IONIC LIQUIDS – CLASSES AND PROPERTIES

Edited by **Scott T. Handy**

Ionic Liquids – Classes and Properties
Edited by Scott T. Handy

Published by InTech

Janeza Trdine 9, 51000 Rijeka, Croatia

Publishing Process Manager Alenka Urbancic

Technical Editor Teodora Smiljanic

Cover Designer InTech Design Team

Additional hard copies can be obtained from orders@intechopen.com

Ionic Liquids – Classes and Properties, Edited by Scott T. Handy
p. cm.
ISBN 978-953-307-634-8

For EU product safety concerns:
IN TECH d.o.o., Prolaz Marije Krucifikse Kozulić 3, 51000 Rijeka, Croatia,
info@intechopen.com or visit our website at intechopen.com.

Meet the editor

Dr. Scott T. Handy received his Ph.D. in Chemistry from Indiana University in 1996 under the direction of Professor Paul Grieco. Following an NIH postdoctoral fellowship at Stanford University in the labs of Professor Paul Wender, he joined the Chemistry Department at Binghamton University in 1999. In 2006 he moved to Middle Tennessee State University as an Associate Professor and was promoted to Professor in 2011. His research interests are in the areas of cross-coupling reactions (particularly regioselectivity in the couplings of polyhaloheteroaromatics) and ionic liquids (as recyclable solvents for transition metal-catalyzed reactions and electrochemistry).

Contents

Preface

Ionic liquids (more specifically, room temperature ionic liquids (RTIL)) have attracted considerable interest over the last few years. Although the specific definition of what an RTIL is varies from person to person, the prevailing definition would be that it is a salt with a melting point below 100 °C. Such a broad definition leaves considerable room for flexibility, which contributed to labeling RTILs as "designer solvents."[1]

The history of ionic liquids (and the closely related molten salts) has a rather ill-defined beginning, although it is most commonly dated back to 1914 and the work of Walden on the use of alkylammonium nitrates.[2,3] The next burst of interest occurred with the discovery of chloroaluminates formed by combining quaternary heterocyclic cations with aluminum chloride. These materials exhibited a great deal of potential for use in a variety of areas, but all suffered from extreme sensitivity to moisture.

A major step forward was made by Wilkes in the early 1990's, with the report of moisture stable ionic liquids created by replacing the aluminum chloride with other anions, such as tetrafluoroborate or hexafluorophosphate.[4]

Since that seminal report by Wilkes and co-workers, the family of RTILs has seen explosive growth. Starting with imidazolium cations, the cationic component has been varied to include pyridinium, ammonium, phosphonium, thiazolium, and triazolium species.[5] In general, these cations have been combined with weakly coordinating anions, although not all weakly coordinating anions result in RTILs (for example, the very weakly coordinating polyhedral borane anions of Reed afford salts with melting points between 45 and 156 °C for a series of imidazolium cations).[6] Common examples include tetrafluoroborate, hexafluorophosphate, triflate, triflimide, and dicyanimide. The first two have been explored the most, and must be treated with the greatest caution as they are fairly readily hydrolyzed to boric acid and phosphate respectively.[7] Indeed, various phosphate and phosphinate anions have been employed to some advantage in RTILs.[8] The list of possible anionic components continues to grow at a rapid rate.

Several chapters in this volume display the increasing variability found in the family of components used to prepare RTILs, notably those of Sindt, Mieloszynki, and Harmand; Yacob and Liebscher; Engel; and Munawar and Nadeem.

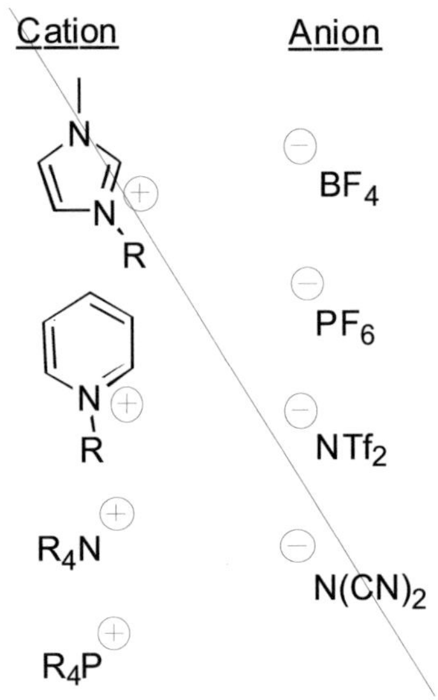

Fig. 1. Representative Ionic Liquid Components

The remaining focus of interest in RTIL research are methods for determining and predicting their physical properties, especially since their unusual and tunable properties are often mentioned as one of the key advantages of RTILs over conventional solvents. Several chapters in this volume focus on this area as well, including those by Shukla, Srivastava, and Saha; Bodo and Migliorati; Forsman, Szparaga, Nordholm, Woodward, and Penfold; Hayamizu; Yoshimura, Takekiyo, Imai and Abe; and Petra, Attila, Katalin, and Laszlo. However, even with this impressive effort, there is still a lot of work to be done before the true power of RTILs as designer solvents (i.e. predictable selection of a particular RTIL for any given application) is effectively harnessed.

Another problem/opportunity is that RTILs are fundamentally synthetic materials. While this does render them highly tunable, it also means that concise, scalable, efficient, and inexpensive synthetic routes to them are of extremely important.

The typical approach is to alkylate the starting nitrogen or phosphorous-containing compound, generating the quaternary halide salt. (Scheme 1) At this point, anion metathesis is employed to introduce the desired anion.

This seemingly simple chemistry has a number of problems associated to it. For the alkylation step, the reactions can require long reaction times (up to several days) in addition to higher temperatures (which can result in a number of side reactions make the halide salt difficult to purify). The anion metathesis is also a generally slow reaction (12+ hours) and can be difficult to push to completion. As a result, a number of different methods have been reported for each of these two steps, including the use of microwaves and sonication.[9] (Scheme 2)

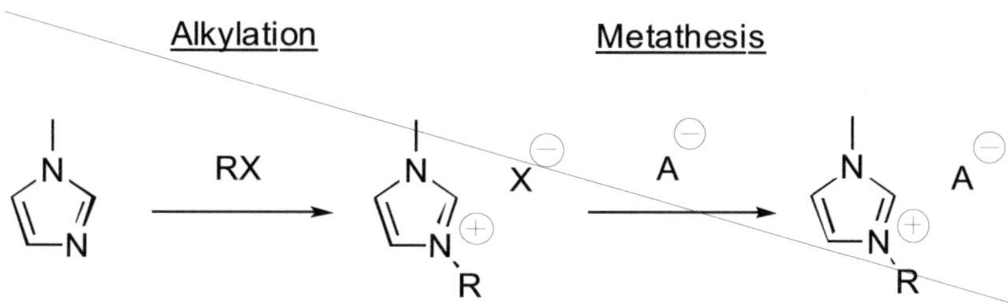

Scheme 1. General Synthetic Route to RTILs

11

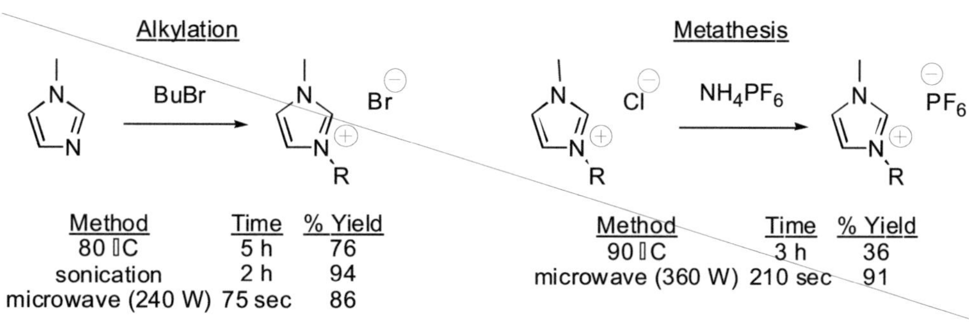

Scheme 2. Non-conventional Methods for the Synthesis of RTILs[10,]

Although this alkylation/metathesis approach is certainly quite flexible and reasonably short, a one-step route to RTILs would be ideal. (Scheme 3) Several different attempts to achieve this goal have been reported, including one-pot reactions (in which alkylation and metathesis are combined),[12] alkylations using species other than alkyl halides (such as alkyl sulfonates, eliminating the need for anion metathesis),[13] Michael reactions with reactive Michael acceptors,[14] and simple protonation using strong acids.[2]

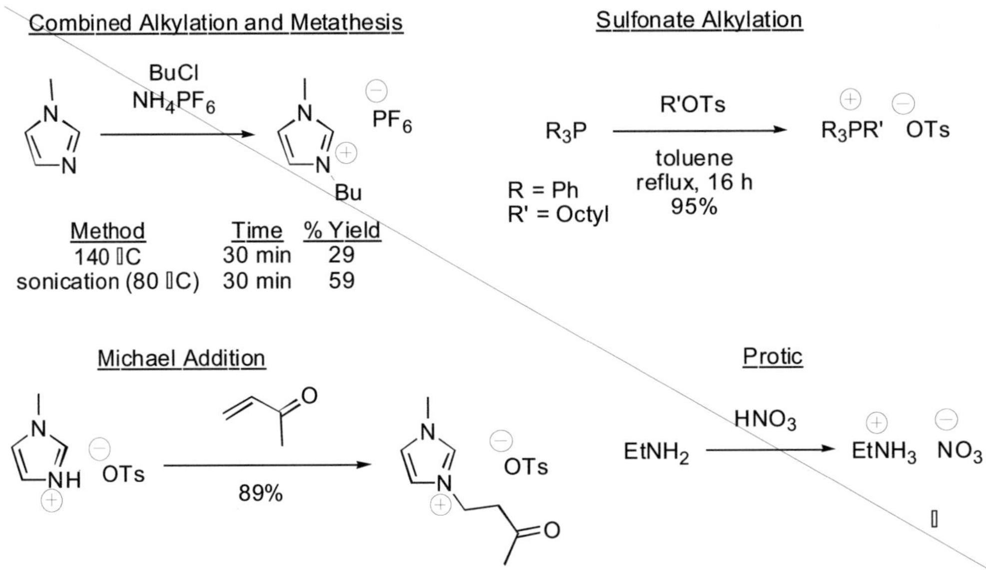

Scheme 3. One-pot Syntheses of RTILs

Even with all of the synthetic advances, one of the biggest limitations to the use of RTILs is still their cost. Common RTILs remain quite expensive, particularly when compared to conventional organic solvents. For example, the price of one liter of THF is under $60, whereas butylmethylimidazolium tetrafluoroborate (one of the least expensive RTILs at the moment) is over $2,000 per liter.[15] The price is an even bigger issue if the most stable of the common RTILs are considered – the triflimides. The triflimide anion is extremely expensive, making it unlikely that the RTILs will ever become inexpensive. As a result, a very important area of study is in the design of equally stable anions that are much less expensive.

Regardless of the method employed for synthesis, a key concern with RTILs is their purity.[16] There are many possible impurities, but the most common (and the most detrimental) are residual halides and undetermined colored impurities. As discussed in the chapter by Abu-Eishah, there are a number of methods to purify contaminated RTILs, and further impurities can often be avoided by use of careful synthetic methods.

Much of the earliest interest in RTILs was as recyclable, "green" solvents for Organic synthesis.[17] Indeed, by now, there is hardly any Organic reaction that has not been reported in RTILs. In many of cases, there is little practical benefit to the use of a RTIL instead of a conventional organic solvent. One clear exception to this statement is the area of transition metal catalyzed reactions.[18] In such cases, there are dual advantages to the use of RTILs. First, the transition metal catalyst is generally retained in the RTIL layer and can thus be reused several times. Second, the RTIL often helps to stabilize the transition metal catalyst and can result in systems which are also more reactive. A few chapters in this volume deal with the use of RTILs in synthesis: Sun, Liu, Fujita, and Arai; Xian, Yong, Hui, and Zhuo; Mancini, Kneeteman, Della Rossa, Bravo and Adam; and Masahiko.

In conclusion, I hope this volume (and its companion, which focuses on the widely varied applications of RTILs in science) will help to inspire further exploration of these fascinating materials and result in many more valuable applications which will improve the quality of life, as well as the environment in the future.

Sincerely,

Scott T. Handy,
Department of Chemistry
Middle Tennessee State University
Murfreesboro,
USA

References

[1] Freemantle, M. *Chem. Eng. News* 1998, 32-38.
[2] Walden, P. *Bull. Acad. Imper. Sci. (St Petersburg)*, 1914, 1800.
[3] For an excellent review regarding the early history of ionic liquids, see Wilkes, J.S. *Green Chem.* 2002, *4*, 73.
[4] Wilkes, J.S.; Zaworotko, M.J. *J. Chem. Soc. Chem. Commun.* 1992, 965.
[5] Handy, S.T.; "Room Temperature Ionic Liquids: Different Classes and Physical Properties." *Curr. Org. Chem.* 2005, *9*, 959-989.
[6] Larsen, A.; Holbrey, J.D.; Tham, F.S.; Reed, C.A. "Designing Ionic Liquids: Imidazolium Melts with Inert Carborane Anions." *J. Am. Chem. Soc.* 2000, *122*, 7264-7272.
[7] Freire, M.G.; Neves, C.M.S.S.; Marrucho, I.M.; Coutinho, J.A.P.; Fernandes, A.M. "Hydrolysis of Tetrafluoroborate and Hexafluorophosphate Counter Ions in Imidazolium-Based Ionic Liquids." *J. Phys. Chem. A* 2010, *114*, 3744-3749.
[8] Lall, S.I.; Mancheno, D.; Castro, S.; Behaj, V.; Choen, J.I.; Engel, R. Polycations. Part X. LIPs, a new category of room temperature ionic liquid based on polyammonium salts." *Chem. Commun.* 2000, 2413-2414.

[9] Leveque, J-M.; Estager, J.; Draye, M.; Cravotto, G.; Boffa, L.; Bonrath, W. "Synthesis of Ionic Liquids Using Non Conventional Activation Methods: An Overview." *Monats. Chemie* 2007, *138*, 1103-1113.

[10] Namboodiri, V.V.; Varma, R.S. "Solvent-Free Sonochemical Preparation of Ionic Liquids." *Org. Lett.* 2002, *4*, 3161-3163. Varma, R.S.; Namboodiri, V.V. "An expeditious solvent-free route to ionic liquids using microwaves." *Chem. Commun.* 2001, 643-644.

[11] Namboodiri, V.V.; Varma, R.S. "An improved preparation of 1,3-dialkylimidazolium tetrafluoroborate ionic liquids using microwaves." *Tetrahedron Lett.* 2002, *43*, 5381-5383.

[12] Estager, J.; Leveque, J-M.; Cravotto, G.; Boffa, L.; Bonrath, W.; Draye, M. "One-pot and Solventless Synthesis of Ionic Liquids under Ultrasonic Irradiation." *Synlett* 2007, 2065-2068.

[13] Narodai, N.; Guise, S.; Newlands, C.; Andersen, J-A. "Clean catalysis with ionic solvents – phosphonium tosylates for hydroformylation." *Chem. Commun.* 1998, 2341-2342.

[14] Wasserscheid, P.; Driessen-Hoelscher, B.; van Hal, R.; Steffens, H.C.; Zimmermann, J. "New, functionalized ionic liquids from Michael-type reactions – a chance for combinatorial ionic liquid development." *Chem. Commun.* 2003, 2038-2039.

[15] Pricing data from the 2010-2011 Aldrich Catalog.

[16] Seddon, K.R.; Stark, A.; Torres, M-J. "Influence of chloride, water, and organic solvents on the physical properties of ionic liquids." *Pure & App. Chem.* 2000, *72*, 2275-2288.

[17] Hallett, J.P.; Welton, T. "Room-Temperature Ionic Liquids: Solvents for Synthesis and Catalysis. 2.." *Chem. Rev.* 2011, *111*, 3508-3576.

[18] Parvulescu, V.I.; Hardacre, C. "Catalysis in Ionic Liquids." *Chem Rev.* 2007, *107*, 2615-2665.

Part 1

Classes of Ionic Liquids

1

1,2,3-Triazolium Salts as a Versatile New Class of Ionic Liquids

Zekarias Yacob and Jürgen Liebscher
Humboldt-University Berlin
Germany

1. Introduction

Among the various classes of ionic liquids (ILs), those containing *N*-heterocyclic cations are most widely used. Imidazolium salts **1** (Figure 1) represent the most prominent subclass in this area and a number of them are commercially available. Their solvent properties such as melting points, solubility, and viscosity can easily be tuned in a wide range by varying the substituents at the nitrogen atoms as well as by varying the counter-ions. This makes ionic liquids in general, real designer solvents. Even if imidazolium salts have found very wide application in organic synthesis and catalysis, they have some limitations. One important limitation is that, they do not behave as innocent solvents under strongly basic conditions, where they suffer from deprotonation at carbon 2 leading to *N*-heterocyclic carbenes. 1,2,3-Triazolium salts **2** (Figure 1) lack an acidic ring-carbon flanked by two *N*-atoms, consequently they should be advantageous in this aspect. Surprisingly, unlike 1,2,4-triazolium salts, 1,2,3-triazolium salts were neglected as potential ionic liquids until recently. Even though the first examples of 1,2,3-triazolium salts had been known since 1887 (Zincke and Lawson 1887) they were not investigated for their ionic liquid properties. Prior to the first report on stable 1,3,4-trialkyl-1,2,3-triazolium-based ionic liquids 1,2,3-triazolium salts were only interesting when they bear 1-amino groups and contain oxygen-rich anions **3** (Figure 1) and thus can be utilized as highly energetic fuels (Drake, Kaplan et al. 2007). Here, a review is provided on the state of the art in the area of 1,2,3-triazolium salts as ionic liquids and ionic liquid tags for organocatalysts.

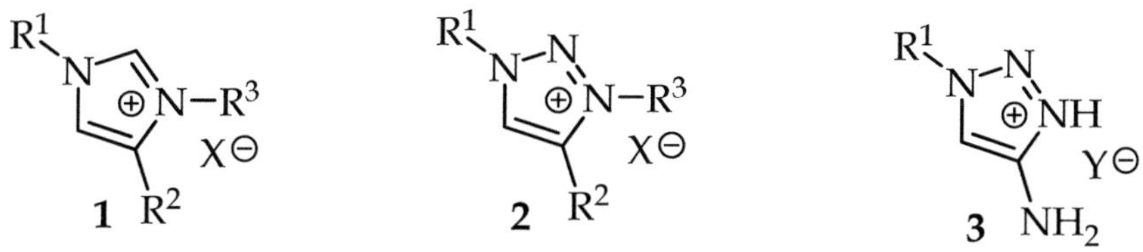

1 **2** **3**

Fig. 1. General structure of common ionic liquids

2. Synthesis of 1,2,3-triazolium ionic liquids

The syntheses of 1,2,3-triazolium salts consists of two major steps, which are: construction of the 1,2,3-triazole ring system and then its *N*-alkylation (Scheme 1). While 1-amino-1,2,3-

triazoles can be obtained by oxidation of glyoxal bishydrazones (Kaplan, Drake et al. 2005), the 1,4-disubstituted 1,2,3-triazoles are mostly synthesized by [3+2] cycloaddition reactions of azides **4** and terminal alkynes **5**. In the classical manner this reaction was performed as thermal 1,3-dipolar cycloaddition where it suffers from lack of regioselectivity. It gives rise to mixtures of 1,4- and 1,5-disubstituted products (Gompper 1957), which are difficult to separate. Nowadays, cycloaddition of azides and terminal alkynes is realized using metal catalysis with high regioselectivtiy and improved yields. Most frequently, Cu-catalysis ("Click"-chemistry) is used to synthesise 1,4-disubstituted 1,2,3-triazoles regioselectively. An alternative copper free methodology has also been developed primarily for the crosslinking of biomolecules. This reaction involves the use of strained cyclooctyne ring substituted with an electron-withdrawing group such as fluorine, which can promote a [3 + 2] dipolar cycloaddition with azides (Baskin, Prescher et al. 2007; Bernardin, Cazet et al. 2010; Debets, van Berkel et al. 2010). Regioselectivity of the cycloaddition of azides to alkynes can be changed to the 1,5-disubstituted 1,2,3-triazoles if Ru-catalysis is applied (Johansson, Lincoln et al. 2011; Zhang, Chen et al. 2005).

So far, only 1,4-disubstituted 1,2,3-triazoles were transformed into ionic liquids by further alkylation. Different systems such as Cu(I) salts with triphenylphosphine, with iminopyridine or with mono or polydentate nitrogen ligands, Cu(I) isonitrile complex in water (Liu and Reiser 2011), Cu(0) nanoclusters (Orgueira, Fokas et al. 2005; Pachon, van Maarseveen et al. 2005) and most commonly $CuSO_4$ with sodium ascorbate (Kolb, Finn et al. 2001) were applied in the click reaction between alkynes and azides to produce 1,4-disubstituted 1,2,3-triazoles regioselectively. Even though this reaction is often termed as "Click-reaction", it is known for its extended reaction times, which can span from several hours to several days. In order to overcome this limitation it is often assisted by suitable ligands such as tris-(benzyltriazolylmethyl)amine (Chan, Hilgraf et al. 2004), triphenylphosphine or other amines.

The outstanding feature of the Cu-catalyzed cycloaddition of azides to alkynes in producing 1,2,3-triazoles is the fact that many functional groups can be tolerated. Hence, interesting functionalities can be incorporated as substituents of the resulting triazole ring. Additionally the access to 1,4-disubstituted 1,2,3-triazoles can be further facilitated by combing the synthesis of the organic azides (from a halogenated precursors and sodium azide) with the copper catalyzed click reaction in a one pot procedure. In this reaction, the copper catalyst facilitates both the replacement of the halide by the azide, and the subsequent [3+2]

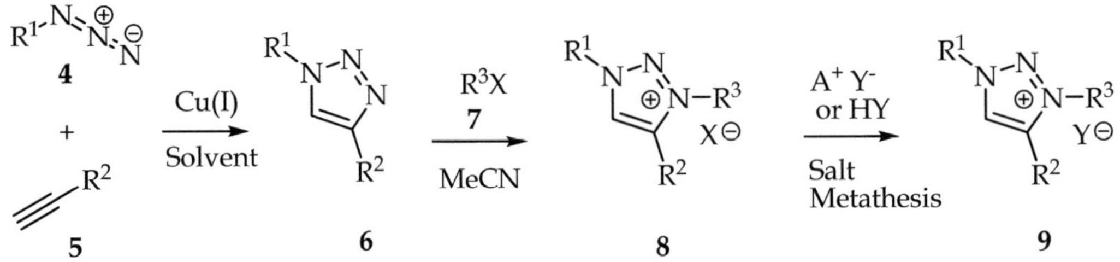

Scheme 1. Synthesis of 1,3,4-trialkylated 1,2,3-triazolium ionic liquids.

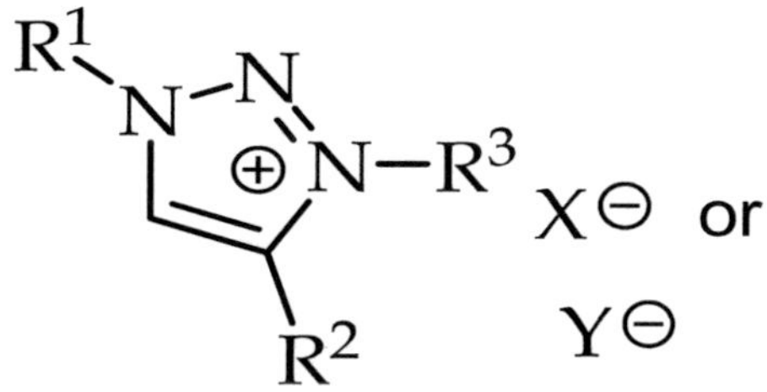

R1	R2	R3	X or Y	Ref
n-Butyl	*n*- Butyl	Me	CF_3,(MeO)PO_3, N(CN)$_2$ or BF_4	1
n-Butyl	*n*-Decyl	Me	$MeSO_4$	1
Benzyl	*n*- Butyl	Me	I	1, 4
Benzyl	*n*-Pentyl	Me	TsO	1
Benzyl	CH_2OH	Me	I	1
Benzyl	$(CH_2)_2OH$	Me	I	1
Benzyl	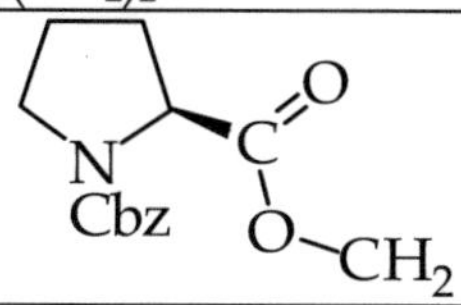	Me	I	1
4-Benzyloxybenzyl	*n*-Pentyl	Me	I	1
3-Methoxybenzyl	3-Phthalimido propyl	Et	Br	1
3-Methoxybenzyl	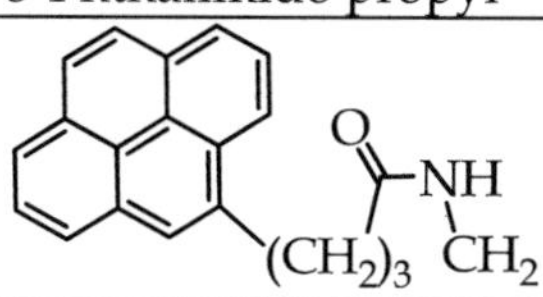	*n*-Pr	I	1
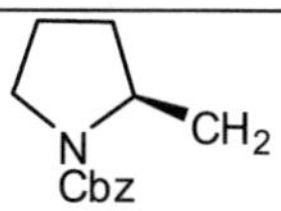	*n*-Pentyl	Me	I	1
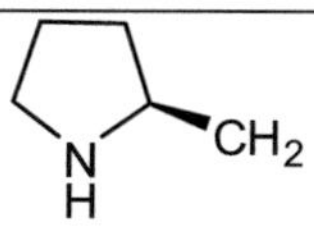	Ph or *n*-Pentyl	Me	I	2
3-Phthalimidopropyl	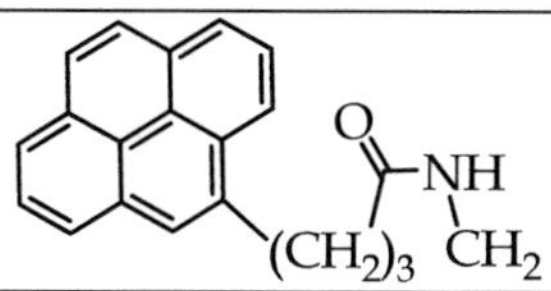	Me	I	1
Benzyl	H	Me	NTf_2, OTf, PF_6 or BF_4	3
n-Butyl	H	Me	NTf_2, OTf, PF_6 or BF_4	3
n-Butyl	H	*n*-Butyl	NTf_2, OTf, PF_6 or BF_4	3
Allyl	*n*-Pentyl	Me	I	4

[1.] (Khan, Hanelt et al. 2009), [2.] (Yacob, Shah et al. 2008) [3.] (Jeong and Ryu 2010) [4.] (Fletcher, Keeney et al. 2010)

Table 1. Some examples of 1,2,3-triazolium ionic liquids **9** cycloaddition reaction. Remarkably, even iodoarenes which normally are unsuitable substrates for a nucleophilic substitution by azides can be transformed to the corresponding arylazides and further converted to 1-aryl-1,2,3-triazoles under these conditions (Fletcher, Keeney et al. 2010; Yan and Wang 2010).

N-Alkylation as the second step in the synthesis of 1,2,3-triazolium salts can produce a 2-alkylated or 3-alkylated 1,4-disubstituted triazolium salt which can present a problem of regioselectivity. However, when 1,4-disubstituted 1,2,3-triazoles are alkylated by soft alkylating reagents **7** (alkyl-, benzyl-, allyl halides, sulphates and sulfonates) only 1,3,4-trisubstituted 1,2,3-triazolium salt products are obtained. Accordingly, both major steps in the synthesis of 1,2,3-triazolium salts can be performed in a highly regioselective manner (Begtrup 1971; Gompper 1957).

Purification of ionic liquids by conventional techniques can be a hideous process. Therefore, obtaining ionic liquids in a pure state from the alkylation reactions is an essential and straightforward alternative. In order to avoid impurities, it is important to start the alkylation reaction with pure 1,2,3-triazoles. Often the alkylation reactions are conducted without solvent. The alkylating agents are used in excess quantities and serve as solvents. The resulting 1,2,3-triazolium salts are obtained in a highly pure state by removal of the excess alkylating agent utilizing high vacuum and/or washing with non-polar solvents (such as diethyl ether or hexane) to remove nonpolar impurities as well. At this stage, when the leaving group of the alkylating reagent is considered as an unsuitable counter ion, it can be exchanged with another anion using salt metathesis (e.g. treatment with $AgBF_4$ or corresponding acids).

A few 4-amino substituted1,2,3-triazolium salts **3** (Figure 1) without any alkyl group at the third *N*-atom were obtained by protonation of 1,4-substituted 1,2,3-triazole by an inorganic acid. These salts however may not be interesting as ionic liquid solvents in chemical transformations due to their extremely low stability. They can rather serve as high energetic substances (Drake, Hawkins et al. 2003).

In general, the synthesis of 1,2,3-triazolium-based ionic liquids has a wide scope and numerous functional groups can be introduced as substituents. When 1,2,3-triazolium ionic liquids with only two substituents namely at positions 1 and 3 are required, trimethylsilylethyne (TMS-ethyne) can be used for the click reaction and the TMS-group can be removed from the resulting 1,2,3-triazole by fluoride (Jeong and Ryu 2010; Khan, Shah et al. 2010). Alternatively calcium carbide (CaC_2) was used for the synthesis of 1-mono substituted 1,2,3-triazoles by click chemistry (Jiang, Kuang et al. 2009).

Ionic liquids in which a 1,2,3-triazolium moiety is linked to other 1,2,3-triazolium or imidazolium units by an alkylene bridge were synthesized (Schemes 2 and 3) utilizing copper catalyzed click reaction between appropriate precursors and final double N-alkylation (Khan, Shah et al. 2010). The compounds **14** and **19** were investigated as useful ligands in Pd-catalyzed Suzuki-Miyaura reaction. Here, the imdiazolium moiety serves as a precursor of *N*-heterocyclic carbene ligand for Pd as in known cases where simple imidazolium salts were used (Mathews, Smith et al. 2000). Interestingly, the combination of an imidazolium and a 1,2,3-triazolium unit demonstrated a better performance than a single azolium salt or a dicationic salt where two imidazolium or two 1,2,3-triazolium units are found. Aryl chlorides, iodides and triflates turned out to be the best substrates for this new catalytic system.

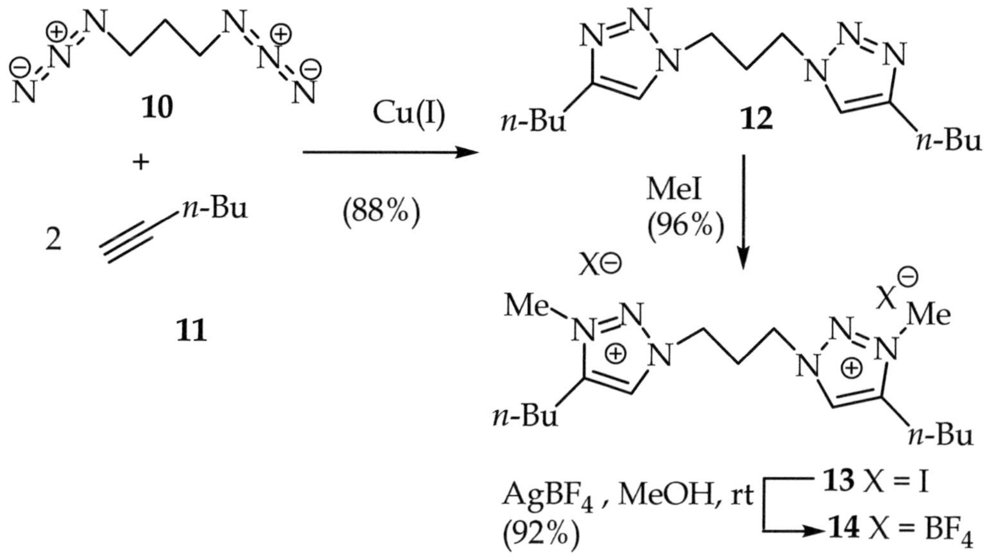

Scheme 2. Synthesis of di-(1,2,3-triazolium) salts **13, 14**

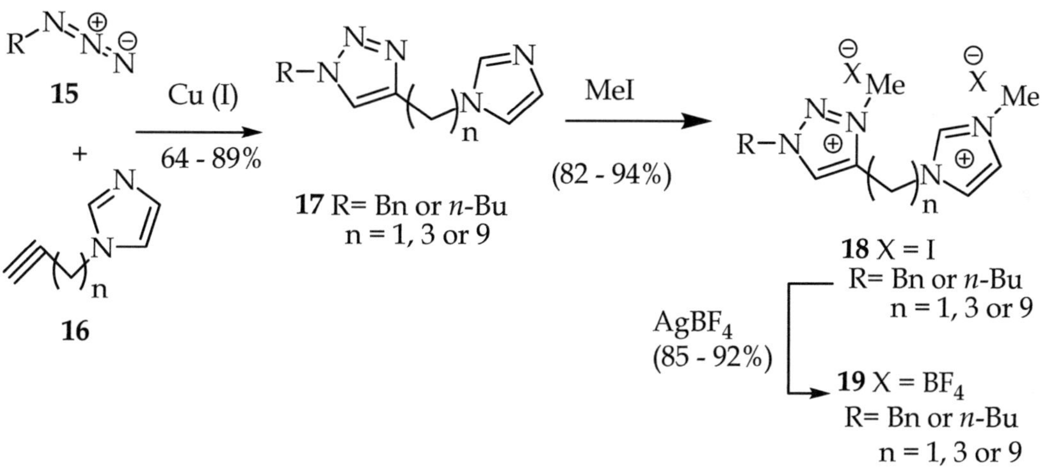

Scheme 3. Synthesis of mixed imidazolium triazolium ionic liquids **18, 19**

3. Properties of 1,2,3-triazolium ionic liquids

Most of the 1,2,3-triazolium based ionic liquids are room temperature ionic liquids (RTILs) (Fletcher, Keeney et al. 2010; Khan, Hanelt et al. 2009). So far, not so many physical constants have been reported. Viscosity measurements and differential thermogravimetry (TGA) measurements of some 1,3,4-trisubstituted 1,2,3-triazolium ionic liquids counter-ions (Jeong and Ryu 2010; Khan, Hanelt et al. 2009). These reports indicated good thermal stability up to 355 °C, which is strongly dependent on several variables such as the kind of counter-ion and the nature of substituents on the triazolium ring. One can tune the stability of 1,2,3-

triazolium ionic liquids by the choice of substituents and the anion. The 4-amino-1,2,3-triazolium salts **3** with oxygen-rich anions can be very unstable and posses high explosive nature. Therefore, they need to be handled with caution. In general 1,2,3 triazolium ionic liquids possess weaker thermal stabilities with the onset of decomposition occurring at about 100 °C when the counter-ions are iodide or TfO^{-} while salts with bulky anions such as bis(trifluoromethylsulfonyl)amide, hexafluorophospate and tetrafluoroborate show much higher stabilities (Jeong and Ryu 2010; Khan, Hanelt et al. 2009).

1,2,3-Triazolium salts are chemically relatively stable. However, nucleophilic displacement of the triazole ring by nucleophilic attack at an alkyl group in position 3 can occur in some cases. Furthermore, deprotonation at position 4 or 5 was observed with strong bases or under H-D exchange conditions. Remarkably, the formation of transition metal complexes was found recently wherein the triazolium unit was transformed into an *N*-heterocyclic carbene acting as a ligand for the corresponding metal. However, this deprotonation is much more difficult than in related imdiazolium salts (Mathew, Neels et al. 2008, Guisado-Barrios, Bouffard et al.2010)

4. Applications of 1,2,3-triazolium ionic liquids

Since 1,2,3-triazolium ionic liquids have not been commercially available, their application as mere solvent is rare. 1,3-Dialkyl-1,2,3-triazolium ionic liquids have been developed as stable and recyclable solvents for the Baylis-Hillman reaction. The Baylis-Hillman reaction between *p*-chlorobenzaldehyde and methyl acrylate was conducted in 1,2,3-triazolium ionic liquids at room temperature in the prescence of DABCO. Interestingly the reaction furnished improved yields within shorter reaction time in the triazolium ionic liquids as compared to analogous imidazolium ionic liquids (Jeong and Ryu 2010).

4.1 Application of 1,2,3-triazolium ionic liquids in catalyst tagging

So far, the major field of application for 1,2,3-triazolium ionic liquids is in ionic liquid tagging of organocatalysts. In this recent methodology, ionic liquid moieties are covalently bonded to organocatalytic structures affording the so-called ionic liquid tagged organocatalysts (ILTOCs) (Sebesta, Kmentova et al. 2008). This strategy aims to better catalytic performance, better solubility and in particular easier recycling of catalysts. Unlike catalysts fixed on solid supports which are often insoluble in reaction media, ionic liquid tagged organocatalysts can be dissolved in various polar solvents and thus act as homogeneous catalysts. Recycling of the ionic liquid tagged organocatalysts is usually achieved by extracting the products from the reaction mixture using non-polar organic solvents. This leaves behind the ionic liquid tagged organocatalyst, which is eventually dissolved in an appropriate solvent or reagent for the next cycle. 1,2,3-Triazolium ionic liquid tagging is in particular useful because the key step in the synthesis of these catalysts (which is a copper-catalyzed cycloaddition of azides to alkynes) can tolerate many functional groups, amongst them are most of the organocatalytic moieties. Either the alkyne or the azido functionality can be introduced into the organocatalytic moiety enabling it to undergo a copper-catalyzed click reaction to establish the triazole tag. Alternatively, an alkylating function can also be tethered with the organocatalyst, which can later be used to alkylate a triazole thus forming the triazolium tag (Scheme 4).

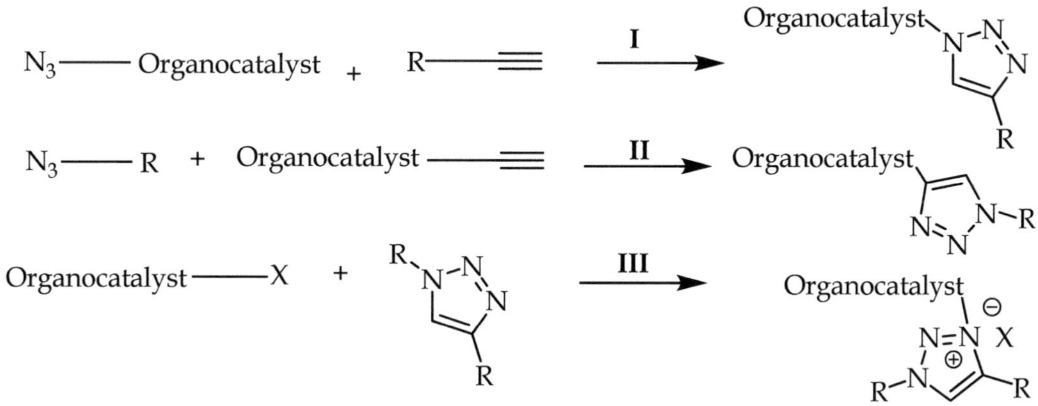

Scheme 4. Various approaches towards 1,2,3-triazolium ionic liquid tagged catalysts

Chiral 1,2,3-triazolium ionic liquid tethered pyrrolidine catalysts built from (*S*)-proline and its derivatives have been successfully applied in various catalytic reactions (Khan, Shah et al. 2010; Khan, Shah et al. 2010; Maltsev, Kucherenko et al. 2010; Yacob, Shah et al. 2008). The 1,2,3-triazolium ionic liquid-tagged organocatalysts derived from proline and its derivatives are mostly viscous liquids at room temperature and are completely miscible with polar solvents such as methanol, chloroform, acetonitrile, dimethylsulfoxide, dimethylformamide and water. They are insoluble in less-polar solvents such as *n*-hexane and diethyl ether. In some cases, the ionic liquid sub-unit serves not only as a phase tag for efficient recycling but also as an effective chiral amplifier through polar interactions and steric shielding.

1,2,3-Triazolium ionic liquid tethered pyrrolidine organocatalysts **25, 26, 30 - 32, 38, 44, 48-50, 58** were synthesised in a simple and versatile way from (*S*)-proline or *trans* 4-hydroxy-(*S*)-proline. The synthesis of these catalysts involves primarily the preparation of the corresponding azide and terminal alkyne derivatives for click cycloaddition reaction. The ionic liquid products were obtained quantitatively with a straightforward two-step procedure by *N*-alkylation of the click reaction products. The protecting groups are finally removed to liberate the target catalyst.

In the first approach (Scheme 5), the carboxyl group of (*S*)-proline was reduced to the corresponding alcohol. The resulting (*S*)-pyrrolidine-2-ylmethanol was transformed into tosylate and was substituted with azide resulting the (*S*)-pyrrolidine-2-ylazide following literature procedures (Luo, Xu et al. 2006). In this case, the (*S*)-proline served as an azide precursor of the click reaction to react with an alkyne. The resulting 1,2,3-triazol-1-yl-methylpyrrolidine **22** was transformed to an ionic liquid tagged catalyst by alkylation with methyl iodide. The exchange of the counter-ion iodide of catalyst **25** to tetrafluoroborate by anion metatheses using $AgBF_4$ furnished the respective ionic liquid-tagged organocatalysts **26** in quantitative yields (Scheme 5) (Yacob, Shah et al. 2008).

Catalytic application of the triazolium ionic liquid tagged organocatalysts **25** and **26** gave very interesting results in the Michael addition of various unmodified aldehydes and ketones to *trans*-β-nitrostyrenes. 10 mol % phenyl substituted ionic liquid tagged organocatalyst **25** provided excellent yields (>99%), diastereoselectivities (>99:1) and

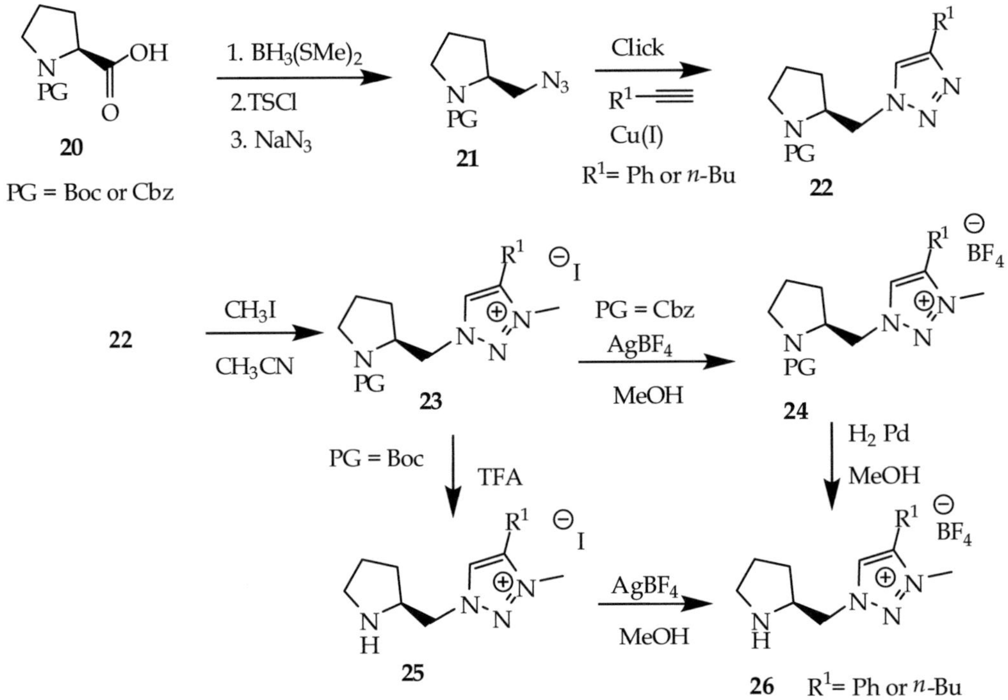

Scheme 5. Synthesis of (*S*)-proline derived 1,2,3-triazolium ionic liquid tagged organocatalysts **25** and **26**

enantioselectivities (ee >99%) for the Michael addition of cyclohexanone to *trans*-β-nitrostyrene. The catalyst was recyclable by extraction of the reaction mixture with diethyl ether and usage of the remainder with a fresh batch of reactants. Recycling provided slightly decreasing yields with rapidly diminishing enantioselectivities and persistently high diastereoselectivities (> 93:7). The yields decreased from 99 % for the first run, to 90 % for the second run, 83 % for the third run and 74 % for the fourth run, but the enantioselectivities rapidly diminished (first run, 99%; second run, 90%; third run, 88 % and fourth run, 58 %) with identical duration of the reaction. Reduction of the amount of catalyst **25** to 5 mol % gave excellent yield (>95%), diastereoselectivity (97:3) and diminished enantioselectivity (ee 82%) after a slightly prolonged reaction duration.

Tetrafluoroborate based catalyst **26** demonstrate improved recyclability than those with iodide counter-ion. The recycling of these catalysts was carried out at least four times without sacrificing the yield or enantioselectivity (Yacob 2010).

The direct application of unmodified aldehydes in catalytic Michael additions can be severely hindered due to the presence of undesirable intermolecular self-aldol reactions (Hagiwara, Komatsubara et al. 2001; Hagiwara, Okabe et al. 2001). Barbas and co-workers achieved the first direct catalytic asymmetric Michael reaction between unmodified aldehydes and nitroolefins. The usage of an (*S*)-2-(morpholinomethyl) pyrrolidine catalyst in 20 % furnished the Michael addition products in 72 % enantioselectivity, 12:1 diastereoselectivity and 78 % yield (Betancort and Barbas 2001; Betancort, Sakthivel et al. 2004; Mosse, Andrey et al. 2006). The utilization of the ionic liquid tagged catalysts **25** and **26** in the Michael reactions of *trans*-β-nitrostyrenes to aldehydes resulted in high yields but lower enantioselectivities. Isobutyraldehyde provided 95% yield, but the enantiomeric excess was 72%. Isovaleraldehyde resulted in high diastereoselectivity of 95:5 but it gave very low enantioselectivities (<45%). Acetone reacted comparatively faster to give high yield and enantioselectivity of 52 %. Comparable results were obtained by Yan et al. with a non ionic trialkyl 1,2,3-triazolylmethylpyrrolidine catalyst (Yan, Niu et al. 2006). The reaction of cyclopentanone with *trans*-β-nitrostyrene gave a good yield of 85% and satisfactory enantioselectivity (ee 82 %) but low diastereoselectivity (2:1). Cyclopentanone is known to be a less striking substrate for Michael additions to *trans*-β-nitrostyrene. It often gives enantioselectivity of around 50 % and low yield products after prolonged reaction times (Betancort and Barbas 2001; Betancort, Sakthivel et al. 2001).

Another instance of 1,2,3-triazolium ionic liquid tagged proline was achieved utilizing *trans*-4-hydroxy-(*S*)-proline as the azide precursor for the click reaction (Shah, Khan et al. 2009). A *cis*-4-azido functionalized proline **28** was prepared by tosylating of a doubly protected *trans*-4-hydroxy (*S*)-proline **27** followed by a nucleophylic substitution using sodium azide. The Cu(I) catalyzed click reaction with 1-hexyne produced the *cis*-4-triazoyl substituted proline **29**. Alkylation with methyl iodide and salt metathesis with $AgBF_4$ resulted the ionic liquid tagged diprotected proline , which was deprotected by palladium catalyzed hydrogenation to give catalyst **30** (Scheme 6).

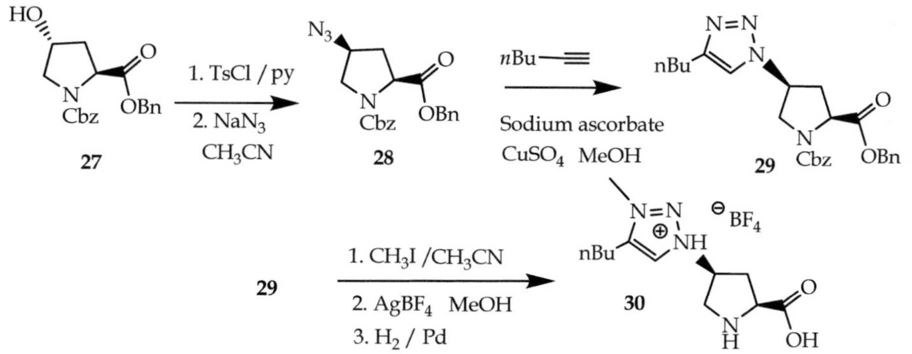

Scheme 6. Synthesis of *trans*-4-hydroxy-(*S*)-proline derived 1,2,3-triazolium ionic liquid tagged organocatalyst **30**

A similar approach was used to prepare α,α-diphenylprolinol derived ionic liquid tagged organocatalysts **31** and **32** (Maltsev, Kucherenko et al. 2010). The synthetic route starts with benzyl protected 4-hydroxydiphenylprolinol and goes through identical reaction conditions as for compound **30** indicated in Scheme 6. The 1,2,3-triazolium ionic liquid tagged catalyst was obtained after final reductive hydrogenation, anion metathesis and treatment with TMS-OTf (Scheme 7). These catalysts were applied in 0.1 equimolar quantities in a domino reaction involving aza-Michael and intramolecular acetalization, between *trans*-cinnamaldehyde and *N*-Cbz protected hydroxyl amines (Maltsev, Kucherenko et al. 2010).

The 4-unsubstituted 1,2,3-triazolium salt **38** was obtained using TMS-ethyne **34** as alkyne component in the reaction with the 4-azidoproline derivative **33** and a consecutive desilylation of the resulting TMS-triazole **35** by fluoride (Scheme 8). Independently of our work this strategy was also used to synthesize 1,3-dialkyl-1,2,3-triazolium salts as ionic liquid solvents useful for Baylis-Hillman reactions (Jeong and Ryu 2010). Catalyst 38, which lack a substituent in position 4 of the 1,2,3-triazole ring and thus is less lipophilic performed poorer as compared to catalyst **30**, resulting in 78% ee and 82% yield (Khan, Shah et al. 2011).

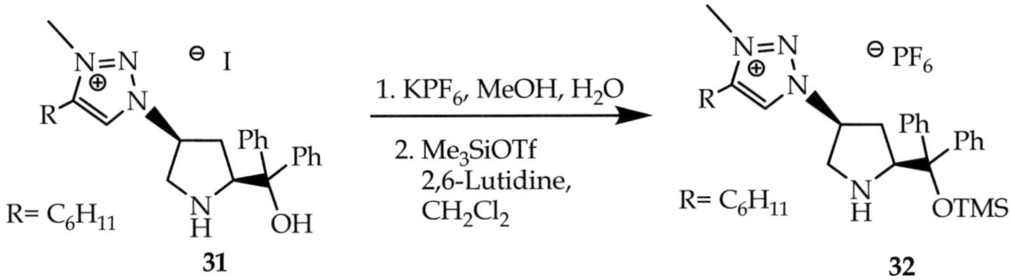

Scheme 7. Synthesis of α,α-diphenylprolinol derived ionic liquid tagged catalyst

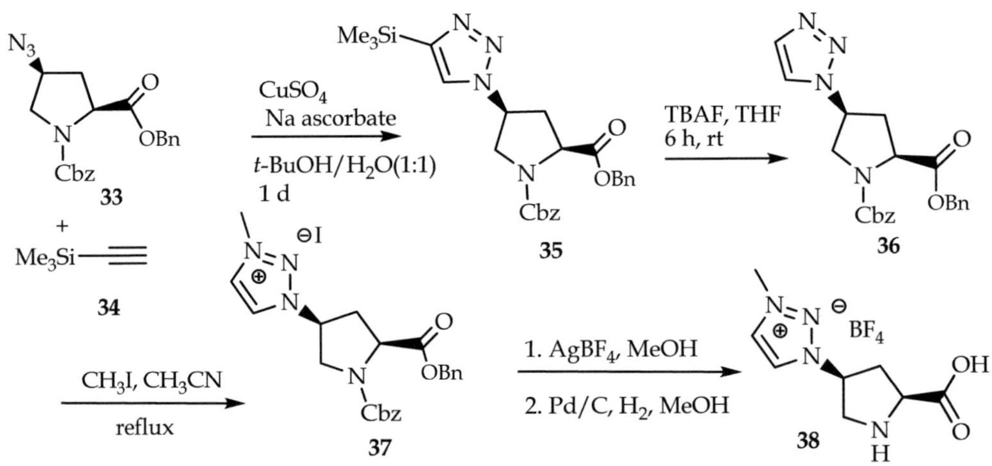

Scheme 8. Synthesis of 4-unsubstituted 1,2,3-triazolium ionic liquid tagged (*S*)proline **38**

In a second approach, the (*S*)-proline derivative was used as the alkyne precursor in the click reaction. A propargyl subunit was introduced at the 4-hydroxy position of a diprotected *trans*-4-hydroxy proline **39** by means of Williamson ether synthesis. Click reaction with an alkyl azide furnished the triazole **41** in good yield. This product was transformed into the desired 1,2,3-triazolium ionic liquid tagged (*S*)-proline catalyst **44** by means of an analogous methylation, salt metathesis and a final removal of the protecting groups (Scheme 9) (Shah, Khan et al. 2009).

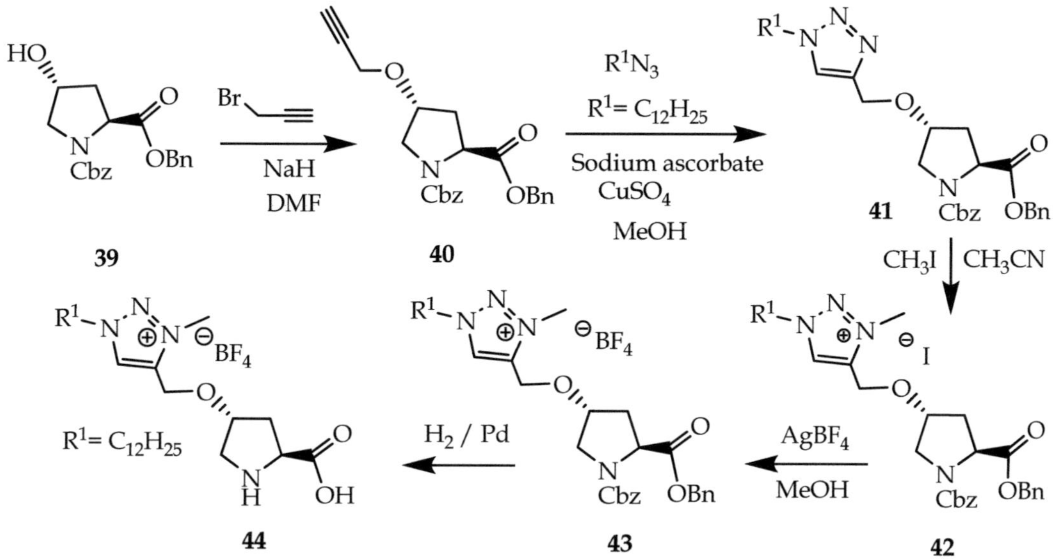

Scheme 9. Synthesis of *trans*-4-hydroxy (*S*)-proline derived 1,2,3-triazolium ionic liquid-tagged organocatalyst **44**

The application of 20 mol% *trans*-4-hydroxy-(*S*)-proline derived triazolium tagged catalysts **30** and **44** in a direct aldol reaction of aromatic aldehydes with acetone, cyclohexanone, or cyclopentanone resulted excellent diastereoselectivities and enantioselectivities (> 90%)(Shah, Khan et al. 2009). Cyclopentanone gave lower diastereoselectivities and enantioselectivities as compared with cyclohexanone. The application of catalyst **44** in the Michael addition of cyclohexanone to *trans*- β-nitrostyrene did not provide satisfactory results. The enantioselectivities found were low (<54 %) even if the reaction proceeded with high yields and good diastereoselectivities.The recycling experiments for the aldol reaction of cyclohexanone with 4-bromobenzaldehyde in the presence of excess cyclohexanone as solvent and 20 mol% of the catalyst were performed by extraction of the reaction mixture with diethyl ether or cyclohexane. The triazolium tagged organocatalyst remained in the reaction flask as an oil, which was combined with fresh reactants for the next cycle (Shah, Khan et al. 2009). High yields (> 80%) were achieved until the 5th cycle and the diastereoselectivity remained excellent. However, the enantioselectivity diminished with recycling, finally reaching 44% ee. Extraction with less polar cyclohexane gave better results in recycling experiments. Under these conditions, 88% yield with 68% ee and a diastereomeric ratio of 97:3 could be achieved in the 5th cycle. Unlike other reported recycling efforts in organocatalysis, there was still a drop in yields and enantioselectivities.

The aldol reactions were also performed in ionic liquid solvents namely N,N-diethyl-N′-hexyl-N′-methyl-N″,N″dipropylguanidinium iodide and the commercially available 1-butyl-3-methylimidazolium tetrafluoroborate [bmim][BF$_4$] (Shah, Khan et al. 2009). Equimolar ratios of cyclohexanone and 4-bromobenzaldehyde were used and the reaction mixture was extracted with diethyl ether. The guanidinium salt gave improved results than the neat reaction. Application of the commercially available [bmim][BF$_4$] resulted in a rapidly diminishing enantioselectivity for the recycling.

The influence of other functional groups (such as a carboxyl moiety or an acyclic amino acid) on the performance of the ionic liquid tagged catalysts was investigated by introducing these functional groups as appendages on the 1,2,3-triazolium ionic liquid tags. Click reaction of diprotected *trans*-4-propargyloxyproline **40** with either azidoacetic acid, azidoacetic acid methylester or a lysine-derived azide give access to the 1,2,3-triazolium ionic liquid tagged catalyst precursors **48** - **50** with additional functional groups. The synthesis involved similar steps as the one indicated in Scheme 9. A final anion metathesis using silver tetrafluoroborate and palladium catalyzed hydrogenative deprotection furnished the required triazolium tetrafluoroborates **48** - **50** as oils. The Cbz-protective group of the side chain in the triazolium iodide **47** was also removed during the final reduction step providing the product **50** which has two unprotected α-amino acids (Scheme 10).

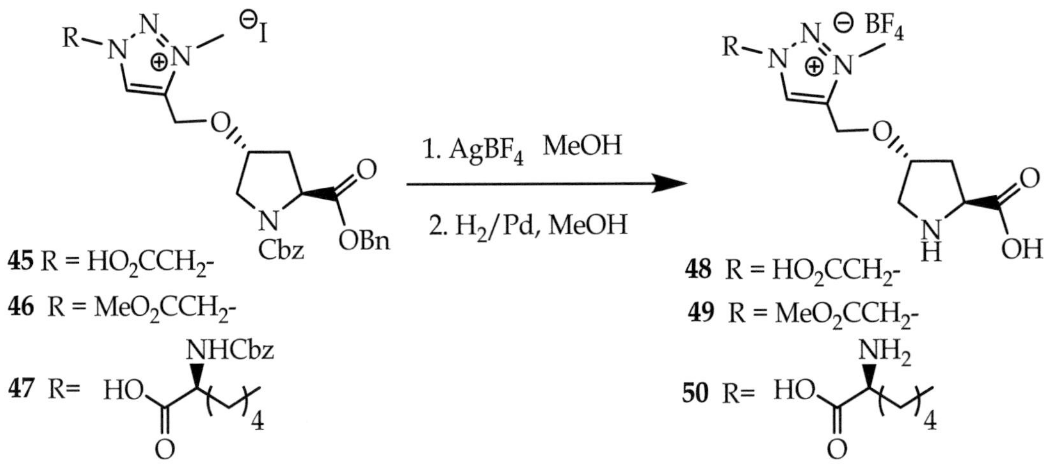

Scheme 10. Synthesis of *trans*-4-hydroxy(*S*)-proline derived functionalized 1,2,3-triazolium ionic liquid tagged catalysts.

In the α-aminoxylation of cyclohexanone with nitrosobenzene using [bmim][BF_4] as an ionic liquid solvent at room temperature, the lipophilic dodecyl 1,2,3-triazolium ionic liquid tagged catalyst **44** gave a high enantioselectivity 96%. However, the short chain carboxylate containing catalyst **49** gave 8% enantioselectivity and the carboxylic acid containing catalyst **48** provided 10% enantioselectivity. Placing an additional α-amino acid moiety in the side chain of the 1,2,3-triazoliumsalt **50** had a disadvantageous effect resulting in only 52% ee.

By means of an analogous reaction sequence an acyclic 1,2,3-triazolium tetrafluoroborate tagged catalyst **53** was synthesised starting from lysine derivative **51** and 1-dodecyne **52** (Scheme 11). The product was obtained as a colour less solid (mp 91 °C). A similar Cu-catalyzed cycloaddition was reported for the synthesis and biochemical application of a lower homologue of **53** (Gajewski, Seaver et al. 2007). Unexpectedly, the triazolium ionic liquid tagged catalyst **53** wherein a noncyclic α-amino acid unit was found instead of proline failed to give any enantioselectivity. However, this catalyst gave good results in aldol reactions (Khan, Shah et al. 2010).

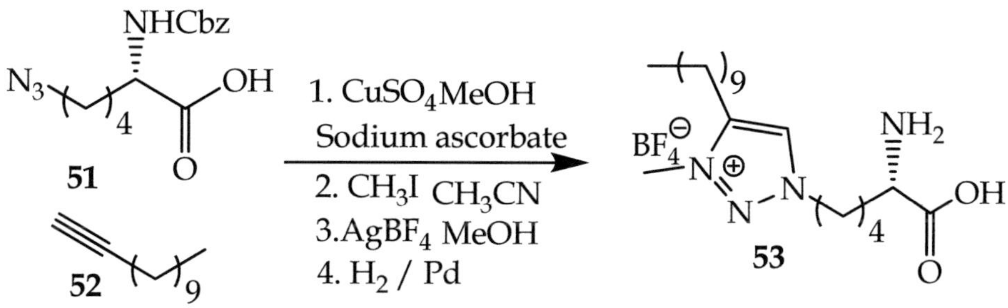

Scheme 11. Synthesis of (*S*)-lysine derived 1,2,3-triazolium ionic liquid tagged catalyst **53**

1,2,3-Triazolium ionic liquid tagged proline in which, the ionic liquid tag is several carbon atoms away from the pyrrolidine sub-unit was realized by transforming the diprotected proline into an alkylating agent **56**. The esterification of the 4-hydroxy group of a diprotected *trans*-4-hydroxy (*S*)-proline **54** with 5-bromovaleric acid **55** resulted in the bromo substituted ester **56.** This molecule was used in the alkylation of 1,4-dibutyl-1,2,3-triazole **57** to obtain the required ionic liquid tagged diprotected proline. Salt metathesis with $AgBF_4$ and hydrogenative deprotection of the triazolium bromide lead to the triazolium tagged proline organocatalyst **58** (Scheme 12).

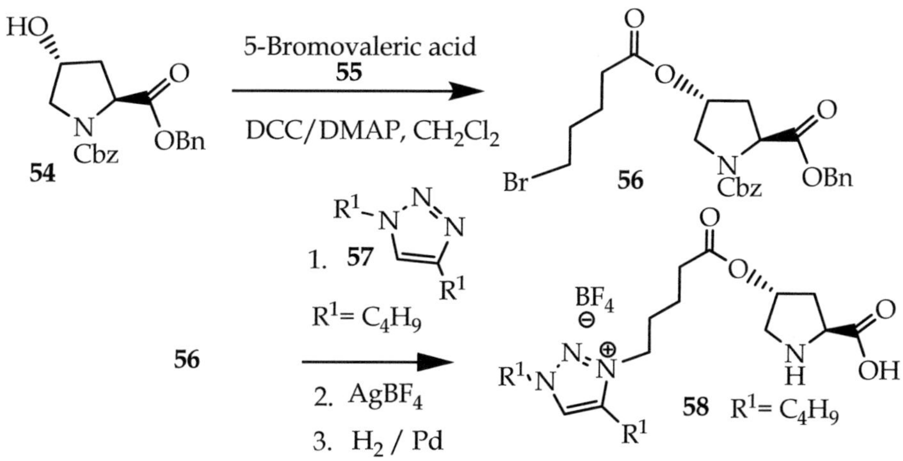

Scheme 12. Synthesis of (*S*)-proline derived 1,2,3-triazolium ionic liquid tagged catalyst **58**

The deprotection of the Cbz-protecting group from some of the ionic liquid tagged catalyst precoursors by reductive hydrogenation using palladium was found to be problematic. In cases where iodide was the counter ion it showed no selectivity between benzyloxycarbonyl protecting groups and benzyl substituents eventually found at the 1,2,3-triazolium rings. The reduction reaction was unsuccessful when iodide is found as counter-ion probably due to poisoning of the palladium surface. In these cases exchange of iodide by tetrafluoroborate was essential for the palladium-catalyzed reductive deprotection to take place. Thus anion metathesis is used not only to vary the physical properties of the ionic liquid such as viscosity and melting points but also to facilitate the deprotection reaction.

The α-aminoxylations of carbonyl compounds with nitrosobenzene was also investigated using catalyst **30** and analogous 4-hydroxy-*(S)*-proline derived catalyst (where the counter-anion is triflate) resulting from alkylation with MeOTf and lysine derived 1,2,3-triazolium tagged catalysts. All the ionic liquid tagged catalysts were screened in the α-aminoxylation of cyclohexanone with nitrosobenzene using [bmim][BF] as ionic liquid solvent at room temperature. The 1,2,3-triazolium tagged proline tetrafluoroborate **32** gave superior enantioselectivity (>99%) and yield (91%). These results are better than reported in cases where simple (*S*)-proline was used (Sunden, Dahlin et al. 2005). Exchanging the anion with triflate resulted in a somewhat lower enantioselectivity (92%). A less lipophilic triazolium-tagged proline lacking a substituent at carbon 4 of the 1,2,3-triazole ring surprisingly demonstrated lower performance of 78% enantioselectivity and 82% yield after prolonged reaction time. Another 4-hydroxy-*(S)*-proline derived catalyst with three ionic liquid-tags provided only 50%ee (Khan, Shah et al. 2011).

Another interesting types of catalysts, which can be tethered with 1,2,3-triazolium based ionic liquids are TADDOLs (α,α,α′,α′-tetraaryl-2,2-dimethyl-1,3-dioxolane-4,5-dimethanol). TADDOLs are among the most prominent chiral scaffolds used as metal ligands in chiral Lewis acid catalysis and as organocatalysts in hydrogen bonding catalysis. In the latter case TADDOLs can activate carbonyl compounds and imines by decreasing the LUMO energy through hydrogen-bonding interaction with the carbonyl oxygen atom or imine nitrogen atom, respectively. The activated carbonyl compounds or imines can undergo highly enantiofacial addition of carbon nucleophiles to furnish the corresponding alcohols and amines with excellent stereoselectivities (Huang and Rawal 2002).

The synthesis of hitherto unknown 1,2,3-triazolium core ionic liquid tagged TADDOL catalysts was achieved using commercially available *L*-(+)-tartaric acid dimethyl ester and relevant carbonyl compounds (Scheme 13) (Yacob 2010). The synthesis starts with the reaction between *L*-(+)-tartaric acid dimethyl ester and a carbonyl compound to furnish the dioxolane ring. This step is crucial in binding the two chiral centres of the TADDOL which render it the ability to induce chirality. The resulting ester is then treated with aryl Grignard reagents to furnish the tetraaryl dimethanol.

As a common practice, TADDOLs are often tethered through position 2 of the dioxolane ring. This keeps the tethered entry far from the two active tetraaryl dimethanol units. The functional groups necessary for the tethering can be introduced relatively easily and at earlier stages of the synthesis. Other sites on TADDOL can influence the catalytic activity by either steric congestion or permanently altering its stereochemistry (Seebach, Beck et al. 2001). Thus the synthesis of 1,2,3-triazolium ionic liquid tagged TADDOL organocatalysts **64 - 65** was achieved starting from 3-hydroxybenzaldehyde (Yacob 2010). Treatment of 3-hydroxybenzaldehyde with propargyl bromide in the suspension of potassium carbonate gave the 3-propargyloxybenzaldehyde **59**. Reaction of **59** with *L*-(+)-tartaric acid dimethyl ester **60** in the presence of excess trimethylorthoformate catalyzed by a small quantity of *p*-toluenesulfonic acid furnished the intended dioxolane **61** (Scheme 13). Grignard reaction was employed to introduce the necessary aromatic sub-units with moderate yields. The Cu (I) catalyzed click reaction of **62** with butyl or benzyl azide furnished the triazole compounds **63** in good yields. The 1,2,3-triazolium ionic liquid tags were finally prepared from these triazoyl tethered TADDOLs **63** by a straight forward alkylation and salt metathesis using methyl iodide and silver tetrafluoroborate, respectively. Investigations of the application of the new ionic liquid tagged TADDOLs **65** are currently underway.

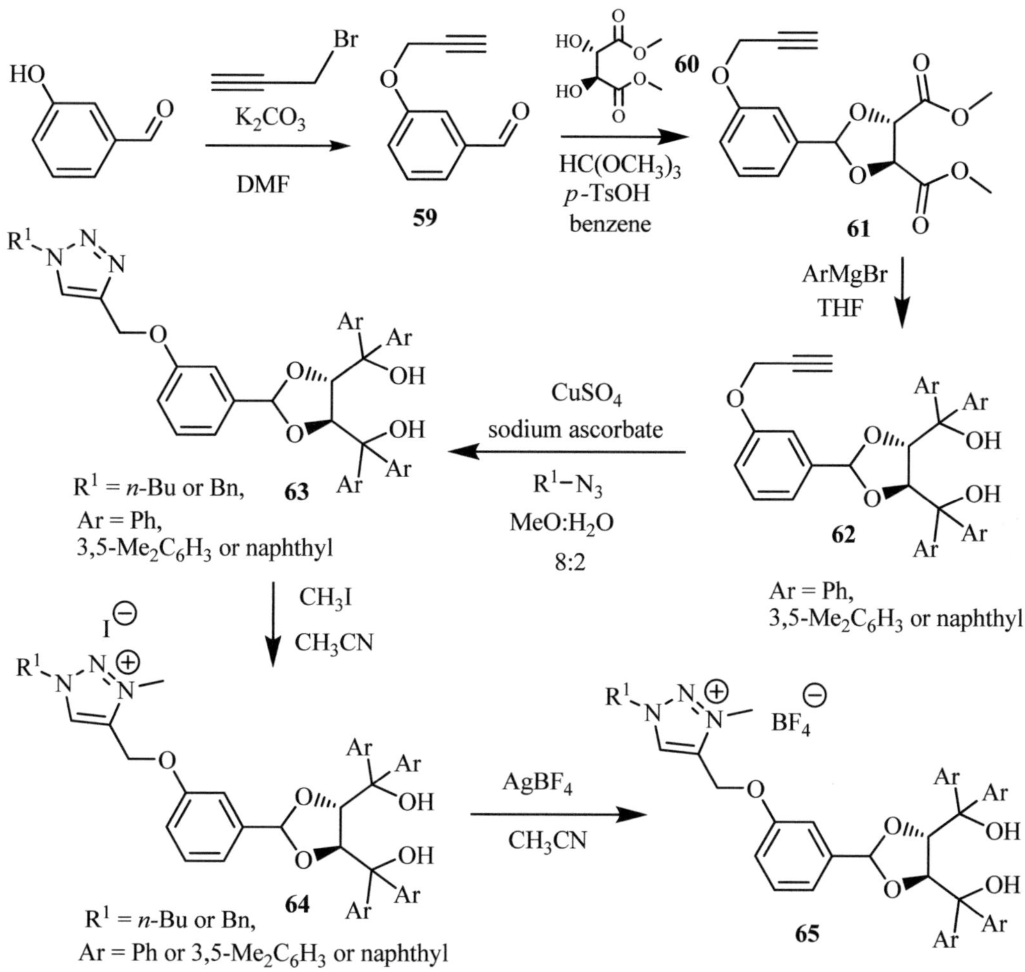

Scheme 13. Synthesis of 1,2,3-triazolium ionic liquid tagged TADDOL catalysts

Although in general the performance of 1,2,3-triazolium tagged organocatalysts in asymmetric syntheses was often excellent their behaviour in recycling experiments still needs some improvement because yields and enantioselectivities decreased after several runs. Leaching of the ionic liquid tagged organocatalysts can cause this effect. In cases of enamine activation by proline-derived ionic liquid tagged organocatalysts also trapping by oxazolidine formation (Khan, Shah et al. 2011) and in the light of recent findings of Maltsev et al. on the deactivation of ionic liquid tagged Jørgensen-Hayashi-type catalysts in asymmetric Michael reactions an oxidation of oxazoline or enamine intermediates can be responsible for the deactivation of the catalyst. While this phenomenon does not give a great effect in a single organocatalyzed reaction, it becomes crucial when the organocatalyst is repeatedly used after recycling. As shown by the same group oxidative poisoning of the ionic liquid tagged organocatalysts could be circumvented by working under oxygen free conditions (Maltsev, Chizhov et al. 2011). It seems to be worthy to examine the applications of 1,2,3-triazolium-based ionic liquid tagged organocatalysts in organocatalyzed reaction

running via intermediate enamines under oxygen free conditions because a better performance can be expected after recycling.
Very recently, transition metal complexes were reported where 1,2,3-triazolium salts served as precursors for 1,2,3-triazolylidene ligands (Mathew, Neels et al. 2008). However, application of these carbene complexes as catalysts in organic synthesis has not been reported so far. We tried to apply an in situ procedure in Suzuki reaction using $Pd_2(dba)_3$, 1,2,3-triazolium salts and $CsCO_3$ as base expecting formation of corresponding palladium-triazolydene complexes as catalytic species. However, the outcome of these reactions was not satisfactory neither when ligand precursors were used wherein two triazolium units were tethered to each other. However, the combination of a 1,2,3-triazolium- and a imidazolium unit **18** and **19** (Scheme 3) provided good results in Suzuki reactions. It can be assumed that the imidazolium unit is deprotonated in this case rather than the triazole moiety thus giving imidazolydene ligands for the Pd. Remarkably, the combination of one 1,2,3-triazolium with one imdidazolium salt performed much better than two imidazolium units in one compound showing the synergistic effect excreted by the triazolium moiety (Khan, Hanelt et al. 2009).

5. Miscellaneous

In a very unique approach cyano substituted triazolate based ionic liquids were prepared by exchanging the anions in common ionic liquids such as [emim][I], [empyrr][I] or [epy][I] with silver 4,5-dicyano-triazolate in water. Here the triazolate ring serves as a counter-anion with delocalized negative charge, and results in a diminished cation-anion interaction due to its decreased charge density. Hence, there is a decrease in the viscosity of the resulting ionic liquids as compared to those with iodide counter-ions (Kitaoka, Nobuoka et al. 2010).

6. Summary

1,2,3-Triazolium salts as a new class of ionic liquid can be prepared with a wide scope by Cu-(I)-catalyzed cycloaddition of terminal alkynes and azides followed by *N*-alkylation and eventual salt metathesis. This synthesis tolerates many functional groups and is therefore in particular important for developing functionalized ionic liquids. In this area already a number of successful applications in asymmetric catalysis were reported wherein organocatalysts with 1,2,3-triazolium tags gave higher yields and stereoselectivities than the bare organocatalyst. Recycling and re-usage of ionic liquid tagged organocatalysts was possible by extraction of reaction products with nonpolar solvents. The catalytic performance was found to be reproducible for a number of cycles.

7. References

Baskin, J. M.; J. A. Prescher; S. T. Laughlin; N. J. Agard; P. V. Chang; I. A. Miller; A. Lo; J. A. Codelli and C. R. Bertozzi (2007). "Copper-free click chemistry for dynamic in vivo imaging." *Proceedings of the National Academy of Sciences of the United States of America,* Vol.104, No.43, (Oct 23, 2007), pp. 16793-16797, ISSN 0027-8424

Begtrup, M. (1971). "Reactions between Azolium Salts and Nucleophilic Reagents .2. Bromo-1,2,3-Triazolium Salts and Sodium Hydroxide." *Acta Chemica Scandinavica,* Vol.25, No.1, pp. 249-259, ISSN 0904-213X

Bernardin, A.; A. Cazet; L. Guyon; P. Delannoy; F. Vinet; D. Bonnaffe and I. Texier (2010). "Copper-free click chemistry for highly luminescent quantum dot conjugates: application to in vivo metabolic imaging." *Bioconjugate Chemistry,* Vol.21, No.4, (Apr 21, 2010), pp. 583-588, ISSN 1520-4812

Betancort, J. M. and C. F. Barbas (2001). "Catalytic direct asymmetric Michael reactions: Taming naked aldehyde donors." *Organic Letters,* Vol.3, No.23, (Nov 15, 2001), pp. 3737-3740, ISSN 1523-7060

Betancort, J. M.; K. Sakthivel; R. Thayumanavan and C. F. Barbas (2001). "Catalytic enantioselective direct Michael additions of ketones to alkylidene malonates." *Tetrahedron Letters,* Vol.42, No.27, (Jul 2, 2001), pp. 4441-4444, ISSN 0040-4039

Betancort, J. M.; K. Sakthivel; R. Thayumanavan; F. Tanaka and C. F. Barbas (2004). "Catalytic direct asymmetric Michael reactions: Addition of unmodified ketone and aldehyde donors to alkylidene malonates and nitro olefins." *Synthesis,* No.9, (Jun 18, 2004), pp. 1509-1521, ISSN 0039-7881

Chan, T. R.; R. Hilgraf; K. B. Sharpless and V. V. Fokin (2004). "Polytriazoles as copper(I)-stabilizing ligands in catalysis." *Organic Letters,* Vol.6, No.17, (Aug 19, 2004), pp. 2853-2855, ISSN 1523-7060

Debets, M. F.; S. S. van Berkel; S. Schoffelen; F. P. J. T. Rutjes; J. C. M. van Hest and F. L. van Delft (2010). "Aza-dibenzocyclooctynes for fast and efficient enzyme PEGylation via copper-free (3+2) cycloaddition." *Chemical Communications,* Vol.46, No.1, (Nov 6, 2010), pp. 97-99, ISSN 1359-7345

Drake, G.; T. Hawkins; A. Brand; L. Hall; M. Mckay; A. Vij and I. Ismail (2003). "Energetic, low-melting salts of simple heterocycles." *Propellants Explosives Pyrotechnics,* Vol.28, No.4, (Aug, 2003), pp. 174-180, ISSN 0721-3115

Drake, G.; G. Kaplan; L. Hall; T. Hawkins and J. Larue (2007). "A new family of energetic ionic liquids 1-amino-3-alkyl-1,2,3-triazolium nitrates." *Journal of Chemical Crystallography,* Vol.37, No.1, (Jan, 2007), pp. 15-23, ISSN 1074-1542

Fletcher, J. T.; M. E. Keeney and S. E. Walz (2010). "1-Allyl- and 1-Benzyl-3-methyl-1,2,3-triazolium Salts via Tandem Click Transformations." *Synthesis,* No.19, (Oct, 2010), pp. 3339-3345, ISSN 0039-7881

Gajewski, M.; B. Seaver and C. S. Esslinger (2007). "Design, synthesis, and biological activity of novel triazole amino acids used to probe binding interactions between ligand and neutral amino acid transport protein SN1." *Bioorganic & Medicinal Chemistry Letters,* Vol.17, No.15, (Aug 1, 2007), pp. 4163-4166, ISSN 0960-894X

Gompper, R. (1957). "Untersuchungen in Der Azolreihe .5. Umsetzungen Der Oxazolone-(2) Mit Metallorganischen Verbindungen." *Chemische Berichte-Recueil,* Vol.90, No.3, pp. 374-382, ISSN 0009-2940

Hagiwara, H.; N. Komatsubara; H. Ono; T. Okabe; T. Hoshi; T. Suzuki; M. Ando and M. Kato (2001). "Diethylamino(trimethyl)silane-mediated direct 1,4-addition of naked aldehydes to electron-deficient olefins." *Journal of the Chemical Society-Perkin Transactions 1,* No.3, (Jan 11, 2001), pp. 316-322, ISSN 1472-7781

Hagiwara, H.; T. Okabe; K. Hakoda; T. Hoshi; H. Ono; V. P. Kamat; T. Suzuki and M. Ando (2001). "Catalytic enamine reaction: an expedient 1,4-conjugate addition of naked aldehydes to vinylketones and its application to synthesis of cyclohexenone from Stevia purpurea." *Tetrahedron Letters,* Vol.42, No.14, (Apr 2, 2001), pp. 2705-2707, ISSN 0040-4039

Huang, Y. and V. H. Rawal (2002). "Hydrogen-bond-promoted hetero-Diels-Alder reactions of unactivated ketones." *Journal of the American Chemical Society,* Vol.124, No.33, (Aug 21, 2002), pp. 9662-9663, ISSN 0002-7863

Jeong, Y. and J. S. Ryu (2010). "Synthesis of 1,3-dialkyl-1,2,3-triazolium ionic liquids and their applications to the Baylis-Hillman reaction." *Journal of Organic Chemistry,* Vol.75, No.12, (Jun 18, 2010), pp. 4183-4191, ISSN 1520-6904

Jiang, Y. B.; C. X. Kuang and Q. Yang (2009). "The Use of Calcium Carbide in the Synthesis of 1-Monosubstituted Aryl 1,2,3-Triazole via Click Chemistry." *Synlett,* No.19, (Dec 1, 2009), pp. 3163-3166, ISSN 0936-5214

Johansson, J. R.; P. Lincoln; B. Norden and N. Kann (2011). "Sequential One-Pot Ruthenium-Catalyzed Azide-Alkyne Cycloaddition from Primary Alkyl Halides and Sodium Azide." *Journal of Organic Chemistry,* Vol.76, No.7, (Apr 1, 2011), pp. 2355-2359, ISSN 1520-6904

Kaplan, G.; G. Drake; K. Tollison; L. Hall and T. Hawkins (2005). "Synthesis, characterization, and structural investigations of 1-amino-3-substituted-1,2,3-triazolium salts, and a new route to 1-substituted-1,2,3-triazoles." *Journal of Heterocyclic Chemistry,* Vol.42, No.1, (Jan-Feb, 2005), pp. 19-27, ISSN 0022-152X

Khan, S. S.; S. Hanelt and J. Liebscher (2009). "Versatile Synthesis of 1, 2, 3-Triazolium-based Ionic Liquids." *Arkivoc,* 2009), pp. 193-208, ISSN 1551-7004

Khan, S. S.; J. Shah and J. Liebscher (2010). "Synthesis of new ionic-liquid-tagged organocatalysts and their application in stereoselective direct aldol reactions." *Tetrahedron,* Vol.66, No.27-28, (Jul 3, 2010), pp. 5082-5088, ISSN 0040-4020

Khan, S. S.; J. Shah and J. Liebscher (2010). "Synthesis of new ionic-liquid-tagged organocatalysts and their application in stereoselective direct aldol reactions (vol 66, pg 27, 2010)." *Tetrahedron,* Vol.66, No.48, (Nov 27, 2010), pp. 9468-9468, ISSN 0040-4020

Khan, S. S.; J. Shah and J. Liebscher (2011). "Ionic-liquid tagged prolines as recyclable organocatalysts for enantioselectiive a-aminoxylations of carbonyl compounds." *Tetrahedron* Vol.67, (Jan 15, 2011), pp. 1812 -1820 ISSN 0040-4020

Kitaoka, S.; K. Nobuoka; N. Yoshiiwa; T. Harran and Y. Ishikawa (2010). "Preparation and Properties of Low Viscous Triazolate-based Ionic Liquids Containing Two Cyano Groups." *Chemistry Letters,* Vol.39, No.11, (Nov 5, 2010), pp. 1142-1143, ISSN 0366-7022

Kolb, H. C.; M. G. Finn and K. B. Sharpless (2001). "Click chemistry: Diverse chemical function from a few good reactions." *Angewandte Chemie-International Edition,* Vol.40, No.11, 2001), pp. 2004-2021, ISSN 1433-7851

Liu, M. N. and O. Reiser (2011). "A Copper(I) Isonitrile Complex as a Heterogeneous Catalyst for Azide-Alkyne Cycloaddition in Water." *Organic Letters,* Vol.13, No.5, (Mar 4, 2011), pp. 1102-1105, ISSN 1523-7060

Luo, S.; H. Xu; X. Mi; J. Li; X. Zheng and J. P. Cheng (2006). "Evolution of pyrrolidine-type asymmetric organocatalysts by "click" chemistry." *Journal of Organic Chemistry,* Vol.71, No.24, (Nov 24, 2006), pp. 9244-9247, ISSN 0022-3263

Maltsev, O. V.; A. O. Chizhov and S. G. Zlotin (2011). "chiral ionic liquid/esi-ms methodology as an efficient tool for the study of transformations of supported organocatalysts: deactivation pathways of jorgensen-hayashi-type catalysts in asymmetric michael reactions." *Chemistry,* Vol.17, No.22, (May 23, 2011), pp. 6109-6117, ISSN 1521-3765

Maltsev, O. V.; A. S. Kucherenko; A. L. Chimishkyan and A. G. Zlotin (2010). *Tetrahedron: Asymmetry,* Vol.21, (Nov 20, 2010), pp. 2659-2670, ISSN 0957-4166

Mathew, P.; A. Neels and M. Albrecht (2008). "1,2,3-triazolylidenes as versatile abnormal carbene ligands for late transition metals." *Journal of the American Chemical Society,* Vol.130, No.41, (Oct 15, 2008), pp. 13534-13535, ISSN 0002-7863

Mathews, C. J.; P. J. Smith and T. Welton (2000). "Palladium catalysed Suzuki cross-coupling reactions in ambient temperature ionic liquids." *Chemical Communications,* No.14, 2000), pp. 1249-1250, ISSN 1359-7345

Mosse, S.; O. Andrey and A. Alexakis (2006). "The use of N-iPr-2,2 '-bipyrrolidine derivatives as organocatalysts for asymmetric Michael additions." *Chimia,* Vol.60, No.4, 2006), pp. 216-219, ISSN 0009-4293

Orgueira, H. A.; D. Fokas; Y. Isome; P. C. M. Chan and C. M. Baldino (2005). "Regioselective synthesis of [1,2,3]-triazoles catalyzed by Cu(I) generated in situ from Cu(0) nanosize activated powder and amine hydrochloride salts." *Tetrahedron Letters,* Vol.46, No.16, (Apr 18, 2005), pp. 2911-2914, ISSN 0040-4039

Pachon, L. D.; J. H. van Maarseveen and G. Rothenberg (2005). "Click chemistry: Copper clusters catalyse the cycloaddition of azides with terminal alkynes." *Advanced Synthesis & Catalysis,* Vol.347, No.6, (May, 2005), pp. 811-815, ISSN 1615-4150

Sebesta, R.; I. Kmentova and S. Toma (2008). "Catalysts with ionic tag and their use in ionic liquids." *Green Chemistry,* Vol.10, No.5, 2008), pp. 484-496, ISSN 1463-9262

Seebach, D.; A. K. Beck and A. Heckel (2001). "TADDOLs, their derivatives, and TADDOL analogues: Versatile chiral auxiliaries." *Angewandte Chemie International Edition,* Vol.40, No.1, 2001), pp. 92-138, ISSN 1433-7851

Shah, J.; S. S. Khan; H. Blumenthal and J. Liebscher (2009). "1,2,3-Triazolium-Tagged Prolines and Their Application in Asymmetric Aldol and Michael Reactions." *Synthesis,* No.23, (Dec 1, 2009), pp. 3975-3982, ISSN 0039-7881

Sunden, H.; N. Dahlin; I. Ibrahem; H. Adolfsson and A. Cordova (2005). "Novel organic catalysts for the direct enantioselective alpha-oxidation of carbonyl compounds." *Tetrahedron Letters,* Vol.46, No.19, (May 9, 2005), pp. 3385-3389, ISSN 0040-4039

Yacob, Z. (2010). 1,2,3-Triazolium Ionic Liquid Tagged Catalysts in Asymmetric Organocatalysis Department of Chemistry and Bioorganic Chemistry. Berlin, Humboldt University of Berlin.

Yacob, Z.; J. Shah; J. Leistner and J. Liebscher (2008). "(S)-pyrrolidin-2-ylmethyl-1,2,3-triazolium salts - Ionic liquid supported organocatalysts for enantioselective Michael additions to beta-nitrostyrenes." *Synlett,* No.15, (Sep 15, 2008), pp. 2342-2344, ISSN 0936-5214

Yan, J. C. and L. Wang (2010). "Synthesis of 1,4-Disubstituted 1,2,3-Triazoles by Use of Copper(I) and Amino Acids Ionic Liquid Catalytic System." *Synthesis,* No.3, (Feb, 2010), pp. 447-452, ISSN 0039-7881

Yan, Z. Y.; Y. N. Niu; H. L. Wei; L. Y. Wu; Y. B. Zhao and Y. M. Liang (2006). "Combining proline and 'click chemistry': a class of versatile organocatalysts for the highly diastereo- and enantioselective Michael addition in water." *Tetrahedron: Asymmetry,* Vol.17, No.23, (Dec 11, 2006), pp. 3288-3293, ISSN 0957-4166

Zhang, L.; X. Chen; P. Xue; H. H. Sun; I. D. Williams; K. B. Sharpless; V. V. Fokin and G. Jia (2005). "Ruthenium-catalyzed cycloaddition of alkynes and organic azides." *Journal of the American Chemical Society,* Vol.127, No.46, (Nov 23, 2005), pp. 15998-15999, ISSN 0002-7863

Zincke, T. and A. T. Lawson (1887). "Azo-derivatives of phenyl-β -naphthylamine." *Ber. deut. chem. Ges,* Vol.20, (1887), pp. 1167-1176, ISSN 0365-9496

2

Thiazolium and Benzothiazolium Ionic Liquids

Munawar Ali Munawar and Sohail Nadeem
Institute of Chemistry, University of the Punjab, Lahore Pakistan

1. Introduction

In a general perspective the term "ionic liquid" includes all classical molten salts (Visser 2002), even those which are composed of more thermally stable ions, such as sodium with chloride or potassium with nitrate. Although the term dates back as early as 1943 (Barrer 1943), in the language/field of chemistry an ionic liquid (IL) is specifically a salt having organic cation and organic/inorganic anion, which is liquid at room temperature or reaction temperature. Wasserscheid and Keim (Wasserscheid&Keim 2000) have proposed that an organic salt having a melting point below 100 °C could be called ionic liquid and this is indeed now one of the most widely accepted definitions of ionic liquids. Some scientists consider this set point as 150 °C (Huddleston *et al.* 2001). The low melting behaviour of ionic liquids is due to the poor coordination of the cations and anions. The strength of coordination depends upon the nature and structure of the cation and anion and a little unsymmetry in the structure may lead to decrease the coordination of the ions.
The most common heteroaromatic based ionic liquids include imidazolium, thiazolium, tetrazolium, pyridinium *etc.* However thiazolium and benzothiazolium based ionic liquids are very scarcely studied. This chapter will describe the synthesis and applications of thiazolium and benzothiazolium based ionic liquids.

2. Thiazolium salts / ionic liquids

The thiazoles are known in chemistry mainly due to their presence in thiamine (Vitamin B_1) in the form of substituted thiazolium salt (Mcguinness *et al.* 2001). Thiazolium salts can be obtained successfully by a modification of the Hanztsch's thiazole synthsis. This method is particularly valuable for those thiazolium compounds in which the substutuent on the ring nitrogen cannot be introduced by direct alkylation, for example, aryl or heteroaryl thiazolium salts.

1 **2** **3**

Scheme 1. Preparation of thiazolium salts.

N-Monosubstututed thioamides (**1**) can be cyclized with α-halocarbonyl compounds (**2**) to give thiazolium salts (**3**) in excellent yields. Quaternization of thiazole strongly enhances the reactivity of the thiazole ring towards nucleophiles. The following reactions have been observed. Addition of a nucleophile to the 2-position to give a pseudobase, followed by ring opening, e.g. with sodium hydroxide solution:

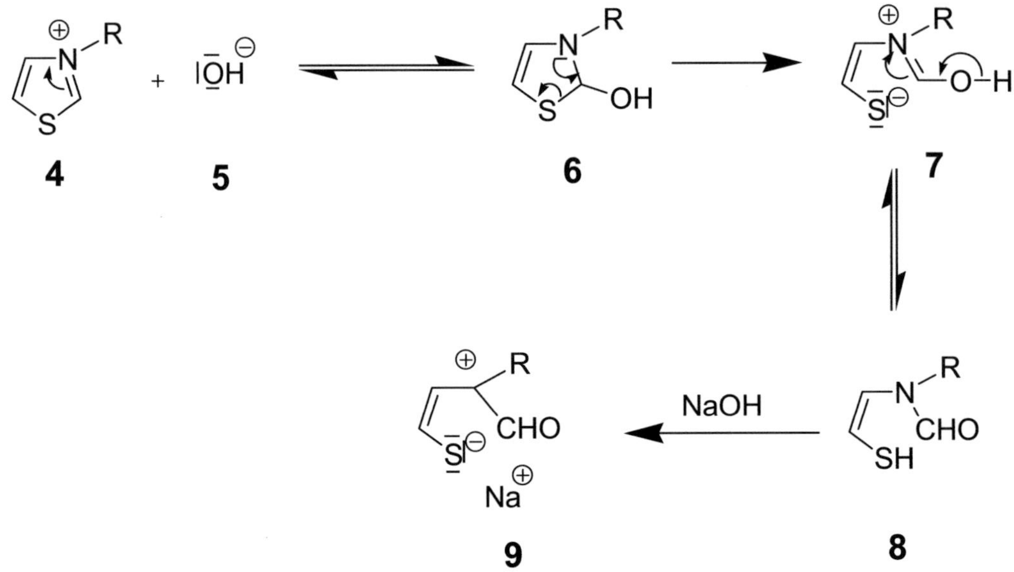

Scheme 2. Reactivity of thiazolium salts

3-Alkyl substituted thiazolium salts react in an analogous way. The deuterodeprotonation of 3-alkylthiazolium salts (*e.g.* **10**) by D_2O proceeds via an *N*-ylide (*e.g.* **11,** sheme-3).

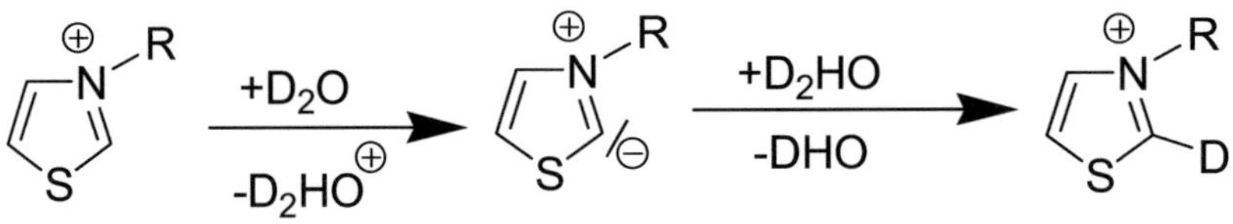

10 **11** **12**

Scheme 3. Dueterium exchange of thiazolium salts.

Todd *et al.* have reported a very convenient method for the preparation of *N*-alkylthiazolium salts. The method of Hantzsch thiazole synthesis was modified to obtain direct formation of thiazolium salts. *N*- substituted thioamides (*e.g.* **13**) were used instead of the unsubstituted thioamides to obtain the desired thiazolium salts (*e.g.* **15**) (Todd *et al.* 1936)

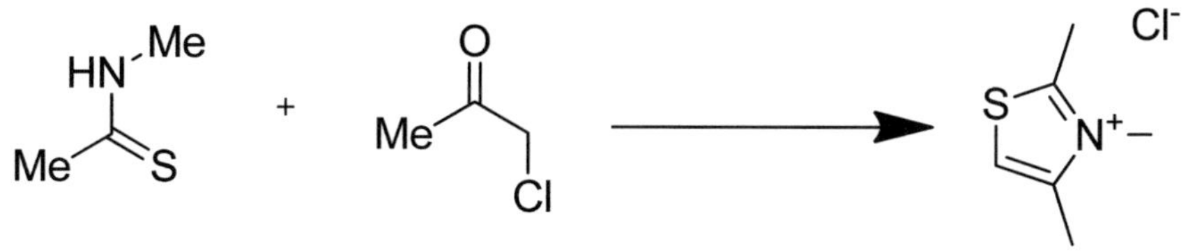

13 14 15

Scheme 4. A convenient method for thiazolium salts.

Schöberl and Stock have reported the synthesis of *N*-Benzyl-2,4-dimethylthiazolium chloride followed by the ion exchange with potassium iodide to get *N*-Phenyl-2,4-dimethylthiazolium iodide (**18**) from *N*-phenylthioacetanilide (**16**) and chloroacetone (**17**) (scheme-5) (Schöberl&Stock 1940).

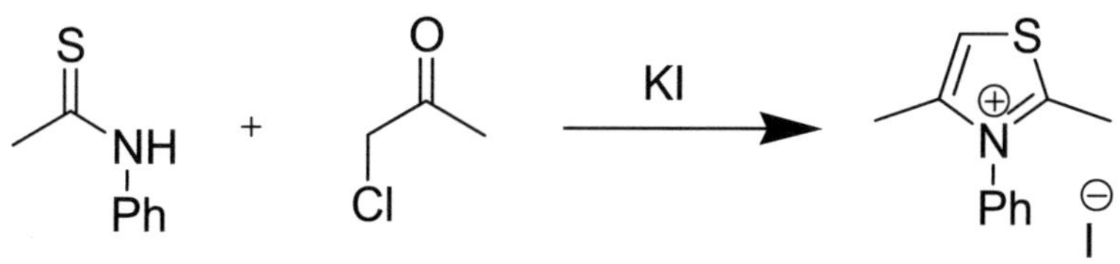

16 17 18

Scheme 5. Preparation of *N*-phenyl-2,4-dimethylthiazolium iodide.

The reaction of *N*-methyl-p-dimethylaminothiobenzamide (**20**) with a number of α-halo ketones (**19a-19d**) and one α -halo aldehyde (**19e**) have been studied by Egan and his coworkers and they were able to get stable 4-hydroxy-2-thiazolinium derivatives , which were isolated as the iodide salts (**21**) (scheme-6) (Egan *et al.* 1968). The 4-hydroxy-2-thiazolinium salts (**21**) were stable in neutral and basic media, but could be dehydrated to the thiazolium salts (**22**) by treating with methanolic hydrogen chloride. Dehydration could also be accomplished, although less conveniently, by treating with methanesulfonyl chloride containing sulfur dioxide in the presence of collidine (Hazen&Rosenburg 1964).
There are many reviews and books regarding the synthesis and applications of thiazolium salts in different fields (Eicher *et al.* 2003). However thiazolium based ionic liquids have not been well documented in the literature. Many researchers have studied the preparation and applications of the thiazole based ionic liquids in different reactions. A brief account of that has been given here.
Chen *et al.* prepared [2-^{13}C]-labeled 3-benzyl and 3-methylthiazolium tetrafluoroborate (may be liquids in nature) and studied the reactions of benzaldehyde with thiazolium salts in Me_2SO. They found that the reactions of thiazolium salts with aromatic aldehydes: p-anisaldehyde and cinnamaldehyde, in MeOH/MeONa, led to the formation of significant amounts of the corresponding dimethyl acetals, rather than to the benzoin products (Chen *et al.* 1994).

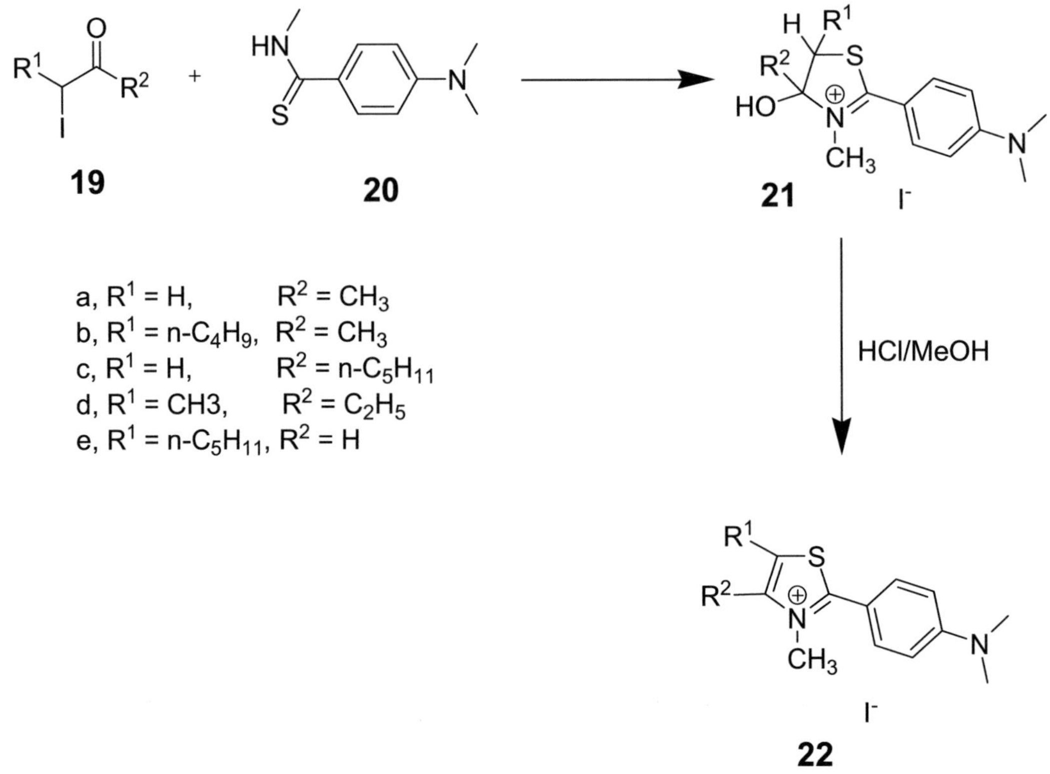

Scheme 6. Thiazolinium derivatives.

Motesharei and Myles prepared thiazolium salts (**e.g. 26**, scheme 7) for the synthesis of self-assembled thiazolium thiolate monolayers on gold (fig-1) which promoted the acyloin condensation. (Motesharei&Myles 1997)

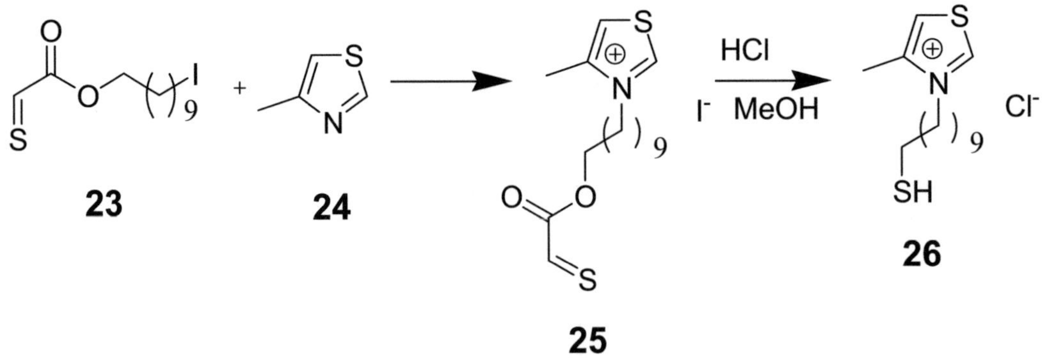

Scheme 7. Thiazolium thiols preparation.

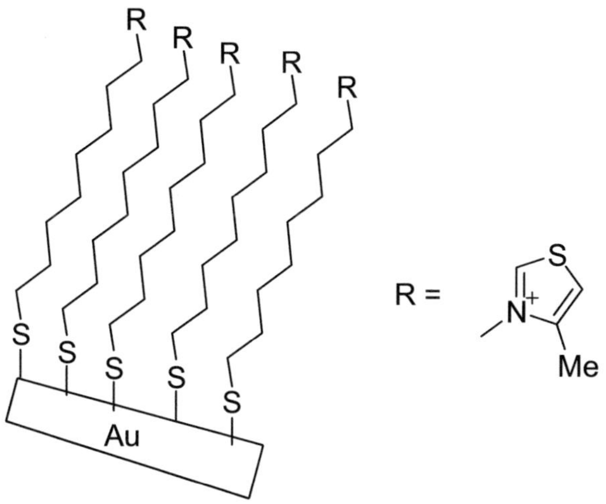

Fig. 1. Thiazolium thiolate monolayers on gold

Davis and Forrester prepared thiazolium based organic ionic liquids (**29**, scheme-8) and used them for catalyzing the benzoin condensation (Davis&Forrester 1999).

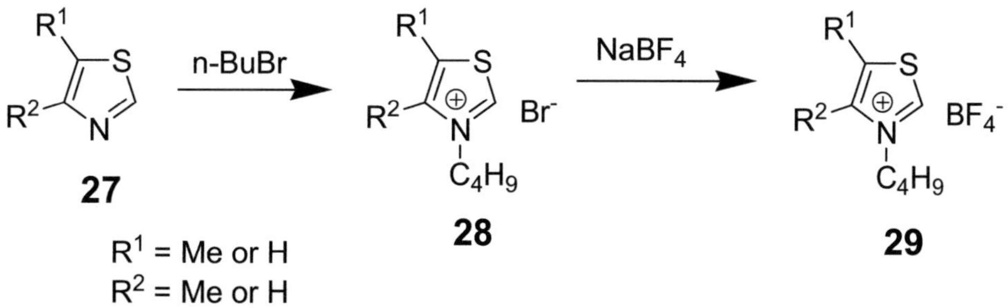

27 **28** **29**

Scheme 8. Tetrafluoroborate thiazolium salts.

The first thiazolium gold (III) compound (**32**) that qualifies as an ionic liquid has been prepared by Deetlefs *et.al.* during the study of stoichiometric and catalytic reactions of gold utilizing ionic liquids (scheme-9) (Deetlefs *et al.* 2002).

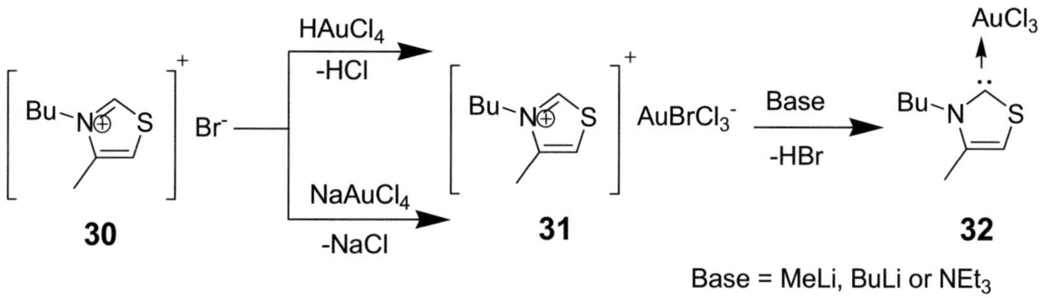

30 **31** **32**

Scheme 9. Thiazolium- gold catalyst.

Yu *et al*. have reported a series of thermoregulated thiazolium based ionic liquids containing polyether moieties attached to thiazolium cation (**36**, scheme-10). These thermoregulated ionic liquids (**36**) were used as catalysts in the Stetter reaction (Yu *et al.* 2010). The ionic liquids were pale yellow, viscous liquids at room temperature. The viscosity was found to increase with the length of the polyether chain. The structures of these ionic liquids were determined by NMR.

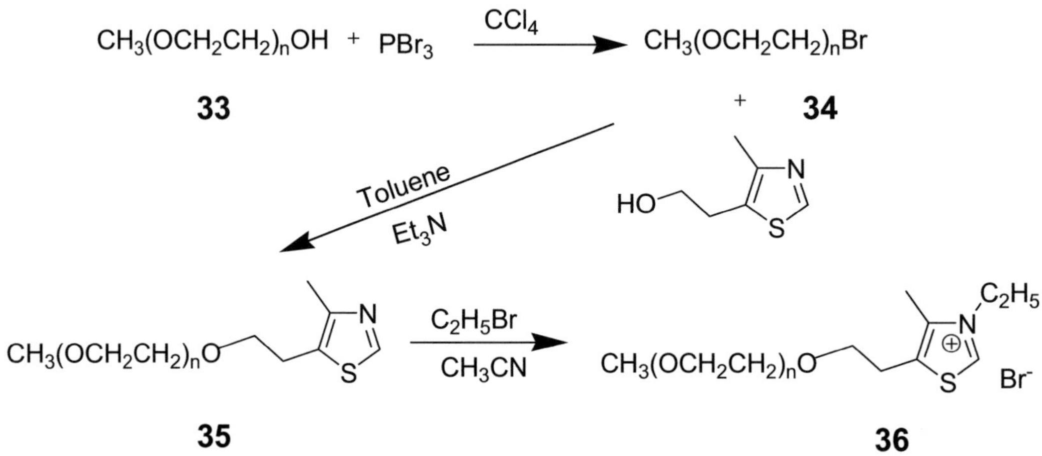

Scheme 10. Thermoregulated thiazolium based ionic liquids.

Li and Xu prepared the 2-bromo-3-ethyl-4-methylthiazolinium tetrafluoroborate (BEMT) (**41**, scheme-11), and found it to be an efficient coupling reagent for hindered amino acids. It was used for the coupling of *N*-alkyl or α,α-dialkyl amino acids (Li&Xu 1999).

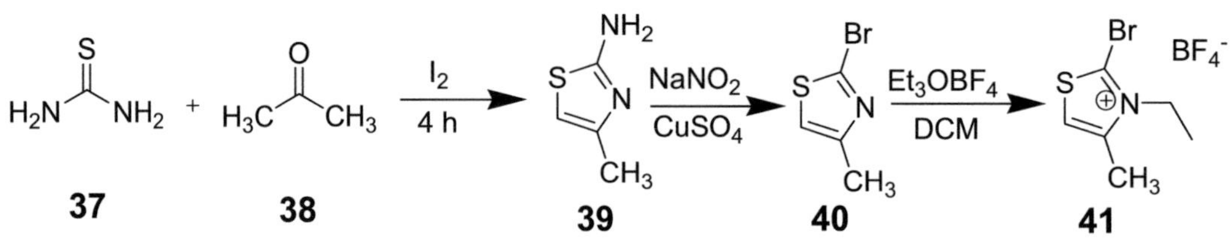

Scheme 11. Preparation of 2-bromo-3-ethyl-4-methylthiazolinium tetrafluoroborate (BEMT).

The dramatic influence of a new tailor-made, task-specific, and stable ionic liquid, 3-[2-(1-butyl-1*H*-imidazol-3-ium-3-yl)ethyl]-4,5-dimethyl-1,3-thiazol-3-ium bromide (**44**, scheme-12), in benzoin condensations have been described by Mohanazadeh and Aghvami (Mohanazadeh&Aghvami 2007).

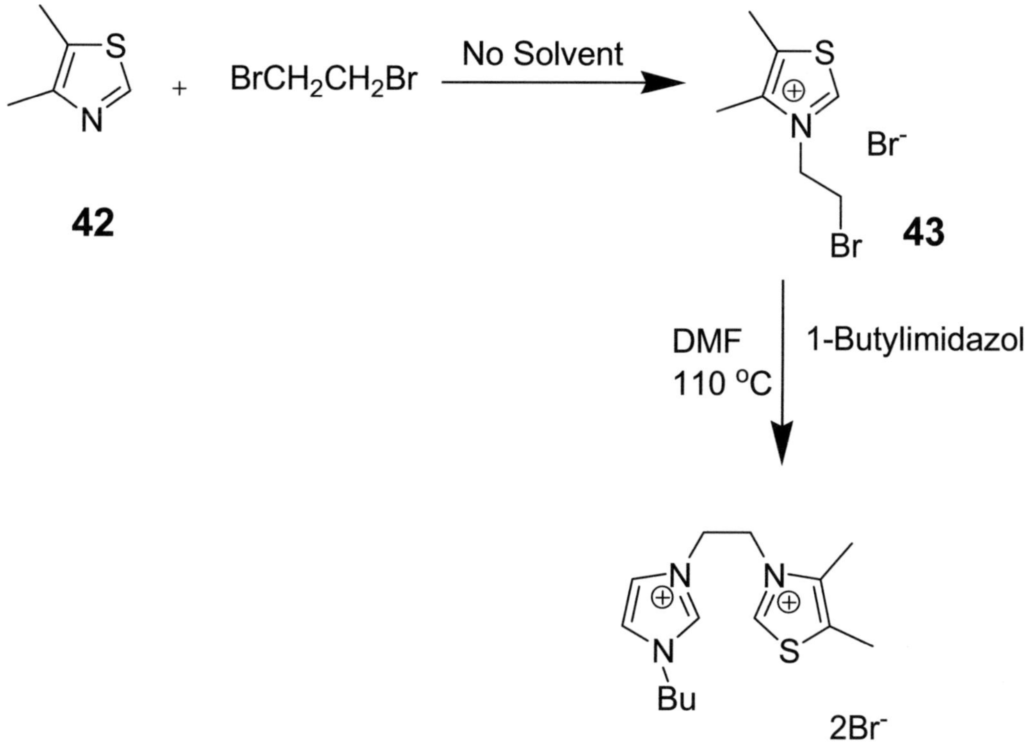

Scheme 12. Preparation of 3-[2-(1-butyl-1*H*-imidazol-3-ium-3-yl)ethyl]-4,5-dimethyl-1,3-thiazol-3-ium bromide.

3. Benzothiazolium salts and ionic liquids

Benzothiazoliums are also a part of many important dyes (Svetlichnyĭ *et al.* 2007) *e.g* Cyanine Dyes, Thioflavin T & S, (Stsiapura *et al.* 2007) (Sabaté&Saupe 2007) Direct yellow 7, Carnotine, Primuline, *etc.* In medicine the benzothiazolium salts play an important role, for example Riluzole a benzothiazole-based drug is used to treat amyotrophic lateral sclerosis. (Kitzman 2009).

During the attempts to prepare the cyanine dyes (**49**) in 1887, Hofmann was able to isolate benzothiazolium salts which were produced by reaction of benzothiazole (**47**) with amyl iodide (Scheme-13) (Hofmann 1887).

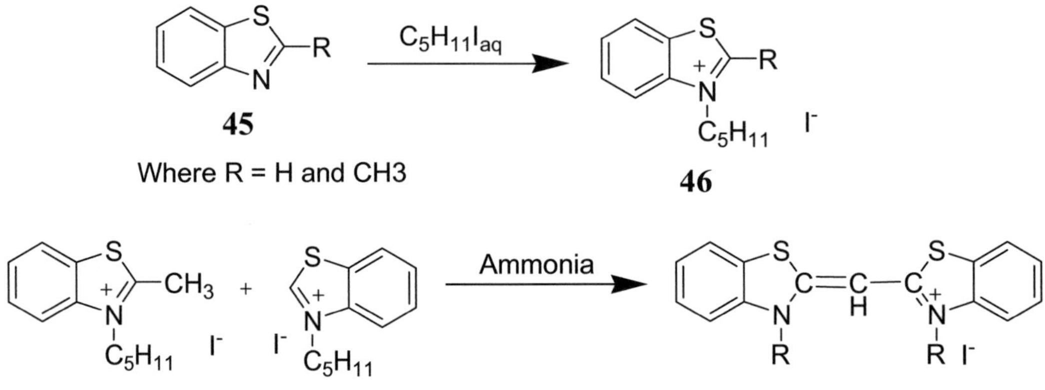

45 46 47 48 49

Scheme 13. Hofmann Benzothiazolium synthesis

Mills (1922) prepared N-Ethylbenzothiazolium iodide (**51**)), to be used in the synthesis of thiocyanine dyes, by heating the equimolar quantities of ethyl iodide and benzothiazole (**50**) (scheme-14) (Mills 1922).

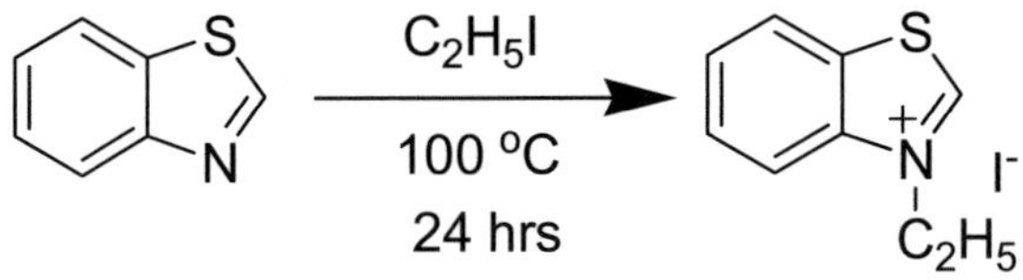

50 51

Scheme 14. Mills benzothiazolium synthesis

Then in 1925 Konig used *N*,2-dimethylbenzothiazolium perchlorate (**52**) for the preparation of methylene base (**53,** scheme-15) which was used in the synthesis of thiocyanine dyes (Konig&Meier 1925). The quaternary salts of benzothiazoles also found recognition in 1928 by Clark (Clark 1928) while studying the absorption sensitivity of the methylene bases from the quaternary salts of 2-methyl benzothiazole.

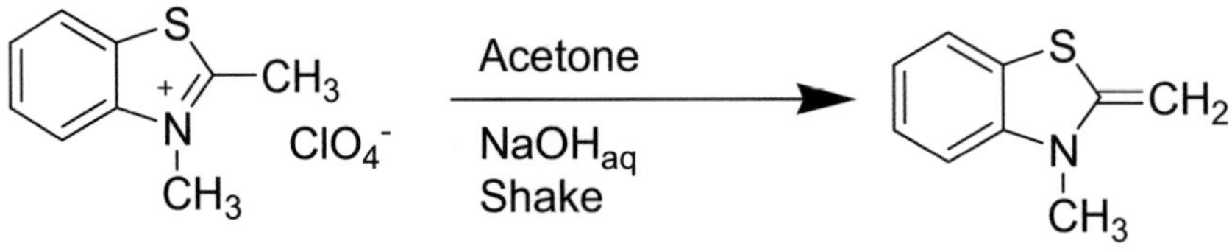

52 53

Scheme 15. Konig preparation of methylene bases from benzothiazolium salts

In the same year Konig (König 1928) used the quaternary salts of benzothiazole for the preparation of the heterocyclopolymethine dyes during investigations on color and constitution. He claimed that a number of 2-methylbenzothiazoles (**56**) can be prepared much more conveniently by heating alkaline solutions of 2-aminobenzenethiol (**55**) with acetic anhydride (**54**). The benzothiazoles (**56**) were converted to quaternary salts (**57**) by reacting with alkyl halides or dimethyl sulfate (scheme-16).

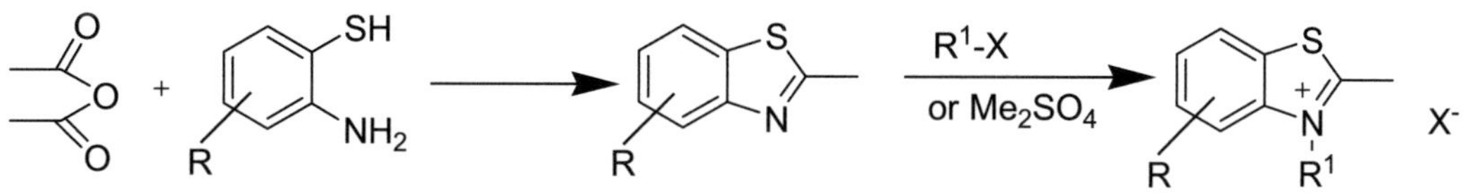

54 **55** **56** **57**

Where X= Halogen or CH_3SO_4

Scheme 16. Konig benzothiazolium synthesis from 2-aminobenzenethiol

Fisher and Hamer prepared benzothiazolinium chlorides (e.g. **59**) from the corresponding iodides (e.g. **58**), these chlorides (e.g. **59**) were then converted to 2-cycnobenzothiazolium chlorides (e.g. **61**) for their onward conversion to thiocyanines (**62**, scheme-17) (Fisher&Hamer 1930).

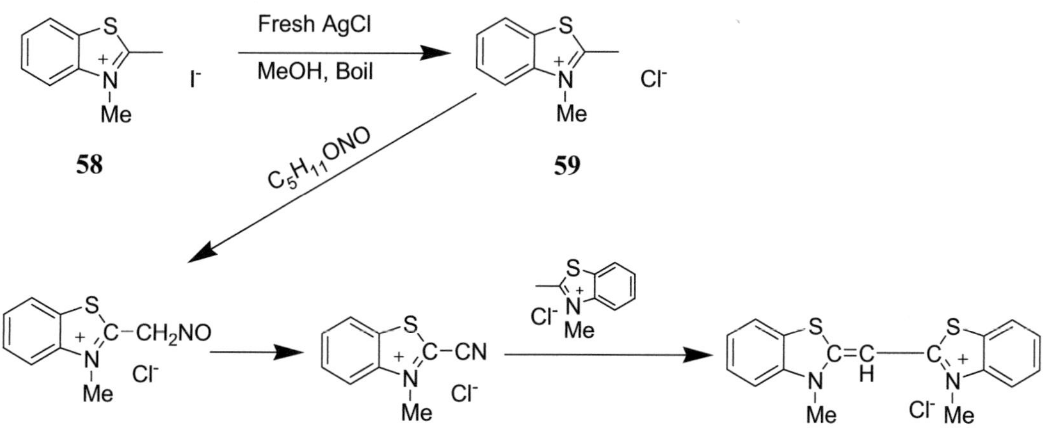

58 **59** **60** **61** **62**

Scheme 17. Fisher and co-workers benzothiazolium synthesis

In 1935 Evans and Smiles obtained 2-methyl-*N*-(2-nitrophenyl)benzothiazolium iodide (**64**) when 2-[*N*-acetyl-*N*-(2-nitrophenylamino)]benzenethiol (**63**) was treated with hydroiodic acid in acetone (scheme-18). The same salt was also isolated when 2-[*N*-acetyl-*N*-(2-nitrophenylamino)]benzenesulfinic acid was reduced with hydriodic and sulfurous acids. The corresponding perchlorate was prepared by treating the benzenethiol with perchloric acid (Evans&Smiles 1935).

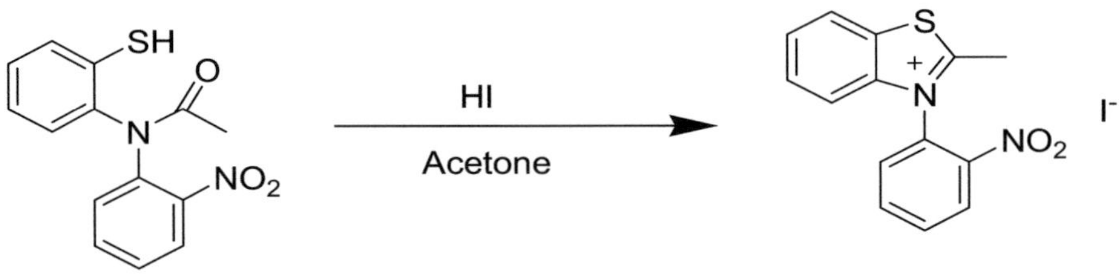

63 **64**

Scheme 18. Evans benzothiazolium synthesis

After one year Beilenson and Hamer prepared the benzothiazoles based quaternary salts (e.g. **66**) by sealed tube reactions of benzothiazoles (e.g. **65**) with methyl iodide at 100 °C (scheme-19) and used them for the preparation of cyanine dyes (Beilenson&Hamer 1936).

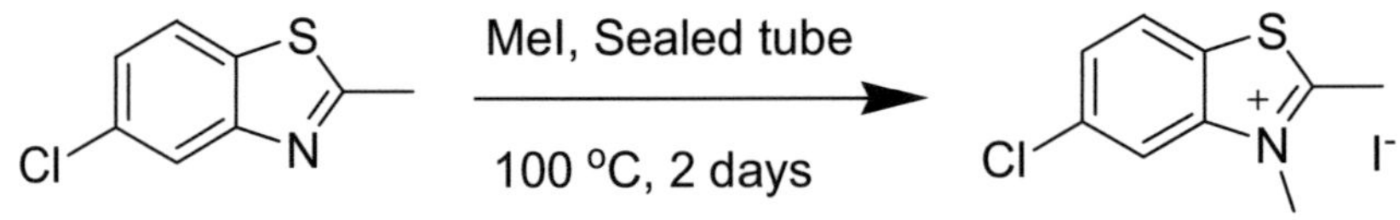

65 **66**

Scheme 19. Beilenson and Hamer benzothiazolium synthesis

In 1941 Brooker and co-workers prepared quaternary salts (**70**) of 2-alkyl/aryl sulfanyl benzothiazoles (**69**) (scheme-20) (Brooker *et al.* 1941).

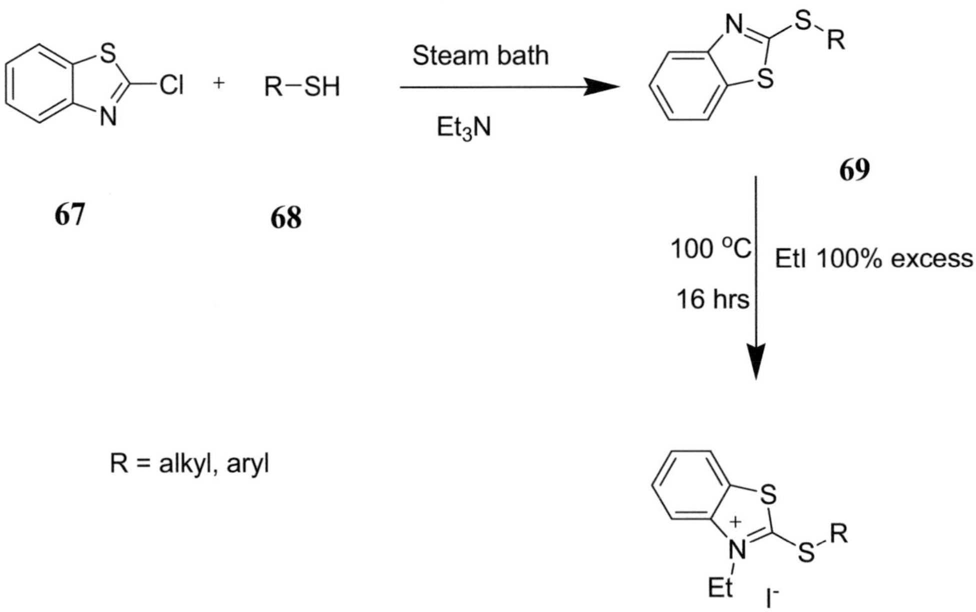

67 **68** **69**

70

Scheme 20. Brooker and co-workers benzothiazolium synthesis

In 1948 Kendall and Majer prepared (Kendall&Majer 1948) 2-(2-ethylsulfanylvinyl)-*N*-methylbenzthiazolium iodide (**74**, R = H) and 2-(2-ethylsulfanyl-1-methylvinyl)-*N*-methylbenzthiazolium iodide (**74**, R = Me) by fusing 2-methylbenzothiazole (**71**, R = H) and 2-ethylbenzothiazole (**71**, R = Me) with methyl-p-toluene sulphonate respectively followed by treating the resultant quaternary salts with tri(ethylsulfanyl)methane and 20% KI solution (scheme-21).

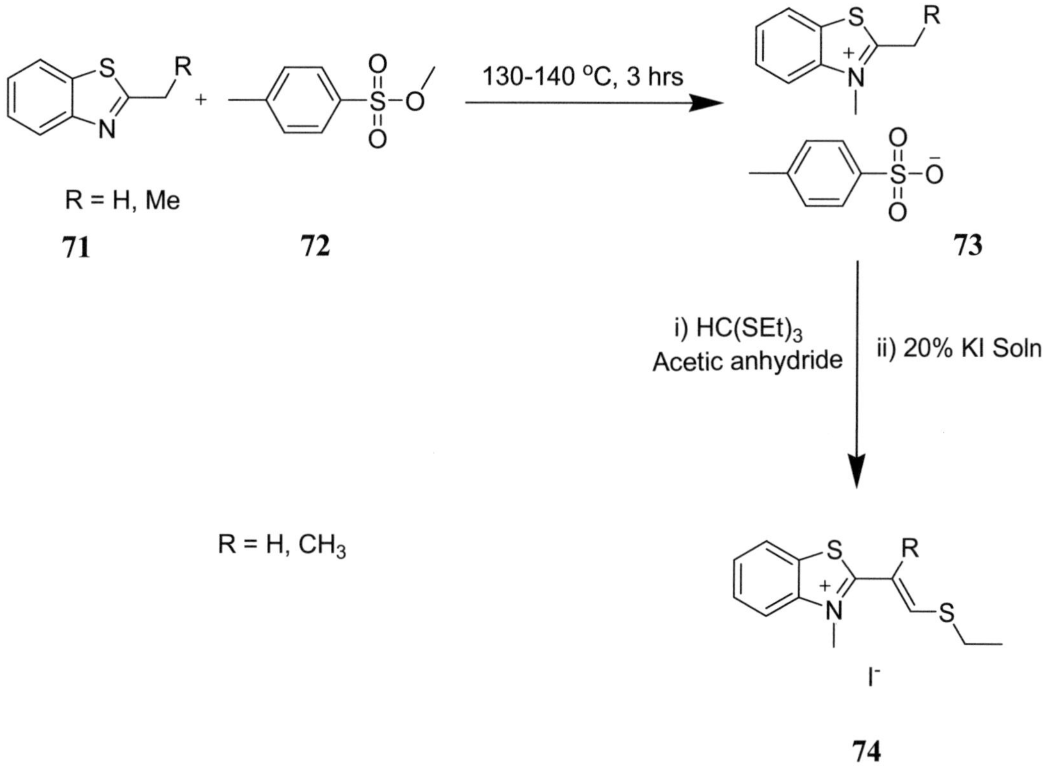

Scheme 21. Kendall and Majer preparation of 2-(2-ethylsulfanylvinyl)-*N*-methylbenzthiazolium iodide

A direct synthesis of quaternary salts (*e.g.* alkylsulfates, iodides, bromides, perchlorates) of benzothiazole from 2-aminobenzenethiol has been narrated in literatue by Kiprianov and Pazenko. Fry and Kendall synthesized quaternary salts of 2-alkylsulfanylbenzothiazole and used them in the synthesis of thiacyanine dyes (Fry&Kendall 1951). Similar type of synthesis of 2-haloalkylsufanyl benzothiazolium salts was carried out by Knott in 1955 (Knott 1955). Brooker (1951) prepared benzothiazolium salts from substituted benzothiazoline and *p*-toluensulfonate for further conversion to cyanine dyes (Brooker *et al.* 1951).

Kiprianov (1957) studied the relative rates of formation of quaternary salts of 2-methylbenzothiazole with methyl iodide, methyl sulfate, and methyl esters of sulfonic acids and *p*-nitrosulfonic acids. He found that the reaction with methyl 2-nitrobenzenesulfonate was six times faster than that with Me_2SO_4, while the methyl esters of the 2,3- and 2,4-dinitrobenzenesulfonic acid reacted sixty times more rapidly than did Me_2SO_4. Such esters afford a method for very mild alkylation of very weak bases (Kiprianov&Tolmachev 1957).

Benzothiazolium salts have great use in the synthesis of different types of cyanine dyes while on the other hand the benzothiazolium salts have also been studied for their activity in different biological systems. In 1959 Pianka and Hall studied the fungi toxicity of 3-ethylbenzothiazolium iodide and 3-ethylbenzoxazolium ethyl sulfate derivatives and activity was found to reside only in the cation (Pianka&Hall 1959). In the same year (1959) Leslie and co-workers patented a method for the preparation of quaternary salts of benzothiazole for making the thiacarbocyanine dyes. In 1961 Horwitz and co-workers prepared a benzothiazolium salt (**76**) from 6-(2-aminocarbonyl)ethoxy-2-methylbenzothiazole (**75**, Scheme-22) for the use in preparation of different dyes.

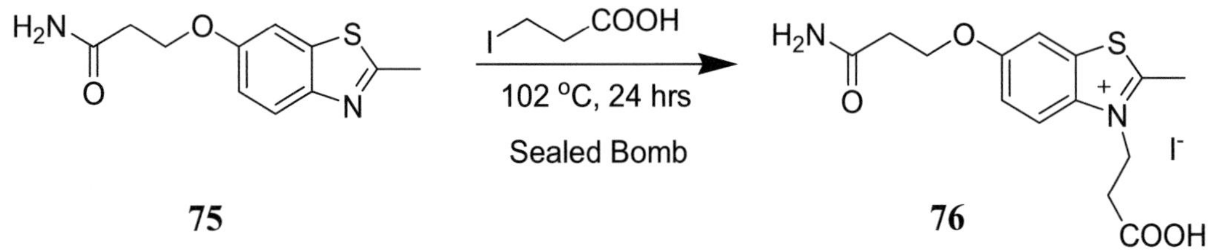

Scheme 22. Horwitz and co-workers preparation of benzothiazolium salts

In 1967 Messmer and Gelleri prepared 1,3-diphenyl[1,2,3]triazolo[5,1-*b*]-benzothiazolium and benzimidazolium bromide salts by the action of NBS on benzothiazol-2-yl-phenyl ketone phenyl hydrazone in ethyl acetate at room temperature (Messmer&Gelléri 1967). Vorsanger (1968) made the spectroscopic studies for the existence of carbenes generated from the heterocyclic bases which were prepared from the quaternary salt (**78**) of the benzothiazole with dietyl sulphate. Similar studies were made with 3-deuteriomethylbenzothiazolium iodide (**79**) prepared by heating the CD_3I with benzothiazole in a sealed ampule at 130 °C for 4 hours. He prepared the dimers of (**78**) and (**79**) by mixing themselves and also by mixing (**80**) and (**81**) separately to get the hybrid dimer (**83**) (Scheme-23) (Vorsanger 1968).

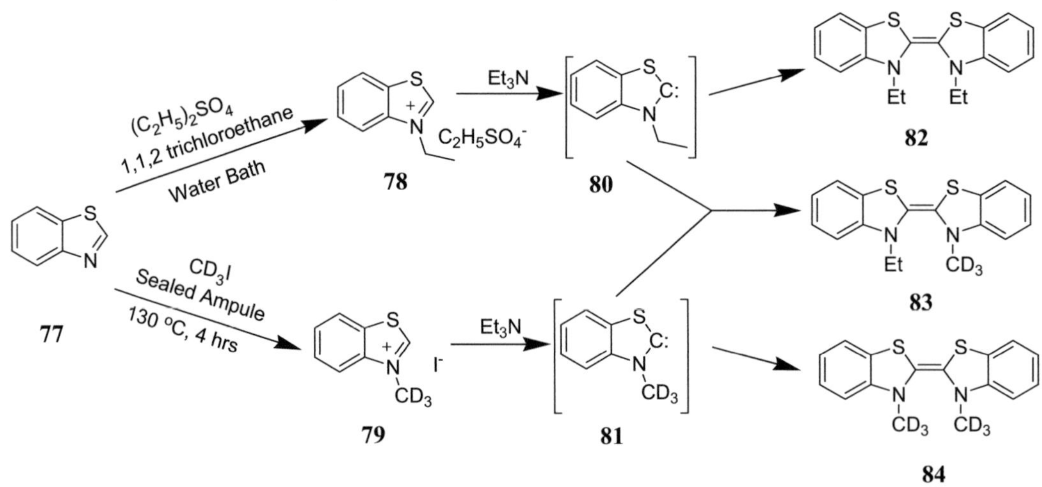

Scheme 23. Vorsanger's synthesis of 3-deuteriomethylbenzothiazolium iodide

In 1969 Garmaise *et al.* prepared a number of benzothiazolium salts (**87**) from 2-alkylaminobenzenethiol (**85**) (Scheme-24). A study of the anthelmintic activity of the benzothiazolium salts in comparison to the dye thioflavin T was made. Compounds which showed activity were all closely related to thioflavin in that they had a 2-(*p*-dialkylaminophenyl) substituent and were quaternized on the heterocyclic nitrogen atom (Garmaise *et al.* 1969).

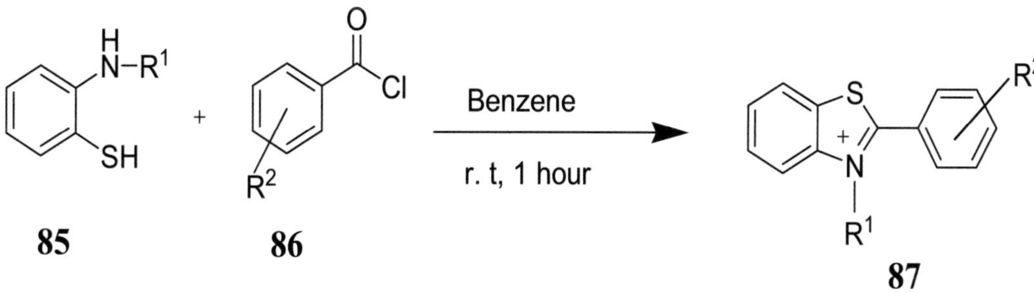

Where $R^1 = CH_3$, *n*-C_3H_7, $CH_2CH{=}CH_2$ and R^2 = 2-Cl, 4-Cl, 4-F *etc.*

Scheme 24. Benzothiazolium salts from 2-alkylaminobenzenethiol

In 1970 Weinhardt *et al.* prepared 6-methoxy-2-oxo-1,2,3,4-tetrahydropyrimido[2,1-*b*]benzothiazol-5-ium chloride (**89**) by heating 2-(3-chloropropanoylamino)-4-methoxybenzothiazole (88) and they observed that the salt with base i.e. diethylamine lead to chloride free product which can be readily converted back to the salt by heating with HCl saturated chloroform. They also claimed that the salt (**88**) can be converted in to (2-amino-6-methoxybenzothiazol-*N*-propanoic acid (**91**, R = H) and their esters (**91**, R = Me or Bu) by reacting with H_2O and alcohol respectively (scheme-25) (Weinhardt&Neumeyer 1970).

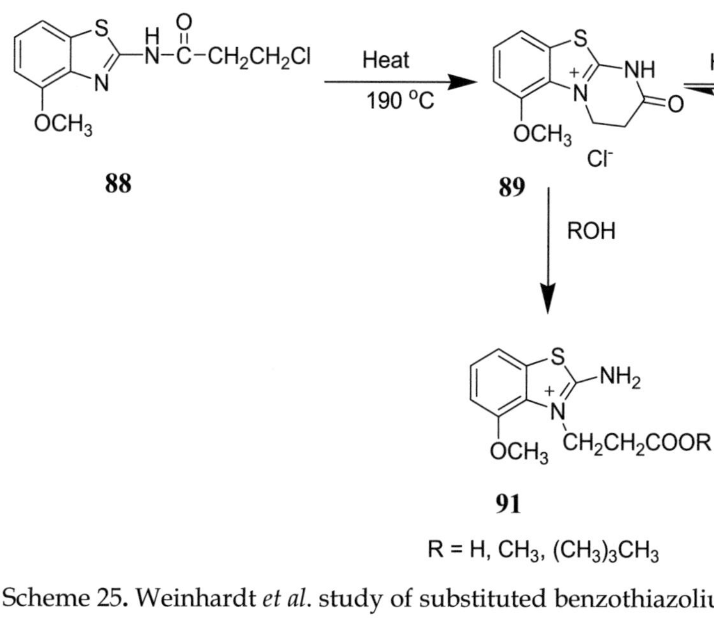

R = H, CH_3, $(CH_3)_3CH_3$

Scheme 25. Weinhardt *et al.* study of substituted benzothiazolium salts

In 1971 Mushkalo *et al.* prepared the benzothiazolium salts (**94**) by the condensation of 2-aminobenzenethiols (**92**) with different α-haloketones (**93**) (scheme-26) (Mushkalo *et al.* 1971).

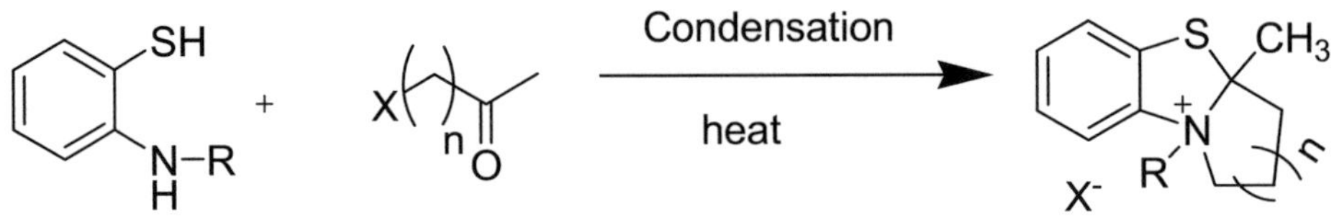

92 R = H, Me

93 X = Cl, Br; n = 3, 4

94 n = 1,2; R = H, Me; X = Cl, Br

Scheme 26. Synthesis of benzothiazolium salts by the condensation of 2-amino benzenethiol with different α-haloketones

Takamizawa *et al.* (1972) narrated that benzothiazine derivatives (**97**) could be obtained by treating the benzothiazolium salts (**95**) with diethyl acylphosphonates followed by basic hydrolysis (Scheme-27) (Takamizawa *et al.* 1972).

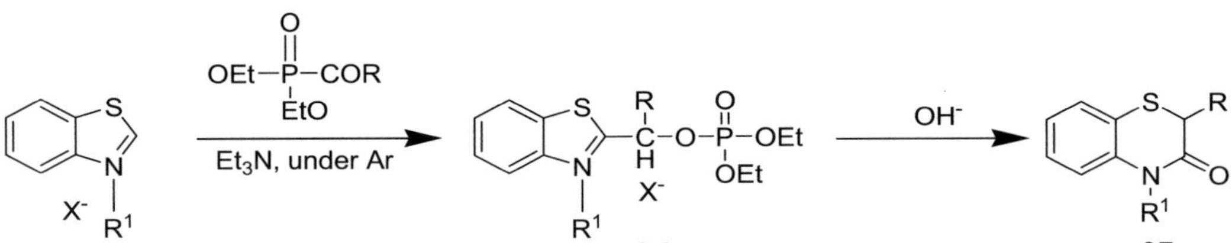

95 **96** **97**

Where R^1 = CH_2Ph, Me, *p*-CH_2PhNO_2
R^2 = Ph, Me, CH_2Ph, *p*-$MePhCH_2$, *p*-$ClPhCH_2$

Scheme 27. Benzothiazine derivatives from benzothiazolium salts

Takahashi *et al.* (1973) studied the aminolysis of the *N*-methyl-2-ethoxycarbonylbenzothiazolium perchlorate **(98)** with a series of amines and obtained *N*-methylbenzothiazolium chlorate **(100)**, carbamates (**101**) in addition to the amides (**99**) (scheme-28) (Takahashi *et al.* 1973).

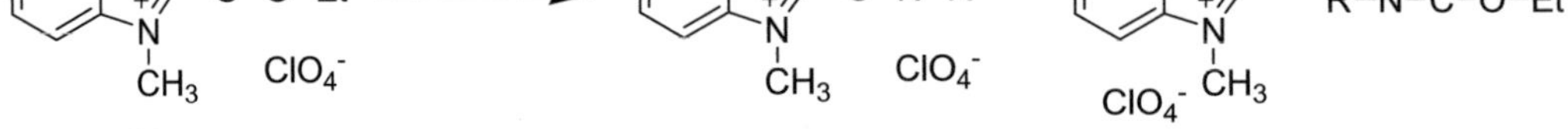

98 **99** **100** **101**

Amine = NH_3, $BuNH_2$, *iso*-$PrNH_2$, *tert*-$BuNH_2$, piperdine

Scheme 28**.** Reaction of *N*-methylo-2-ethoxycarbonylbenzothiazolium perchlorate with the amine

Chapman *et al.* (1974) found that protonated heterocycles including benzthiazolium bromide (**102**), possessing an alkyl substituent (*e.g.* benzyl) on the carbon adjacent to the positive nitrogen reacted with methylvinylketone to form adduct (**103**) which can be converted into fused pyridinium salts (**104,105**) (Scheme-29) (Chapman *et al.* 1974).

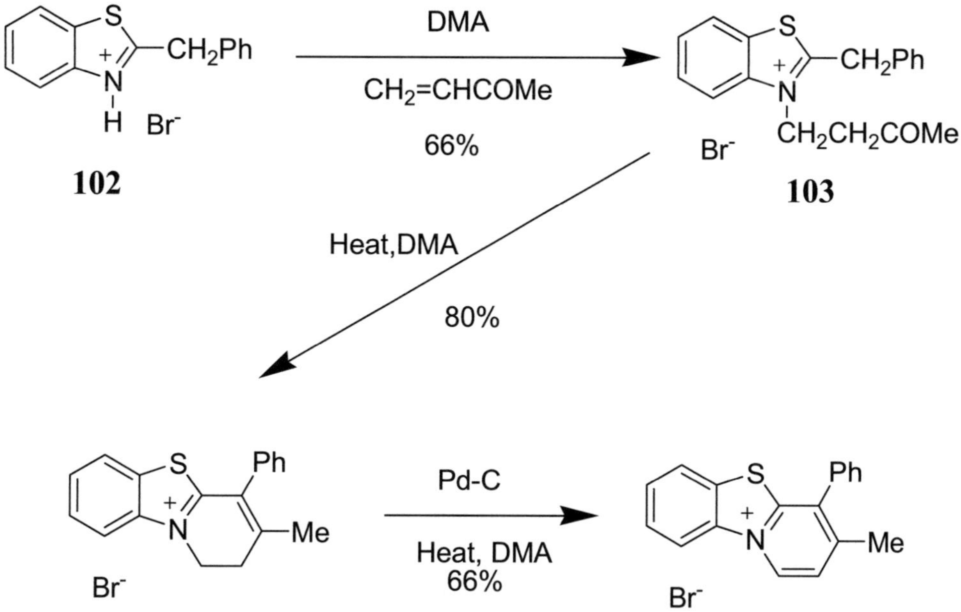

102 **103** **104** **105**

Scheme 29. Benzothiazolium based fused pyridinium salts

Baldwin and Walker (1974) observed that *N*-benzyl and *N*-allyl benzothiazolium salts form bis benzothiazolidine (**106**) on treatment with triethylamine in DMF at 0 °C. These dimers were found to undergo 1,3 or 3,3 sigmatropic rearrangements (scheme-31) depending upon the reaction conditions and the nature of the *N*-alkyl group to provide the more stable 2,3-dialkyl-2-(benzothiazol-2-yl)benzothiazoline (**107**) (Scheme-30) (Baldwin&Walker 1974).

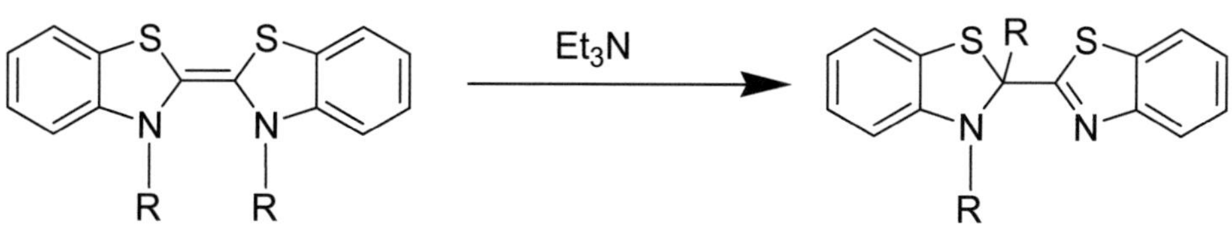

106 **107**

Scheme 30. Sigmatropic shift in benzothiazolium salts

Mukaiyama and Hojo have described the interconversion of enantiomeric secondary alcohols (**112**) by the treatment with an optically active 2-alkoxybenzothiazolium salt (**110**) followed by reaction with trichloroacetic acid and base hydrolysis subsequently (scheme-31) (Mukaiyama&Hojo 1976).

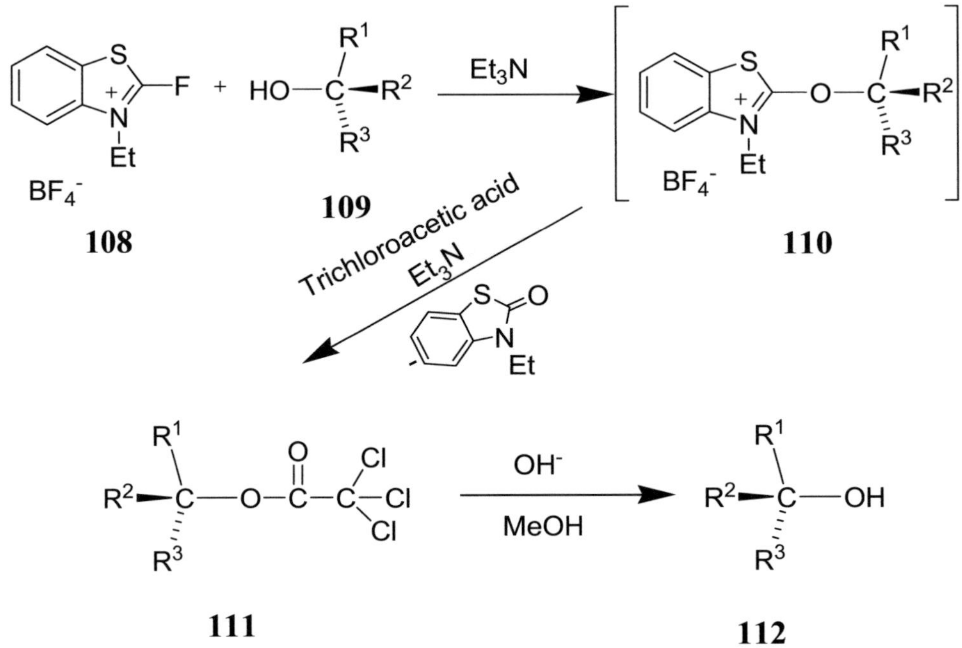

Scheme 31. Interconversion of enantiomeric secondary alcohols with benzothiazolium salts

Sawada *et al.* (1977) synthesized 3-(substituted-phenyl)thiazolo[2,3-*b*]benzothiazolium perchlorates (**116**) by the acid catalysed cyclization of the ketosulfides which were prepared by the alkylation of 2- sulfanylbenzothiazole sodium salt with substituted phenacyl halides (scheme-32).

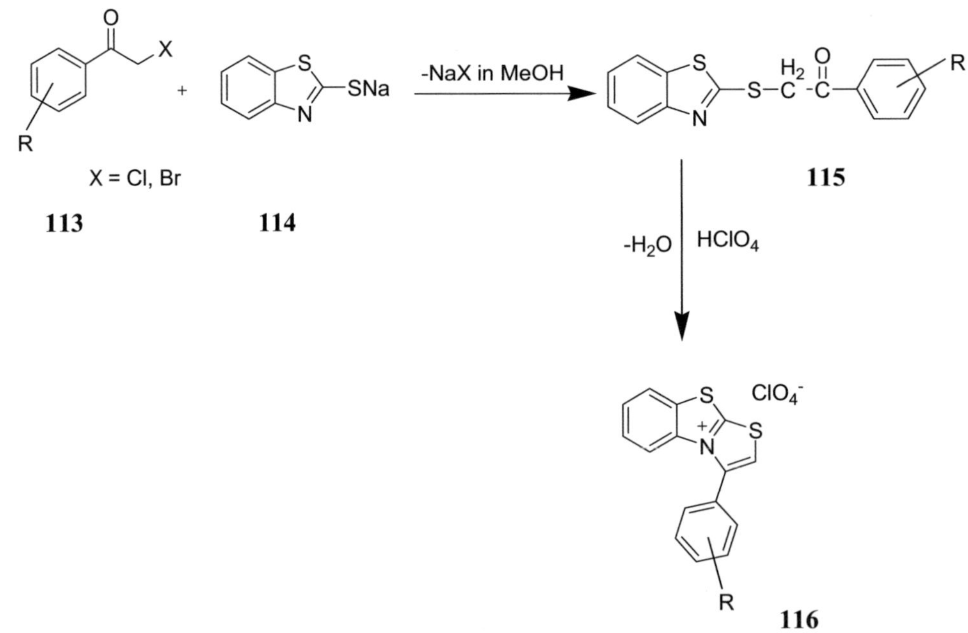

Scheme 32. Acid catalysed cyclization of ketosulfides to benzothiazolium

Kossmehl (1979) *et al.* have described the polymerization of the benzothiazolium salts by the Knoevenagel condensation of 2,3,6-trimethylthiazolo[4,5-*f*]benzothiazolium perchlorate (**117**, X = ClO_4^-) and iodide (**117**, X = I^-) with dialdehydes such as *p*-phenylenedicarbaldehyde, 2,5-dimethoxy-*p*-phenylenedicarbaldehyde, thiophene-2,3 dicarbaldehyde (scheme-33) (Kossmehl *et al.* 1979). In the same year Mukaiyama (1979) studied that the benzothiazolium compounds can be used to activate the carboxylic acids or alcohols to give 2-acyloxy or 2-alkoxy intermediates, which can be converted into esters, thioesters, amides, lactones, acid fluorides, isocyanates, *etc* (Mukaiyama 1979).

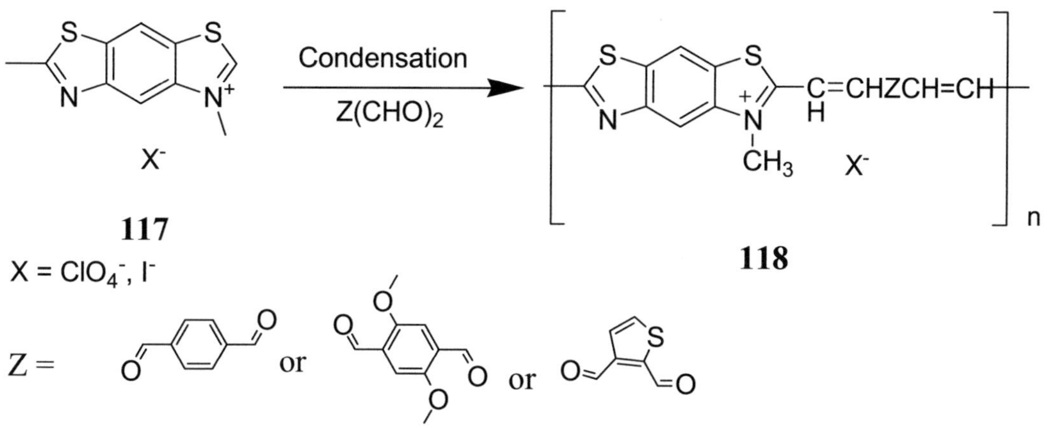

Scheme 33. Knoevenagel condensation of benzothiazolium salts with dialdehydes

Federsel *et al.* have devised the method for the preparation of 6-, 7- and 8-membered heterocycles (**120**) by base-induced ring expansion of quaternized benzothiazolium salts. The treatment of the quaternized benzothiazolium salt (**119**) with NaOH at 0-40 °C gave benzothiazine and benzothiazepine (Scheme-34) (Federsel&Bergman 1980).

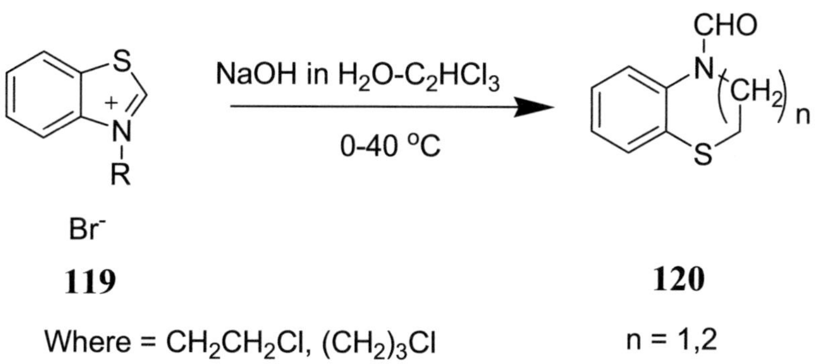

Scheme 34. 6, 7 and 8 membered hetrocycles from benzothiazolium salts

Halgas *et al.* (1983) prepared a series of 3,4,6-substituted benzothiazolium salts and studied their plant growth regulatory and antimicrobial activities. They observed the stimulation and inhibitory effects of the prepared benzothiazolium salts on plants. The highest stimulation activity was found with 6-methyl-3-propoxycarbonylmethylbenzothiazolium bromide and 4-chloro-3-methylbenzothiazolium bromide, and the highest inhibitory effect was observed with 4-chloro-3-methylbenzothiazolium methyl sulfate, while all these benzothiazolium salts did not show any noticeable antimicrobial activity (Halgas *et al.* 1983). Other reports for the similar studies by the same group can be found in the literature (Sutoris *et al.* 1983).

Bryce *et al.* (1984) have reported the preparation of a highly conjugated bisbenzothiazoline (**124**) by the reaction of dimethyl cyclohexa-1,3-diene-1,4-diacetate (**121**) with 2-(methylamino)benzenethiol (**122**) in the presence of triphenylcarbenium tetrafluoroborate and the subsequent treatment of the resultant bisbenzothiazolium tetrafluoroborate (**129**) with triethylamine at 20 °C (Scheme-35) (Bryce *et al.* 1984).

Xu *et al.* (1988) have described a method for the preparation of polystyrene based benzothiazolium salts and their use to study their catalytic activity in the Michael addition reactions of α,β-unsaturated compounds (Xu *et al.* 1988). He also studied the catalytic activity of the benzothiazolium salts for the Michael addition reactions in the presence of triethylamine (Xu *et al.* 1988).

Sutoris *et al.* (1988) studied the effects of benzothiazolium compounds on the growth of sugar beet and vetch (Vicia sativa) as well as their sugar and chlorophyll contents. They found that most of the benzothiazolium salts possessed auxin-like activity. *N*-Methylbenzothiazolium bromide was found to stimulate sugar production in vetch at different concentrations and among the *N*-benzyl and substitutedbenzyl benzothiazolium salts, *N*-benzylbenzothiazolium bromide was found to be the most active compound. *N*-(methoxycarbonylmethyl)benzothiazolium bromide was found to be the most active compound for increasing chlorophyll contents in sugar beet leaf after treatment of the seed. Changing the methyl group for ethyl or propyl in benzothiazolium salt decreased chlorophyll contents (Sutoris *et al.* 1988).

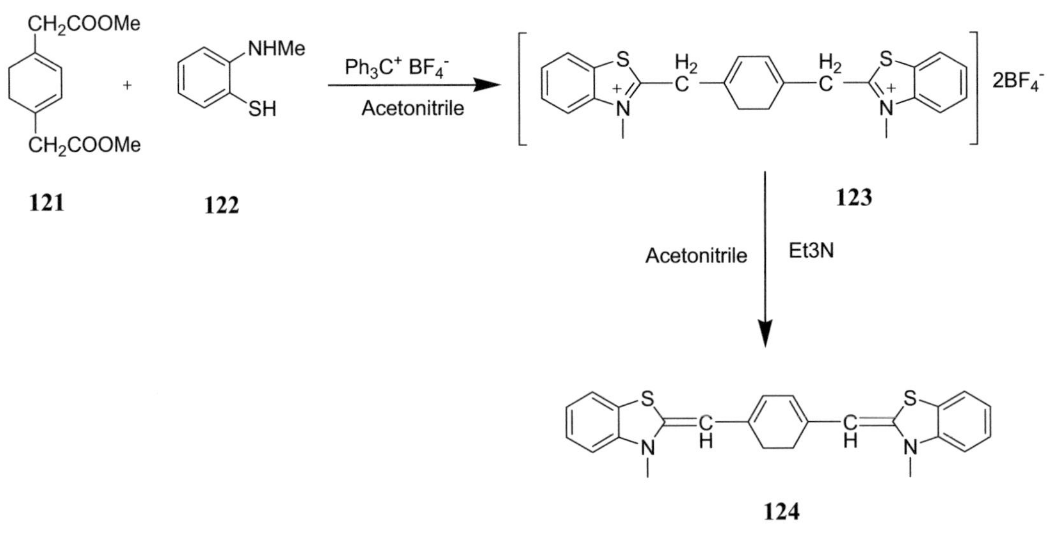

Scheme 35. Preparation of a highly conjugated bisbenzothiazoline

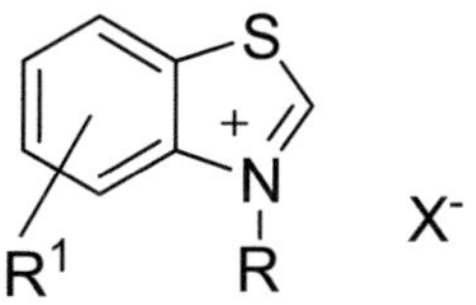

125

Chikashita *et al.* (1991) have described a synthetic and systematic use of benzothiazole ring system as an "on-off" type of leaving group for the preparation of ketones (**128**) and carboxylic acid derivatives (**129-132**) from a variety of benzothiazoles, *e.g.* (R = H, Me, ph; R^1 = propyl, ph, cyclohexyl, CH:CHPh) (off state) via the corresponding benzothiazolium salts (on state) obtained by quaternization (on switch). A variety of such compounds (R ≠ H) underwent a carbon-carbon bond cleavage at the 2-position to give the corresponding ketones $RCOR^1$ on simple treatment with a base under mild reaction conditions. The similar reaction (R^1 = H) to aldehydes proceeded less efficiently. The oxidative reaction of a variety

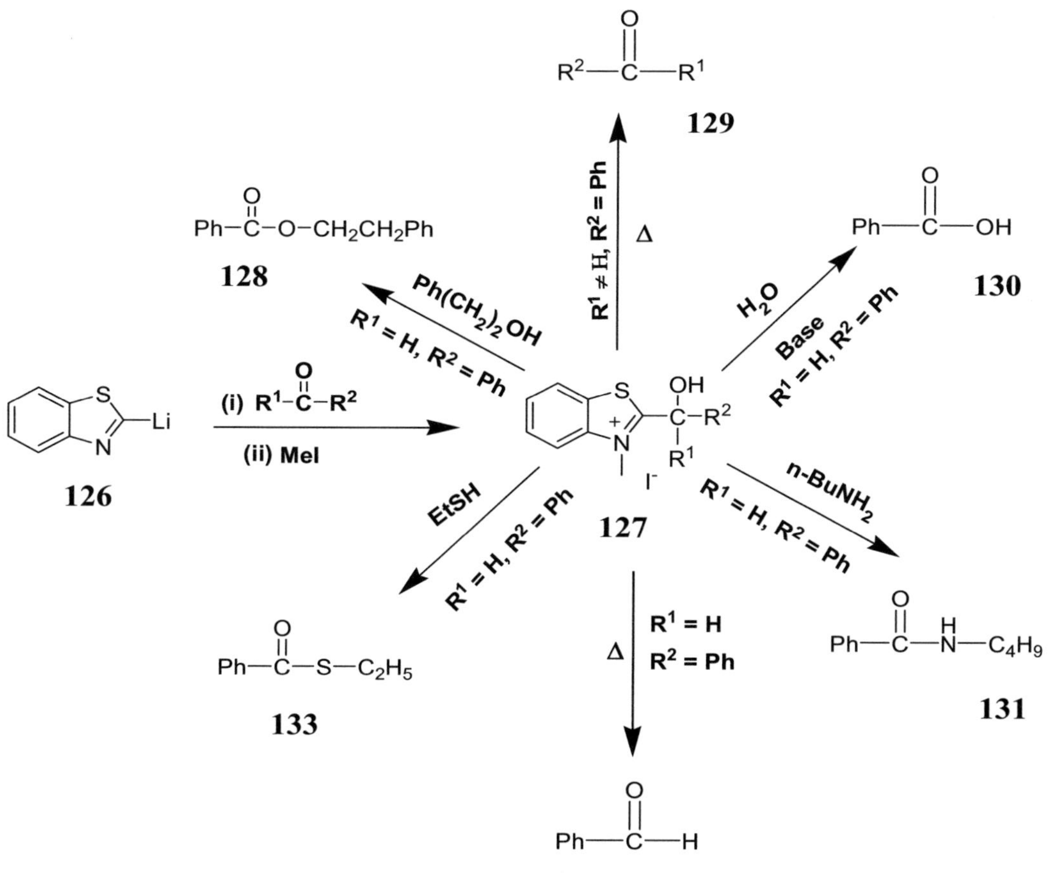

Scheme 36. Use of benzothiazole ring system as an "on-off" type of leaving group **for** the preparation of ketones and carboxylic acid derivatives.

(R^1 = H) with base and nonactivated MnO_2 in ethanol afforded the corresponding ethyl esters. This type of oxidative reaction could be also achieved in THF with different nucleophiles, such as alcohol, water, thiol, and amine, to give the corresponding ester, carboxylic acid, thioester, and amide, respectively (scheme-36) (Chikashita *et al.* 1991).
Riemer and Liebscher (1993) have described the synthesis of pyrimido[2,1-*b*]benzoxazolium, pyrimido[2,1-*b*]benzothiazolium salt (**134**) by the reaction of 3-isothiocyanato-2-propeniminium salts with 2-aminothiophenol followed by cyclization using methyl iodide by a new condensation reaction (Riemer&Liebscher 1993).

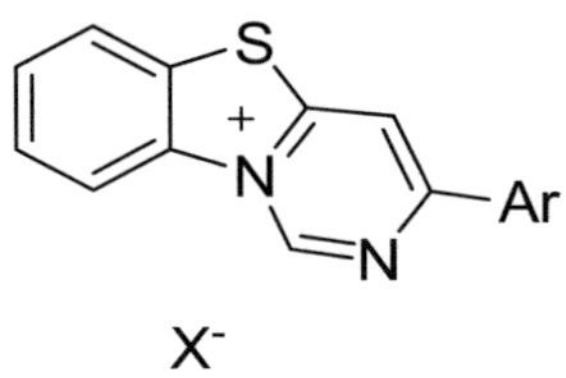

Ar = Ph, 4-MePh, 4-ClPh
X = I^-, ClO_4^-

134

Lopez-Celahorra *et al.* (1994) have narrated that the 3,3'-polymethylene-bridged benzothiazolium and thiazolium salts could be used as pre-catalysts for the benzoin condensation and that catalytic activity depends strongly on the methylene bridge length. The authors claimed that in aprotic medium, the catalytic activity was due to the bis(thiazolin-2-ylidene)s or bis(benzothiazolin-2-ylidene)s species (Lopez-Celahorra *et al.* 1994).
Zimmermann and Klaus (1996) have reported a method in which the ring of the pyrylium and thiopyrylium salts (**135**) have been transformed to substituted benzene ring (**136, 137**) by treatment with anhydrobases derived from 1*H*-benzimidazolium and benzothioazolium salts (Scheme-37) (Zimmermann&Schmidt 1996).
Hatrik and Zahradnik (1996) have claimed that the toxicity of benzothiazolium salts can be predicted by the neural network (NN) method. Results were found to be in good agreement with the previously used Free-Wilson method. They have used a number of benzothiazolium salts to calculate their toxicity by the NN method (Hatrík&Zahradník 1996).
Kumar *et al.* prepared aza-enediyne analogues by the incorporation of *N*-propargyl moiety to 2-alkynylbenzothiazolium salts (**139**) and the aza-enediynes (**140**) were proven to be the modest DNA cleavage agents. The mechanism probably involves the formation of an adduct (**141**) prior to cleavage of DNA. They also observed DNA cleavage with the *N*-methyl-2-alkynylbenzothiazolium salt, which lacks the aza-enediyne moiety (scheme-38) (Kumar *et al.* 2001).

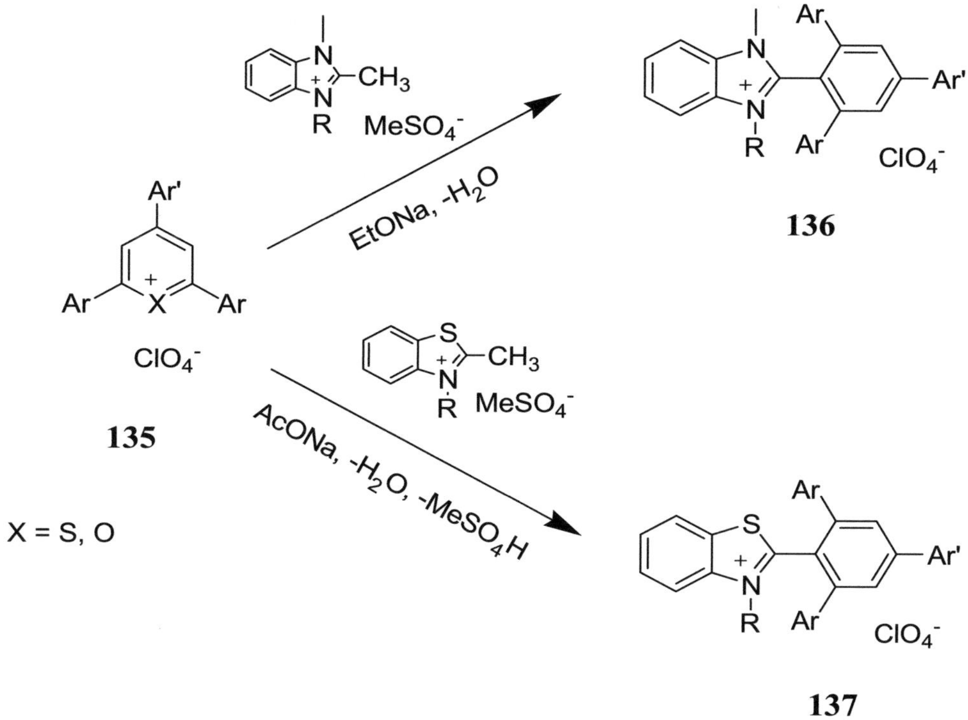

Where R = Me, Et
Ar = Ph or p-Me, p-MeO, p-Cl, p-Br phenyl
Ar' = may be one of above but not similar

Scheme 37. Conversion of pyrylium and thiopyrylium salts to substituted benzene ring

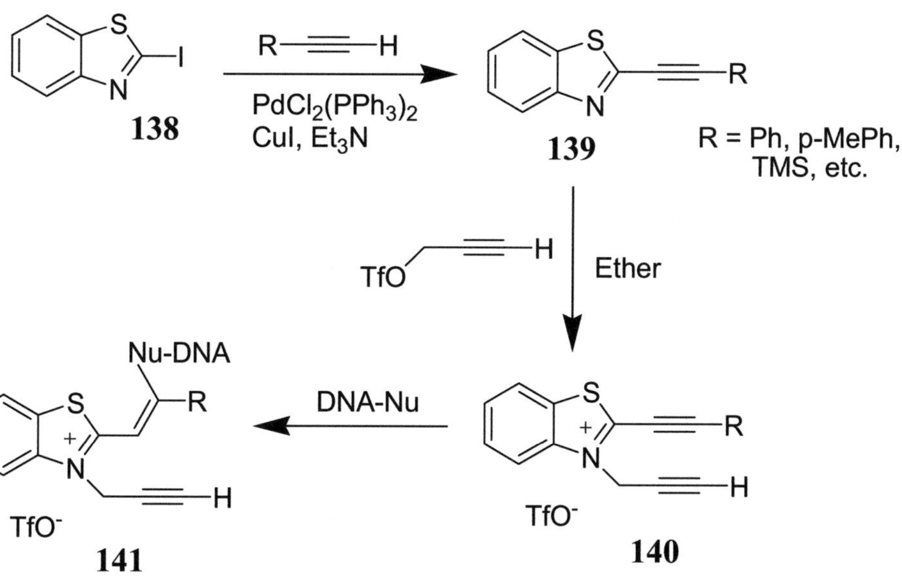

Scheme 38. *N*-propargyl alkynylbenzothiazolium salts as DNA cleavage agents

Puterová *et al.* (2004) have prepared different benzothiazolium compounds by reaction of 5-arylfuran-2-carboxaldehydes and furo[*b*]pyrrole type aldehydes with 2-methyl-3-methyl or ethyl benzothiazolium bromide. The new compounds were based on highly conjugated systems that have potential biological activity. Cyanine dye precursors can also be obtained by the reaction of furo[*b*]pyrrole type aldehydes with benzothiazolium salts (Puterová *et al.* 2004).

Tseng *et al.* (2005) have used different benzothiazolium compounds (**142** & **143**) to study their activity for the inhibition of nitric oxide production in a cell culture system. They found that benzothiazolium salts have been the better inhibitors than *N*-methyl arginine (L-NMMA) a known inhibitor. They have demonstrated these results by the correlation of in *vivo* and in *vitro* activities using mouse paw edema (Tseng *et al.* 2005).

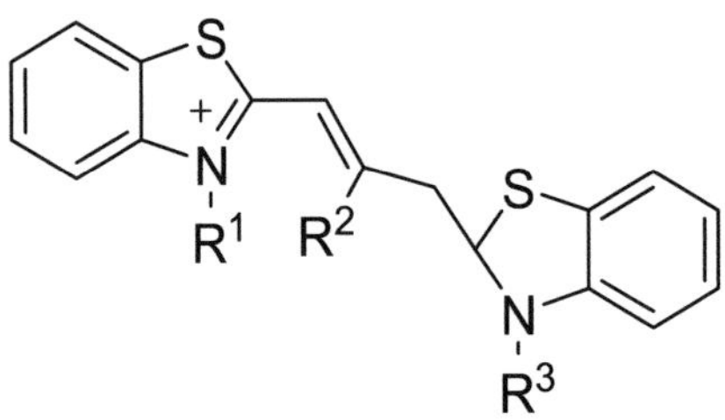

142

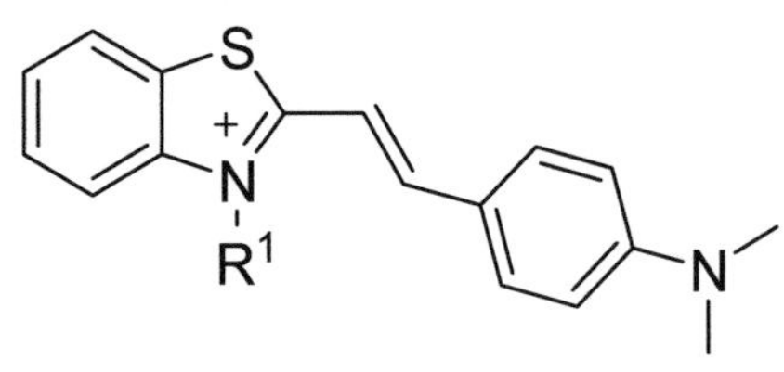

143

Yang *et al.* (2007) synthesized a series of benzothiazolium compounds (**144** and **145**) which showed inhibitory activity against gastroenteritis virus (TGEV). They used swine testicle (ST) cells infected with transmissible gastroenteritis virus (TGEV) and an indirect immunofluorescent assay with antibodies against TGEV spike and nucleocapsid proteins to screen the benzothiazolium compounds that inhibit TGEV replication. The benzothiazolium compounds were found to have inhibitory activity against TGEV 3CL(pro) (Yang *et al.* 2007).

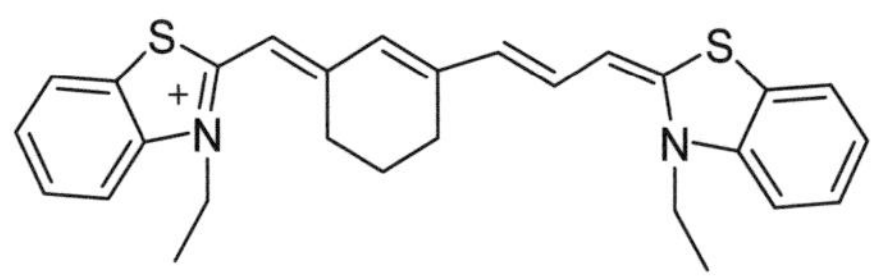

144

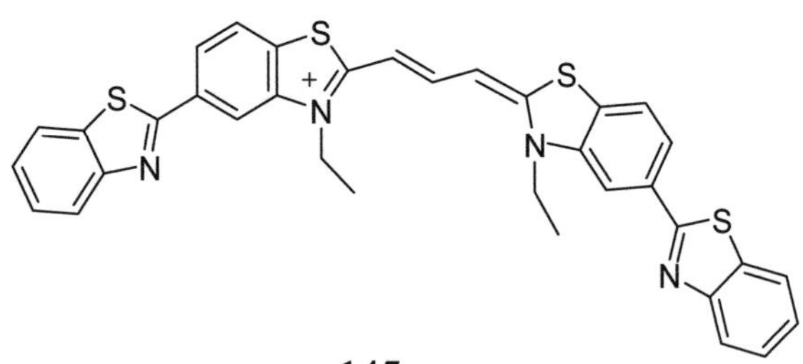

145

Wanzlick *et al.* have prepared the Bis(3-methylbenzothiazolin-2-yliden) which was used by Castells *et al.* for the catalysis of benzoin condensation (Castells *et al.* 1988).

Due to the unavailability of data about the liquid range of the above benzothiazolium salts, it is not certain whether the salts should be classified as ionic liquids or not. Recently our group have prepared a series of benzothiazolium iodides in absence of any solvent, studied their physical properties and performed metathesis reactions. The benzothiazolium bistriflimides showed a good liquidity range near ambient temperatures and have wide stability range of 274-366 °C in comparison to the other benzothiazolium salts having high melting points and low stabilities (Nadeem *et al.* 2010) (scheme-39).

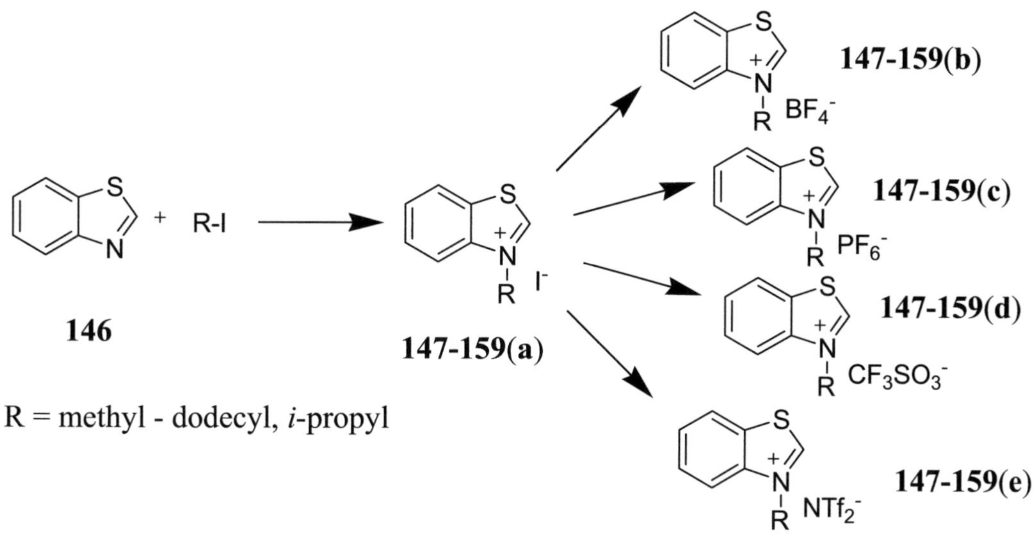

Scheme 39.

No.	R	Yield %	Temperature °C	Time h
147a	CH_3	82, 99.8*	20-25	**9.0**
148a	C_2H_5	75	50 - 60	**4.0**
149a	C_3H_7	68	80 - 90	**6.0**
150a	*iso*C_3H_7	62	70 - 80	**7.0**
151a	C_4H_9	80	110 - 120	**5.5**
152a	C_5H_{11}	79	125 - 135	**1.0**
153a	C_6H_{13}	77	160 - 170	**3.0**
154a	C_7H_{15}	75	180 - 190	**4.0**
155a	C_8H_{17}	80	210 - 220	**4.0**
156a	C_9H_{19}	73	150 - 160	**2.0**
157a	$C_{10}H_{21}$	70	110 - 120	**4.5**
158a	$C_{11}H_{23}$	75	95 - 105	**3.0**
159a	$C_{12}H_{25}$	72	140 - 150	**5.0**

*Sonicated at room temperature.

Table 1. Specific temperatures, reaction times, and yields for compounds **147a-159a** (Scheme 39).

4. Thermal studies of benzothiazolium ionic liquids

The benzothiazolium salts/ionic liquids have been comprehensively studied by our group (Nadeem *et al.* 2010). The thermal studies such as TGA and DSC reveal important information about the stability and thermal behavior of these compounds.

Decomposition temperatures of benzothiazolium iodide salts were determined by TGA, heating at 10 °C min^{-1} under dried air atmosphere and are reported (table-2) as (i) onset to 5 wt% mass loss ($T_{5\%dec}$) and (ii) onset to total mass loss (T_{dec}) (in parentheses).

Sr No.	R	mp (°C)	$T_{5\% dec}$ (T_{dec}) (°C)
147a	CH_3	221-222[c] *(lit 216)*	194 (214)
148a	C_2H_5	136-137[c] *(lit 140)*	177 (205)
149a	C_3H_7	158-159[c] *(lit 158)*	179 (205)
150a	*iso*C_3H_7	131.1[a,b] *(lit 131)*	181 (205)
151a	C_4H_9	116.3 *(lit 114)*	175 (201)
152a	C_5H_{11}	117.8[a] *(lit 119)*	172 (199)
153a	C_6H_{13}	126.7	173 (199)
154a	C_7H_{15}	**65.1**[a,b]	164 (193)
155a	C_8H_{17}	**74.5**[a,b]	166 (188)
156a	C_9H_{19}	**82.1**[a,b]	162 (186)
157a	$C_{10}H_{21}$	**82.4**[a,b]	171 (194)
158a	$C_{11}H_{23}$	**80.4**[a,b]	161 (188)
159a	$C_{12}H_{25}$	**90.5**[a,b]	162 (187)

Melting points and thermal stabilities.§

Table 2. [§] Melting (mp) or glass transition (T_g) points (°C) were measured from the transition onset temperature and determined by DSC from the second heating cycle at 10 °C min^{-1}, after initially melting and then cooling samples to -100 °C. Decomposition temperatures shown were determined by TGA, heating at 10 °C min^{-1} under dried air atmosphere and are reported as (i) onset to 5 wt% mass loss ($T_{5\%dec}$) and (ii) onset to total mass loss (T_{dec}) (in parentheses). Salts meeting the definition of ionic liquids (mp < 100 °C) are in **bold**. [a] Sample exhibits additional glass transition temperatures (°C) of supercooled liquids (i) with consecutive crystallization and melting on heating; **148b**, -42.7; **149b**, -48.1; **149c**, -17.3; **150c**, -5.2; **151c**, -4.4;**152a**, 2.3; **152b**, -51.2; **153b**, -47.6; **154b**, -42.8; **155b**, -43.5; **156b**, -49.1; **163b**, -58.2; **164b**, -45.2; **165b**, -48.1; (°C); (ii) with no crystallization and melting under e experimental conditions: **148c**, -5.2; **150a**, 2.4; **152c**, -3.2; **153c**, -6.5; **154a**, -5.6; **154c**, -10.1; **155a**, -2.3; **155c**, -7.2; **156a**, 5.5; **156c**, -9.1;**157a,** -5.6; **163c**, -11.2; **158a**, -5.8; **164c**, -11.1;**158a**, -6.0; **159c**, -10.3 (°C); [b] Irreversible transition, from first heating; [c] melting point measured visually due to close proximity to the decomposition temperature and possibility of contamination of the DSC cell.

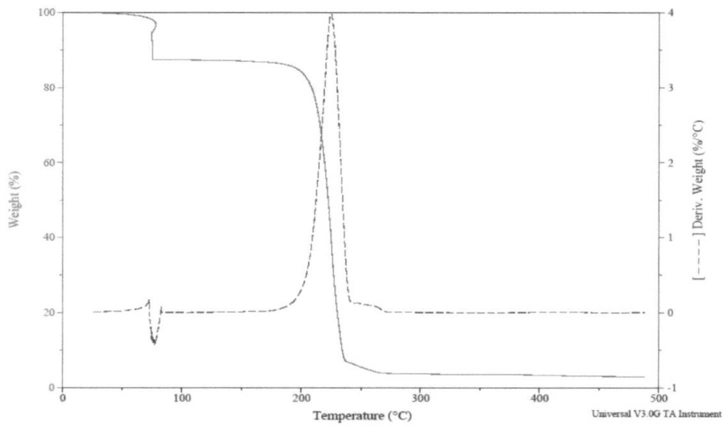

Fig. 2. TGA of *N*-Methyl benzothiazolium iodide

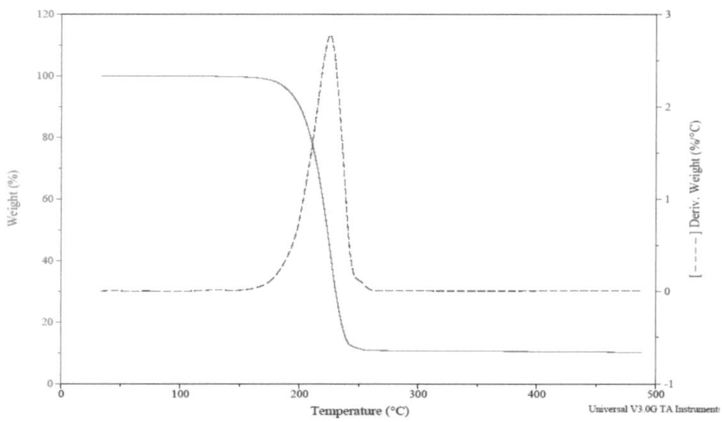

Fig. 3. TGA of *N*-Ethyl benzothiazolium iodide.

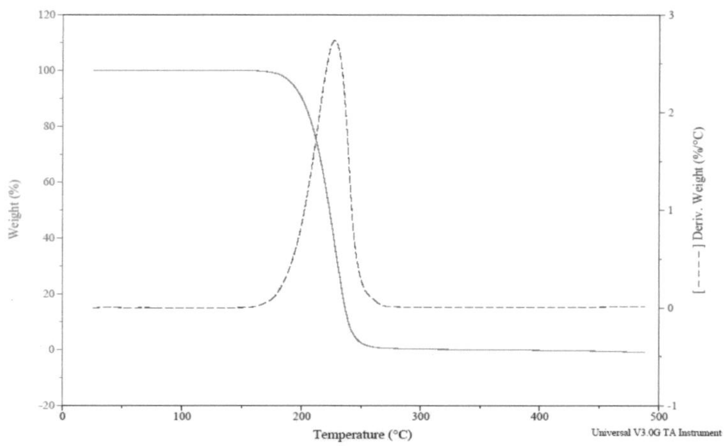

Fig. 4. TGA of N-Propyl benzothiazolium iodide.

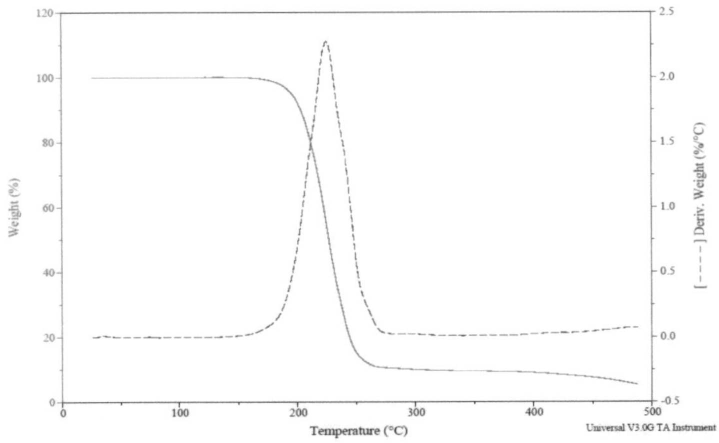

Fig. 5. TGA of N-iso-Propyl benzothiazolium iodide.

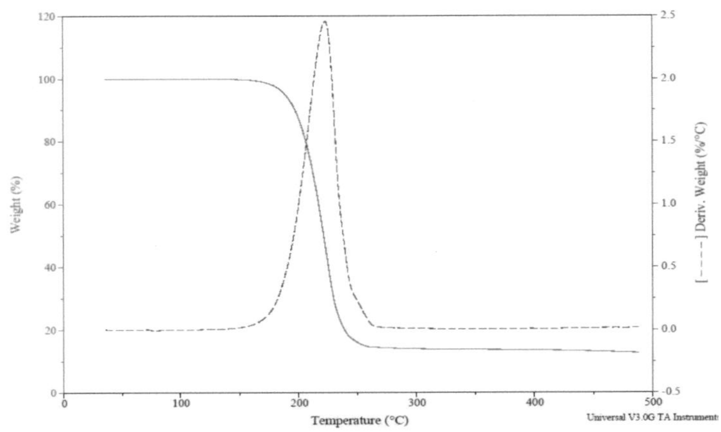

Fig. 6. TGA of N-Butyl benzothiazolium iodide.

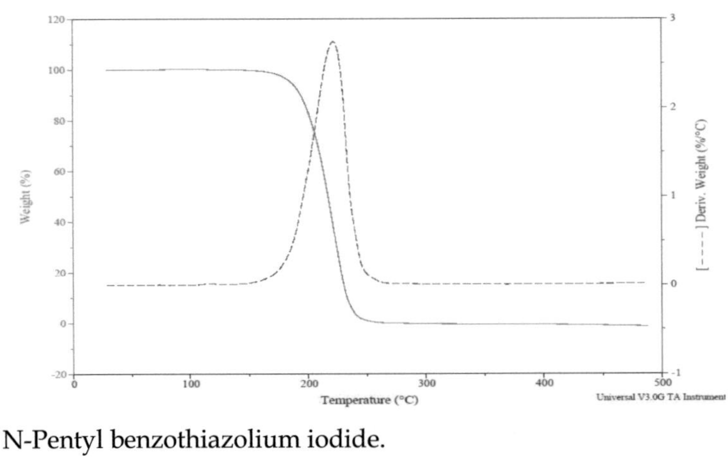

Fig. 7. TGA of N-Pentyl benzothiazolium iodide.

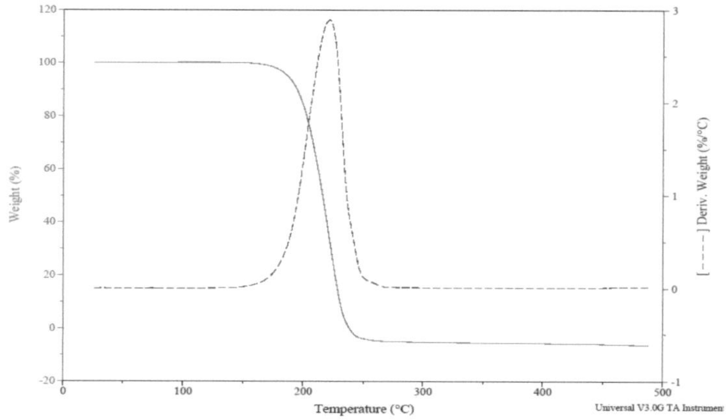

Fig. 8. TGA of N-Hexyl benzothiazolium iodide.

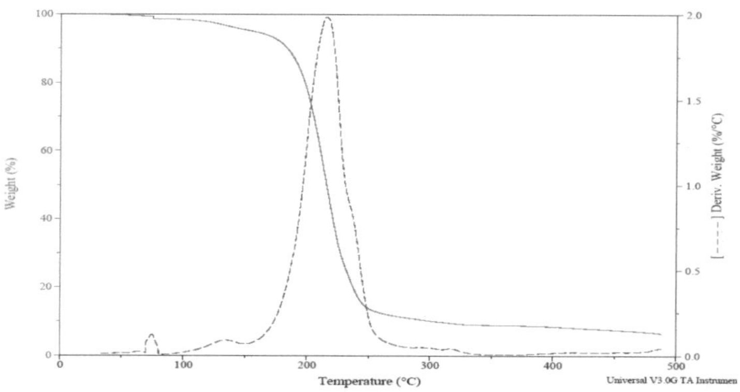

Fig. 9. TGA of N-Heptyl benzothiazolium iodide.

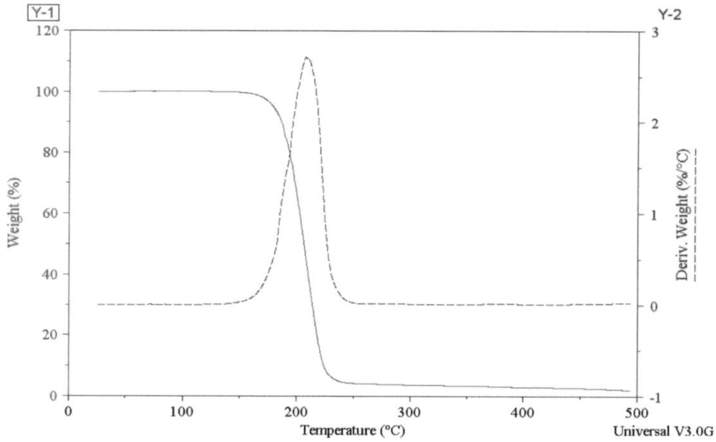

Fig. 10. TGA of N-Octyl benzothiazolium iodide.

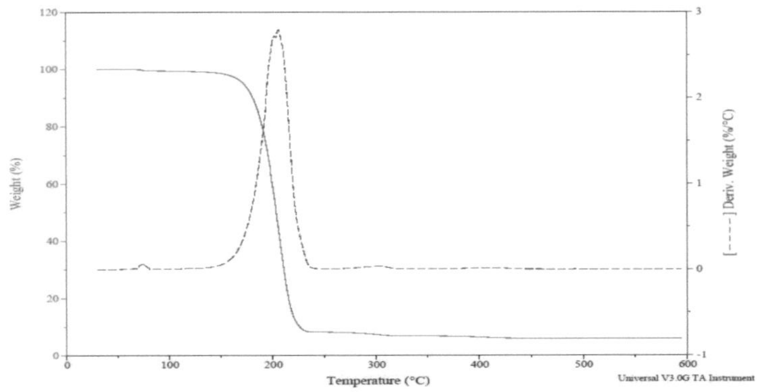

Fig. 11. TGA of N-Nonyl benzothiazolium iodide.

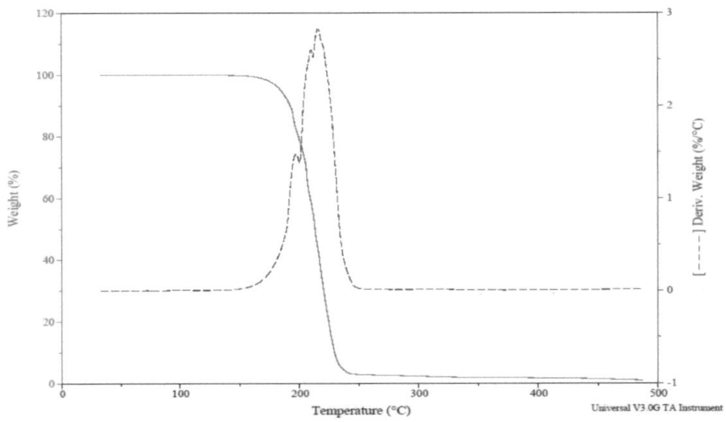

Fig. 12. TGA of N-Decyl benzothiazolium iodide.

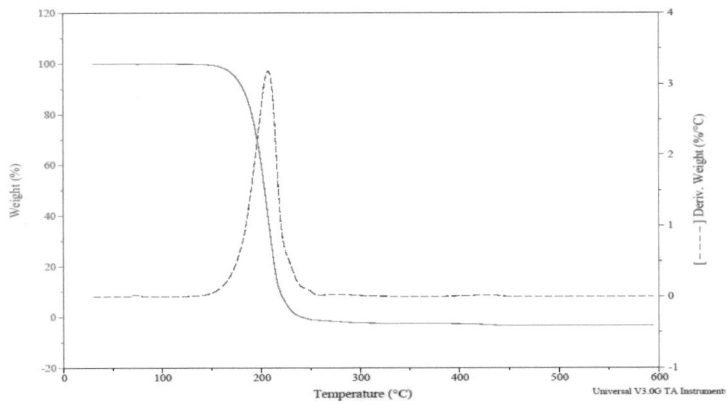

Fig. 13. TGA of N-Undecyl benzothiazolium iodide.

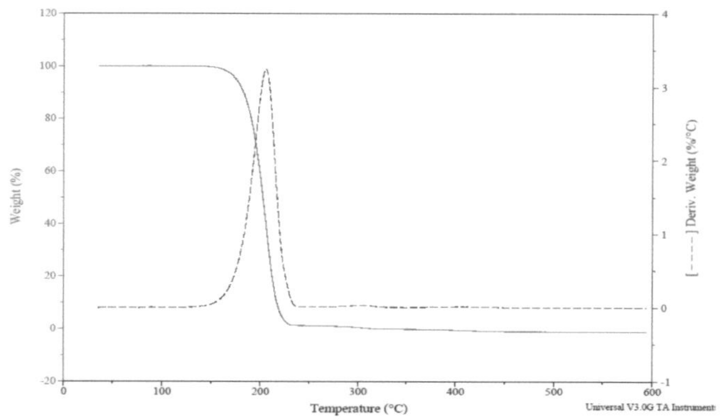

Fig. 14. TGA of N-Dodecyl benzothiazolium iodide.

Melting (mp) or glass transition (Tg) points (°C) were measured from the transition onset temperature and determined by DSC from the second heating cycle at 10 °C min^{-1}, after initially melting and then cooling samples to -100 °C. mThe melting points of *N*-methyl, *N*-ethyl and *N*-propyl were measured visually due to close proximity to the decomposition temperature and possibility of contamination of the DSC cell.

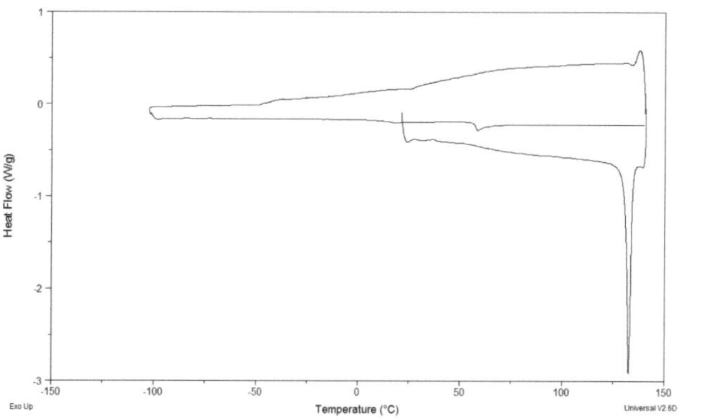

Fig. 15. DSC graph for *N*-IsopropylBTI

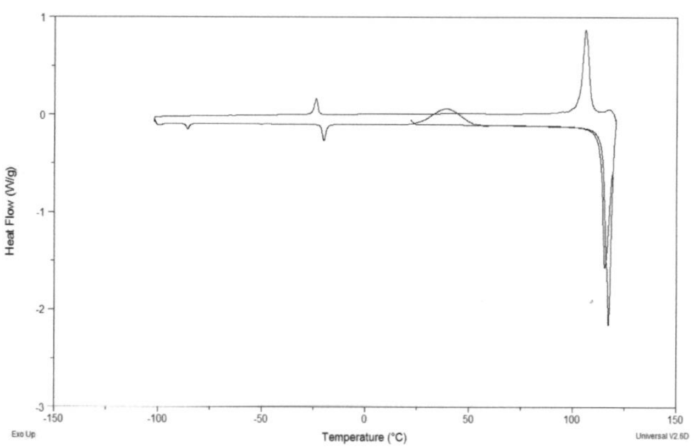

Fig. 16. DSC graph for *N*-butylBTI

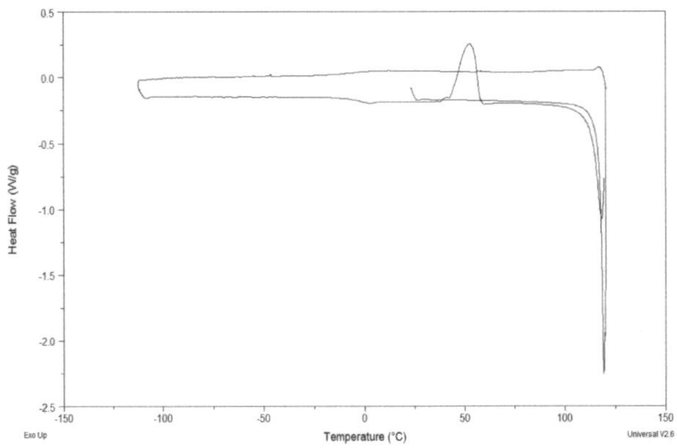

Fig. 17. DSC graph for *N*-pentylBTI

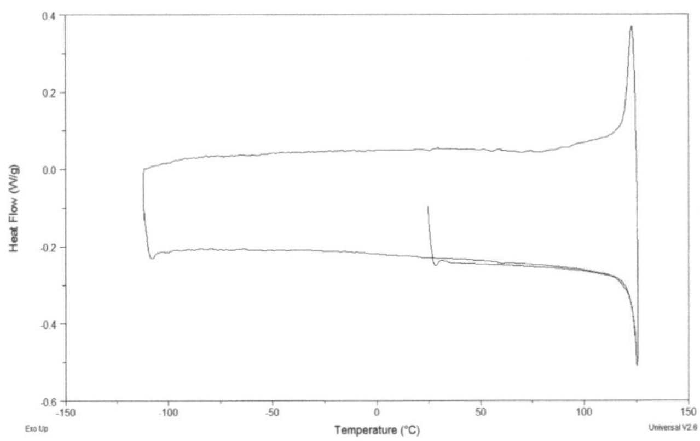

Fig. 18. DSC graph for *N*-hexylBTI

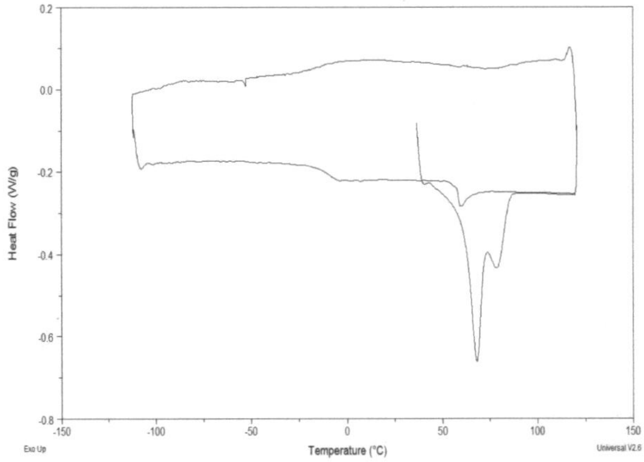

Fig. 19. DSC graph for *N*-heptylBTI

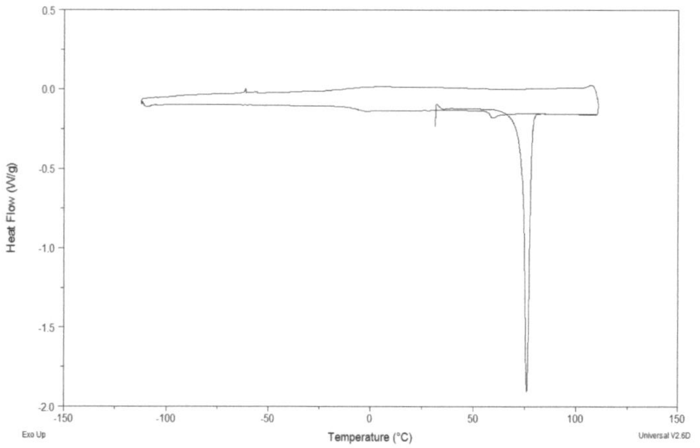

Fig. 20. DSC graph for *N*-octylBTI

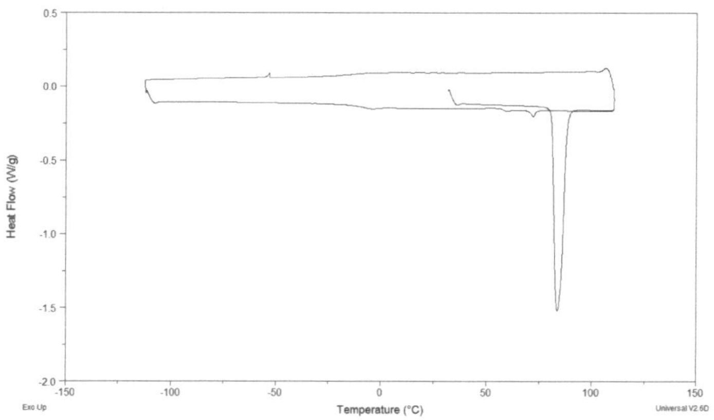

Fig. 21. DSC graph for *N*-nonylBTI

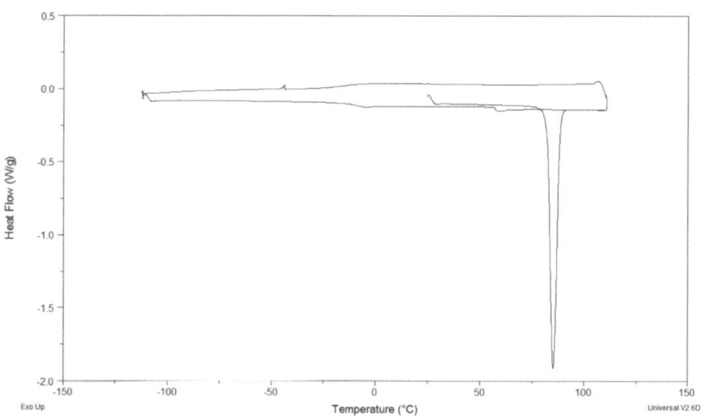

Fig. 22. DSC graph for *N*-decylBTI

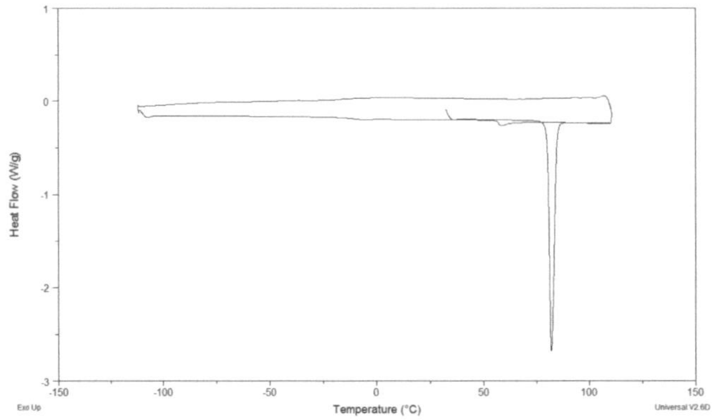

Fig. 23. DSC graph for *N*-undecylBTI

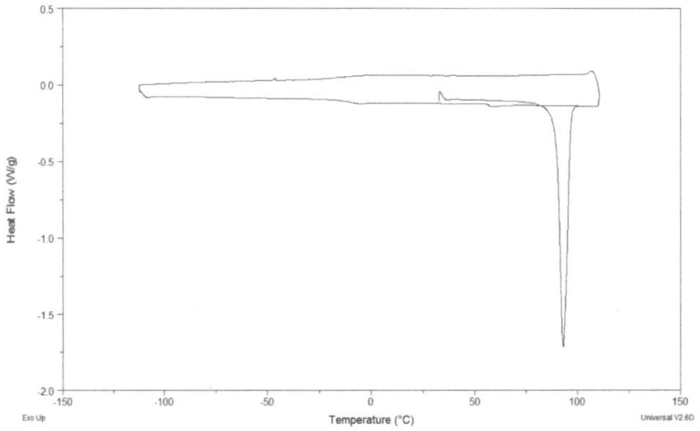

Fig. 24. DSC graph for *N*-dodecylBTI

5. Benzothiazolium iodide-chloroaluminate ionic liquid

The phase behaviour study of $[C_{11}BT]I\text{-}AlCl_3$ mixtures have been carried out with the help of DSC to find the appropriate mole ratio of the aluminum chloride to the $[C_{11}BT]I\text{-}AlCl_3$ mixture (Figure-25). From figurr-25, it is evident that at mole ratio 0.5% the mixture of $[C_{11}BT]I\text{-}AlCl_3$ gave lowest melting point (22.5 ^{o}C) which was checked by DSC. Figure-24 represents the phase diagram of the $[C_2mim]Cl\text{–}AlCl_3$ system as a function of composition, $X(AlCl_3)$ and figure-25 gives an account of the phase study of the *N*-undecylbenzothiazolium iodide-aluminum chloride system. It is obvious that the most distinctive feature of the chloroaluminate(III) system is its dependence on the apparent mole fraction of aluminium(III) chloride, the ionic liquid is acidic (Visser 2002), basic $[X(AlCl_3) < 0.5]$, or neutral $[X(AlCl_3) = 0.5]$, referring to the Franklin acidity and basicity (Franklin 1905, 1924). This is frequently assumed to refer to the Lewis acidity and basicity.

From figures 25 and 26, it is immediately clear that the liquid range of both the systems is very different from each other. The $[C_2mim]Cl\text{–}AlC13$ system is a low viscosity liquid at room temperature from $X(AlCl_3) = 0.33$ to $X(AlCl_3) = 0.67${where $X(AlCl_3)$ is the mole fraction of the nominal aluminium(III) content. While the room temperature liquidity of the

C_{11}BTI-$AlCl_3$ system starts at $X(AlCl_3)$ = 0.45 and ends up to 0.55. Although the liquidus range of the C_{11}BTI-$AlCl_3$ system is very low as compared to the imidazolium system, there is a eutectic point, which may be useful for different applications.

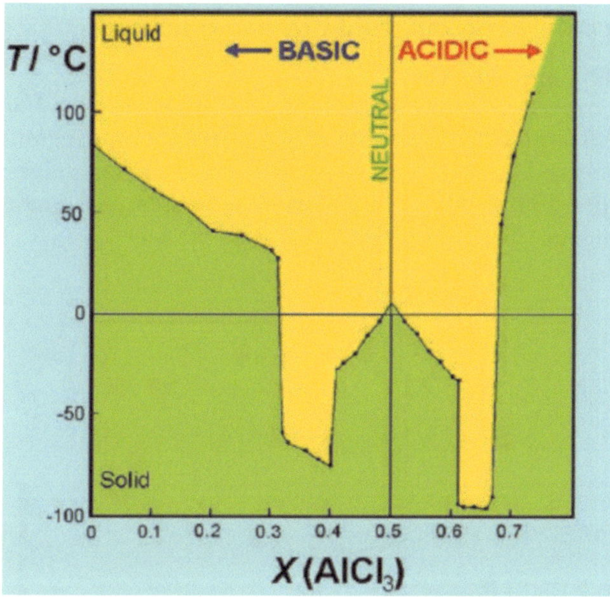

Fig. 25. The phase diagram for the [C_2mim]Cl–$AlCl_3$ system a function of composition, $X(AlCl_3)$

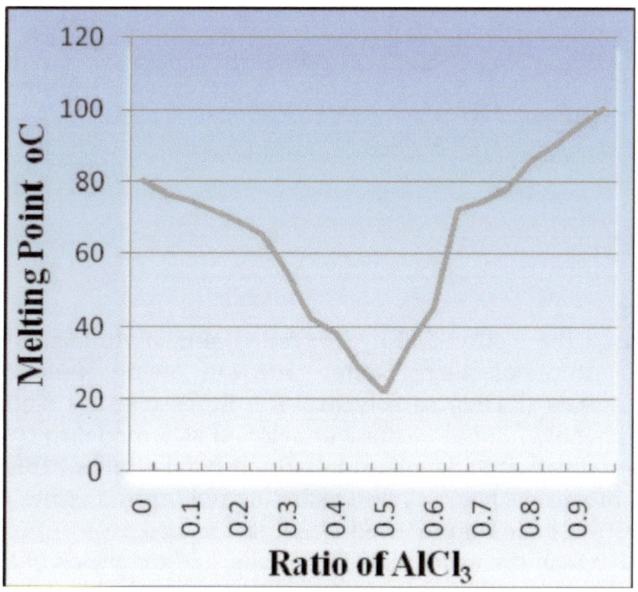

Fig. 26. The phase diagram for the [C11BT]I-AlCl3 system as a function of composition, X(AlCl3)

The [C_2mim]Cl–$AlCl_3$ system have been considered the first genuine example of an ionic liquid system that was liquid at room-temperature. So the C_{11}BTI-$AlCl_3$ system would open up a field for electrochemistry specialists, and laid the foundations for the exploration in this field.

6. Comparison in catalyzing different organic reactions

6.1 Acyloin / benzoin condensation

Davis and Forrester have studied the benzoin condensation promoted with a small (~ 5 mol %) quantity of triethylamine and thiazolium salts (**160**). The reaction was accomplished when the thiazolium based organic ionic liquid (OIL) was stirred under nitrogen as a clearly heterogeneous mixture with a toluene solution of benzaldehyde. The reaction gave about 80% conversion to benzoin (**163**) (scheme-40) (Davis&Forrester 1999).

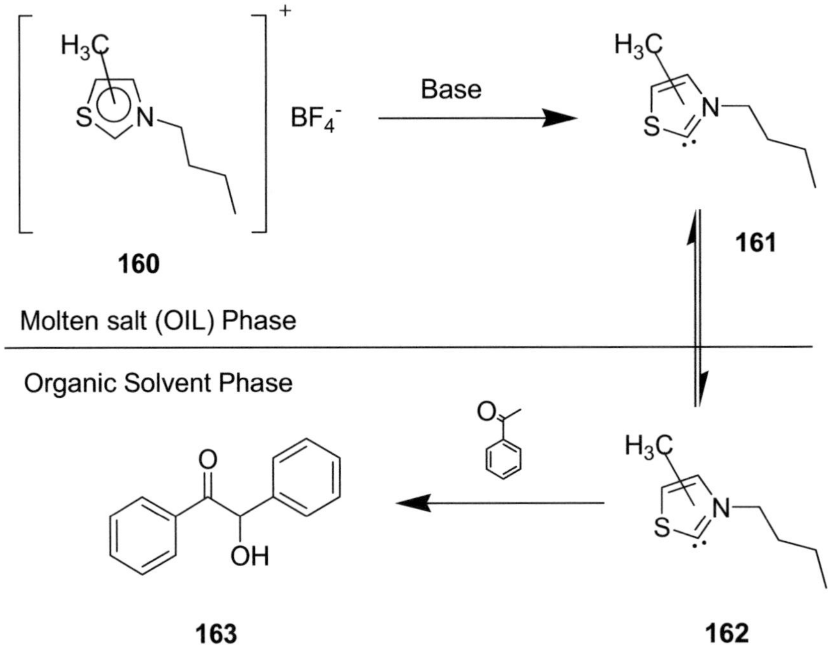

Scheme 40.

6.2 Stetter reaction

Zhou *et al.* have used the 3-ethyl-5-(2-hydroxyethyl)-4-methylthiazolium bromide for the microwave-assisted intramolecular Stetter reaction using imidazolium-type room temperature ionic liquids (RTILs) as solvents. The intramolecular Stetter reaction of (*E*)-methyl 4-(2-formylphenoxy) but-2-enoate was selected as a model to optimize the reaction conditions, using butylmethylimidazolium tetrafluoroborate $[bmim]^+[BF_4]^-$ as the solvent. It is known that salts of various heterocycles, including imidazolium salts, can also be used as catalysts. Therefore, the ionic liquids used as solvent is also able to function as a catalyst, even if it is less active than the usual thiazolium salts, and the yields of the desired product in that case are very poor. Subsequently, it was observed that when the quantity of thiazolium salt increased from 5 to 15 mol%, the yield improved accordingly to 96%. Under microwave irradiation, a variety of aromatic substrates undergo the intramolecular Stetter reaction in imidazolium-type RTILs as solvents, with thiazolium salts and Et_3N as catalysts. Under these conditions the reactions were finished in 5–20 min and the products could be isolated in good to excellent yields, usually higher than those obtained under conventional heating conditions. Furthermore, it was possible to recycle and reuse the ionic liquid and the catalyst thiazolium salts (Zhou *et al.* 2006).

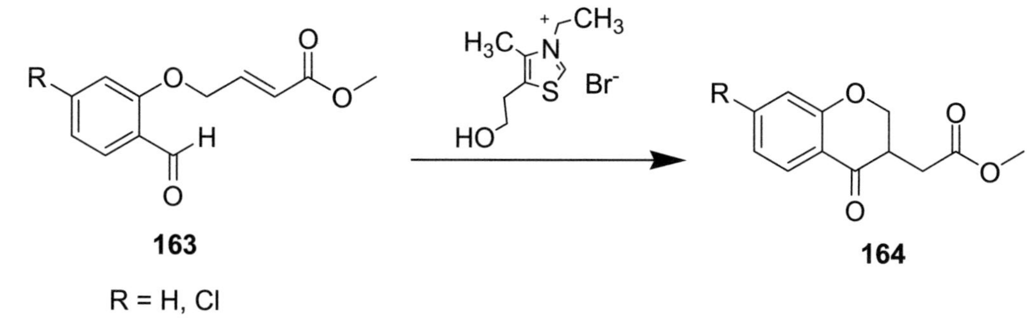

Scheme 41.

Imidazolium-type room temperature ionic liquids (RTILs) have been used for the Stetter reaction, affording the desired 1,4-dicarbonyl compounds (e.g. **167**) in good yields together with the benzoin (e.g. **168**). Thiazolium salts and Et_3N are efficient catalysts for this reaction performed in ionic liquid. The possibility to recycle and reuse the solvent has been demonstrated, although it was not possible to recycle the thiazolium catalyst (Anjaiah *et al.* 2004).

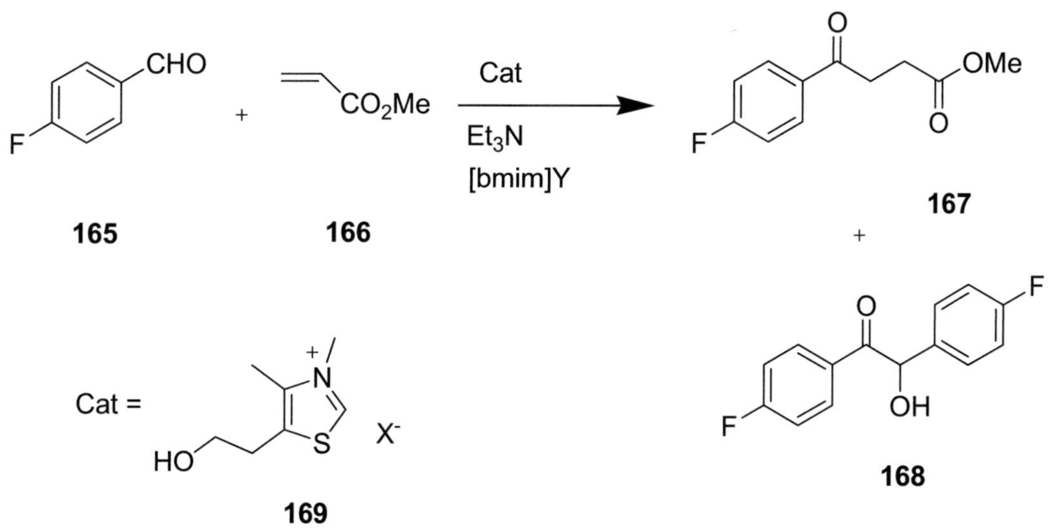

Scheme 42.

6.3 As coupling reagent

Li and Xu prepared the 2-Bromo-3-ethyl-4-methylthiazolinium tetrafluoroborate (BEMT), which is a crystalline solid, and have been reported to be an efficient coupling reagent for hindered amino acids. It was used for the coupling of *N*-alkyl or α,α-dialkyl amino acids (Li&Xu 1999). Mechanism of action of BEMT

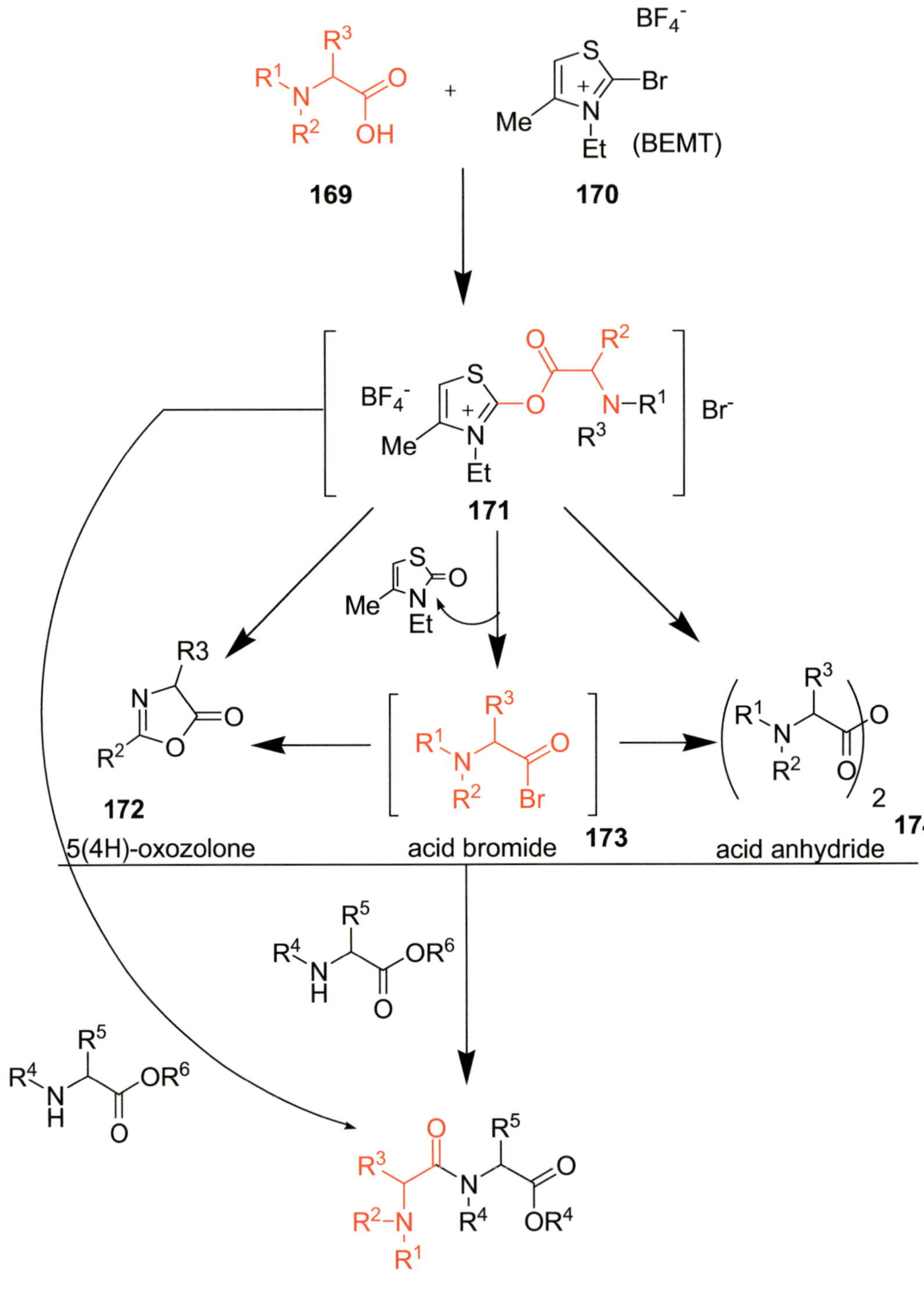

BF4-
R3
R1
N
O
R2
OH
+
S
Br
Me
N
Et
(BEMT)
169
170
BF4-
O
R2
S
O
N–R1
Br-
Me
N
R3
Et
171
S
O
Me
N
Et
R3
N
O
R2
O
R3
R1
N
O
R2
Br
R3
R1
N
O
R2
O
2
172
5(4H)-oxozolone
acid bromide
173
174
acid anhydride
R5
R4
NH
OR6
H
O
R5
R4
N
OR6
H
O
O
R5
R3
N
O
R2–N
R4
OR4
R1
175

Scheme 43.

6.4 Use of thiazolium-gold ionic liquid as catalyst for the hydration of phenylacetylene
Deetlefs *et al.* have used the thiazolium-gold ionic liquids for the catalytic hydration of phenylacetylene to acetophenone using imidazolium ionic liquids as the reaction media.

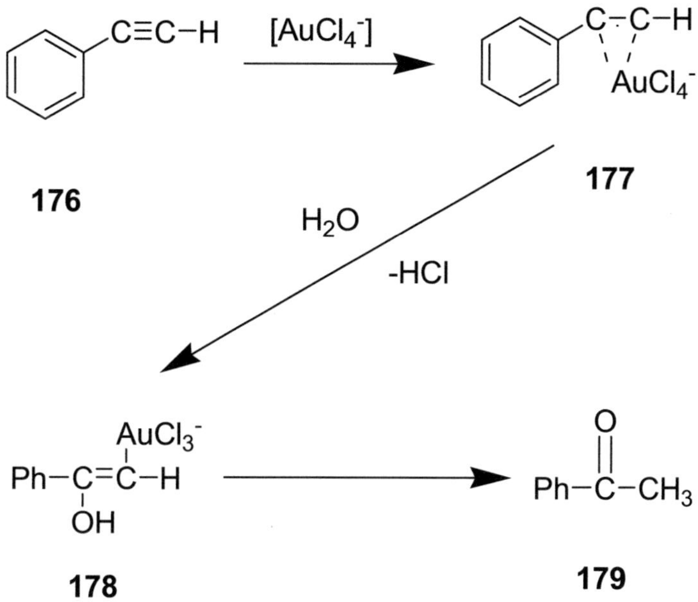

Scheme 44.

7. Conclusion

This chapter provides a comprehensive information about the synthesis of thiazolium and benzothiazolium ion based salts and ionic liquids. it is evident that thiazolium and benzothiazolium ionic liquids can be as efficient solvents/catalysts as the other heterocyclic cation based ionic liquids could be (*e.g.* imidazolium). Although the quaternary salts of thiazolium and benzothiazolium have been studied since 18th century but their exploitation as ionic liquids remained limited to a few reports in the literature. Based on the Scifinder information the only article on the benzothiazolium ionic liquids has been published by our group. Moreover, the phase behaviour study of $[C_{11}BT]I\text{-}AlCl_3$ mixtures have been described where the appropriate mole ratio of the aluminum chloride to the $[C_{11}BT]I\text{-}AlCl_3$ mixture (Figure-25) have been found with the help of DSC. From the figur-25 it is clear that at mole ratio 0.5% the mixture of $[C_{11}BT]I\text{-}AlCl_3$ gave lowest melting point (22.5 ^{o}C) which was checked by DSC. Here the use of thiazolium ionic liquids in different reactions (*e.g.* Acyloin condensation, Stetter reaction) as solvent/catalyst have also been included.

8. Acknowledgment

We gratefully acknowledge the assistance of Prof. Dr. Misbahul Ain Khan for the preparation of manuscript. We particularly thank Prof. Dr. Makshoof Athar for facilities at Institute of Chemistry, University of the Punjab Lahore, Pakistan. Special thanks to Prof. Khizar Iqbal Malik for helpful comments and additional information.

9. References

Anjaiah, S.; Chandrasekhar, S. & Gree, R. (2004). Stetter reaction in room temperature ionic liquids and application to the synthesis of haloperidol. *Adv. Synth. Catal.* Vol. 346, No. 13-15, pp. 1329-1334, ISSN. 1615-4150

Baldwin, J. E. & Walker, J. A. (1974). Competing [1-3]- and [3,3]-sigmatropic rearrangements of electron-rich olefins. *J. Am. Chem. Soc.* Vol. 96, No. 2, pp. 596-597, ISSN. 0002-7863

Barrer, R. M. (1943). The viscosity of pure liquids. II. Polymerised ionic melts. *J. Chem. Soc. Faraday. Trans.* Vol. 39, pp. 59-67, ISSN. 0014-7672

Beilenson, B. & Hamer, F. M. (1936). 260. 5-Chloro- and 5-bromo-1-methylbenzthiazole and cyanine dyes prepared from them. *J. Chem. Soc.* pp. 1225-1231, ISSN. 0368-1769

Brooker, L. G. S.; White, F. L.; Keyes, G. H.; Smyth, C. P. & Oesper, P. F. (1941). Color and Constitution. II.1 Absorptions of Some Related Vinylene-Homologous Series. *J. Am. Chem. Soc.* Vol. 63, No. 11, pp. 3192-3203, ISSN. 0002-7863

Brooker, L. G. S.; Keyes, G. H.; Sprague, R. H.; Vandyke, R. H.; Vanlare, E.; Vanzandt, G. & White, F. L. (1951). Studies in the Cyanine Dye Series. XI.1 The Merocyanines. *J. Am. Chem. Soc.* Vol. 73, No. 11, pp. 5326-5332, ISSN. 0002-7863

Bryce, M. R.; Coates, H. M.; Cooper, J. & Murphy, L. C. (1984). A new route to 1,4-disubstituted cyclohexa-1,3-diene derivatives: the synthesis of a highly conjugated bisbenzothiazoline derivative. *J. Org. Chem.* Vol. 49, No. 18, pp. 3399-3401, ISSN. 0022-3263

Castells, J.; Lopez-Calahorra, F. & Domingo, L. (1988). Postulation of bis(thiazolin-2-ylidene)s as the catalytic species in the benzoin condensation catalyzed by a thiazolium salt plus base. *J. Org. Chem.* Vol. 53, No. 19, pp. 4433-4436, ISSN. 0022-3263

Chapman, D. D.; Elwood, J. K.; Heseltine, D. W.; Hess, H. M. & Kurtz, D. W. (1974). Method for annelation of pyridinium rings on to nitrogen heterocycles. *J. Chem. Soc. Chem. Commun.* No. 16, pp. 647-648, ISSN. 0022-4936

Chen, Y.-T.; Barletta, G. L.; Haghjoo, K.; Cheng, J. T. & Jordan, F. (1994). Reactions of Benzaldehyde with Thiazolium Salts in Me2SO: Evidence for Initial Formation of 2-(.alpha.-Hydroxybenzyl)thiazolium by Nucleophilic Addition, and for Dramatic Solvent Effects on Benzoin Formation. *J. Org. Chem.* Vol. 59, No. 25, pp. 7714-7722, ISSN. 0022-3263

Chikashita, H.; Ishihara, M.; Takigawa, K. & Itoh, K. (1991). Synthetic application of benzothiazole ring system as an on-off type of leaving group. Synthesis of ketones and carboxylic acid derivatives from 2-(1-substituted 1-hydroxyalkyl)- and 2-(1-hydroxyalkyl)benzothiazoles. *Bull. Chem. Soc. Jpn.* Vol. 64, pp. 3256-3261, ISSN. 0009-2673

Clark, L. M. (1928). CCCIII.-The methylene bases from 1-methylbenzthiazole and 1-methylbenzselenazole methiodides; with a note on the preparation of 1-substituted benzthiazoles. *J. Chem. Soc.* pp. 2313-2320, ISSN. 0368-1769

Davis, J. H., Jr. & Forrester, K. J. (1999). Thiazolium-ion based organic ionic liquids (OILs). Novel OILs which promote the benzoin condensation. *Tetrahedron Lett.* Vol. 40, No. 9, pp. 1621-1622, ISSN. 0040-4039

Deetlefs, M.; Raubenheimer, H. G. & Esterhuysen, M. W. (2002). Stoichiometric and catalytic reactions of gold utilizing ionic liquids. *Catal. Today*. Vol. 72, No. 1-2 pp. 29-41, ISSN. 0920-5861

Egan, R. S.; Tadanier, J.; Garmaise, D. L. & Gaunce, A. P. (1968). Intermediates in the Hantzsch thiazole synthesis. *J. Org. Chem.* Vol. 33, No. 12, pp. 4422-4426, ISSN. 0022-3263

Eicher, T.; Hauptmann, S. & Speicher, A. (2003), *Five-Membered Heterocycles* (2), Wiley-VCH Verlag GmbH & Co. KGaA, 9783527601837, Grünstadt.

Evans, W. J. & Smiles, S. (1935). A rearrangement of o-acetamido-sulphones and -sulphides. *J. Chem. Soc.* pp. 181-188, ISSN. 0368-1769

Federsel, H.-J. & Bergman, J. (1980). A novel base-induced ring expansion of quaternized heterocycles. *Tetrahedron Lett.* Vol. 21, No. 25, pp. 2429-2432, ISSN. 0040-4039

Fisher, N. I. & Hamer, F. M. (1930). A general method for the preparation of thiocyanine dyes. Some simple thiocarbocyanines. *J. Chem. Soc.* pp. 2502-2510, ISSN. 0368-1769

Franklin, E. C. (1905). Reactions in Liquid Ammonia. *J. Am. Chem. Soc.* Vol. 27, No. 7, pp. 820-851, ISSN. 0002-7863

Franklin, E. C. (1924). Symtems of Acids, Bases and Salts. *J. Am. Chem. Soc.* Vol. 46, No. 10, pp. 2137-2151, ISSN. 0002-7863

Fry, D. J. & Kendall, J. D. (1951). The formation and fission of quaternary salts of heterocyclic bases containing reactive alkylthio-groups. *J. Chem. Soc.* pp. 1716-1722, ISSN. 0368-1769

Garmaise, D. L.; Paris, G. Y.; Komlossy, J.; Chambers, C. H. & Mccrae, R. C. (1969). Anthelmintic quaternary salts. III. Benzothiazolium salts. *J. Med. Chem.* Vol. 12, No. 1, pp. 30-36, ISSN. 0022-2623

Halgas, J.; Sutoris, V.; Foltinova, P. & Sekerka, V. (1983). Benzothiazole compounds. XX. Synthesis of 3,4,6-substituted benzothiazolium salts as plant growth regulators and their antimicrobial activity. *Chem. Zvesti.* Vol. 37, No. 6, pp. 799-808, ISSN. 0366-6352

Hatrík, Š. & Zahradník, P. (1996). Neural Network Approach to the Prediction of the Toxicity of Benzothiazolium Salts from Molecular Structure. *J. Chem. Inf.Comput. Sci.* Vol. 36, No. 5, pp. 992-995, ISSN. 0095-2338

Hazen, G. G. & Rosenburg, D. W. (1964). Preparation of 9(11)-Unsaturated Steroids. A Novel Reagent System. *J. Org. Chem.* Vol. 29, No. 7, pp. 1930-1932, ISSN. 0022-3263

Hofmann, A. W. (1887). Noch einige weitere Beobachtungen über das o-Amidophenylmercaptan und seine Abkömmlinge. *Ber. Deut. Chem. Ges.* Vol. 20, No. 2, pp. 2251-2265, ISSN. 1099-0682

Huddleston, J. G.; Visser, A. E.; Reichert, W. M.; Willauer, H. D.; Broker, G. A. & Rogers, R. D. (2001). Characterization and comparison of hydrophilic and hydrophobic room temperature ionic liquids incorporating the imidazolium cation. *Green. Chem.* Vol. 3, No. 4, pp. 156-164, ISSN. 1463-9262

Kendall, J. D. & Majer, J. R. (1948). 136. Preparation of unsymmetrical carbocyanines. *J. Chem. Soc.* pp. 687-690, ISSN. 0368-1769

Kiprianov, A. I. & Tolmachev, A. I. (1957). Preparation of quaternary salts of weak organic bases. *Zh. Obshch. Khim.* Vol. 27, pp. 142-150, ISSN. 0044-460X

Kitzman, P. H. (2009). Effectiveness of riluzole in suppressing spasticity in the spinal cord injured rat. *Neurosci. Lett.* Vol. 455, No. 2, pp. 150-153, ISSN. 0304-3940

Knott, E. B. (1955). Compounds containing sulphur chromophores. Part III. Experiments on the synthesis of ionic dyes. *J. Chem. Soc.* pp. 933-937, ISSN. 0368-1769

Konig, W. & Meier, W. (1925). Cyanine dyes. VI. Thio- and oxocyanines. *J. Prakt. Chem. (Leipzig).* Vol. 109, pp. 324-344, ISSN. 0021-8383

König, W. (1928). Über ein vereinfachtes Verfahren zur Gewinnung substituierter μ-Methyl-benzthiazole und deren Umwandlung in neue Heterocyclo-Polymethinfarbstoffe. *Ber. Deut. Chem. Ges. (A and B Series).* Vol. 61, No. 9, pp. 2065-2074, ISSN. 1099-0682

Kossmehl, G.; Bocionek, P. & Manecke, G. (1979). Polyarylenalkenylenes and polyheteroarylenalkenylenes, 9. Syntheses and Characterization of poly(5-methylthiazolo[4,5-f]benzthiazolium-2,6-diylvinylenarylenevinylene) s and some model compounds. *Makromol. Chem.* Vol. 180, pp. 1441-1463, ISSN. 0025-116X

Kumar, D.; David, W. M. & Kerwin, S. M. (2001). N-propargyl-2-alkynylbenzothiazolium aza-enediynes: role of the 2-alkynylbenzothiazolium functionality in DNA cleavage. *Bioorg. Med. Chem. Lett.* Vol. 11, No. 22, pp. 2971-2974, ISSN. 0960-894X

Li, P. & Xu, J. C. (1999). A novel thiazolium type peptide coupling reagent for hindered amino acids. *Tetrahedron Lett.* Vol. 40, No. 47, pp. 8301-8304, ISSN. 0040-4039

Lopez-Celahorra, F.; Castells, J.; Domingo, L.; Marti, J. & Bofill, J. M. (1994). Use of 3,3'-polymethylene-bridged thiazolium salts plus bases as catalysts of the benzoin condensation and its mechanistic implications: proposal of a new mechanism in aprotic conditions. *Heterocycles.* Vol. 37, No. 3, pp. 1579-1597, ISSN. 0385-5414

Mcguinness, D. S.; Cavell, K. J.; Yates, B. F.; Skelton, B. W. & White, A. H. (2001). Oxidative Addition of the Imidazolium Cation to Zerovalent Ni, Pd, and Pt: A Combined Density Functional and Experimental Study. *J. Am. Chem. Soc.* Vol. 123, No. 34, pp. 8317-8328, ISSN. 0002-7863

Messmer, A. & Gelléri, A. (1967). Synthesis of 1,3-Triaryl[1,2,3]triazolo[5,1-b]-benzothiazolium and -benzimidazolium Salts, and of 1,3-Diphenyl[1,2,3]triazolo[1,5-a]benzimidazole. *Angew. Chem. Intl. Ed.* Vol. 6, No. 3, pp. 261-262, ISSN. 1521-3773

Mills, W. H. (1922). LIV.-The cyanine dyes. Part IV. Cyanine dyes of the benzothiazole series. *J. Chem. Soc. Trans.* Vol. 121, pp. 455-466, ISSN. 0368-1645

Mohanazadeh, F. & Aghvami, M. (2007). Thiazolium Salt Immobilized on Ionic Liquid: An Efficient Catalyst and Solvent for Preparation of α-Hydroxyketones. *Phosphorus, Sulfur Silicon Relat. Elem.* Vol. 182, No. 10, pp. 2467-2475, ISSN. 1042-6507

Motesharei, K. & Myles, D. C. (1997). Multistep Synthesis on the Surface of Self-Assembled Thiolate Monolayers on Gold: Probing the Mechanism of the Thiazolium-Promoted Acyloin Condensation. *J. Am. Chem. Soc.* Vol. 119, No. 28, pp. 6674-6675, ISSN. 0002-7863

Mukaiyama, T. (1979). Neue Synthesen mit Oniumsalzen von Aza-arenen. *Angew. Chem.* Vol. 91, No. 10, pp. 798-812, ISSN. 1521-3757

Mukaiyama, T. & Hojo, K. (1976). Optical interconversion of enantiomeric secondary alcohols using 2-fluorobenzothiazolium salt. *Chem. Lett.* Vol. 5, No. 9, pp. 893-896, ISSN. 0366-7022

Mushkalo, L. K.; Chuiguk, V. A. & Annamuradov, C. (1971). Condensation of o-aminothiophenol with γ- and δ-halo ketones. *Ukr. Khim. Zh.* Vol. 37, No. 7, pp. 568-572, ISSN. 0041-6045

Nadeem, S.; Munawar, M. A.; Ahmad, S.; Smiglak, M.; Drab, D. M.; Malik, K. I.; Amjad, R.; Ashraf, C. M. & Rogers, R. D. (2010). Solvent-free synthesis of benzothiazole-based quaternary ammonium salts: precursors to ionic liquids. *Arkivoc.* No. 7, pp. 19-37, ISSN. 1551-7012

Pianka, M. & Hall, J. C. (1959). Studies in fungitoxicity. II.—Fungitoxicity of certain anilinovinyl quaternaries. *J. Sci. Food Agric.* Vol. 10, No. 7, pp. 385-388, ISSN. 1097-0010

Puterová, Z.; Krutošíková, A.; Lyčka, A. & Ďurčeková, T. (2004). Reactions of Substituted Furan-2-carboxaldehydes and Furo[b]pyrrole Type Aldehydes with Benzothiazolium Salts. *Molecules.* Vol. 9, No. 4, pp. 241-255, ISSN. 1420-3049

Riemer, B. & Liebscher, J. (1993). Synthesis of Pyrimido[2,1-b]benzoxazolium, Pyrimido[2,1-b]benzothiazolium and Pyrimido[1,2-a]benzimidazolium Salts by Reaction of 3-Isothiocyanato-2-propeniminium Salts with o-Functionalized Anilines. *Synthesis.* Vol. 1993, No. 1, pp. 75-76, ISSN. 0039-7881

Sabaté, R. & Saupe, S. J. (2007). Thioflavin T fluorescence anisotropy: An alternative technique for the study of amyloid aggregation. *Biochem. Biophys. Res. Commun.* Vol. 360, No. 1, pp. 135-138, ISSN. 0006-291X

Schöberl, A. & Stock, M. (1940). Über Ringspaltungen bei Thiazoliumsalzen. *Ber. Deut. Chem. Ges. (A and B Series).* Vol. 73, No. 11, pp. 1240-1252, ISSN. 1099-0682

Stsiapura, V. I.; Maskevich, A. A.; Kuzmitsky, V. A.; Turoverov, K. K. & Kuznetsova, I. M. (2007). Computational Study of Thioflavin T Torsional Relaxation in the Excited State. *J. Phys. Chem. A.* Vol. 111, No. 22, pp. 4829-4835, ISSN. 1089-5639

Sutoris, V.; Bajci, P.; Sekerka, V. & Halgas, J. (1988). Benzothiazole compounds. XXXIII. Benzothiazolium salts which increase sugar and chlorophyll contents in plants. *Chem. Pap.* Vol. 42, No. 2, pp. 249-261, ISSN. 0366-6352

Sutoris, V.; Halgas, J.; Sekerka, V.; Foltinova, P. & Gaplovsky, A. (1983). Benzothiazole compounds. XVIII. Preparation of benzothiazolium salts, their growth regulation effects, and antimicrobial activities. *Chem. Zvesti.* Vol. 37, No. 5, pp. 653-662, ISSN. 0366-6352

Svetlichnyĭ, V.; Bazyl', O.; Artyukhov, V.; Kopylova, T.; Derevyanko, N. & Ishchenko, A. (2007). Specific features of the two-photon absorption of cationic symmetric polymethine dyes. *Opt. Spectrosc.* Vol. 103, No. 5, pp. 753-760-760, ISSN. 0030-400X

Takahashi, S.; Hashimoto, S. & Kano, H. (1973). Heteroaromatic N-oxides. XI. Aminolysis of esters of benzazole N-oxides and related quaternary salts. *Chem. Pharm. Bull.* Vol. 21, No. 2, pp. 287-295, ISSN. 0009-2363

Takamizawa, A.; Sato, H. & Sato, Y. (1972). Pyrimidine derivatives and related compounds. LXXV. Reactions of thiazolium salts with diethyl acylphosphonates and hydroxylation of some 3-oxo-2,3-dihydro-4H-1,4-thiazine derivatives. *Chem. Pharm. Bull.* Vol. 20, No. 5, pp. 892-900, ISSN. 0009-2363

Todd, A. R.; Bergel, F. & Karimullah. (1936). Über Aneurin, II. Mitteil.: Über die Synthese von N-Arylthiazoliumsalzen; über Einzelheiten in der Konstitution des Aneurins und Thiochroms. *Ber. Deut. Chem. Ges. (A and B Series)*. Vol. 69, No. 1, pp. 217-223, ISSN. 1099-0682

Tseng, H.-Y.; Wu, S.-H.; Huang, W.-H.; Wang, S.-F.; Yang, Y.-N.; Mahindroo, N.; Hsu, T.; Jiaang, W.-T. & Lee, S.-J. (2005). Benzothiazolium compounds: novel classes of inhibitors that suppress the nitric oxide production in RAW264.7 cells stimulated by LPS/IFN[gamma]. *Bioorg. Med. Chem. Lett.* Vol. 15, No. 8, pp. 2027-2032, ISSN. 0960-894X

Visser, A. E. S., R.P.; Reischert, W.M.; Mayton, R.; Sheff, S.; Wierzbicki, A.; Davis, J.H. Jr.; Rogers, R.D. (2002). Task-specific ionic liquids incorporating novel cations for the coordination and extraction of Hg^{2+} and Cd^{2+}: synthesis, characterization, and extraction studies. *Environ. Sci. Technol.* Vol. 36, No. 11, pp. 2523–2529, ISSN.

Vorsanger, J. J. (1968). Bases derived from 3-methyl benzothiazolium ions. Existence of a heterocyclic carbene. *Bull. Soc. Chim. Fr.* No. 3, pp. 971-980, ISSN. 0037-8968

Wasserscheid, P. & Keim, W. (2000). Ionic Liquids—New "Solutions" for Transition Metal Catalysis. *Angew. Chem. Intl. Ed.* Vol. 39, No. 21, pp. 3772-3789, ISSN. 1521-3773

Weinhardt, K. K. & Neumeyer, J. L. (1970). Preparation and ring opening of 1,2,3,4-tetrahydro-2-oxopyrimido[2,1-b]benzothiazol-5-ium chloride. *J. Org. Chem.* Vol. 35, No. 4, pp. 1176-1178, ISSN. 0022-3263

Xu, L.; Tao, F.; Sun, G. & Yu, T. (1988). Preparation and catalytic reactions of benzothiazolium salts supported on polymers. *Huaxue Xuebao*. Vol. 46, No. 7, pp. 706-710, ISSN. 0567-7351

Xu, L.; Tao, F.; Shen, X. & Yu, T. (1988). Catalytic synthesis of 1,4-dicarbonyl compounds with benzothiazolium salts. *Huaxue Xuebao*. Vol. 46, No. 7, pp. 663-668, ISSN. 0567-7351

Yang, C.-W.; Yang, Y.-N.; Liang, P.-H.; Chen, C.-M.; Chen, W.-L.; Chang, H.-Y.; Chao, Y.-S. & Lee, S.-J. (2007). Novel Small-Molecule Inhibitors of Transmissible Gastroenteritis Virus. *Antimicrob. Agents Chemother.* Vol. 51, No. 11, pp. 3924-3931, ISSN.

Yu, F.; Zhang, R.; Xie, C. & Yu, S. (2010). Polyether-substituted thiazolium ionic liquid catalysts - a thermoregulated phase-separable catalysis system for the Stetter reaction. *Green Chem.* Vol. 12, No. 7, pp. 1196-1200, ISSN. 1463-9262

Zhou, Z.-Z.; Ji, F.-Q.; Cao, M. & Yang, G.-F. (2006). An efficient intramolecular Stetter reaction in room temperature ionic liquids promoted by microwave irradiation. *Adv. Synth. Catal.* Vol. 348, No. 14, pp. 1826-1830, ISSN. 1615-4150

Zimmermann, T. & Schmidt, K. (1996). Ring transformations of heterocyclic compounds. XIV. Ring transformations of pyrylium and thiopyrylium salts with anhydrobases derived from 1H-benzimidazolium and benzothioazolium salts: an easy access to 2-(2,4,6-triarylphenyl)-1H-benzimidazolium and -benzothiazolium derivatives. *J. Heterocycl. Chem.* Vol. 33, No. 6, pp. 1717-1721, ISSN. 0022-152X

3

Glycoside-Based Ionic Liquids

Robert Engel
Department of Chemistry and Biochemistry
Queens College of the City University of New York
USA

1. Introduction

A significant portion of our efforts over the past decade has been dedicated to the design and synthesis of polycationic organic salts, both for their use as metabolic regulators in a variety of settings, and for their use in the construction of ionic liquids. Following from this body of work we present this summary of efforts toward the development of a particular category of ionic liquid, those related to carbohydrates and their derived compounds, glycosides and related polyols.

Our synthetic endeavours have focused on the generation of a library of polycationic derivatives of carbohydrates and their derivatives. There is a fundamental rationale in contemplating such structures; if one is to fulfill the promise of ionic liquids as "designer solvents," (Plechkova & Seddon 2007) then one must consider structures with which significant and subtle variation in properties can be achieved with relatively minor changes in their structures. Carbohydrates and their related compounds (glycosides, polyols) clearly meet this requirement through the presence of multiple functionalities through which structural variation can be accomplished with relative ease.

Thus we turn our attention to the design of cationic (and particularly polycationic) derivatives related to glycosides, simple carbohydrates, and the reduced polyols. It is necessary that such parent structures bear functionalities that may be used readily for the incorporation of cationic sites. Most prominently, in all of these types of compounds, the presence of primary hydroxyl group(s) readily permits such cationic site incorporation. We will contemplate the reaction systems allowing such transformations in a later section of this treatment. Initially, however, we should note the limitations on the nature of the anion in ionic liquids generated from such structures if one is to allow facile production of the ionic liquid without side reactions that could modify the inherent organic (carbohydrate) structure.

Finally, we will contemplate the experimental strategies most amenable for such ionic liquids, as well as their particular applications that would appear to be significant.

2. Anionic choices

There has been a wide range of anions that have been used for the construction of ionic liquids. Historically, anions such as tetrachloroaluminate, which are highly reactive with water and other hydroxylic species, were involved for the construction of ionic liquids that were reasonably fluid at temperatures below 100°C. Unfortunately, tetrachloroaluminate is itself quite reactive toward both water and other hydroxylic solvents and reagents, and therefore would be quite unsuitable for use with cations themselves bearing hydroxylic functionalities (Kubisa 2004).

Other commonly used anions for ionic liquids, such as tetrafluoroborate and hexafluorophosphate, also undergo hydrolysis at sufficiently rapid rates that would cause difficulties for their use with either aqueous solvents or hydroxylic cations. While the hexafluorophosphate anion undergoes hydrolysis most rapidly only under acidic conditions (Freire et al. 2010), the preparation of the ionic liquids using hexafluorophosphoric acid can lead to significant formation of the (hydrolyzed) phosphate salt (Lall et al. 2000). Similarly, the alternative approach for the preparation of the phosphate salts (liquid ionic phosphates, LIPs) involving the use of 98% phosphoric acid (Lall et al. 2002) similarly experiences difficulty when cations are used that bear hydroxylic sites. Presumably the "anion exchange" method (Lall et al. 2002) would not experience this difficulty, but this process is extremely inefficient for the preparation of any sizable quantity of material. Similarly, the tetrafluoroborate anion undergoes hydrolysis sufficient rapidly (Freire, et al. 2010) and can be anticipated to react with free hydroxylic functions such that it would not be a useful anion for ionic liquids based on carbohydrate and carbohydrate-related materials.

While these reactivity characteristics eliminate such anions from practical use with the polyhydroxylic cationic species of current interest, there remain a number of other anions that would be quite suitable. Of particular interest are those that remain unreactive in the presence of hydroxylic solvents (and cations) and may be incorporated by simple metathesis processes. For example, the bis(trifluoromethyl)sulfonylimide anion can be introduced through the addition of an equivalent amount of lithium bis(trifluoromethyl)sulfonylimide to an aqueous solution of the halide salt of the desired cationic (ammonium) species, which causes separation of the desired liquid salt from the aqueous remainder (Bonhôte et al. 1996). The ionic liquid can then be isolated by decantation and drying. In addition, use of the lithium bis(oxalato)borate salt in similar ionic exchange processes produces the "BOB" salts that are useful as anions for ionic liquids (Lishka et al. 1999; Xu & Angell 2001; Xu et al. 2002). The "BOB" anion is quite intriguing in that it has no hydrogen associated with it and exhibits a wide electrochemical stability window (Xu & Angell 2001). This intriguing anion has recently been used for the construction of ionic liquids bearing traditional cations (Lall-Ramnarine et al. 2009). In complement to the hydrophobic solubility characteristics presented by these anions, the phosphate salts provide aqueous solubility without dissolution in organic solvents. The anion structures are illustrated in Figure 1.

3. Cationic systems

The interest for us in the construction of new ionic liquids involved cationic components bearing more than one quaternary ammonium ion site within a given cationic species. These "polycations" were of particular interest given our involvement in the construction and use for a variety of purposes of polyammonium and polyphosphonium salts.

Polyammonium and polyphosphonium salts in several geometric categories were prepared, including dendrimers (Rengan & Engel 1990; Rengan & Engel 1991; Engel et al. 1991; Engel 1992; Rengan & Engel 1992; Engel et al. 1993a; Engel et al. 1993b; Engel 1995; Cherestes & Engel 1994; Cherestes et al. 1998), linear cationic arrays, "strings" (Fabian et al. 1997; Shevchenko & Engel 1998; Strekas et al. 1999; Engel et al. 1999; Cohen et al. 1998; Gordon et al. 2006), and cyclic systems, "rings" (Cohen et al. 1999; Cohen & Engel 2000; Cohen et al.

2000a; Cohen et al. 2000b). Examples of these categories of polycationic species are illustrated in Figures 2-4.

BF_4^- tetrafluoroborate

PF_6^- hexafluorophosphate

PO_4^{3-} phosphate

$(CF_3SO_2)_2N^-$ bis(trifluoromethyl)sulfonylimide

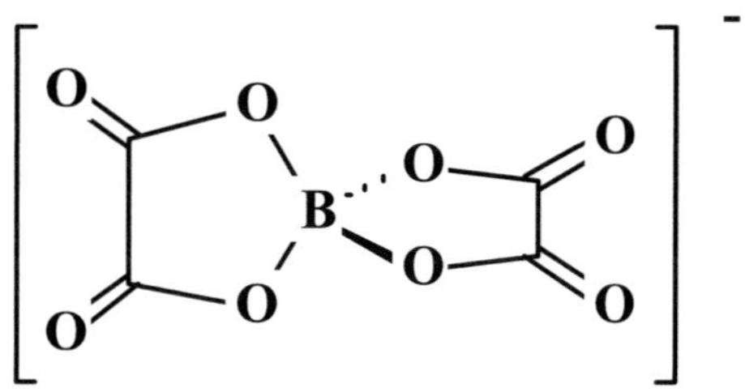

bis(oxalato)borate

Fig. 1. Common anions for ionic liquids

Efforts using the cyclodextrins led to increasing interest in a variety of carbohydrate mono- and polycationic species. The polycationic cyclodextrin derivatives themselves exhibited most intriguing physical characteristics, particularly with regard to their serving as inclusion agents for a variety of anions (Cohen et al. 2000b). It was determined that the host:guest association was of a 1:1 nature, with association constants for guest nucleoside monophosphate anions being generally in the range 5-500; similarly, 1:1 association with acylated amino acids was observed, but with association constants found to be as great as 10^5. Clearly the ring of charges about the top of the cyclodextrin torus held quite rigidly the charged species with the hydrophobic interior of the torus providing a "comfortable" space for the organic portion of the guest species. It is also intriguing that aqueous solubility of the polycationic cyclodextrin salts is significantly dependent on the associated anion. As the reaction is performed, a mixture of halide and tosylate anions would be anticipated to be associated with the product salts. However, a selectivity is observed, strictly halide forms of the salt precipitating readily from aqueous solution while tosylate forms of the salts remain in solution, the tosylate being at least partially associates as a guest species within the hydrophobic interior of the torus.

Carbohydrate derivatives bearing charged sites (as ammonium or phosphonium components) clearly held significant potential not only with the cyclodextrins, but also for other carbohydrate species owing to the wealth of functionality present within the structure and the possibilities for interaction with solvents and other external anionic components. Owing to the size of the cationic portion and the wealth of associated anionic species in a multitude of possible locations relative to the individual cationic sites, the possibilities of cationic carbohydrate derivatives for serving in ionic liquids are quite sizable.

In this regard a series of polyhydroxylic polyammonium species have been prepared, derived from carbohydrate precursors (Thomas et al. 2009a) and ionic liquids generated from them using phosphate and tosylate anions (Engel et al. 2007). Of particular note are the syntheses of chiral tosylate associated ionic liquids derived from methyl α-D-glucopyranoside and mannose, as illustrated in Figure 5.

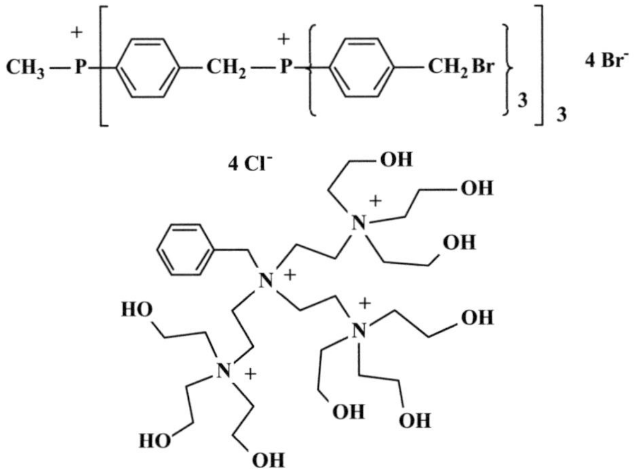

Fig. 2. Examples of polycationic dendrimers. top: a tetraphosphonium dendrimer; bottom: a tetraammonium dendrimer

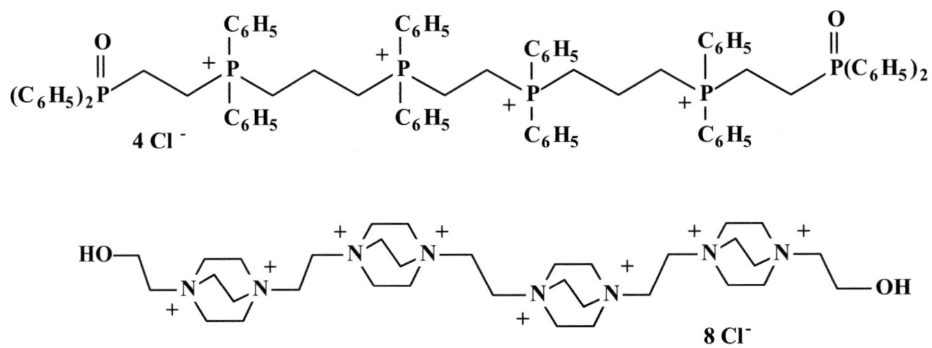

Fig. 3. Examples of polycationic "strings". top: a tetraphosphonium "string" species bottom: an octaammonium "string" species

It is notable that simple tosylate salts of these polyhydroxylic cations produce ionic liquids. This is an indication of the nature of the polyhydroxylic species bearing one or more cationic sites possessing several arrangements of cation and anion of relatively similar energies leading to an inability to form crystalline structures readily. Preliminary MM2 calculations concerning the relative energy minima available for polycationic salts of several common anions indicate a host of such relative minima to exist of relatively close energies (Engel 2011). In such a situation their formation of a simple crystalline structure at room temperature is disfavored and a liquid or glass material is to be anticipated. Indeed, one of the first ionic liquids generated in our laboratory was that from a derivative of a polyhydroxylic species as a chloride (Cherestes et al. 1998). In this effort, formation of a chiral ammonium salt derived from (S)-malic acid by reaction of the primary chloride with 1,4-diazabicyclo[2.2.2]octane produced the chloride salt as a viscous liquid that resisted all attempts at crystallization (Figure 6).

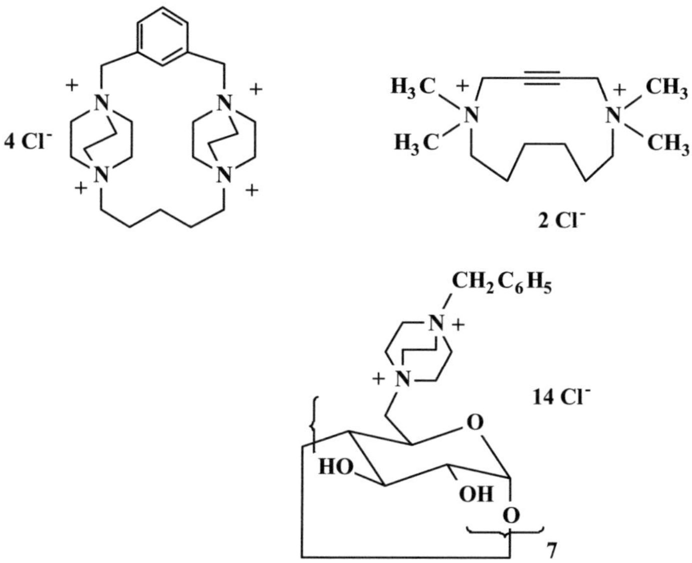

Fig. 4. Examples of polycationic rings. top left: a tetraammonium salt derived from 1,4-diazabicyclo[2.2.2]octane; top right: a cyclic dicationic alkyne; bottom: a β-cyclodextrin derivative bearing fourteen ammonium sites.

As noted in Figure 5 (*vide infra*) the formation of the tosylates of polyhydroxylic species is a relatively simple matter. Tosyl chloride, in early efforts allowed to react in pyridine solution with hydroxylic substrates, reacts selectively with primary hydroxylic sites under aqueous conditions reminiscent of the simple Schotten-Baumann procedure for ester formation (Frantz et al. 2002), as has been used for a wide range of ester preparations. The selectivity of the tosylation reaction and the ability to perform it in aqueous medium (Thomas et al. 2009a) provide major advantages to this procedure, allowing facile formation of the tosylate ester and the equally facile subsequent displacement of the tosylate leaving group by an incoming tertiary amine nucleophile.

Using the simple approach of selective tosylation of primary hydroxyl sites followed by nucleophilic substitution by a tertiary amine reagent, quaternary ammonium salts of a wide range of structures derived from carbohydrate precursors have been prepared. for the cyclodextrin derivatives, per-tosylation of the available hydroxylic sites was established by early efforts at generating cyclodextrin derivatives (Cramer et al. 1969) with later modification to the procedure using aqueous medium.

Using the method of selective tosylation of primary hydroxyl sites, a variety of simple carbohydrate and their glycosidic and reduced derivatives have been generated and investigated (Thomas et al. 2009). Of particular concern were cationic and polycationic derivatives of methyl α-D-glucopyranoside, examples of which are illustrated in Figure 7.

It will be noted that a variety of tertiary amine species were used as nucleophiles in generating the cationic glycoside species from the appropriate tosylate. These included simple mono-basic reagents as had been used previously in the preparation of cationic components of ionic liquids, as well as a series of mono-basic/di-nitrogen reagents derived from 1,4-diazabicyclo[2.2.2]octane. These reagents are illustrated in Figure 8.

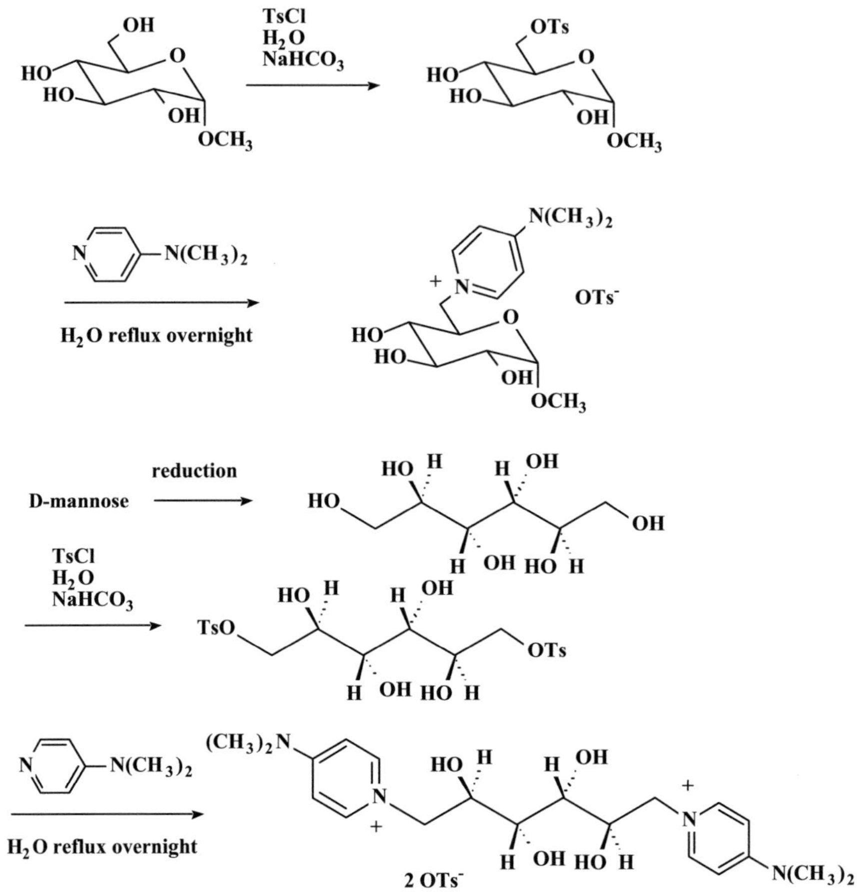

Fig. 5. Chiral ionic liquids prepared from carbohydrate derived precursors with tosylate anions. top: chiral tosylate ionic liquid based on methyl α-D-glucopyranoside; bottom: chiral tosylate ionic liquid based on derivative of D-mannose.

Several di-basic reagents were used for the linking of two glycosidic units, as illustrated in Figure 9. These types of reagents were utilized in the construction of several di-cyclodextrin species bearing a most intriguing structural feature (*vide infra*).

In addition to using methyl α-D-glucopyranoside as a scaffolding for the cationic components, reduced carbohydrates were used for the construction of materials bearing two cationic sites attached directly to the carbohydrate backbone (Thomas et al. 2009b). There are present in several of these species, examples being illustrated in Figure 10, additional cationic sites as components of integrated 1,4-diazoniabicyclo[2.2.2]octane components. All of these species bear potential for the construction of ionic liquids as the association of a variety of anions would be anticipated to exhibit numerous thermodynamic minima of close energy rendering the formation of crystalline species unlikely. The presence of the numerous hydrophilic oxygen sites and hydroxyl groups in such materials would provide intriguing solubilizing characteristics with both classically hydrophobic and hydrophilic anions. The variety of solubilizing characteristics is particularly numerous using the 1,4-diazoniabicyclo[2.2.2]octane adjuncts as the distal unit of these materials can be varied widely, from long and hydrophobic to multifunctional hydrophilic. As such, these materials hold significant promise to fulfill the aim of ionic liquids as being "designer solvents" (Plechkova & Seddon 2007).

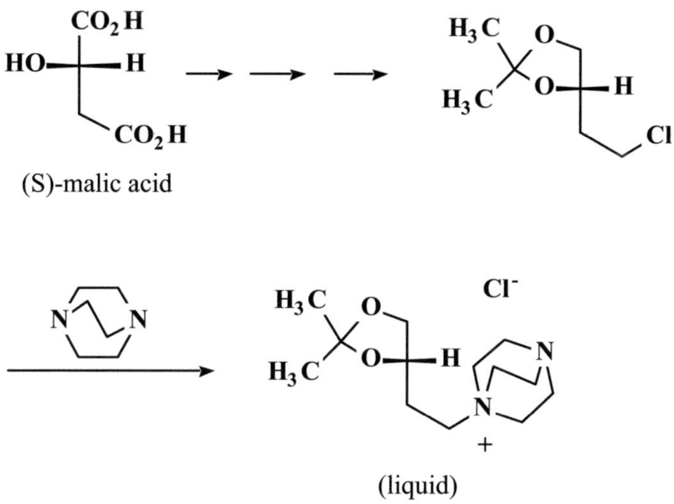

Fig. 6. Formation of an ionic liquid chloride salt derived from a polyhydroxylic precursor

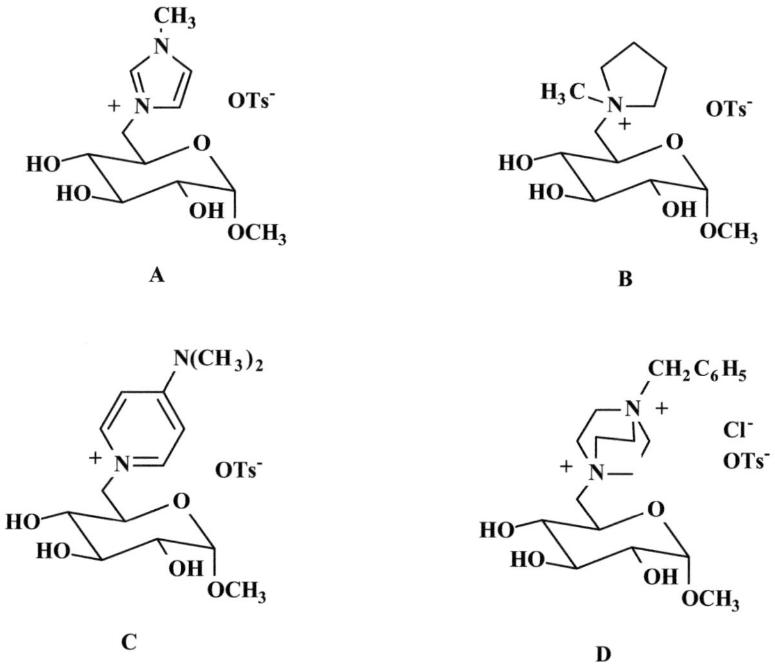

Fig. 7. Examples of cationic derivatives of methyl α-D-glucopyranoside as reported in Thomas et al. 2009b: **A**, derivative of *N*-methylimidazole; **B**, derivative of *N*-methylpyrrolidine; **C**, derivative of 4-dimethylaminopyridine; **D**, derivative of 4-benzyl-1-aza-4-azoniabicyclo[2.2.2]octane

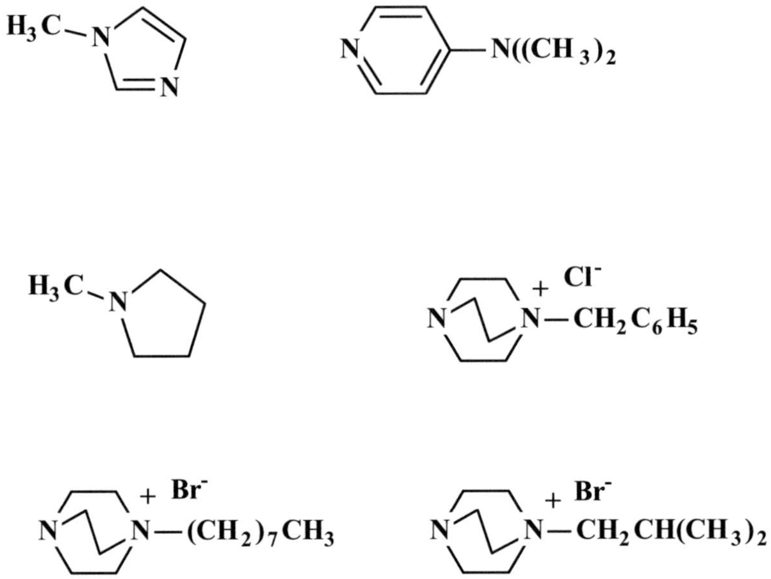

Fig. 8. Tertiary amines used in the formation of cationic derivatives of α-D-glucopyranoside as reported in Thomas et al. 2009b: these include (from top, left to right) *N*-methylimidazole, 4-dimethylaminopyridine, *N*-methylpyrrolidine, 4-benzyl-1-aza-4-azoniabicyclo[2.2.2]octane chloride, 4-octyl-1-aza-4-azoniabicyclo[2.2.2]octane bromide, and4-isobutyl-1-aza-4-azoniabicyclo[2.2.2]octane bromide

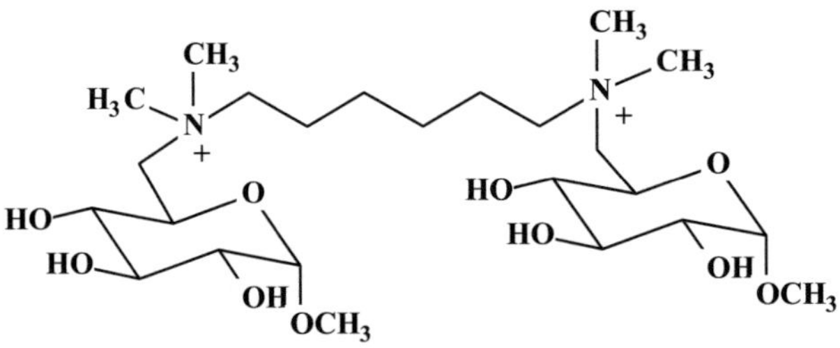

derived from 1,6-bis(dimethylamino)hexane

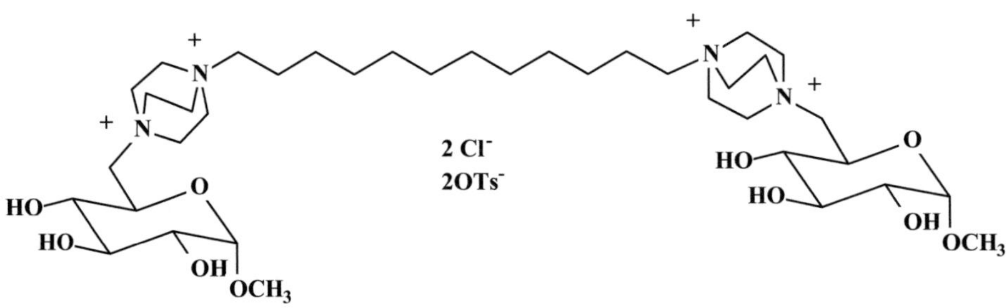

derived from 1,12-bis-(1'-azonia-4'-azabicyclo[2.2.2]octane) dichloride

Fig. 9. Constructs with two α-D-glucopyranoside units connected through linking di- and tetracationic bridges (Thomas et al 2009b)

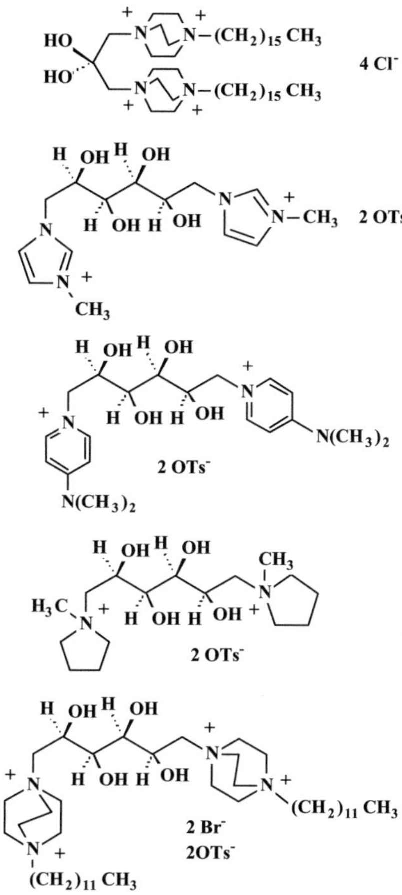

Fig. 10. Constructs of polycationic salts derived from reduced carbohydrate scaffoldings, including glycerol(top) and mannitol (lower structures). A host of potential structures bearing a variety of functionalities and subsequent solubilizing characteristics are available.

As noted (*vide supra*) a series of salts derived from cyclodextrins have been prepared in which the α,ω-bis(basic) reagents were used for the construction of a most intriguing category of host species. Both the α,ω-bis(dimethylamino)alkanes as well as the α,ω-bis(1'-azonia-4'azabicyclo[2.2.2]octane salts were used as bridging reagents linking two cyclodextrin units. These particular host molecules bear hydrophobic cyclodextrin interiors at each end as well as in the middle, with cationic sites near each end at the "top" portion of each of the cyclodextrin units. This structural category is shown in cartoon form in Figure 11. The particular significance of these structures is that they can serve as host for anions bearing a hydrophobic region (*e.g.* fatty acid anions) that holds them virtually permanently. Using MM2 calculations it is indicated that once such an anion is trapped inside the tube, it remains there perpetually, unable to "escape" to the surrounding solution (Engel et al. 2004).

While the cyclodextrins are most intriguing polysaccharides in that six, seven, or eight glucose unites are combined in a useful cyclic shape, other polysaccharides are also of interest. Polycationic derivatives of sucrose, a non-reducing disaccharide, have recently been reported (Engel et al. 2011). The compounds of interest, each bearing six cationic sites as well as lipophilic chains, are notable as being significantly antibacterial in character. The sucrose structure provides three primary hydroxyl sites within each molecule through which the cationic sites can be introduced easily using the tosylation procedure noted previously (Thomas et al 2009). Several structures of interest are illustrated in Figure 12.

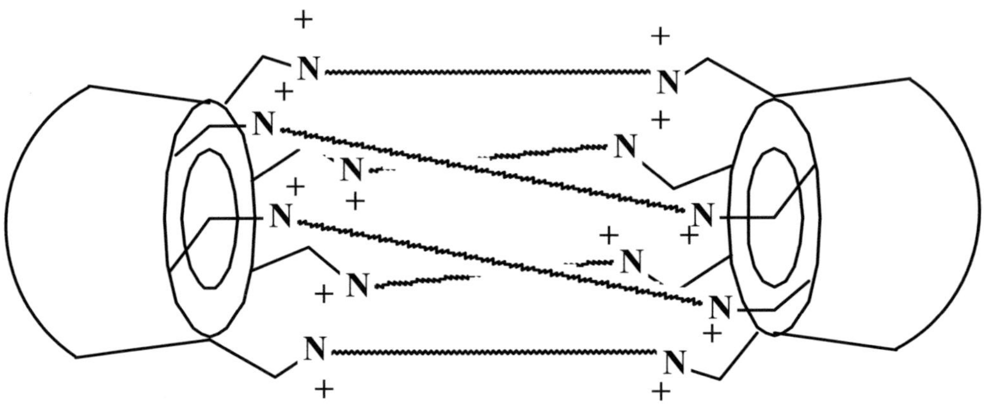

Fig. 11. Cartoon representation of a molecular version of the "Chinese Finger Puzzle" constructed from α,ω-bis(basic)alkane reagents and cyclodextrins modified by tosylation at each of their primary hydroxyl sites.

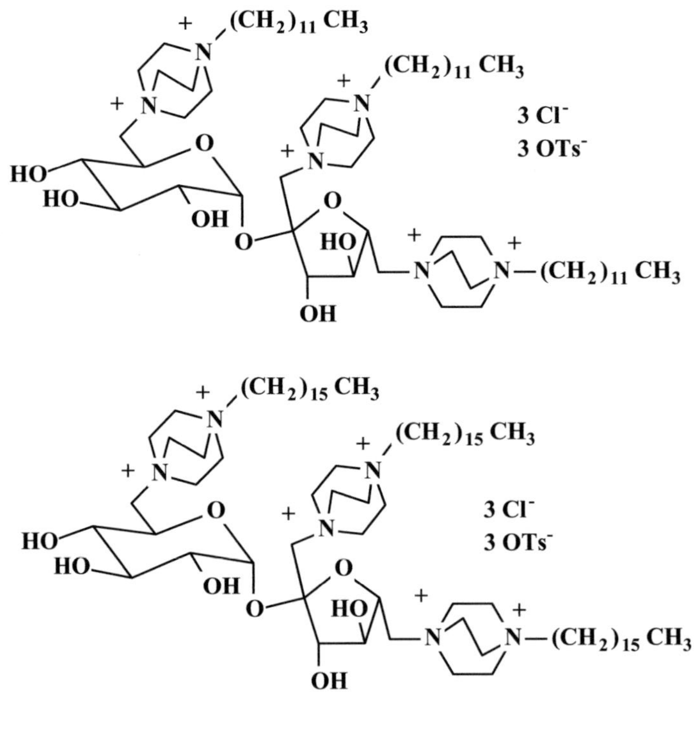

Fig. 12. Polycationic derivatives of sucrose

The significant antibacterial effect of the polycationic derivatives of sucrose is not unexpected. Antibacterial (and antifungal) characteristics are noted for even simple cationic derivatives of lipophilic materials, increasing as the density of charge and tethered sites of charge increase. Even greater antibacterial activity is noted for polycationic derivatives of oligosaccharide species such as soluble starch.

There remain to be exploited numerous variable structures based on carbohydrate and glycosidic systems to which are tethered cationic linkages. For examples are, simple carbohydrates of shorter chain length and their reduced derivatives, simple nucleosides, alginates, carrageenan, amino sugars, chitosan, and other nitrogen-containing carbohydrate species (see Figure 13).

These carbohydrate derivatives, including the oligosaccharides, are intriguing potential structures for cationic systems and ultimately ionic liquids owing not only to their particular structural characteristics, but also to their ready availability for large scale production.

4. Outlook

While only a portion of the carbohydrate and derived glycosidic species considered in the previous section have to date been converted into ionic liquids, they present a major source of materials for the generation of ionic liquids with tunable physical and chemical characteristics, among these being: solubilizing capabilities, fluidity, melting point, and reactivity, to itemize a few. If one is looking to actualize the potential of ionic liquids to serve as "designer solvents" (Plechkova & Seddon 2007) the glycosidic materials, along with their parent carbohydrates and their reduced derivatives provide readily available scaffoldings with facile conversion routes.

Caution need be taken in exploring these possibilities. The multifaceted functionality of the glycoside and related carbohydrate components place limitations on the nature of the associated anions and on the approaches used to incorporate them with the cations. For example, while it was quite reasonable to use crystalline phosphoric acid for the generation of liquid ionic phosphates from simple quaternary ammonium ion chloride salts bearing only alkyl groups on the ammonium species (Lall et al 2002), this approach will not be feasible with the carbohydrate-based systems; there would be anticipated to occur significant dehydration of the carbon structure rendering yields low and purification extremely difficult. Similarly, the approach toward such materials using HPF_6 would result in reaction of the free hydroxyl groups. It would appear to be necessary to generate the phosphate salts through the use of an ion-exchange approach.

Other classical systems for ionic liquid construction would also require care, or might possibly be excluded from consideration owing to the presence of the reactive hydroxyl groups. Of particular note here would be the classical tetrachloroaluminates and the tetrachloroborates, both of which would be anticipated to undergo reaction with the polyhydroxylic cations in undesirable ways.

The major options thereby for ionic liquid construction using glycosides and their related carbohydrate and reduced-carbohydrate species are those that are unreactive in general, and particularly with hydroxylic reagents. These would include the bisoxalato borate and bis(trifluoromethyl)sulfonylimide anions (illustrated in Figure 1), both of which can be incorporated into ionic liquid materials by simple metathesis processes or ion-exchange. Other anions that can be incorporated into ionic liquid materials *via* ion-exchange processes include phosphate (albeit with low efficiency for large-scale conversions) or the substituted phosphate system, di(phenyl)phosphate anion (Thomas et al 2009b). The use of ion-exchange for incorporating such phosphate diesters is tedious, but has the advantages of

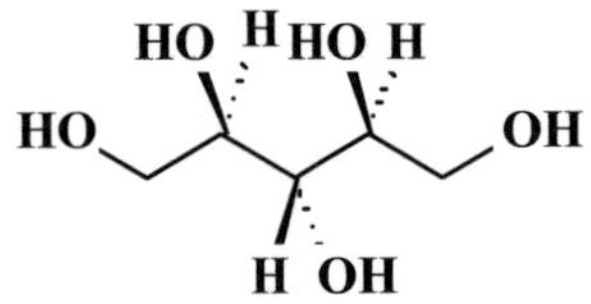

ribitol - reduced ribose

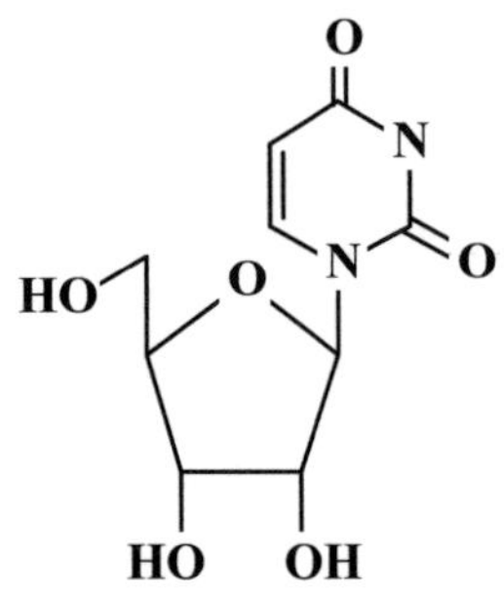

uridine - simple nucleoside

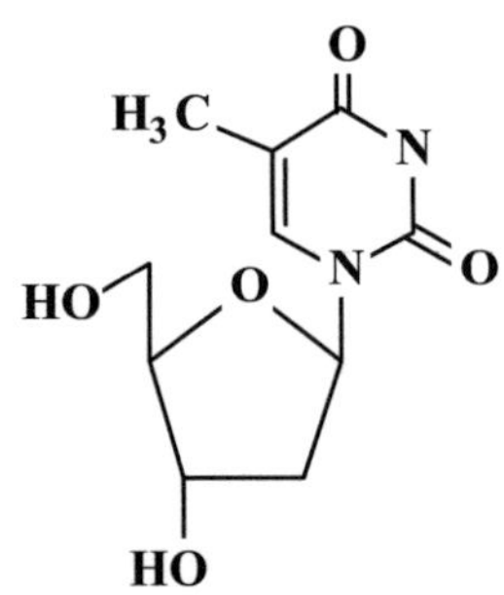

deoxythymidine - simple nucleosid

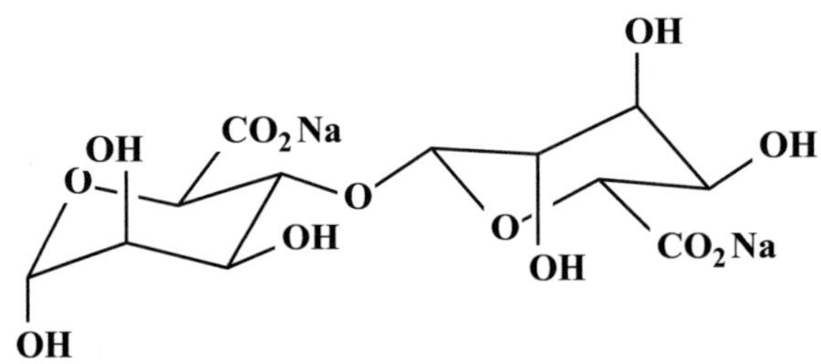

alginate portion bearing one glycosidic linkage - the carboxylate linkages are reducible to primary hydroxyl sites to accommodate facile incorporation of cationic sites

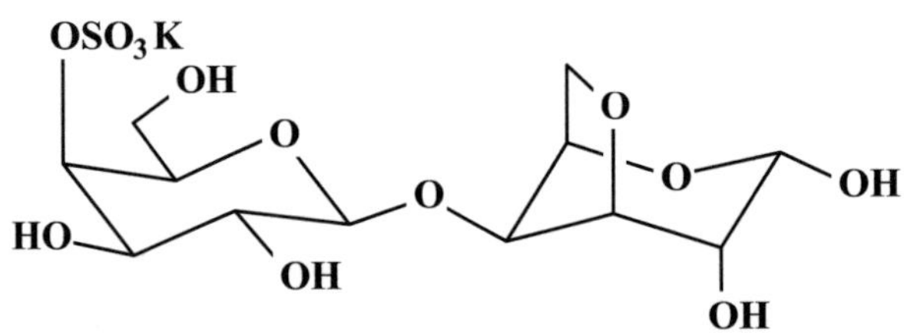

carrageenan portion bearing one glycosidic linkage - the primary hydroxyl group is amenable to tosylation in the presence of the sulfate ester linkage

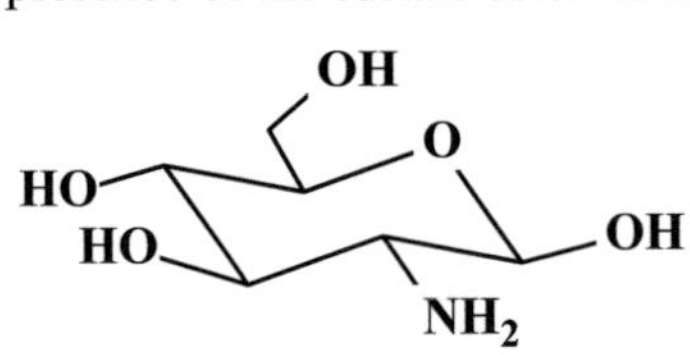

D-2-glucosamine - component of chitosan

Fig. 13. Glycoside and related structures suitable for facile preparation of cationic derivatives with ultimate conversion to ionic liquids. In each instance a primary hydroxyl is available or readily obtained by simple reduction for introduction of a quaternary ammonium (or phosphonium) site for salt construction.

precluding reaction of the target anion supplying reagent from undergoing reaction with a functionally sensitive cation, and allows further structural variation for tuning properties of the target ionic liquid. The variability of this type of anion, being an organic species in itself, is illustrated in Figure 14.

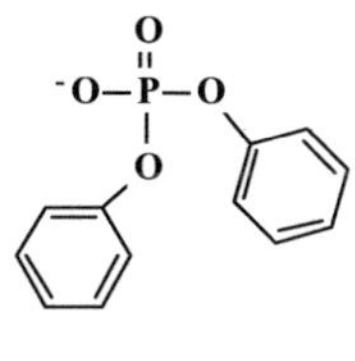

diphenyl phosphate anion

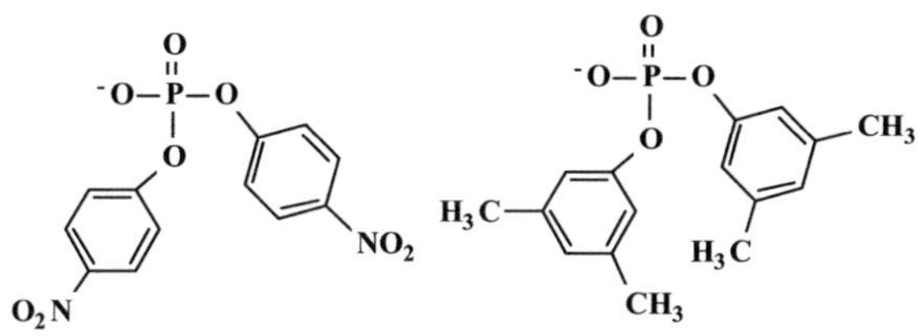

substituted diphenyl phosphate anions readily available for incorporatioı using ion-exchange techniques

Fig. 14. Phosphate and substituted phosphate anions for ionic liquids.

Of course, the potential for ionic liquids being generated from glycosides and related carbohydrate-derived cations with tosylate anions remains. These have been observed in several instances, and the possibility remains for additional species to have similar behavior and characteristics. There are also possibilities, including the tunability of characteristics, through the use of mono-esters of phosphoric acid or sulfuric acid (see Figure 15), again

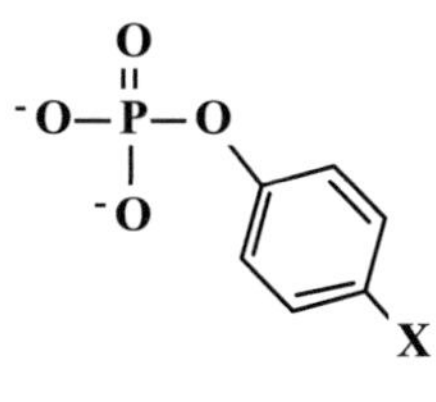

substituted phenyl phosphate dianion

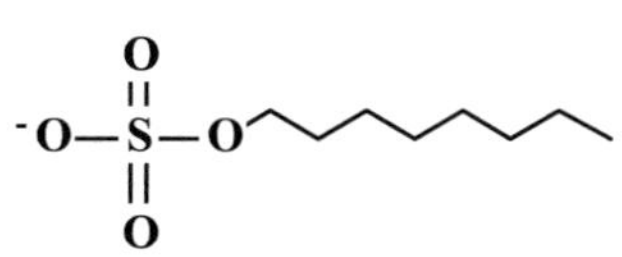

octyl sulfate anion

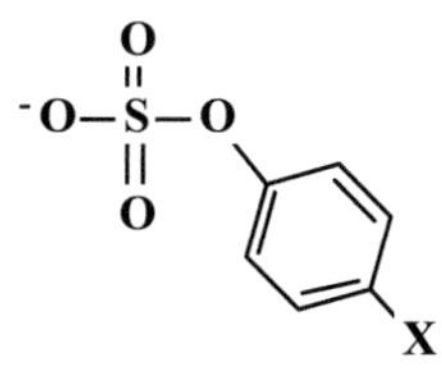

substituted phenyl sulfate anioı

Fig. 15. Phosphate dianion and sulfate anions with structural and conformational variability for ionic liquid construction.

incorporated into the ionic liquid through ion-exchange or metathesis techniques. These again may be generated easily with a variety of substituents that will provide a broad range of chemical and physical characteristics to the materials. The fact that such anions as the partial esters of such inorganic acids are relatively bulky in nature with possibilities for variation in shape *via* bond rotation among a selection of conformations with closely lying relative minima also increases their potential to provide salts that are liquid at room temperature. The construction of these salts with glycosides and related carbohydrate derived cations can provide us with a significantly large selection of ionic liquids for a broad range of purposes.

5. Acknowledgements

The author wishes to thank his numerous students who have generated over the years most of the results discussed in this presentation that emanate from his laboratory. He also wishes to thank Queens College of The City University of New York for providing him the laboratory for his work and the Fellowship Leave during which this presentation was prepared.

6. References

Bonhôte, P., Dias, A.-P., Papageorgiou, N., Kalyanasundaram, K., & Grätzel, M. (1996). Hydrophobic, Highly Conductive Ambient-Temperature Molten Salts. *Inorg. Chem.*, Vol. 35, No. 5, pp. 1168-1178, ISSN 0020-1669.

Cherestes, A., & Engel, R. (1994). Dendrimeric Ion Exchange Materials. *Polymer,* Vol. 35, No. 15, pp. 3343-3344, ISSN 0032-3861.

Cherestes, A., October, T., & Engel, R. (1998). Polycations. II. Chiral Ammonium Dendrimer Synthesis. *Heteroatom Chem.*, Vol. 9, No. 5, pp. 485-494, ISSN 1042-7163.

Cohen, J.I., Traficante, L., Schwartz, P.W., and Engel, R. (1998). Polycations. 4. Synthesis and Antihydrophobic Effect of Polycationic Strings. *Tetrahedron Lett.,* Vol. 39, pp. 8617-8620, ISSN 0040-4039.

Cohen, J.I., Rusinowski, A., Strekas, T.C., & Engel, R. (1999). Polycations. 6. Polycationic Heterocyclic Salts: Their Synthesis and Effect on Double Stranded DNA, *Heteroatom. Chem.*, Vol. 10, No. 7, pp. 559-565, ISSN 1042-7163.

Cohen, J.I. & Engel, R. (2000). Polycations. 7. Polycationic Cyclic Alkynes, *Synth. Commun.,* Vol. 30, No. 12, pp. 2161-2164, ISSN 0039-7911.

Cohen, J.I., Shteto, V., & Engel, R. (2000a). Polycations. 8. The Synthesis of Polycationic Rings, *Synthesis*, pp.1263-1268, ISSN 1088-4327.

Cohen, J.I., Castro, S., Han, J.-a., Behaj, V., & Engel, R. (2000b). Polycations. IX. Polyammonium Derivatives of Cyclodextrins: Synthesis and Binding to Organic Oxyanions, *Heteroatom Chem.*, Vol. 11, No. 7, pp. 546-555, ISSN 1042-7163.

Cramer, F., Mackensen, G., & Kensse, K. (1969). Über Einschlussverbindungen, XX. ORD-Spektren und Konformation der Glucose-Einhalten in Cyclodextrinen, *Chem. Ber.*, Vol. 102, No. 2, pp. 494-508, ISSN 1099-6082

Engel, R. (1992). Cascade Molecules. *Polymer News*, Vol. 17, pp. 302-305, ISSN 0032-3918.

Engel, R., Rengan, K., & Milne, C. (1991). Cationic Cascade Molecules. *Polym. Prepr.*, Vol. 32, p. 601, ISSN 0332-3934.

Engel, R., Rengan, K., & Chan, C.-s. (1993a). Phosphorus Cascade Molecules, *Phosphorus and Sulfur*, Vol. 77, p. 221, ISSN 1042-6507.

Engel, R., Rengan, K., & Chan, C.-s. (1993b). Cascade Molecules Centered About Phosphorus, *Heteroatom Chem.*, Vol. 4, No. 2/3, pp. 181-184, ISSN 1042-7163.

Engel, R. (1995). Ionic Dendrimers and Related Materials, *Advances in Dendritic Macromolecules*, Vol. 2, pp.73-99, ISBN 1-55938-939-7.

Engel, R., Shevchenko, V. Lall, S., Tropp, B., Lau, N., and Strekas, T. (1999). Synthesis and Biological Activities of Phosphorus-Containing Polycationic Strings. *Phosphorus and Sulfur*, Vol. 147, p. 83, ISSN 1042-6507.

Engel, R., Coleman, D., & Thomas, M. (2004). Molecular Version of the Chinese Finger Puzzle, presented at the 228th Meeting of the American Chemical Society, Philadelphia, PA, August 2004.

Engel, R., Lall-Ramnarine, S., Coleman, D., & Thomas, M. (2007). New cations for ionic liquids, including chiral adjuncts with phosphate and sulfonylimide anions. *Proceedings of the Ionic Liquids for Organic Synthesis Symposium*, American Chemical Society Symposium Series #950, pp. 259-266, ISBN 13:9780841274075.

Engel, R. (2011). Unpublished work of this laboratory.

Engel, R., Ghani, I., Montenegro, D., Thomas, M., Klaritch-Vrana, B., Castaño, A., Friedman, L., Leb, J., Rothman, L., Lee, H., Capodiferro, C., Ambinder, D., Cere, E., Awad, C., Sheikh, F., Rizzo, J., Nesbitt, L.-M., Testani, E., & Melkonian, K. (2011). Polycationic Glycosides. *Molecules*, Vol. 16, pp.1508-1518, ISSN 1420-3049.

Fabian, J., October, T., Cherestes, T., & Engel, R. (1997). Polycations: Syntheses of Polyammonium Strings as Antibacterial Agents. *Synlett*, pp. 1007-1010, ISSN 0936-5214.

Frantz, D.E., Weaver, D.G., Carey, J.P., Kress, M.H., & Dolling, U.H. (2002). Practical Synthesis of Aryl Triflates under Aqueous Conditions. *Org. Lett.*, Vol. 4, No. 26, pp. 4717-4718, ISSN 1523-7060.

Freire, M., Neves, C., Marrucho, I., Coutinho, J., & Fernandes, A. (2010). Hydrolysis of Tetrafluoroborate and Hexafluorophosphate Counter Ions in Imidazolium-Based Ionic Liquids. *J. Phys. Chem.*, Vol. 114, No. 11, pp. 3744-3749, ISSN 1089-5647.

Gordon, E., Cohen, J., Engel, R., & Abbott, G.W. (2006). 1,4-Diazabicyclo[2.2.2]octane Derivatives: A Novel Class of Voltage-Gated Potassium Channel Blockers. *Mol. Pharmacol.*, Vol. 69, No. 3, pp. 718-726, ISSN 0026-895X.

Kubisa, P. (2004). Application of ionic liquids as solvents for polymerization processes. *Prog. Polym. Sci.*, Vol. 29, (January 2004), pp. 3-12, ISSN 0079-6700.

Lall, S., Mancheno, D., Castro, S., Behaj, V., Cohen, J. & Engel, R. (2000). Polycations. X. LIPs, a New Category of Ionic Liquid Based on Polyammonium Salts. *J. Chem. Soc., Chem. Commun.*, (November 2004), pp. 2413-2414, ISSN 1364-548X.

Lall, S., Behaj, V., Mancheno, D., Casiano, R., Thomas, M., Rikin, A., Gaillard, J., Raju, R., Scumpia, A., Engel, R., & Cohen, J. (2002). Polycations-12. The Synthesis of Liquid Ionic Phosphates (LIPs) from Mono- and Polycationic Ammonium Halides. *Synthesis*, No. 11, pp.1530-1540, ISSN 0039-7881.

Lall-Ramnarine, S., Castaño, A. Subramaniam, G., Thomas, M.F., Wishart, J. (2009). Synthesis, characterization and radiolytic properties of bis(oxalato)borate containing ionic liquids. *Radiation Phys. Chem.*, Vol. 78, No. 12. pp. 1120-1125, 0969-806X

Lishka, U., Wietelmann, & Wegner, M. (1999). German Patent DE 19829030.

Plechkova, N., & Seddon, K. (2007). Ionic Liquids: "Designer" Solvents for Green Chemistry, In: *Methods and Reagents for Green Chemistry: An Introduction*, Tundo, P., Perosa, A., & Zecchini, F., Chapter 5, John Wiley & Sons, Inc., wiley.com/WileyCDA/WileyTitle/productCd-0470124075.html.

Rengan, K. and Engel, R. (1990). Phosphonium Cascade Molecules, *J. Chem. Soc., Chem. Commun.*, pp. 1084-1085, ISSN 1364-548X.

Rengan, K. & Engel, R. (1991). The Synthesis of Phosphonium Cascade Molecules, *J. Chem. Soc., Perkin Trans. I*, pp. 987-990, ISSN 0300-922X.

Rengan, K. & Engel, R. (1992). Ammonium Cascade Molecules, *J. Chem. Soc. Chem. Commun.*, p. 758, ISSN 1364-548X.

Shevchenko, V. & Engel, R. (1998). Polycations. III. Synthesis of Polyphosphonium Salts for Use as Antibacterial Agents. *Heteroatom Chem.*, Vol. 9, No. 5, pp. 495-502, ISSN 1042-7163.

Strekas, T.C., Engel, R., Locknauth, K., Cohen, J., & Fabian, J. (1999). Polycations. 5. Inducement of Ψ-DNA Circular Dichroism Signals for Duplex Deoxyribonucleotide Homopolymers by Polycationic Strings. *Arch. Biochem. Biophys.*, Vol. 364, No. 1, pp. 129-131, ISSN 0003-9861.

Thomas, M., Montenegro, D., Castaño, A., Friedman, L., Leb, J., Huang, M.L., Rothman, L., Lee, H., Capodiferro, C., Ambinder, D., Cere, E., Galante, J., Rizzo, J., Melkonian, K., & Engel, R. (2009a). Polycations. 17. Synthesis and properties of polycationic derivatives of carbohydrates, *Carbohydrate Res.*, Vol. 344, No. 13, pp. 1620-1627, ISSN 0008-6215.

Thomas, M., Rothman, L., Hatcher, J., Agarkar, P., Ramkirath, R. Lall-Ramnarine, S., & Engel, R. (2009b). Synthesis and Thermochemical Properties of Stereoisomeric Dihydroxy- and Tetrahydroxyalkylammonium Salts. *Synthesis*, No. 9, pp. 1437-1444, ISSN 0039-7881.

Xu, K., Zhang, S., Jow, T.R., Xu, W., & Austen, C. (2002). A, LiBOB as Salt for Lithium-Ion Batteries - A Possible Solution for High Temperature Operation. *Electrochem. and Solid-State Letters*, Vol. 5, No. 1, pp. A26-A29, ISSN 0013-4651.

Xu, W., & Angell, C.A. (2001). LiBOB and Its Derivatives: Weakly Coordinating Anions, and the Exceptional Conductivity of Their Nonaqueous Solutions. *Electrochem. and Solid-State Letters*, Vol. 4, No. 1, pp. E1-E4, ISSN 0013-4651.

4

Ionic Liquids from (Meth) Acrylic Compounds

Sindt M., Mieloszynski J.L. and Harmand J.
Université Paul Verlaine-Metz
France

1. Introduction

Ionic Liquids (ILs) are known and used for the past several decades (Davis J.H. et al, 2003; Wasserscheid & Welton, 2008). In general, there are salts made up of an organic cation and inorganic anion. Their ionic structure gives them particular properties: they are usually liquid at room temperature, non volatile even at high temperature, capable of solubilizing organic and inorganic compounds… Because of these properties, enthusiasm for ILs was initially focused on their use as a solvent and at the beginning of 1990, most of the researches consisted in reproducing conventional reactions in this new reaction medium (Earle & Seddon 2000; Rogers & Seddon, 2003; Baudequin C. et al., 2005; Biswas A. et al., 2006). Compared to conventional solvents, ILs present a great advantage: they are tuneable. In modifying the side chain length of the cation or in exchanging the anion for a bigger or a smaller one, you are able to obtain ILs with specific chemical properties and make them miscible in the aqueous or organic phase or allow them to dissolve a particular compound. As a logical course of events, the early 2000's, a new class of ILs was synthesized to meet special needs such as catalyzing a particular reaction or enabling a reaction which is typically not achievable in a conventional organic solvent. Task Specific Ionic Liquids (TSILs) results from the covalent tethering of a functional group to the cation or the anion or both of them of an ordinary IL, giving the latter the capacity to interact with dissolved substrates in specific ways.

The scope of their applications has expanded over the years. They are attracting much interest in many fields of chemistry and industry due to their properties cited above. First seen as an alternative to volatile solvent, they are now used in electrochemistry, catalysis, asymmetric synthesis, extraction, chromatography... (Seddon, 1997; Sheldon, 1993; Cull et al, 2000; Wilkes, 2004; Wasserscheid & Keim, 2000; Welton, 1999; Brennecke & Maginn, 2001; Fredlake et al., 2004; Henderson & Passerini, 2004).

One of the most recent uses of ionic liquids is as an electrolyte in batteries of the new generation. Ionic liquids are introduced in a polymer matrix and leads to better performance and reliability system, thus allowing no electrochemical reactions which could cause damage (electrochemical window), low pressure steam, and a high ionic conductivity even at room temperature.

The cost and availability of ILs, especially those with an atypical function in their side chain are two issues curbing the use of TSILs in protocols on an industrial scale. One of our objectives is to develop an efficient and rapid synthesis of new TSILs with a function in their side chain. To achieve our goal, we chose to use acrylic compounds which are industrial products easily accessible, available in large quantity and relatively inexpensive. Moreover, our laboratory has a great knowledge of acrylic compounds since we worked in this field for years (Caye et al., 1998; Cerf et al., 1992, Cochez et al., 2000; Curci et al., 1993; Gentilhomme et al., 2005; Jullien et al., 1996a, 1996b; Mancardi et al., 2007; Pees et al., 2001, 2002, 2003, 2004)

The reactivity of acrylic compounds can provide extensive opportunities to access to various original ILs. As shown on Fig. 1, on an acrylic type compound, it is possible to introduce an Ionic Part (IP) on two different sites: via a nucleophilic substitution at the end of the esterifying chain (Way 1) or via a Michaël type addition on the activated double bond (Way 2). As a consequence it's possible to obtain cationic or anionic polymerisable monomers using Way 1, depending on the type of ionic part.

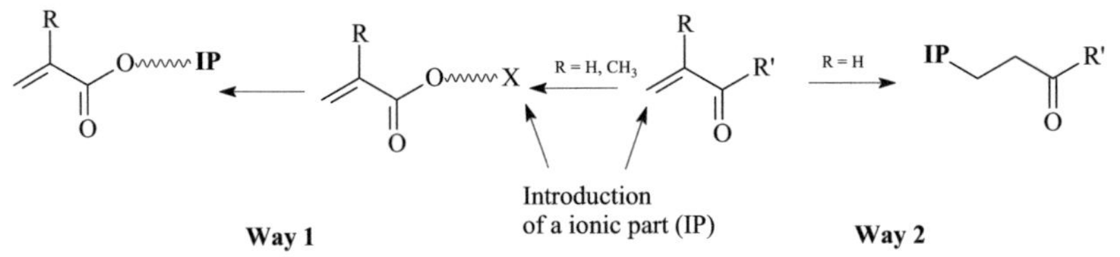

Fig. 1. Reactivity ways of (meth)acrylic compounds

A variety of cations can potentially be used to produce ILs. However later in this chapter, we will only give description of syntheses of ILs with a cation-type N-alkylimidazolium.

1.1 General synthetic method

A mass of methods exist to synthesize ILs. However there is a common denominator among all of them: the first step which lies in the formation of the cation. In the case of an N-alkylimidazole, it consists in quaternization of the molecule's nitrogen atoms. This step is often followed by the metathesis of the resulting counter ions, meaning the exchange of the anion of the molecule by another one, providing different or more interesting properties. The basis of ILs synthetic path is summarized in Figure 2.

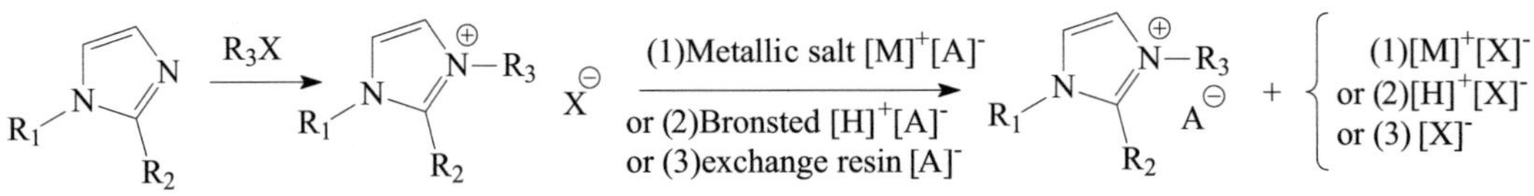

Fig. 2. Typical steps for the ionic liquid (type N-alkylimidazolium)

Depending on haloalkanes and temperature used in the first step, reaction time can vary from one to several hours. Nevertheless, the use of microwaves to perform the nucleophilic substitution can significantly reduce reaction time (Varma & Namboodiri, 2001a; Levêque et al., 2006). Alkyl triflates (Bonhôte et al., 1996) or tosylates (Karodia et al., 1998) may also be used.

It is also possible to easily obtain imidazolium ionic liquids, by performing a one-pot synthesis of 2 steps, using Michael addition. First the counter-ion is introduced via the protonation of N-alkylimidazole by an acid (*p*-toluenesulfonic acid, methanesulfonic acid, tetrafluoroboric acid ...). Then, in a second step, the addition of the N-protonated alkylimidazole on a α, β unsaturated compound (methyl vinyl ketone, methyl acrylate, acrylonitrile ...) leads to the obtention of an imidazolium type IL with a functionalized side (Fig. 3):

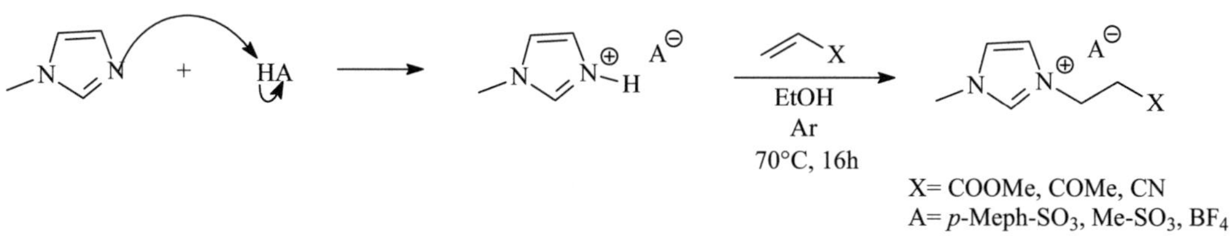

Fig. 3. One pot synthesis of N-methyl-N-alkylimidazolium ionic liquid

In reversing the two steps, i.e. first imidazole addition on a conjugated double bond, then a N-alkylation, the same derivatives may be obtained. The Michael addition of imidazole or its derivatives on an activated double bond can be catalyzed by various reagents: enzyme catalysis (Cai et al., 2004), Montmorillonite catalysis (Martin Aranda et al., 1997, 2002) and activated microwave or ultrasound (Zaderenko et al., 1994), or molecular sieve graft (Blasko-Jimenez et al., 2009), KF(Yang et al., 2005), Cu (Acac)$_2$(Lakshmi Kantam et al., 2007), liquid ion ([bMIM] OH) (Xu et al., 2007).

1.2 Acrylic monomers

One of the latest applications of ILs is the synthesis of polymer electrolytes with high ionic conductivity. Polymer electrolytes are an area of active research since the early 1980s and have applications ranging from rechargeable lithium electrochromic flexible displays and to smart windows (Ratner & Shriver, 1988). Most of the time they are made of polymers like polyethylene oxide (PEO) filled with diluted alkali salts leading to conductive solutions. However, their conductivity remains relatively low at room temperature, so the researchers are trying to find ways to improve it. For this they usually carry out doping of the polymer structure using compounds such as ion-NTF$_2$ (Christie et al., 2005; Reiche et al., 1995) or with salts as plasticizers $LiClO_4$, and $NaCF_3SO_3$ $LiCF_3SO_3$ (MacFarlane et al., 1995; Forsyth et al., 1995) Recently, polymers have also been doped with ionic liquid type imidazolium and pyridinium (Noda & Watanabe, 2000): the conductivity of these compounds is located around 1mS.cm^{-1}. The team of H. Ohno (Ohno, 2001), has chosen to synthesize polymers bearing ionic function, rather than using the ionic liquid as a dopant. The molecular weights of synthesized polymers were not measured, the authors only report the stickiness and rubbery compounds obtained as proof of proper functioning of the reaction (Yoshizawa & Ohno, 2001; Ohno, 2001). The team of Chen (Chen et al., 2009) studied the effect of random polymer composition of ionic liquid polymerize, while recent publication of Matsumoto (Matsumoto et al., 2010) report the synthesis of Met-IL (1-(2-methacryloloxyethyl)-3-methylimidazolium bis(trifluoromethanesulfonyl)imide), and evaluate molecular weight of homopolymer, after transformation in PMMA. As expected the ionic conductivity of ILs decrease during their polymerization: it's reported that they lose 20% to 33% of their initial conductivity. The authors show that the chain length between the imidazolium cation and the polymeric chain seems to have an influence on the polymer conductivity: if the string is longer, it should have a greater flexibility and as a consequence a higher conductivity. Furthermore the length of the side chain has more influence on the conductivity of a cationic polymer than that of an anionic polymer.

Since the beginning of our researches, one of our main concerns has been to prepare new monomers having an element that can confer flexibility in their esterifying chain: we chose the sulfur atom, known for this property.

2. Synthetic methods modifying the esterifying chain: Way 1

Herein, we wish to report our recent work on the syntheses of new ILs containing (meth) acrylic functions. Looking for interest of potential applications of ILs we focused our attention in two directions:

- Synthesis of new ILs possessing a high conductivity and a polymerisable double bond which can lead, after polymerization, to a novel family of polymer electrolytes,
- Development of new synthetic methods from (meth) acrylic compounds which provide simple and efficient ILs containing a carboxylate function.

The (meth) acrylic compounds are of some interest. They are highly reactive due to the presence of the activated double bond. This latter can undergo nucleophilic attack (Michael reaction) or can be subject to radical reaction. The esterifying chain can also be modified by simple (trans) esterification reaction. Moreover they are cheap industrial compounds synthesized on very large scale. Figure 4 describes two aspects of the synthetic methods (process 1 -process 2):

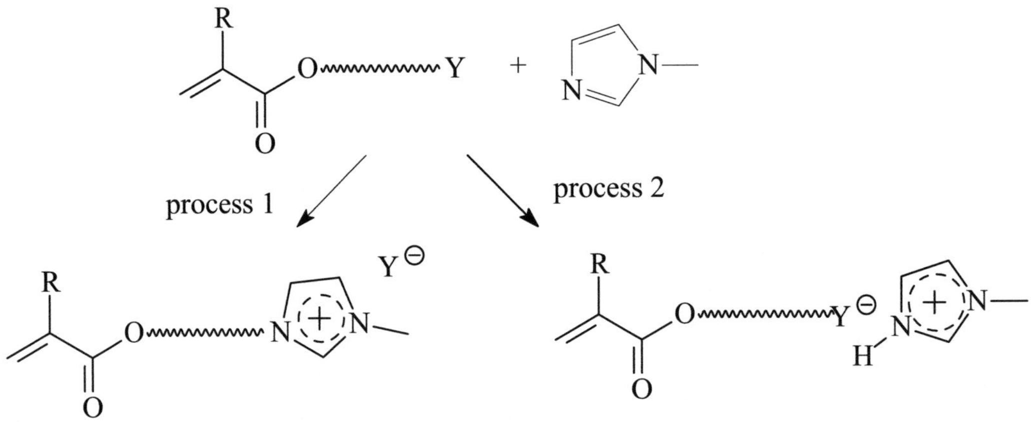

Fig. 4. General scheme of ionic acrylic monomers synthesis

The process consists in modifying the esterifying chain of the (meth) acrylic system thus leading to two types of ILs :

- Cationic ILs, which are (meth)acrylic monomers with a cationic part grafted at the end of their esterifying chain (Fig.5a).
- Anionic ILs, which have this time an anionic part attached to the (meth)acrylate function (Fig.5b).

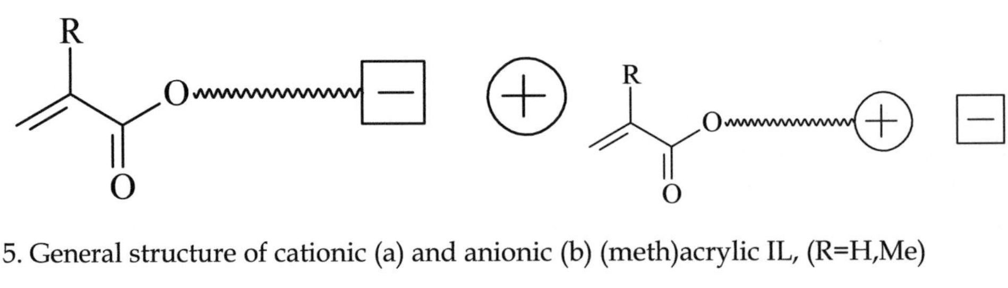

Fig. 5. General structure of cationic (a) and anionic (b) (meth)acrylic IL, (R=H,Me)

2.1 Process 1: Synthesis of cationic monomers

2.1.1 Synthesis

The starting material, halogeno alkyl (meth)acrylate, can easily be prepared through the esterification of a (meth) acrylate type compound with various halogenoalcohols. Two methods are then available: the use of (meth)acrylic acid refluxing in cyclohexane or the one of (meth)acryloyl chloride in the presence of a tertiairy amine (pyridine or triethylamine). In order to avoid polymerization during the process, hydroquinone methyl ether (HQME) is added in the reaction medium. Yield depends on the halogenoalkyl chain length and also on the (meth)acrylic system nature. The best results are obtained with n=6 and acrylic acid (Fig.6).

Note: halogenoalcohols were purchased from Aldrich chemical or synthesized using the described methods (Ford-Moore & Perry, 1963; Suk-Ku et al., 1985)

At first, we adapted a classical well described method consisting in the reaction of the halogeno(meth)acrylates with N-methylimidazole (Nakajima & Ohno, 2005) (Fig. 6).

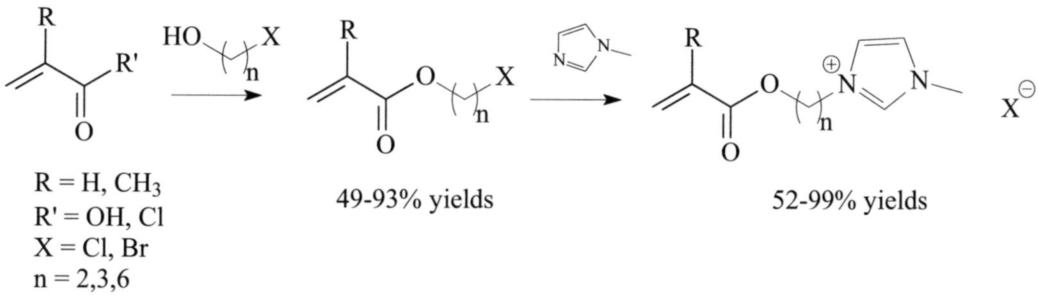

Fig. 6. Synthesis of cationic monomers (R = Me, [MAlkylMim][X] and R = H, [AAlkylMim][X])

Unlike Ohno's procedure (i.e. an excess of N-methylimidazole) we used an equivalent amount of N-methylimidazole and halogeno(meth)acrylate which enabled us to avoid a purification step since no N-methylimidazole was left in the final product. Some amount of HQME was added to the other reactants in order to avoid polymerization reaction. The reaction was followed by ^1H NMR and its completeness was unambiguous since the signals of N-methylimidazole moieties at 3.64 (CH_3), 6.86 (HCN(Me)), 7.01 (NCH) and 7.38 (NCHN) ppm have disappeared in favour of those from N-methylimidazolium respectively at 3.74, 7.35, 7.41, 8.5-10 ppm.

We observed that when n = 6, the product is well synthesized, both with Br and Cl halides. When n = 3, the reaction is really slower. As a consequence, the reaction time must be increased when the less reactive halide (X = Cl) is present and consequently some polymerization products appears. Nevertheless the final product is easily purified by an ether washing.

All synthesized products are viscous liquids or solids having melting below 130°C which can be considered as ILs. They are also watersoluble. As the reaction is carried out without any solvent, the work up is very easy and yields are generally around 90%. However, as reaction time is somewhat too long, several trials to reduce it were attempted. Sonochemistry (Namboodiri & Varma, 2002; Estager et al., 2007; Levêque et al., 2007) and microwave chemistry (Levêque et al., 2007; Varma & Namboodiri, 2001b) which are well described in the literature and often named as a good way to shorten reaction times, were used to try to improve our reaction. In the previous synthetic method the quaternarisation of 3-bromopropyl methacrylate with N-methylimidazole leads to 99 % yield in 24 hours. The same reaction was realized using ultrasonic amplitude of 40 %(apparatus: 20 kHz ultrasonic processor, power: 130 W). In a few minutes temperature raised and ^{1}H NMR follow-up revealed the disappearance of N-methylimidazole protons at the favour of those of imidazolium salt. The final product is obtained with a 99% yield in only 35 minutes (Fig. 7). Still the will to decrease the reaction time of the quaternarisation step, an experiment was realized in a microwave oven (150 W delivered, power is a function of temperature) determinate temperature (75°C) to avoid the polymerization of the methacrylate. After 45 seconds, the quaternarisation is achieved with a 99% yield (Fig. 7). :

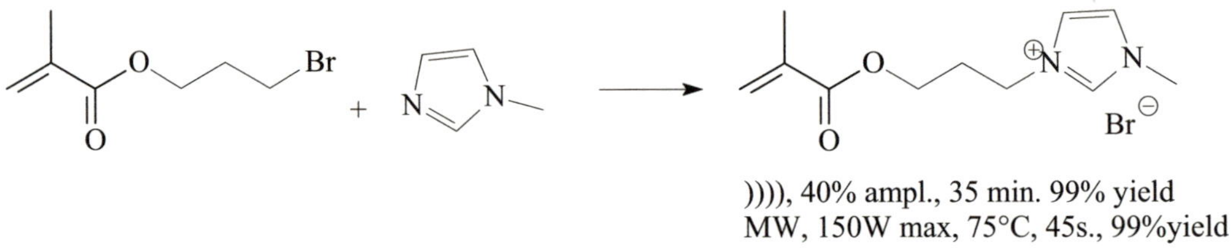

Fig. 7. Synthesis of [MPMIm][Br] using ultrasons or microwave activation

Remark: the reaction with n=2 has not been studied. Indeed The halide in β-position for the ester function is somewhat poor reactive referring to some other works (Dubosclard-Gottardi et al, 1995; Fort et al., 1993; Cerf, 1991) which clearly explain the non-reactivity of the β-C in halogeno(meth)acrylates.

In conclusion, we have demonstrated that synthesis of imidazolium acrylates are possible varying length chain. Moreover, N-(methacryloyloxypropyl)-N'methylimidazolium bromide [MPMIm][Br] was successfully obtained via quaternarisation using alternative methods: under ultrasounds and under microwave irradiation in a microwave oven. Compared with thermal heating, we observed a rate acceleration and drastically reduced reaction times.

2.1.2 Metathesis of halogenoimidazolium salts

The properties of IL are widely dependent on the nature of the counter-ion: its size or structure can affect the physical properties of the resulting IL, especially its melting point and its viscosity (Handy S.T, 2005). When the anion is small, the structure tends to act as a classical crystal salt like sodium chloride (Fig 8.A). But when the anion becomes more bulky, the structure of the IL is then disorganized and, becomes unsymmetrical: so it acts like a liquid (Fig. 8.B):

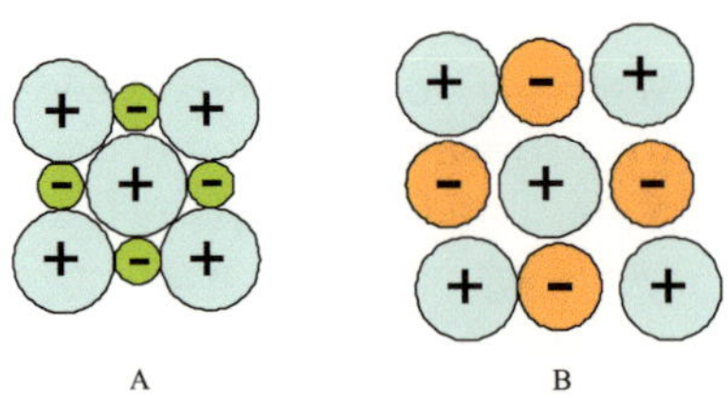

Fig. 8. Ionic structure of crystal salt (A) and liquid (B)

On this basis we explored several ways to realize the counter-ion metathesis (Fig. 9).

- Metathesis of anion with hydracids: at first we applied a well-established method consisting of a reaction between an IL and an hydracids (HY) in an aqueous medium (Gordon et al., 1998). The IL (Br counter-ion) reacts with HY in water: the reaction is very exothermic and HY must be carefully added. The novel IL is obtained by filtration or extraction with an organic solvent, depending on the compound state (solid or liquid). In all cases IL are not (or only lightly) water-soluble. Yields (table 1, entry 1, 3, 5, 7, 9, 11, 13, 15) are depending on the solubility of ILs. When they are partially water-soluble, a small quantity of them is lost and yields decrease.
- Metathesis of anion with ammonium and lithium salts: we try to perform a counter-anion metathesis of the IL with NH_4BF_4 salt in order to try to improve the reaction yield. The solvent used during the reaction is acetone which allows the solubilisation of final product but not the one NH_4Br, by-product of the reaction, which makes the purification easier. Moreover the solvent is simple to remove. The reaction is carried out with one molar equivalence of IL ammonium salt which are soluble in acetone. The by-product is removed by filtration and the filtrate is concentrated under vacuum to get a liquid or a solid. Yields are quite improved compare to the first metathesis (table 1, entry 2, 6, 10, 14).

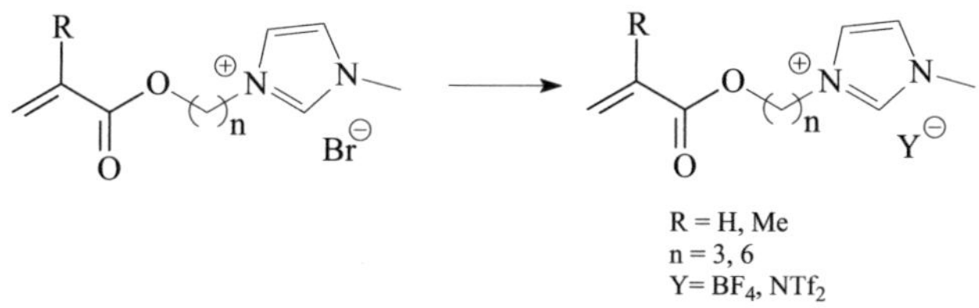

Fig. 9. Metathesis of bromide in [MAlkylMim][Br] or [AAlkylMim][Br]

Entry	Product	n	R	Reagent	Yield	Solubility in water
1 2	[MPMim][BF_4]	3	Me	HBF_4 NH_4BF_4	33% 91%	Partial
3 4	[MPMim][NTf_2]	3	Me	$HNTf_2$ $LiNTf_2$	34% 91%	No
5 6	[APMim][BF_4]	3	H	HBF_4 NH_4BF_4	34% 82%	Partial
7 8	[APMim][NTf_2]	3	Me	$HNTf_2$ $LiNTf_2$	85% 68%	No
9 10	[MHMim][BF_4]	6	Me	HBF_4 NH_4BF_4	56% 87%	Partial
11 12	[MHMim][NTf_2]	6	Me	$HNTf_2$ $LiNTf_2$	78% 89%	No
13 14	[AHMim][BF_4]	6	H	HBF_4 NH_4BF_4	51% 79%	Partial
15 16	[AHMim][NTf_2]	6	Me	$HNTf_2$ $LiNTf_2$	38% 63%	No

Table 1. Results of Metathesis of bromide in [MAlkylMim][Br] or [AAlkylMim][Br]

In a similar way, we tried the same reaction but this time with a lithium salt, $LiNTf_2$, which is far less expensive than $HNTf_2$. We chose $CHCl_3$ as solvent reaction because the reaction by-product, LiBr, isn't soluble in it and so, this latter is easy removable from the reaction mixture. The reaction was performed under the same conditions than those used with the tetrafluoroborateammonium salt. The IL is obtained by extraction and the solvent is removed under vacuum to get a liquid. We observed that yields are generally improved compare to those from the first method (table 1, entry 4, 8, 12, 16)
In conclusion, the metathesis of anion can be performed in different ways. Laborious reactions with hydracids can be satisfactorily avoided using the corresponding salts.

2.1.3 Special IL and new development

In cationic ILs, viscosity and conductivity are largely dependant on the flexibility of the esterifying chain of the (meth) acrylic moiety. From this point of view, we tried to insert a sulfur atom in the esterifying chain. The synthesis is realised in three steps and the overall yields are 50% and 58% when R is respectively Me and H (Fig. 10).

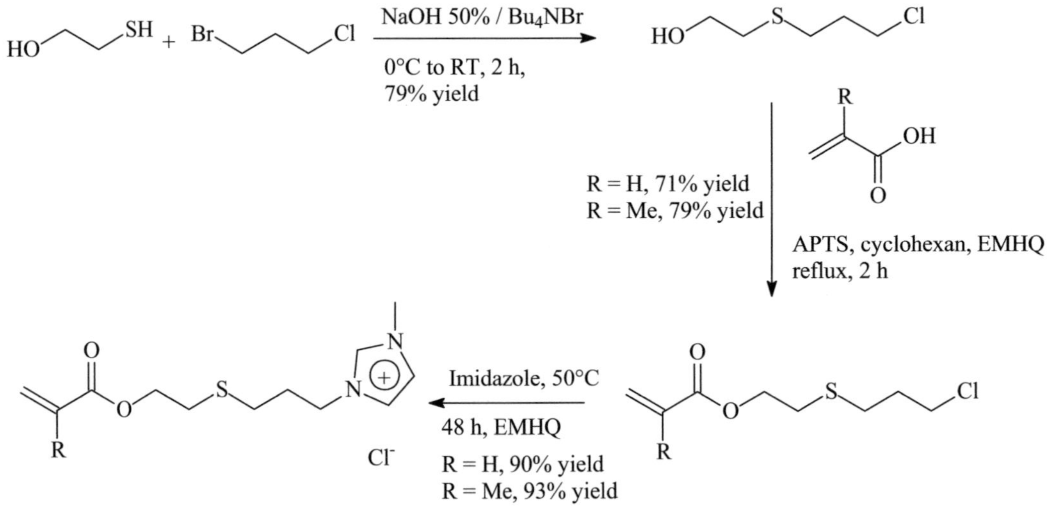

Fig. 10. Synthesis of sulfurated ionic (meth)acrylic monomers

These new compounds are identified without ambiguity with the NMR spectra:
R = H :^{1}H NMR, δ ppm (250 MHz; $CDCl_3$): 8.25 (NCHN, s), 7.86 (2 CHN,d), 6.39 (CH=, dd, J_{gem}=1,5Hz,J_{trans}=16,7Hz), 6.14 (CH=, dd), 5.87 (CH=, dd, J_{gem}=1,5Hz, J_{cis}=10Hz), 4.32 (CH_2O, t), 4.25 (CH_2N, t), 3.74 (CH_3N, s), 2.81 (2 CH_2S, m), 1.98 (CH_2, q). ^{13}C NMR, δ ppm (250 MHz; $CDCl_3$): 165.8 (CO), 136.7 (NCHN), 131.0 (CH_2=), 128.1 (CH=), 122.9 (CHN), 122.4 (CHN), 63.6 (CH_2O), 40.9 (CH_2N), 35.8 (CH_3N), 33.2 (CH_2S), 30.4 (CH_2S), 28.0 (CH_2).
R = Me: ^{1}H NMR, δ ppm (250 MHz; $CDCl_3$): 8.25 (NCHN, s), 7.42 (2 CHN,d), 6.12 (CH=, d), 5.55 (CH=, d), 4.36 (CH_2O, t), 4.21 (CH_2N, t), 3.74 (CH_3N, s),2.80 (2 CH_2S, m), 1.92 (CH_2, q), 1.85 (CH_3, s). ^{13}C NMR, δ ppm (250 MHz; $CDCl_3$): 167.1 (CO), 136.7(NCHN), 136.0 (C=), 125.8 (CH_2=), 122.9 (CHN), 122.4 (CHN), 63.8 (CH_2O), 41.0 (CH_2N), 35.8 (CH_3N), 33.0 (CH_2S), 30.5 (CH_2S), 28.2 (CH_2).
This work is currently under study.

2.2 Process 2: Synthesis of anionic monomers

As mentioned above, anionic ILs generally have a better conductivity and a lower viscosity than cationic ILs: This is due to the higher lability of the cation in an anionic IL which induces a greater flexibility of the structure. In order to compare physicochemical properties of cationic and anionic IL, we focused our attention on two ionic ILs, we chose to synthesize two anionic ILs which structures are alike with those from cationic ILs described above: one has three carbon atoms in the side chain and the other exhibits a sulfur atom.

2.2.1 Synthesis of 4-(meth)acryloyloxybutanoate N-methylimidazolium salt

1,4 butanediol, in large excess, was reacted with (meth)acryloyloxy chloride to prepare 4-hydroxybutyl(meth)acrylate (Chaudron, 1999). Resulting monomers were reacted with Jones reagent ($CrO_3/H_2SO_4/H_2O$) in order to prepare corresponding carboxylic acids. Filtration on silica and a liquid-liquid extraction allowed the obtention of pure acids. Finally an acido-basic reaction between carboxylic acids and N-methyl imidazole to led the formation of the corresponding anionic IL (Fig. 11).

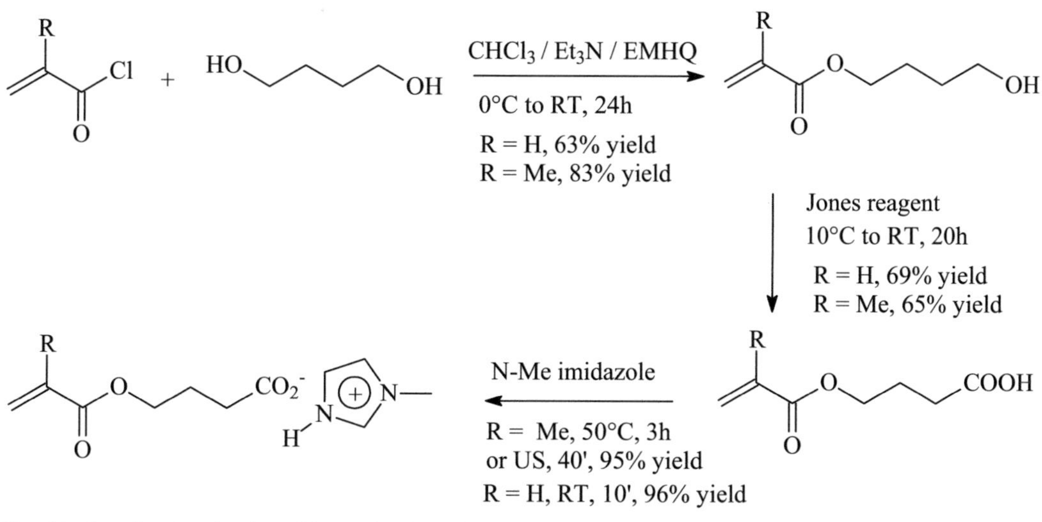

Fig. 11. Synthesis of 4-(meth)acryloyloxybutanoate N-methylimidazolium salt

In a first step, the reaction (R = Me) was performed at 50°C during 48h. Along with the desired IL, we observed the presence of a by-product resulting from the Michael addition of N-methyl imidazole on the activated double bound of the (meth)acrylate compound(Fig.12).

It seemed that the reaction time was too long. In fact we observed that the reaction was complete after 3h of heating. After this period, if heating continues, IL decomposes to give back initial products which perform a Michael addition. As a consequence it is preferable to stop the reaction as soon as the IL is synthesized. In order to avoid formation of the Michael addition by-product and to decrease reaction time, sonochemistry was tested on this synthesis. In these conditions, it only took 40 minutes to prepare the desired IL. When R = H, the conditions have to be less harsh since the acrylic compounds is more reactive than the methacrylic one and is more inclined to give the Michael by-product. 10 minutes room temperature are enough to obtain the acrylic type IL.

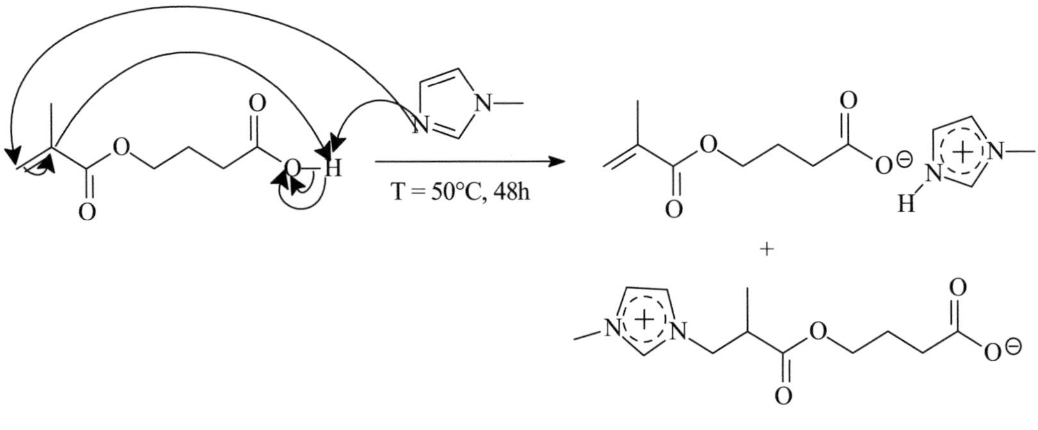

Fig. 12. Scheme for the formation of expected IL an by-product (Michael reaction)

Finally the overall yields are 42% and 54% for the acrylate and methacrylate compounds respectively. These new compounds are identified with the NMR spectra:
R = H :^{1}H NMR, δ ppm (250 MHz; D_2O): 8.40 (NCHN, s), 7.21 (2 CHN, m), 6,34 (CH=, dd, J_{gem}=1Hz,J_{trans}=17.5Hz) ; 6,07 (CH=, dd), 5.82 (CH=, dd, J_{gem}=1Hz, J_{cis}=10Hz), 4.16 (CH_2O, t), 3.72 (CH_3N, s), 2.44 ($CH_2C(O)$, t), 1.98 (CH_2, m). ^{13}C NMR, δ ppm (250 MHz; D_2O): 178 (CO), 166.3 (CO acryl.), 136 (NCN), 130.9 (CH_2=), 128.3 (CH=), 123 (CN), 120 (CN), 65 (CH_2O), 35 (CH_3N), 34 ($\underline{C}H_2CO$), 25 (CH_2).
R = Me :^{1}H NMR, δ ppm (250 MHz; D_2O): 8.41 (NCHN, s), 7.21 (2 CHN, m), 6.06 (CH=,d), 5.5 (CH=, d), 4.17 (CH_2O, t), 3.73 (CH_3N, s), 2.11 ($CH_2C(O)$, t), 1.73 (CH_3 and CH_2, m). ^{13}C NMR, δ ppm (250 MHz; D_2O): 182 (CO), 170 (CO acryl), 136 (NCN), 135 (C=), 126.5 (CH2=), 123 (CN), 120 (CN), 65 (CH_2O), 35 (CH_3N), 34 ($\underline{C}H_2CO$), 25 (CH_2), 18 (CH_3)

2.2.2 Synthesis of 6-(meth)acryloyloxy-4-thiahexanoate N-methylimidazolium salt

Tertiobutylacrylate was first reacted with mercaptoethanol in the presence of triton B (N-benzyltrimethyl ammonium hydroxyde (Kharasch & Fuchs, 1948; Melwig, 1995) resulting in a Michael reaction with a 99% yield. These conditions tend to be more effective than those used in classical radical reaction which only give 54% yield (Curci, 1992). Resulting alcohol was esterified with (meth)acryloyl chloride under usual conditions. Nevertheless, the choice of the acidic medium during the workup must be taken really carefully. Indeed HCl 10% leads easily to the desired products, while H_2SO_4, hydrolyzes the two ester functions and which recovers $HO(CH_2)_2S(CH_2)_2COOH$. Tertiobutyl group was then removed with formic acid 98%. In a final step, the quaternarization of the acid was realized with N-methyl imidazole using the conditions described above. The reaction time depends on the type of compound: 10 minutes for an acrylate compound and 15 minute for a methacrylate one. Finally the overall yields are 31% and 37% for the acrylate and methacrylate compounds respectively (Fig.13).
We identified these new compounds with the NMR spectra:
8.45 (NCHN, s), 7.21 (2 CHN, m), 6,35 (CH=, dd, J_{gem}=1.5Hz,J_{trans}=17Hz), 6,08 (CH=, dd), 5.79 (CH=, dd, J_{gem}=1.5Hz, J_{cis}=10.2Hz), 4.30 (CH_2O, t), 3.72 (CH_3N, s), 2.82 ($CH_2C(O)$, t), 2.71 (CH_2S, t), 2.58 (CH_2S, t)

MHz; $CDCl_3$): 178 (CO), 168 (CO acryl.), 136 (NCN), 130.6 (CH_2=), 128.3 (CH=), 123 (CN), 120 (CN), 64.6 (CH_2O), 35 (CH_3N), 34 (CH_2CO), 31.5 (CH_2S), 26.9 (CH_2S).
8.42 (NCHN, s), 7.21 (2 CHN, m), 6.07 (CH=,d), 5.62 (CH=, d), 4.27 (CH_2O, t), 3.71 (CH_3N, s), 2.83 (2 CH_2S, t), 2.82 ($CH_2C(O)$, t), 1.92 (CH_3, s)

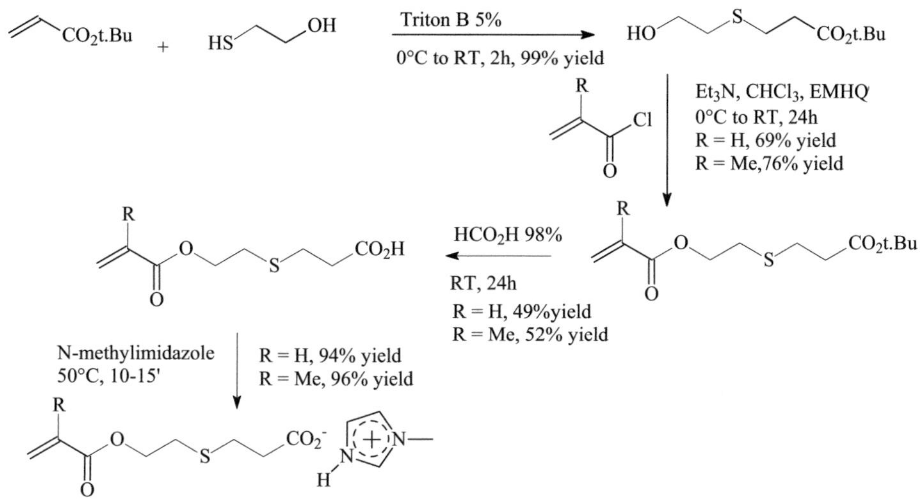

Fig. 13. Synthesis of 6-(meth)acryloyloxy-4-thiahexanoate N-methylimidazolium salts

3. Synthetic methods resulting from aM addition: Way 2

3.1 Synthesis of (N-(2-carboxyethyl)methylimidazolium salts by Michael addition of NMI on acrylic acid

In literature, the high cost of ILs preparation is repeatedly reported. This is mainly due to multi-step syntheses and special counter-ions introduction (NTf_2^-, $OSO_2CF_3^-$...). We developed a "one-pot" reaction, between acrylic acid (AA), N-methylimidazole (NMI) and an fluorinated acid (HBF_4, HPF_6, $HNTf_2$) (FA), to obtain a fluorine anion. The same reaction was performed several times by switching everytime the introduction order of reactants to see if that parameter has an impact on the result of the reaction. In all cases, two reagents were stirred together and the third one was added after to the mixture. The whole was then stirred (24h) at 50°C (Fig. 14).

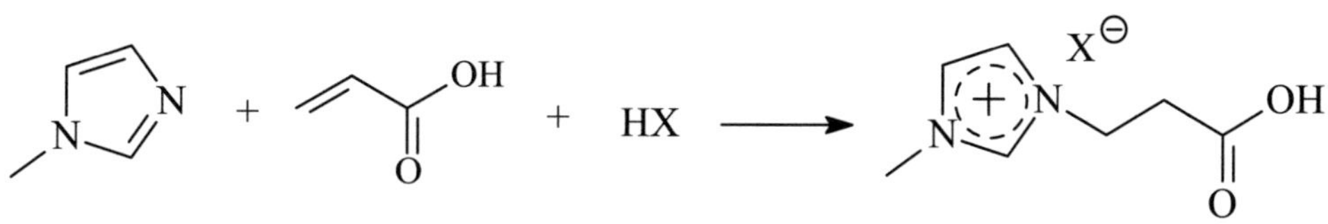

Fig. 14. General scheme of one pot synthesis of carboxylic acid functionalized imidazolium salts by Michael addition

We observed that whatever reactants addition order is, we only obtained the expected product, with a yield of 85% when we use bis(trifluoromethanesulfone)imide acid ($HNTf_2$). With HBF_4 and HPF_6, when NMI and acrylic acid react first, acrylic acid plays the role of an

acid and the imidazolium salt is obtained. According to the work of Wasserscheid (Wasserscheid et al., 2003), it seems important, first, to mix NMI and the fluorinated acid, then to add acrylic acid and heat at 70°C. Following those conditions, we obtained the expected product, respectively with 38 and 33% yield.
The difference of reactivity of these two acids and $HNTf_2$ can be explained by the fact that HBF_4 and HPF_6 are diluted in water, while $HNTf_2$ is pure. As water can be a limiting factor, we added a drying step, carried out by microwave heating, before the addition of acrylic acid. Thus, performance can be improved and yields are respectively 83% and 77% for HBF_4 and HPF_6 (Fig. 15):

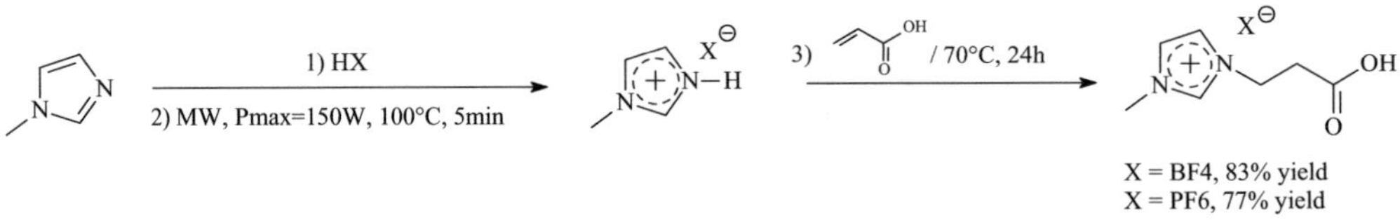

Fig. 15. Preparation of carboxylic acid functionalized imidazolium salts using HBF_4 and HPF_6

However, even if those two acids are highly reactive, they have the disadvantage of being very corrosive. Under certain conditions (high temperature of heating in the presence of water) they decompose in particularly dangerous HF. During a reaction we observed the attack of inner walls of the glass reactor by a small amount of acid formed. To avoid such inconvenience, we developed another path to synthesize these ionic liquids, which does not require the use of HBF_4 and HPF_6, but the one of corresponding ammonium salt. In the first step of this method we used bromohydric acid in order to realize the quaternization of NMI. The bromide anion is replaced by a fluorinated one in a second step (Fig. 16):

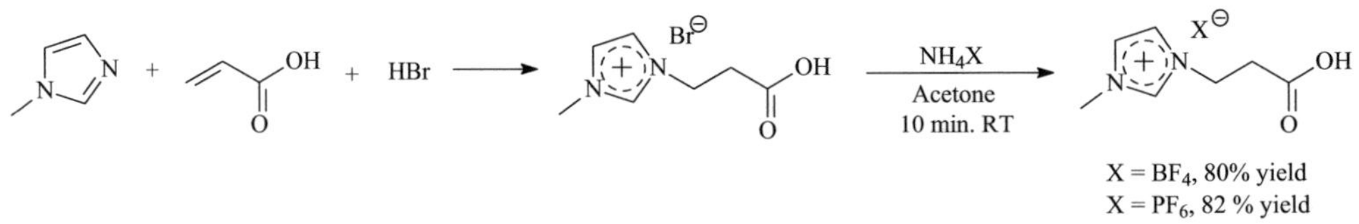

Fig. 16. Preparation of carboxylic acid functionalized imidazolium salts using NH_4BF_4 and NH_4PF_6

The first step can be realized both by reaction of AA with HBr or with NMI (the reaction of NMI with HBr leads to the stable methyl ammonium bromide salt which doesn't react with AA), followed by elimination of water using microwave technical. The exchange of anion using ammonium salt is rapid (10 min in acetone). The ammonium bromide resulting from the counter-anion exchange is insoluble in acetone and, as a consequence, is easily removed from reaction mixture by a simple filtration.

3.2 Synthesis of IL by Michael addition on esters

In order to broaden the series of product available, Michael addition can be realized on acrylic esters, using imidazole (Fig. 17). Quaternization of intermediate products by nucleophilic substitution of the halogen leads to the corresponding ILs. The use of ultrasound is very effective in the 2 steps of this synthetic path since it reduces impressively reaction times.

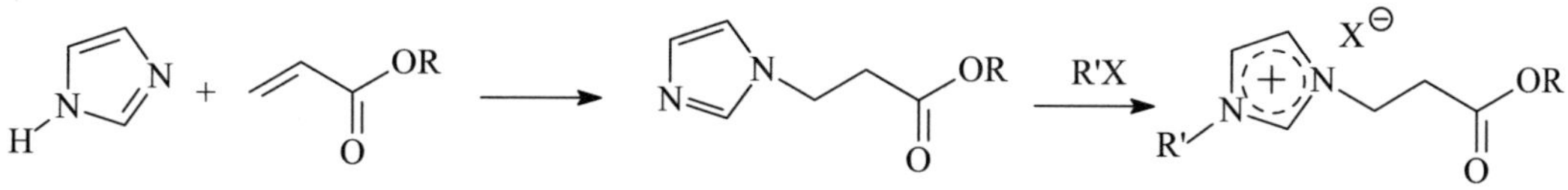

Fig. 17. Synthesis of ester functionalized imidazolium salts

Depending of R substituent (methyl, ethyl, CH_2CH_2Cl, Butyl, *t*-butyl, octyl), this reaction can take between 30 min. and 3 h. under ultrasonic activation instead of 24h. by eating at 50°C. The N-alkylation is easy with bromide derivatives but far less with chlorides ones. So far, we realized nucleophilic substitution with 3-bromopropanol, and other radicals are under study.

4. Thermodynamics data

Interactions between IL cations and anions are the consequence of energetic and geometric factors leading to a variety of strongly organized and oriented structures. These features confer to ILs numerous applications in organic synthesis, separation processes, and electrochemistry (Seddon, 1997; Sheldon, 1993; Cull et al, 2000; Wilkes, 2004; Wasserscheid & Keim, 2000; Welton, 1999; Brennecke & Maginn, 2001; Fredlake et al., 2004; Henderson & Passerini, 2004). The aim of this work is to determine the influence of the (meth)acrylic moiety on the thermodynamic properties of the ionic liquid. For this purpose, the study of (*N*-methacryloyloxyhexyl-*N*-methylimidazolium bromide [MHMim][Br] and *N*-acryloyloxypropyl-*N*-methylimidazolium bromide [APMim][Br] permitted to determine its selectivity toward a hexane/benzene mixture and its interactions toward other compounds using the Linear solvation Energy Relationship (LSER) descriptor.
All of these measurements were performed using the inverse gas chromatography technique. Column packing of 1 m length containing from 7 to 26% of stationary phases (RTIL) on Chromosorb W-AW (60-80 mesh) were prepared using the rotary evaporator technique. After evaporation of the dichloromethane under vacuum, the support was equilibrated at 323 K over 6 h.
LSER characterization: Ionic liquids can easily adsorb onto solid surfaces and may form a strongly structured interface at the support surface. This interface may induce the adsorption of polar solutes. For this reason, it was decided to use the above-described experimental procedure that allows the separation of the adsorption contribution. To quantify intermolecular solute-IL interactions, we used the LSER equation developed by Abraham et al. (Abraham et al., 1987, 1990, 1991, Abraham, 1993). This method allows one to correlate thermodynamic properties of phase transfer processes. The most recent representation of the LSER model is given by (Mutelet, 2008):

$$\log SP = c + eE + sS + aA + bB + lL \quad (1)$$

Both ionic liquids studied in this work were analyzed using the above-described (eq 1) LSER approach. Coefficients c, e, s, a, b, and l of the ILs were obtained by multiple linear regression of the gas-liquid partition coefficients logarithm log K_L of 30 solutes (Mutelet et al., 2008). The system constants for the two ionic liquids studied in this work at 313.15 K and other ionic liquid stationary phases are summarized in Table 2.

Ionic liquid	System constant					
	e	s	a	b	l	c
n-acryloylpropyl-N-methylimidazolium bromide	0	2.88	5.5	0	0.48	-1.03
n-methacryloylhexyl-N-methylimidazolium bromide	0	2.46	5.36	0	0.57	-0.87
1-propenyl-3-methylimidazolium bromide	0	2.16	5.19	0	0.53	-1.86
1-propenyl-3-octylimidazolium bromide	0	1.72	4.96	0	0.57	-1.60
1-propenyl-3-decylimidazolium bromide	0	1.73	4.89	0	0.66	-1.58
1-propenyl-3-dodecylimidazolium bromide	0	1.44	4.87	0	0.7	-1.51
1-butyl-3-methylimidazolium octyl sulfate	0	1.47	4.05	0	0.68	-0.237
1-ethyl-3-methylimidazolium tosylate	0.54	2.40	4.81	0.17	0.48	-0.84
n-butylammonium thiocyanate	0.14	1.65	2.76	1.32	0.45	-0.75
di-*n*-propylammonium thiocyanate	0.3	1.73	2.66	0.68	0.47	-0.6
Ethylammonium nitrate	0.27	2.21	3.38	1.03	0.21	-0.87
n-propylammonium nitrate	0.25	2.02	3.50	0.9	0.36	-0.97

Table 2. LSER descriptors of ionic liquids imidazolium type, determined at 313.5K

LSER coefficients of both ILs studied in this work are slightly different from those obtained with other ionic liquids of the imidazolium bromide type.

- The (c + lL) term gives information on the effect of the cohesion of the ionic liquids on solute transfer from the gas phase. In general, the ionic liquids are cohesive solvents (Poole, 2004).
- The ionic liquid interacts weakly via nonbonding and π-electrons (e system constant is zero) and is not much different from other polar nonionic liquids.
- Dipolarity/ polarizability of ionic liquids is slightly higher than most of dipolar/polarizable nonionic stationary phases ones. The polarizability decreases slightly when the alkyl chain length is increased on the imidazolium ring. But introducing a (meth)acryloyloxyalkyl chain in imidazolium bromide-based ionic liquids considerably increases its dipolarity/polarizability (s system constants).

The hydrogen-bond basicity of the ionic liquid (a system constants) is considerably larger than values obtained for nonionic phases (0 to 2.1) (Poole, 2004), wheras it hydrogen-bond acidity is inexistant (b=0). Ionic liquids have structural features that would facilitate hydrogen-bond acceptor basicity interactions (electron-rich oxygen, nitrogen, and fluorine atoms). Imidazolium bromide based ionic liquids containing a (meth)acryloyloxyalkyl chain have the highest hydrogen-bond basicity, with an a constant of 5,36 and 5,50. This is great support for the idea that the interactions between the -OH group and ionic liquids I and II are very strong.

Selectivity determination: Ionic liquids are solvents that may have great potential in chemical analysis. Specifically, the applications of RTILs are found in chromatography or

extraction chemistry. These compounds, thermally stable, non-toxic with low vapor pressure, can act as stationary phases in inverse gas chromatography (Pacholec, 1982, 1983). Numbers of organic molten salts were evaluated as stationary phases for gas chromatography. Nevertheless only ionic liquids containing imidazolium cation lead to high selectivity toward polar and non-polar solutes.
The selectivity, S_{12}^{∞}, which indicates the suitability of a solvent for separating mixtures of components 1 and 2 by extractive distillation is given by Tiegs (Tiegs et al., 1986)

$$S_{12}^{\infty} = \frac{\gamma_{1/LI}^{\infty}}{\gamma_{1/LI}^{\infty}}$$

$\gamma_{1/LI}^{\infty}$ is the activity coefficient at infinite dilution of compound 1 (hexane) relative to the IL $\gamma_{1/LI}^{\infty}$ is the activity coefficient at infinite dilution of compound 2 (benzene) relative to the IL The selectivity values, S_{12}^{∞}, relative either to the IL studied in this work and other liquid solvents used in industry for the separation of benzene and n-hexane, are reported in Table 3.

Entry	Solvent	S_{12}^{∞}	Reference
1	[MAHMIm][Br]	50,4	**Mutelet et al., 2008**
2	[APMIm][Br]	27,6	**Mutelet et al., 2008**
3	Dichloroacetic acid	6,1	**Tiegs et al., 1986**
4	sulfolane	30,5	**Tiegs et al., 1986**
5	1-propenyl-3-methylimidazolium bromide	6,96	**Mutelet et al., 2006**
6	1-propylboronic acid-3-octylimidazolium bromide	9,91	**Mutelet et al., 2006**
7	1-ethyl-3-methylimidazolium tetrafluoroborate	61,6	**Foco et al., 2006**

Table 3. Selectivity value for the solute Hexane (1) and benzene (2) in solvents

The selectivity of both ionic liquids studied in this work at 313.15 K is very large as compared to the value for classical solvents (entry 1,2). These values are largely higher than those of usual industrial compounds (entry 3,4).Compared to similar IL containing bromide ion we can suspected than (meth)acrylic alkyl chain is of some importance because usually the selectivity increases with decreasing length of the alkyl chain (entry 1,2,4,6). In the same way the proximity of (meth)acrylic function decreases the selectivity (enter 1,2). Only 1-ethyl-3-methylimidazolium tetrafluoroborate has a higher selectivity (entry 7) but it is obvious that the chemical nature of the cation and the anion play an important role in separation of mixtures of aromatic and aliphatic compounds
Through this work, we demonstrated that interfacial adsorption could play a significant role in the retention mechanism of organic compounds. Results indicate that the introduction of (meth)acryloyle substituents on the IL imidazolium cation affects strongly the behavior of organic compounds in mixtures with this IL. For instance, the IL used in the separation of aliphatic hydrocarbons from aromatic hydrocarbons, shows a higher selectivity than the one found by previous workers using classical organic solvents

5. Catalytic properties of IL adsorbed on nanoparticles

One of the recent developments of IL is their use in catalysis (Olivier-Bourbigou et al., 2010). Their catalytic properties are well known, and a broad range of reactions have already been studied (oxidation, polymerization, enantioselective reactions…). Moreover, the capability to support these catalytic species is an attractive alternative to classic use of ionic liquids, because these latter can be recycled. These SILCAs (Supported Ionic Liquids CAtalystS) are formed of different IL immobilized on several supports: active carbon cloth (Mikkola et al. 2007; Virtanen et al. 2007; Maki-Arvela et al. 2006) or silica gel (Riisager et al. 2003; Mehnert et al. 2002). They may also contain metal species like rhodium (Riisager et al. 2003; Mehnert et al. 2002) or palladium (Mikkola et al. 2007; Virtanen et al. 2007; Maki-Arvela et al. 2006). In our case, we realized the adsorption of our monomers on aluminum oxide nanopowder. The use of nanoparticules is really interesting since they are more likely to provide a wide and homogeneous dispersion of the catalyst in the reaction medium. The preparation of these SILCAs is usually based on alumina saturation. After dilution of IL in an organic solvent (Komulski et al. 2005; Zilkova et al. 2006; Li et al. 2007), it is then adsorbed on alumina nanoparticles surface. This procedure is easy to carry out and does not involve the use of metal species in opposition to other methods found in literature (Mikkola et al. 2007; Virtanen et al. 2007; Maki-Arvela et al. 2006; Riisager et al. 2003; Mehnert et al. 2002).
In order to evaluate the catalytic properties of our SILCAs, we chose to test them on a classic and simple reaction. As our laboratory has a great interest in heteroatomic compounds especially sulfurs (Robert et al. 1996; Pees et al. 2001; Thomas et al. 2006), we chose a thioéther synthesis, and benzyl phenyl thioéther in particular (Harmand, 2009). Conversion rates of reaction between thiols and halides depend on reaction conditions which are in most cases not really environment friendly (extraction difficulties, catalysts loss and so on)
To prepare SILCAs, three steps are necessary; first the solubilization of IL in chloroform, then the addition of the alumina nanoparticules in the solution and finally the solvent removal evaporation. This method allows the obtention of a supporting material coated with a thin and uniform IL layer. We first compare the reaction between phenyl bromide and thiophenol in a basic aqueous solution in methylene chloride, with (table 4, entry 2–21) or without (table 4, entry 1) prepared SILCAs (Fig. 18).

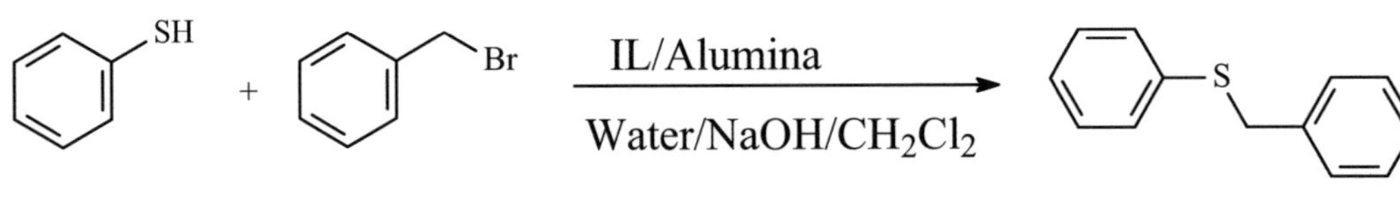

Fig. 18. Synthesis of benzylphenyl thioether, using SILCA as catalysts.

The principal advantage of this method is the simplified working up: after 4h reflux, the solution is filtered to recover the supported catalyst, and then the filtrate is decanted to allow organic layer recovery. Without catalyst the conversion rate is very low as predicted in a heterogeneous system (table 4, entry 1). Rates are improved in the presence of SILCA and results of this preliminary study are satisfactory. Nevertheless depending on the molecular structure, results are different. Generally, conversion rates are best when n is equal to 6 whenever R is H or Me. For similar chain length, rates are higher when R is a methyl. We are currently studying this phenomenon, which could be due to polymerization of monomers on the alumina area during SILCAs preparation. Finally, we can note that the highest conversion rates are reached with n = 6 and are quantitative when R = Me (table 4, entry 12–16).

Entry	IL	Conversion (%)
1	none	25
2	[MPMim][Cl]	96
3	[MPMim][Br]	96
4	[MPMim][PF_6]	100
5	[MPMim][BF_4]	73
6	[MPMim][NTf_2]	75
7	[APMim][Cl]	90
8	[APMim][Br]	68
9	[APMim][PF_6]	65
10	[APMim][BF_4]	73
11	[APMim][NTf_2]	53
12	[MHMim][Cl]	100
13	[MHMim][Br]	100
14	[MHMim][PF_6]	100
15	[MHMim][BF_4]	100
16	[MHMim][NTf_2]	100
17	[AHMim][Cl]	90
18	[AHMim][Br]	99
19	[AHMim][PF_6]	92
20	[AHMim][BF_4]	100
21	[AHMim][NTf_2]	84

Table 4. Results of the synthesis of benzylphenylsulfide using SILCA

In conclusion, after adsorption on alumina nanoparticles surface, these new polymerisable compounds, presented in this work, are used as a new kind of catalysts in the synthesis of benzyl phenyl thioether. These SILCAs give better results than a traditional phase transfer agent. Moreover, they are totally recoverable at the end of the reaction. As a consequence they can be considered as environmentally friendly catalysts.

6. Conclusion and perspectives

We described here the synthesis of several new molecules deriving from acrylates, common industrial compounds which are really reactive and available in large quantities. These molecules exhibit characteristics of ILs like their melting points which are below 100°C and others, such as viscosity and conductivity are being investigated. Moreover they have demonstrated some physical chemical applications such as in catalysis or in separation process.

In the future our laboratory will develop the work described herein, in three directions.

First, our attention will be focused on new heteroatomics IL synthesis. We already synthesized some sulfurated IL and we wish to develop this class of compounds and incorporate some novel heteroatom (P, Si…).

Many IL synthesized have a polymerizable double bound, so as a second way of action, we planned to polymerize these molecules, in order to obtain electrolytes polymers type. These

highly ion-conducting electrolytes could have potential applications in a wide range of areas such as in lithium battery and dye-sensitized solar cell (Choi N.S. et al., 2004; Shibata Y. et al., 2003). ILs have good chemical and thermal stability, low vapor-pressure and high conductivity. These properties could be used to product new polymer with particularly good properties as some described by Ohno (Washiro S. et al., 2004).
In a third way we would like to use our work to develop some potential applications in green chemistry. We already began to synthesize some compounds using glycerol as starting material (Fig. 19). Glycerol is a cheap industrial product which is, for example, currently used in acrylic acid synthesis. From glycerol we intend to build dendritic structures presenting terminal groups functionalized with the type of function described in this work, in order to obtain new green ion-conducting electrolytes.

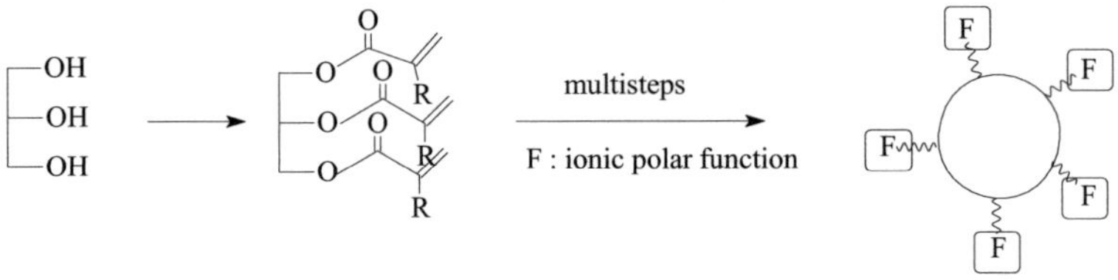

Fig. 19. Dendritic structure starting from glycerol and (meth)acrylic compounds.

7. References

Abraham M.H., Grellier P.L. & McGill R.A., (1987). Determination of olive oil–gas and hexadecane–gas partition coefficients, and calculation of the corresponding olive oil–water and hexadecane–water partition coefficients, *J. Chem. Soc. Perkin Trans II*, n°6, pp.797-803 (June 1987), ISSN 0300-9580

Abraham M.H., Whiting G.S., Doherty R.M. & Shuely W.J. (1990). Hydrogen bonding. Part 13. A new method for the characterisation of GLC stationary phases—the laffort data set, *J. Chem. Soc. Perkin Trans II*, n°8, pp.1451-1460 (August 1990), ISSN 0300-9580

Abraham M.H., Whiting G.S., Doherty R.M. & Shuely W.J. (1991). Hydrogen bonding: XVI. A new solute salvation parameter, π2H, from gas chromatographic data, *J. Chromatogr. A*, vol.587, n°2, pp.213-228, (December 1991), ISSN 0021-9673

Abraham M.H. (1993). Scale of solutes hydrogen bonding: their construction and application to physicochemical and biochemical processes, *Chem. Soc. Rev.*, vol.22, n°2, pp.73-83 (March 1993), ISSN 0306-0012

Baudequin C., Brégeon D., Levillain J., Guillen F., Plaquevent J.C. & Gaumont A.C. (2005). Chiral ionic liquids, a renewal for the chemistry of chiral solvents? Design, synthesis and applications for chiral recognition and asymmetric synthesis. *Tetrahedron: Asymmetry*, vol.16, pp.3921–3945 ISSN 0957-4166

Biswas A., Shogren R.L., Stevenson D.G., Willett JL. & Bhowmik P.K. (2006), Ionic liquids as solvents for biopolymers: Acylation of starch and zein protein, Carbohydrate *Polymers*, vol.66, n°4, pp.546–550, (November 2006), ISSN 0144-8617

Blasco-Jimenez D., Lopez-Peinado A.J., Martin-Aranda A.M., Ziolek M. & Sobczak I. (2009). Sonocatalysis in solvent-free conditions: an efficient eco-friendly methodology to prepare N-alkyl imidazoles using amino-grafted NbMCM-41, *Catalysis today, 6th*

International Symposium on Group Five Elements, Poznań, Poland, vol.142, n°3-4 pp.283-287, (April 2009), ISSN 0920-5861

Bonhôte P., Dias A.-P., Papageorgiou N., Kalyanasundaram K. & Grätzel M. (1996). Hydrophobic, Highly Conductive Ambient-Temperature Molten Salts, *Inorg.Chem.*, vol.35, n°5, pp.1168-1178, (February 1996), ISSN 0020-1669

Brennecke J.F. & Maginn E. (2001). Ionic liquids: Innovative fluids for chemical processing, *J. AIChE*, vol.47, n°11, pp.2384-2389, (November 2001), Online ISSN: 1547-5905

Cai,Y., Yao S-P., Wu Q. & Lin X.F. (2004). Michael addition of imidazole with acrylates catalysed by alkaline protease from Bacillus subtilis in organic media, *Biotechnology letters*, vol.26, n°6, pp.525-528, (March 2004), ISSN 0141-5492

Caye F., Sindt M., Mieloszynski J.L. & Paquer D. (1998). A convenient high yield synthesis of functional methacrylates via dethioacetalization. Synthesis of methacrylate S,S-acetals derivatives as intermediate, *Phosphorus, Sulfur, and Silicon*, vol.143, n°1, pp. 197-220 (December 1998) ISSN 1042-6507

Cerf M. (1991). Etude de monomères acryliques et thioacryliques soufres et phosphores, University Paul Verlaine Metz, Thesis

Cerf M. , J.L. Mieloszynski & D. Paquer (1992). Préparation de monomères (méth)acryliques soufrés à l'aide d'une catalyse par transfert de phase, *Sulfur letters*, vol 15, pp. 61 ISSN 0278-6117

Chaudron P. (1999). Synthèse de monomers methacryliques phosphorés. Application à l'ignifugation. University Paul Verlaine, Metz, Thesis

Chen H., Choi J.H., Salas-de -la Cruz D., Winey K.I. & Elab Y.A. (2009). Polymerized Ionic Liquids: the effect of random copolymer composition on ion conduction, *Macromolecules*, vol.42, n°13, pp.4809-4816, (May 2009), ISSN 0024-9297

Choi N.S., Lee Y.M., Lee B.H., Lee J.A. & Park J.K. (2004). Nanocomposite single ion conductor based on organic–inorganic hybrid, *Solid State Ionics*, vol.167, n°3-4, pp. 293–299 (February 2004), ISSN 0167-2738

Christie A.M, Lilley S.J., Staunton E., Andreev Y.G. & Bruce P.G. (2005). Increasing the conductivity of crystalline polymer electrolytes, *Nature*, vol.433, n°7021, pp. 50-53 (January 2005), ISSN 0028-0836

Cochez M., Ferriol M., Weber J. V., Chaudron P., Oget N. & Mieloszynski J.L. (2000). Thermal degradation of methyl methacrylate polymers functionalized by phosphorus-containing molecules I. TGA/FT–IR experiments on polymers with the monomeric formula CH2C(CH3)C(O)OCHRP(O)(OC2H5)2 (R=H, (CH2)4CH3, C6H5Br, C10H7), *Polym. Degrad. Stab.*, vol.70, n° 3, pp. 455-462 (November 2000) ISSN 0141-3910

Cruickshank A.J.B., Windsor M.L., & Young C.L. (1966). The Use of Gas-Liquid Chromatography to Determine Activity Coefficients and Second Virial Coefficients of Mixtures. I. Theory and Verification of Method of Data Analysis, *Proc. R. Soc. London, A*, vol.295, n°1442, pp.259-270. ISSN 1471-2954

Cull S.G., Holbery J.D., Vargas-Mora V., Seddon K.R. & Lye G. J. (2000). Room-temperature ionic liquids as replacements for organic solvents in multiphase bioprocess operations, *Biotechnol. Bioeng.*, vol.69, n°2, pp.227-233 (July 2000), Online ISSN: 1097-0290

Curci M. (1992). Synthèse de monomères (meth)acryliques soufrés et/ou phosphorés comportant des fonctions pouvant générer des sels, Université Paul Verlaine, Metz, Thesis

Curci M. (1993). Synthesis of functionalized acrylates, *Organic Preparations and Procedures international*, vol.25, n°6, pp.649-657 (December 1993), ISSN 0030-4948

Davis J.H. Jr, Gordon C.M., Hilgers C. & Wasserscheid P. (2003). *Ionic Liquids in Synthesis*, Wiley-VCH Verlag GmbH & Co. KGaA, pp. 7-21, ISBN 978-3-527-30515-7, Weinheim, Germany

Dubosclard-Gottardi C., Caubère P. & Fort Y. (1995). New selective syntheses of (meth)acrylic monomers: Isocyanates, isocyanurates, carbamates and ureas derivatives, *Tetrahedron*, vol.51, n°9, pp.2561-2572, (February 1995), ISSN 0040-4020

Earle M.J & Seddon K.R. (2000). Ionic liquid. Green solvents for the future, *Pure Appl. Chem.*, vol.72, n°7, pp1391-1398, (July 2000), ISSN 0033-4545

Estager J., Levêque J-M., Cravotto G., Boffa L., Benrath W. & Draye M. (2007). One-Pot and Solventless Synthesis of Ionic Liquids under Ultrasonic Irradiation. *Synlett* vol.13, pp. 2065-2068, (2007), ISSN 0936-5214

Foco G.M., Bottini S. B., Quezada N., De La Fuente J.C. & Peters C. J. (2006). Activity Coefficients at Infinite Dilution in 1-Alkyl-3-methylimidazolium Tetrafluoroborate Ionic Liquids, *J. Chem. Eng. Data*, vol.51, n°3, pp.1088-1091, (March 2006), ISSN 0021-9568

Ford-Moore A.H. & Perry B.J. (1963), Triethylphosphite, *Organic Syntheses IV*, Coll. Vol 4, pp.955

Forsyth M., MacFarlane D.R., Meakin P.,.Smith M.E & Bastow T.J. (1995) An nmr investigation of ionic structure and mobility in plasticized solid polymer electrolytes, *Electrochimica Acta, International symposium on polymer electrolytes*, vol.40, n°13-14, pp. 2343-2347, (October 1995), ISSN 0013-4686

Fort Y., Gottardi C. & Caubère P. (1993). New selective syntheses of (meth)acrylic monomers: Isocyanates, carbamates, ureas and isocyanurates derivatives, *Tetrahedron. Letters*, vol.34, n°24, pp.3857-3860, (June 1993), ISSN 0040-4039

Fredlake C.P., Crosthwaite J.M., Hert D.G., Aki S.N.V.K. & Brennecke J.F.J. (2004). Thermophysical Properties of Imidazolium-Based Ionic Liquids, *Chem. Eng. Data*, vol.49, n°4, pp.954-964, (June 2004), ISSN 0021-9568

Gentilhomme A., Cochez M., Ferriol M., Oget N. & Mieloszynski J.L. (2005). Thermal degradation of methyl methacrylate polymers functionalized by phosphorus-containing molecules. III: Cone calorimeter experiments and investigation of residues, *Polym. Degrad. Stab.*, vol.88, n°1, pp. 92-97 (April 2005), ISSN 0141-3910

Gordon C.M., Holbrey J.D., Kennedy A.R. & Seddon K.R. (1998). Ionic liquid crystals : hexafluorophosphate salts, *J. Mater. Chem.*, vol.8, n°12, pp.2627-2636, (December 1998), ISSN 0959-9428

Grant D.W. (1971). *Gas-Liquid Chromatography*, van Nostrand Reinhold company, London.

Handy S.T. (2005). Room temperature ionic liquids: Different classes and physical properties, *Current Organic Chemistry* (2005), vol.9, n°10, pp.959-988, (July 2005), ISSN 1385-2728

Harmand J., Rogalski M., Sindt M. & Mieloszynski J.L. (2009). Preparation of molten salt-type (meth)acrylic monomers and their application as catalysts in the synthesis of benzyl phenyl sulfide: an alternative to the usual toxic sulfides synthesis. *Environ. Chem. Letters*, vol.7, n°3, pp.255-260, (September 2009), ISSN 1610-3653

Henderson W.A. & Passerini S. (2004). Phase Behavior of Ionic Liquid−LiX Mixtures: Pyrrolidinium Cations and TFSI$^-$ Anions, *Chem. Mater.*, vol.16, n°15, pp.2881-2885. ISSN 0897-4756

Jullien Y., Melwig J.Y., Wilhelm J.C., Curci M., Mieloszynski J.L & Paquer D. (1996a). Réaction particulières de cyanoalkyle sulfures. Application à l'obtention de dérivés acryliques, *Phosphorus, Sulfur, and Silicon*, vol. 118, n°1, pp. 95-104 (November 1996), ISSN 1042-6507

Jullien Y., Melwig J.Y., Curci M., Mieloszynski J.L & Paquer D. (1996b). Nouvelle voie de synthèse de dérivés acryliques via des réactions de protection/déprotection, *Phosphorus, Sulfur, and Silicon*, vol.118, pp.105-111 (November 1996). ISSN 1042-6507

Karodia N., Guise S., Newlands C. & Andersen J.-A. (1998). Clean catalysis with ionic solvents - phosphonium tosylates for hydroformylation, *Chem.Comm.*, n°21, pp.2341-2342, (November 1998), ISSN 1359-7345

Kharasch M.S. & Fuchs C.F. (1948). The peroxide effect in the addition of reagents to unsaturated compounds. XXXVIII. The addition of mercaptans to methyl acrylate, *J. Org. Chem.*, vol. 13, n°1, pp.97-100, (January 1948), ISSN 0022-3263

Kosmulski M., Szafran M., Saneluta C. & Marczewska-Boczkowska M. (2005). Low-temperature ionic liquids immobilized in porous alumina. *J. Colloid. Interface Sci.* vol.291, n°1, pp.214–217, (November 2005), ISSN 0021-9797

Lakshmi Kantam M., Neelima B., Venkat Reddy C. & Chakravarti R. (2007). Aza-Michael addition of imidazoles to α,β-unsaturated compounds and synthesis of β-amino alcohols via nucleophilic ring opening of epoxides using copper(II) acetylacetonate ($Cu(acac)_2$) immobilized in ionic liquids. *Ind. Eng. Chem. Res.*, vol.46, n°25, pp.8614-8619, (June 2007), ISSN 0888-5885

Levêque J-M., Desset S., Suptil J., Fachinger C., Draye M., Benrath W. & Cravotto G. (2006). A general ultrasound-assisted access to room-temperature ionic liquids, *Ultrason. Sonochem.*, vol.13, n°2, pp.189-193, (February 2006), ISSN 1350-4177 (ref aussi dans partie 2)

Levêque J-M., Estager J., Draye M., Cravotto G., Boffa L. & Benrath W. (2007) Synthesis of Ionic Liquids Using Non-Conventional Activation Methods: An Overview, *Monatsh. Chem.*, vol.138, n°11, pp.1103–1113 (August 2007), ISSN 0026-9247

Li DY, Lin YS, Li YC, Shieh DL, Lin JL (2007) Synthesis of mesoporous pseudoboehmite and alumina templated with 1-hexadecyl-2,3-dimethyl-imidazolium chloride. *Microporous Mesoporous Mater.*, vol.108, n°1-3, pp.276-282, (February 2008), ISSN 1387-1811

MacFarlane D.R., Sun J., Meakin P., Fasoulopoulos P., Hey J. & Forsyth M. (1995). Structure-property relationships in plasticized solid polymer electrolytes, *Electrochimica Acta, International symposium on polymer electrolytes*, vol.40, n°13-14, pp. 2131-2136, (October 1995), ISSN 0013-4686

Maki-Arvela P., Mikkola J.P., Virtanen P., Karhu H., Salmi T. & Murzin D.Y. (2006) Supported ionic liquids catalyst (SILCA) in the hydrogenation of citral. *Stud. Surf. Sci. Catal.*, vol.162, pp.87–94, (september 2006), ISSN 0-444-52827-X

Mancardi D., Sindt M., Paul J.M. & Mieloszynski J.L. (2007). Convenient synthesis of tributylsilyl methacrylate, *Synth. com.*, vol.37, n°21, pp. 3873-3878 (January 2007) ISSN: 0039-7911

Martin-Aranda R.M., Vicente-Rodriguez M.A., Lopez-Pestana J.M., Lopez-Peinado A.J., Jerez A., Lopez-Gonzalez J.deD. & Banares-Munoz M.A. (1997). Application of basic clays in microwave activated Michael addition : Preparation of N-substituted imidazoles, *J. Mol. Catal. A : Chem.*, vol. 124, n° 2-3, pp.115-121, (October 1997), ISSN 1381-1169

Martin-Aranda R.M., Ortege-Cantero E., Rojas-Cervantes M.L, Vicente-Rodriguez M.A. & Banares-Munoz M.A. (2002). Sonocatalysis and basic clays. Michael addition between imidazole and ethyl acrylate. *Catalysis Letters*, vol.84, n°3-4, pp.201-204, (December 2002), ISSN 1011-372X

Matsumoto K, Talukdar B. & Endo T. (2010). Methacrylate-based ionic liquid: radical polymerization/copolymerization with methyl methacrylate and evaluation of molecular weight of the obtained homopolymers, *Polym. Bull*, vol.66, n°2, pp.199-210, (January 2011), ISSN 0170-0839

Mehnert C.P., Cook R.A., Dispenziere N.C. & Afeworki M. (2002). Supported ionic liquid catalysis—a new concept for homogeneous hydroformylation catalysis. *J. Am. Chem. Soc.*, vol.124, pp.12932–12933, (October 2002), ISSN 0002-7863

Melwig J.Y. (1995). Etude de la synthèse de composés acryliques via une réaction de protection-déprotection de la fonction polymérisable. Université Paul Verlaine, Metz, Thesis

Mikkola J.P.T., Virtanen P.P., Korda`s K., Karhu H. & Salmi T.O. (2007). SILCA—supported ionic liquid catalysts for fine chemicals. *Appl. Catal. A* vol.328, n°1, pp.68–76, (August 2007), ISSN 0926-860X

Mutelet F., Jaubert J.N., Rogalski M., Harmand J., Sindt M . & Mieloszynski J.L. (2008). Activity Coefficients at Infinite Dilution of Organic Compounds in 1-(Meth)acryloyloxyalkyl-3-methylimidazolium Bromide Using Inverse Gas Chromatography, *J. Phys. Chem. B*, vol.112, n°12, pp.3773-3785, (March 2008), ISSN 1520-6106

Mutelet F., Jaubert J.N., Rogalski M., Boukherissa M. & Dicko A. (2006). Thermodynamic Properties of Mixtures Containing Ionic Liquids: Activity Coefficients at Infinite Dilution of Organic Compounds in 1-Propyl Boronic Acid-3-Alkylimidazolium Bromide and 1-Propenyl-3-alkylimidazolium Bromide Using Inverse Gas Chromatography, *J. Chem. Eng. Data*, vol.51, n°4, pp.1274-1279, (April 2006), ISSN 0021-9568

Nakajima H., Ohno H. (2005). Preparation of thermally stable polymer electrolytes from imidazolium-type ionic liquid derivatives, *Polymer*, vol. 46, n°25 pp.11499-11504, (December 2005), ISSN 0032-3861

Namboodiri V.V.& Varma R.S. (2002). Solvent-Free Sonochemical Preparation of Ionic Liquids, *Org. Lett.*, vol.4, n°18, pp.3161-3163, (August 2002), ISSN 1523-7060

Noda A. & Watanabe M. (2000). Highly conductive polymer electrolytes prepared by in situ polymerization of vinyl monomers in room temperature molten salts, *Electrochimica Acta*, vol.45, n°8-9, pp.1265-1270, (January 2000), ISSN 0013-4686

Ohno H. (2001). Molten salt type polymer electrolytes, *Electrochimica Acta*, vol.46, n°10-11, pp.1407-1411, (March 2001), ISSN 0013-4686

Olivier-Bourbigou H., Magna L. & Morvan D. (2010). Ionic liquids and catalysis : Recent progress from knowledge to applications, *Appl. Catal. A* vol.373, n°1-2, pp.1–56, (January 2010), ISSN 0926-860X

Pacholec F., Butler H. T. & Poole C. F. (1982). Molten organic salt phase for gas-liquid chromatography, *Anal. Chem.*, vol.54, n°12, pp.1938-1941, (October 1982), ISSN 0003-2700

Pacholec F. & Poole C. F., (1983). Stationary phase properties of the organic molten salt ethylpyridinium bromide in gas chromatography, *Chromatographia*, vol.17, n°7, p.370-374 (July 1983), ISSN 0009-5893

Pees B., Cahuzac A., Sindt M., Ameduri B., Paul J.M., Boutevin B., & Mieloszynski J.L. (2001). Synthesis of fluorinated acrylic monomers bearing a fonctionalized lateral chain. Part 1 : Preparation of sulfurated monomers, *J.Fluor. Chem.*, vol.108, n°2, pp. 133-142 (May 2001) ISSN 0022-1139

Pees B., Sindt M., Paul J.M. & Mieloszynski J.L. (2002). Synthèse et caractérisation physico-chimiques de polyacrylates d'ω-perfluorooctyl-alkyle : effet pair- impair. *Eur. Polym. Jour*, vol.38, n°5, pp. 921-931 (May 2002), ISSN 0014-3057

Pees B., Paul J.M., Oget N., Sindt M., & Mieloszynski J.L. (2003). Synthesis of fluoro-substituted monomers bearing a functionalised lateral chain. Part 2: Preparation of sulfoxides and sulfones containing monomers, *J. Fluor.Chem.*, vol.124, n°2, pp.139-146 (December 2003) ISSN 0022-1139

Pees B., Paul J.M., Corpart J.M., Sindt M., & Mieloszynski J.L. (2004). Surface properties of ω-perfluorooctyl-alkyl polyacrylates : odd-even effect, *Eur. Polym. Jour.*, vol.40, n°12, pp. 2727-2730 (December 2004) ISSN 0014-3057

Poole C.F. (2004). Chromatographic and spectroscopic methods for the determination of solvent properties of room temperature ionic liquids, *J. Chromatogr., A,* vol.1037, n°1-2, pp.49-82 (May 2004), ISSN 0021-9673

Ratner M.A., Shriver D.F. (1988). Ion transport in solvent-free polymers, *Chem.Rev.*, vol.88, pp. 109-124, (January 1984), ISSN 0009-2665

Reiche A., Steurich T., Sandner B., Lobitz P. & Fleischer G. (1995). Ion transport in gel electrolytes, *Electrochimica Acta, International symposium on polymer electrolytes,* vol.40, n°13-14, pp. 2153-2157, (October 1995), ISSN 0013-4686

Riisager A., Wasserscheid P., Van Hal R. & Fehrmann R. (2003). Continuous fixed-bed gas-phase hydroformylation using supported ionic liquid phase (SILP) Rh catalysts. *J. Catal.*, vol.219, n°2, pp.452–455, (October 2003), ISSN 00219517

Robert D., Schneider M., Born M., Mieloszynski J.L. & Paquer D. (1996). Influence of heteroatomic systems on anti-wear and extreme pressure properties of organo-sulfur compounds. *C. R. Acad. Sci., série IIb*, vol.323, pp.127–132, (1996), ISSN 0320-8437

Rogers R.D. & Seddon K.R. (2003). Ionic Liquids--Solvents of the Future?, *Science*, Vol. 302, n°5646, pp. 792-793 (October 2003), ISSN 0036-8075

Seddon K.R. (1997). Ionic Liquids for Clean Technology, *J. Chem. Technol. Biotechnol.*, vol. 68, n°4, pp.351-356, (April 1997), ISSN 0268-2575

Sheldon R.A. (1993). The role of catalysis in waste minimization. *Precision Process Technology, Perspectives for Pollution PreVention;* Weijnem, M. P. C.; Drinkenburg, A. A. H. Editors; Kluwer: Dodrecht.

Shibata Y., Kato T., Kado T., Shiratuchi R., Takashima W. & Kaneto K. (2003). Quasi-solid dye sensitized solar cells filled with ionic liquid-increase in efficiencies by specific interaction between conductive polymers and gelators, *Chem. Commun.*, n°21, pp. 2730–2731 (November 2003), ISSN 1359-7345

Suk-Ku K., Won-Sup K. & Byoung-Ho M. (1985). An Effective Method for the Preparation of ω-Bromoalkanols from α,ω-Diols, *Synthesis*, n°12, pp.1161-1162, (December 1985), ISSN 0039-7881

Thomas H., Oget N. & Mieloszynski J.L. (2006). A fast and efficient route to 5-chloromethyl-1,4,7,10-tetrathiacyclotridecane and its behavior in a chloroform solution. *J. Incl. Phenom. Macrocyl. Chem.* vol.54, n°3-4, pp.139–145, (April 2006), ISSN 0923-0750

Tiegs D., Gmehling J., Medina A., Soares M., Bastos J., Alessi P. & Kikic I. (1986). Activity Coefficients at Infinite Dilution, *Chemistry Data Series, vol.IX*, Part 1, DECHEMA. ISBN 3-921567-79-3, Main, Frankfurt

Varma R.S. & Namboodiri V.V. (2001a). An expeditious solvent-free route to ionic liquids using microwaves, *Chem. Comm.*, n°7, pp.643-644, (April 2001), ISSN 1359-7345

Varma R.S. & Namboodiri V.V (2001b). Solvent-free preparation of ionic liquids using a household microwave oven , *Pure Appl. Chem.*, Vol. 73, No. 8, pp. 1309-1313, ISSN 0033-4545

Virtanen P., Karhu H., Kordas K. & Mikkola J.P. (2007). The effect of ionic liquid in supported ionic liquid catalysts (silca) in the hydrogenation of α,β-unsaturated aldehydes. *Chem. Eng. Sci.*, vol.62, n°14, pp.3660–3671, (July 2007), ISSN 0009-2509

Washiro S., Yoshizawa M., Nakajima H. and Ohno H. (2004). Highly ion conductive flexible films composed of network polymers based on polymerizable ionic liquids, *Polymer* vol.45, n°5, pp. 1577–1582 (March 2004), ISSN 0032-3861

Wasserscheid P. & Keim W.A. (2000). Ionic Liquids—New "Solutions" for Transition Metal Catalysis, *Ang. Chem. Int. Ed.*, vol.39, pp.3772-3789, (November 2000), Online ISSN: 1521-3773

Wasserscheid P., Drießen-Hölscher B., Van Hal R., Steffens H.C. & Zimmermann J. (2003). New functionalised ionic liquids from Michael-type reactions—a chance for combinatorial ionic liquid development, *Chem. Comm.*, vol. 39, n°16, pp.2038-2039, (August 2003), ISSN 1359-7345

Wasserscheid P. & Welton T. (2008). *Ionic liquid in synthesis*, 2nd ed., Wiley-CH, ISBN 978-3-527-31239-9, Weinheim.

Welton T. (1999). Room-Temperature Ionic Liquids. Solvents for Synthesis and Catalysis, *Chem. Rev.*, vol.99, n°8, pp.2071-2083 (July 1999), ISSN 0009-2665

Wilkes J.S. (2004). Properties of ionic liquid solvent for catalysis, *J. Mol. Catal. A: Chem.*, vol.214, n°1, pp.11-17, (May 2004), ISSN 1381-1169

Xu J.M., Qian C., Liu B.K., Wu Q. & Lin X.F. (2007). A fast and highly efficient protocol for Michael addition of N-heterocycles to α,β-unsaturated compound using a basic ionic liquid [bmIm]OH as catalyst and green solvent. *Tetrahedron*, vol.63, n°4, pp.986-990, (January 2007), ISSN 0040-4020

Yang L., Xu L.W. & Xia C.G. (2005). Highly efficient KF/Al_2O_3-catalyzed versatile hetero-Michael addition of nitrogen, oxygen, and sulfur nucleophiles to α,β-ethylenic compounds. *Tetrahedron Letters*, vol.46, n°19, pp.3279-3282, (May 2005), ISSN 0040-4039

Yang L., Xu L.W., Zhou W., Li L. & Xia C.G. (2006). Highly efficient aza-Michael reactions of aromatic amines and N-heterocyles catalysed by basic ionic liquid under solvent-free conditions. *Tetrahedron Letters*, vol.47, n°44, pp.7723-7726, (October 2006), ISSN 0040-4039

Yoshizawa M. & Ohno H. (2001). Synthesis of molten salt-type polymer brush and effect of brush structure on the ionic conductivity, *Electrochimica Acta*, vol.46, n°10-11, pp.1723-1728, (March 2001), ISSN 0013-4686

Yu B., Zhou F., Wang C. & W. Liu (2007). A novel gel polymer electrolyte based on poly ionic liquid 1-ethyl 3-(2-methacryloyloxy ethyl) imidazolium iodide, *Eur. Polym. J.*, vol.43, n°6, pp.2699-2707 (June 2007), ISSN 0014-3057

Zaderenko P., Soledad Gil M. & Ballesteros P. (1994). Synthesis and regioslective hydrolysis of 2-imidazol-1-ylsuccinic ester, *J. Org. Chem*, vol.59, n°21, pp.6268-6273, (October 1994), ISSN 0022-3263

Zilkova N., Zukal A. & Cejka J. (2006). Synthesis of organized mesoporous alumina templated with ionic liquids. *Microporous Mesoporous Mater.*, vol.95, n°1-3, pp.176–179, (October 2006), ISSN 1387-1811

Part 2

Theoretical Studies

5

Theoretical Description of Ionic Liquids

Enrico Bodo and Valentina Migliorati
University of Rome "La Sapienza", Chemistry Department
Italy

1. Introduction

Ionic liquids are universally considered to be materials of the future. Their peculiar properties appeal to the most diverse technological areas such as chemical industry, electrochemistry, optics, environmental chemistry, medicine and nanotechnology. Their extraordinary macroscopic properties originate from the peculiar ionic nanoscopic structure and the possibility of modulating these properties depends ultimately on small modifications of the material at the molecular level.

Theoretical simulations of ionic liquids are becoming increasingly important in predicting the properties of these new materials and in helping to interpret the experimental determinations. From the theoretical point of view, ionic liquids are peculiar materials because they are entirely composed of ions. They can be defined as pure electrolytes. All the technologically interesting properties that are proper to this class of compounds are due to the strong electrostatic interactions that govern the fluid behavior. Though these interactions are quite strong, the size and nature of the cationic partners (more often) are such that the fluid remains liquid even at room temperature. Its behavior, from a microscopic point of view, is fairly similar to the more conventional high temperature molten salts.

The theoretical framework in which it is possible to provide high quality studies of the microscopic structure of the ionic liquid is mainly represented by classical molecular mechanics and, only very recently, by ab-initio molecular dynamics. While the employed theoretical techniques are not very different from those used for conventional fluids, many difficulties arise because of the microscopic nature of ionic liquids. In particular these substances are extremely viscous and simulation times become quickly prohibitive if one wants to describe dynamical properties, even as simple as diffusion coefficients. Recent technological advances such as the introduction of GPU clusters might allow unprecedented possibilities in the simulation of these material opening the route to the simulation of rare events and long time scale phenomena.

The choice and validation of the many force fields which have been established in the last twenty years, some of which are explicitly tailored for a specific class of ionic liquid, represent a crucial step in performing classical trajectory studies. This task has to be performed by taking into account many different sources of experimental data: on the one hand structure studies with X-ray or neutron diffraction provide indications on the short-range structure, on the other hand measurements of bulk properties such as evaporation enthalpies point to the overall energetic behavior of the fluid.

In this chapter we would like to summarize and review the recent developments and possible future opportunities of theoretical simulations of ionic liquids. In particular we will review some of the issues connected to the theoretical simulations of ionic liquids, focusing on our recent work on a few compounds. We shall also review the theoretical methods in general and the peculiarities in their application to ionic liquids.

2. Ionic liquids

Ionic liquids are one of the most promising classes of new materials investigated in the last decade. They do not easily fit the conventional description of molecular fluids therefore promoting a necessary exploration of their physical properties at a microscopic level. Conventionally, ionic liquids are chemicals entirely made by ions (Freemantle, 2009; Gaune-Escard & Seddon, 2010; Rogers & Seddon, 2005a) that show a melting point lower than 100°C, therefore they are liquid under conventional ambient conditions. Their negligible vapor pressure, high thermal stability, properties tunability upon slight changes in the chemical architecture (including polarity, hydrophobicity, density, solvating activity etc.) have made these materials tailored for a constantly increasing range of applications (Plechkova & Seddon, 2008a; Rogers et al., 2009; Rogers & Seddon, 2005b). Among these applications we find: catalysis, synthesis, sensoristics, medicine, electrochemistry and green chemistry in general (Fischer et al., 1999; Galinski et al., 2006; Garcia et al., 2004; Hough & Rogers, 2007; Hough et al., 2007; Welton & Wasserscheid, 2008). Their versatility has opened the possibility of their employment as functional advanced materials, mediums for materials production and components for preparing highly functional materials. Even if ionic liquids are commonly utilized as solvents, their specific composition where ions solely exist provides amazing functional properties such as dissolution of bio-related materials that never dissolve in conventional solvents. The dissolution of biopolymers as cellulose (Fujita et al., 2009) is one of such examples. Chloride based ionic liquids dissolve cellulose better than other solvents because of the hydrogen bonding between chloride anions with the hydroxyl groups of the polymer. The use of ionic liquids allows a simple, benign system for the processing of cellulose and has potential environmental and cost advantages over current processing methodologies. Ionic liquids are made entirely by ionic couples. The anion is generally inorganic as for example $[PF6]^-$, $[NTf2]^-$, $[BF4]^-$, Br^-, Cl^-. The cation, instead, is an asymmetric organic cation such as alkylammonium, alkylphosphonium, N,N-dialkylimidazolium, and N-alkylpyridinium. Due to the freedom in designing the organic cation (for example by variation of the side chain length or by varying the substituents on the ring and/or on the chain) and to the possible combination of cation and anion, one can generate a huge number of different ionic liquids. Unlike the about six hundred conventional solvents that are extensively used in industrial and synthetic processes, ionic liquids exists in at least a million pure forms and a trillion ternary mixtures. This liberty allows one to design these materials to optimize a specific task such as a solvent for a reaction. This is the reason why these liquids have been termed "designer solvents" (Fei et al., 2006; Freemantle, 1998; Rogers & Seddon, 2003).

In recent years, ionic liquids have emerged as possible "green solvents", that is environmentally benign substances mainly because they have negligible vapor pressure (Liu et al., 2005). One of the primary driving forces behind research into ionic liquids is the perceived benefit of substituting traditional industrial solvents, most of which are volatile organic compounds (VOCs), with nonvolatile ionic liquids. Replacement of conventional solvents by ionic liquids would prevent the emission of VOCs, a major source of environmental pollution (Polshettiwar & Varma, 2008).

One of the fields in which ionic liquids have more to offer in terms of new opportunities is electrochemistry. While there are many properties that individual members of the ionic liquids family exhibit, the single ubiquitous property that we expect to find in all of them is ionic conductivity (Galinski et al., 2006; Garcia et al., 2004). Unfortunately while the ionic liquids are intrinsic ion conductors, their ion conductivity often falls short of that of solvent-based electrolytes because of the high viscosities (Hu & Margulis, 2007). Of course viscosity is the result of the electrostatic interactions that are intrinsic to the nature of these substances. Untangling this "catch 22" situation has been the leading theme of research in electrochemistry (Wilkes, 2002). Many of the structural features that make this substances attractive for electrochemistry are also at the origin of the difficulties in using them in the electrochemical processes (Forsyth et al., 2004; MacFarlane et al., 2007; Xu & Angell, 2003). One potential application of ionic liquids in the electrochemical field concerns their use in supercapacitors also known as electrochemical double-layer capacitors (EDLC). Another intriguing application of ionic liquids concerns the construction of dye-sensitized solar cells (DSSCs) (Wang et al., 2002; Wolfbauer et al., 2001). These are novel photovoltaic devices, also known as Gratzel cells, consisting of a layer of a nanoporous semiconductor, often TiO_2, impregnated with a light sensitive dye and confined between two metallic electrodes. The microporous structure of the semiconductor is embedded in a liquid electrolyte that contains a redox couple, for instance I^-/I_3^-. Different organic solvents have been tested in the manufacture of DSSCs but, as a result of evaporation and leaking, most of them deteriorate the performance of the cell after some time of operation. Ionic liquids have been proposed as an alternative in the design of DSSCs because of their high viscosity and low vapor pressure that minimize solvent leaking and losses.

Molecular simulations are playing an increasing role in developing an understanding of the physical chemistry of ionic liquids. Unlike conventional organic liquids that have been studied for decades, the ionic liquid research field is still young and much is still unknown about their properties and behavior, and, in this respect, simulations are on an equal footing with experimental investigations (Lipkowitz et al., 2009). This means that molecular simulation will be used both for property prediction and to provide qualitative insight into the nature of these substances. Property predictions is of paramount importance under conditions where experiments are difficult to be conducted and thus cannot determine the structural and dynamic properties of the system under investigation.

3. A brief history

The history of ionic liquids is a recent one even if one can trace back the origins of the materials to the early 19th century, by examining the molten salt research (Wilkes, 2002). An ionic liquid can be loosely defined, indeed, as a molten salt with a melting temperature below the boiling point of water. In general an ionic liquid differs from a traditional molten salt because is made by an organic cation and an inorganic anionic partner. Although ionic liquids can be handled as ordinary solvents, they present many specific features that are not seen in ordinary organic solvents or even in high temperature molten salts. These features are mainly due to the strong ion-ion interactions that are peculiar to these materials. One of the most important consequences of the network of strong ionic interactions is the negligible vapor pressure and the huge liquidus range i.e. the span in temperatures between the freezing and boiling point. The consequence for green chemistry is that the ionic liquids represent the ultimate non-volatile solvents (Wilkes, 2002).

As stated before, the roots of nowadays ionic liquids research is firmly planted in the traditional high temperature molten salts development and use. Molten salts have long been established as solvents for particular applications, mainly electrochemical: their main advantages are that they present a high electrical conductivity due to ion mobilities, they have a wide liquidus range due to their thermal stability and they are relatively cheap because they are derived from naturally occurring minerals. Their main drawback is that they operate as liquids at very high temperatures so that their use as a reaction media can be considered unknown to most chemists. Ionic liquids present the advantages of molten salts but avoid the problem of the high temperatures.

The first report for the existence of room temperature ionic liquids dates back to 1914 (Walden, 1914a; Walden., 1914b) when Walden investigated ethylammonium nitrate. After this work, many years passed without a substantial understanding of the considerable potential of this class of materials and only in the 1970's the first generation of ionic liquids (composed of water-unstable organic chloro-aluminates) was studied. As soon as air and water stable ionic liquids have been synthesized in the early nineties (Wilkes & Zaworotko, 1992), ionic liquids have attracted the interest of a considerable part of the material science, electrochemical and organic chemistry community. At the present moment, more than 8000 papers have been published in the last decade and review papers are appearing every two or three days (Plechkova & Seddon, 2008b).

From the theoretical point of view the research on ionic liquids lays its foundations in the simulations of molten salts of the late '60. Although between 1990 and 1997 there have been various theoretical studies on ammonium salt derived ionic liquids, the first study of a dimethy-limidazolium compound traces back to 2001 when Hanke et al. (Hanke et al., 2001) studied the [C_1mim][Cl] and [C_1mim][PF_6] at high temperature (mostly because of computational limitations). Another study followed with the dissolution in the same compounds of water, methanol and propane (Lynden-Bell et al., 2002). Many other theoretical investigations of ionic liquids then followed and helped elucidating some of the key properties and features of this class of compounds. These studies have also highlighted the difficulties of obtaining reliable results on ionic liquids from molecular simulations and the challenges of the next generation of computational simulations. One major problem is that the ionic liquid dynamics is slow, much slower than in conventional liquids and this makes computational method more difficult to be adopted unless very high temperatures or very long simulation times are used. Another common feature of ionic liquids, that can be studied through the use of simulations and that was recognized in these systems from the beginning (Morrow & Maginn, 2002), is the spatial etherogenity due to a long range alternating ordering effect between cations and anions which persists up to 20 Å around each ion center. As an example we report here a comparison we have made in ref. (Bodo et al., 2010) between the measured total distribution functions and those predicted by means of molecular dynamics simulations for the series of compounds [$C_n$mim][Tf_2N] with n =1, 4, 6 and 8. Long range ordering effects can easily be seen from Figure 1 up to 20-25 Å. It has to be pointed out that the theoretical predictions have been obtained by using two different force field: the one from Canongia-Lopes (Lopes & Pádua, 2004) and another one from Köddermann (Ködderman et al., 2007). Differences are very small between the two simulations and both force fields are able to reliably reproduce the structural features of the liquid. These study showed that relatively simple force fields as those adopted in our work that are based on the OPLS one (Jorgensen et al., 1996) are reasonably accurate in describing the structural features of ionic liquids. It turns out to be much more difficult to reliably predict dynamical features as we will show below.

4. Present challenges

A bulk ionic liquid can be considered as a pure electrolyte (Rogers & Seddon, 2005a). It is made entirely of molecular ions and besides its actual composition or the actual nature of the molecular ions that it is made of, many of the appealing properties which have been exploited in the technological applications are due to this feature.

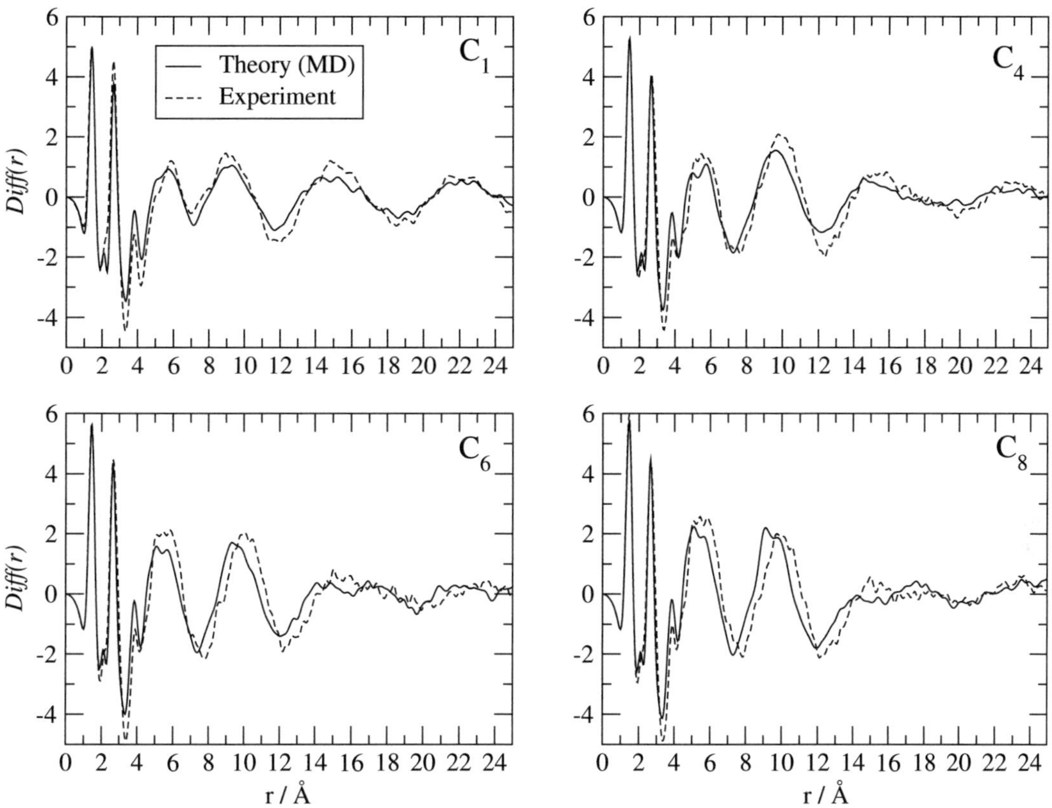

Fig. 1. Experimental and theoretical (MD) $G(r)$ for [$C_n$mim][NTf_2] with n=1 (upper-left corner), n=4 (upper-right), n=6 (lower-left), n=8 (lower right).

The structure of an ionic liquid at the molecular level is therefore characterized by very strong electrostatic interactions which give rise to many of its macroscopic properties: viscosity, negligible vapor pressure, poor conductivity[1]. The relatively low melting points, on the other hand, are due to the complexity of the molecular components and to their asymmetry which defy the natural tendency of ionic compounds to form crystals.
Another aspect that is important to keep in mind is that, unless the halogen anions are used, the charge is not localized on an atom but is distributed all over the molecular structure, therefore making even more difficult to describe the ionic liquid as made by point charges.
From the theoretical point of view, the description of an ionic liquid in its bulk liquid state still represents a formidable challenge under many point of view. Molecular dynamics simulations are the main (if not the only possible) choice for the molecular, atomistic description of an ionic liquid in the bulk or nanoconfined state. As a first point, consider that if one wants to perform a statistically meaningful simulation of a liquid, one has to provide at least few ns of dynamics using a cell large enough to capture the long range structural correlations of the system under investigation. This means that cells of the order of 50-70 Å have often to be used which generally contain few hundreds of ionic couples. This substantially rules out the possibility of

[1] Actually all the ionic liquids are intrinsically conductors, but since the ion drift or diffusion is very low it turns out that only in a limited number of mixture, it has been possible to improve the intrinsic ionic conductivity(Forsyth et al., 2004)

using ab-initio molecular dynamics although it may represent a promising route for the near future or for the simplest (in terms of molecular structure) inhabitants of the vast zoology of ionic liquids. Despite the fact that ab-initio molecular dynamics on such systems is slowly getting within the grasp of modern supercomputer, it still can not represent a viable choice for general applications. This limitation is due to the high computational cost of methods such as Car-Parrinello or even Born-Oppenheimer techniques. The reason is simple: most of the ionic liquids present a very high viscosity, due to the strong electrostatic interactions, and therefore an ab-initio molecular dynamics simulation requires both equilibration and production simulation times that are unaffordable in terms of computational costs. The application of ab-initio techniques, i.e. quantum mechanical methods (QM) is limited, in most cases, to the study of small portion of materials such as clusters or even gas phase isolated molecules. Other issues may arise if one wants to calculate dynamical properties, such as the self diffusion coefficients either from the velocity correlation functions or mean square displacements. In this case the simulation time should be in the hundreds of nanosecond scale and this can be achieved only for small simulation cells or for the simplest compounds.
Molecular dynamics simulations can exploit the fact that they bypass the solution of the electronic quantum problem and use instead a very crude representation of the force acting on nuclei generally obtained by means of mixed QM and empirical data. The choice of the classical force field is of crucial importance for the reliability of the theoretical results. Even if the force fields have been validated against various experimental quantities (viscosity, densities, evaporation enthalpy, diffraction data), their transferability to other systems, not directly involved in the validation stage, can often represent a problem. Given the huge number of molecular variations that are possible with ionic liquids, the use of a given force field has to be experimentally verified anyway, even if the same force field has been successfully used elsewhere. Another challenge that awaits the molecular dynamics simulations is the lack of validated force fields for many of the compounds of interest for experimental applications. For example, while imidazolium based compounds have been extensively investigated and the series of force fields provided by Canongia-Lopes and coworkers (Canongia Lopes & Pádua, 2006b) have been used successfully in predicting the physical properties of the relevant compounds, much less has been done on other compounds such as ammonium based salts although they represent the prototypical ionic liquid. The task of developing and validating a force field is tedious, but it is essential for the theoretical community because, especially if the validation is performed by means of different experimental sources, it provides the basis on which one is able to obtain reliable theoretical data.
It is our strong belief that in the next years theory will not only serve as a post-experiment interpretation tool or a simple "black box" able to predict experimental quantities (even though these already represent highly complex duties), but it will provide a sophisticated tool able to obtain complementary information with respect to experimental quantities and to see many things that are precluded to experiments. In order to do so theory must rely on a well established approximation to the true forces governing the nuclear motion. In this sense, the validation of force fields is crucial.

5. The force field

The problem of choosing, creating or improving a force field for molecular dynamics calculations is obviously non trivial and requires a careful validation of the field with the existing experimental observables. There are even many different paradigms which one may choose from in order to build a force field: for example the "atomistic" level and the electrostatic model. The former can be assessed in the following terms: in principle the most accurate simulations can be obtained from an all-atom calculation in which, as the name says, all of the atoms in the system are explicitly treated. Moreover, all of the bonds among atoms in the molecules can be allowed to vibrate. However, it is well known that some of the vibrational motions of the involved bonds cannot be correctly described by classical mechanics at least in the room temperature range. The frequencies associated to vibrational motions are often too high to allow for the classical equipartition theorem to work correctly and quantization effects should be considered or included. This is especially valid for the stretching of bonds involving the hydrogen atom. As a first step, it is therefore a common practice to freeze the C-H bonds (or the O-H and N-H ones). For this reason it may become computationally convenient to fuse, for example, the three/four Lennard-Jones entities of a CH_2/CH_3 group into a new L-J entity that behaves as a rigid unit. This leads to the construction of United Atoms (UA) force fields that reduce the computational complexity by adopting the above scheme. This approximation can be moved one step forward into considering entire subunits (more or less rigid) of the molecular system as unique entity: the resulting coarse-graining models eliminate out those degrees of freedom that enter into macroscopic properties only through their cooperative effect and substitute them by effective degrees of freedom. Building a coarse grained model implies a rather complex fitting procedure which is based on the knowledge of some previous molecular dynamics run. In a first step one has to decide how to map an all-atom system into a reduced complexity system which has a different structure. One then has to build a reliable force field acting between the newly defined subunits. Needless to say, the second step is the most complex one (Wang & Voth, 2005).

Coarse grained models have been successfully employed in the description of ionic liquids (Wang & Voth, 2005; 2006). It has been shown that for sufficiently long side chains in imidazolium based liquids, the apolar alkali chains tend to aggregate. This important result is in agreement with many experimental observations.

From the point of view of electrostatic interactions (that are obviously of crucial importance for the description of ionic liquids), there are two main choices: fixed partial charges and polarizable models.

Among the many possible variants of force fields the OPLS has been the most popular in this field and has been suitably taylored for the description of a large number of classes of ionic liquids by A. Padua and coworkers (Canongia Lopes & Pádua, 2006a; Canongia Lopes et al., 2005; Canongia Lopes & Pádua, 2006b; Lopes et al., 2004; Lopes & Pádua, 2004)

The non polarizable OPLS force fields can be written by using the following functional form (Jorgensen et al., 1996).

$$V = V_{\text{bonded}} + V_{\text{non-bonded}} = \sum_{ij} k_{ij}(r_{ij} - r_{ij}^0)^2 + \sum_{ijk} a_{ijk}(\theta_{ijk} - \theta_{ijk}^0)^2 + \sum_{ijkl}\sum_{n=1}^{3} \frac{1}{2} V_n \cos(n\phi_{ijkl}) + \sum_{ij}\left[\left(\frac{\sigma_{ij}}{r_{ij}}\right)^{12} - \left(\frac{\sigma_{ij}}{r_{ij}}\right)^{6}\right] + \frac{q_i q_j}{4\pi\varepsilon_0 r_{ij}} \tag{1}$$

where the first term is the harmonic vibrational energy of the bonded pairs, the second term is the bending energy, the third term is due to 1-4 torsional interactions and the last two terms are the Lennard-Jones short range term and the electrostatic interaction energy. The latters are normally referred to as non-bonded interactions. Other force fields use different forms of

each specific term. For example the MM3 one, has a more complex expression for the bonded terms that includes higher order terms in addition to the quadratic one.
The electrostatic term is very simple in the OPLS form but may be more complex in other force fields. This term is actually crucial in the simulation especially if one wants to calculate energetic quantities such as vaporization enthalpies or diffusion coefficients (see below). If we add a polarization term, this has usually the following form:

$$V^{\mathrm{pol}} = -\frac{1}{2}\boldsymbol{\mu}_i \cdot \boldsymbol{E}_1^0 \tag{2}$$

where $\boldsymbol{\mu}_i = \alpha_i \boldsymbol{E}_i$ is the induced dipole at force center i, with polarizability α_i (isotropic) and $\boldsymbol{E}^0$ is the force field due only to the partial atomic charges. Atomic polarizabilities are found by fitting the gas-phase polarizability as determined by quantum mechanics calculations. In general it is found that polarizable force fields, when accurately parametrized, (Borodin, 2009) lead to a realistic description of many important properties of ionic liquids such as densities and transport properties.
In any case a force field has to be "validated" against experimental data. The best way to obtain the parameters for the bonded terms such as bond and bending is that of producing fairly high quality ab-initio calculations in the gas phase for the isolated cation and anion and to obtain equilibrium distances and force constants from an optimization followed by a normal mode analysis. The dihedral angle potential is normally obtained by calculating the torsional profile and fitting the energy expression to the raw points. Electrostatic is generally obtained by assigning a partial charge to each atom (see below). Lennard-Jones parameters are much more difficult to obtain from a quantum calculation and generally the parameters adopted in widely available force fields (AMBER, OPLS) are collected and used.

6. A combined approach

The theoretical simulation of ionic liquids requires a combined approach that involves the use of more than one techniques. It is customary to perform a series of preliminary ab-initio calculations before moving to molecular dynamics simulations. The ab-initio simulations are dictated by the necessity of obtaining the "true" charge distribution in order to check whether it matches the partial atomic charges of the force field and, if necessary, to provide new partial atomic charges.
The first step in the study of a molecular ionic liquid is the modeling of the gas-phase structure of its ionic constituents. This approach is often necessary because the partial atomic charges are not known for many particular structures. Two approaches can be followed: the first one which is more apt to allow for the construction of a general force field is to optimize the isolated cationic or anionic moieties and to calculate the partial atomic charges. The second one is to consider the ionic couple in the gas phase and optimize the two partners altogether. In the former case the net charge on each of the partner will be entire while in the latter case the net charge of the partner can be fractional.

6.1 Charges determination

The simplest way to calculate partial atomic charges is that of fitting the electric potential generated by the electronic and nuclear charge distribution by point charges placed on the various atomic centers. The charge distribution can be easily obtained by ab-initio calculations using well tested DFT or MP2 methods. The more traditional way of producing atomic charges based on wavefunction atomic orbital analysis (Mulliken or Lodwin population) should be avoided because they are not reliable especially when an augmented basis has

been used (which is the normal case for the anionic partner). There are various fitting recipes that can be found implemented in the most common ab-initio packages, but the CHELPG procedure (Frisch et al., 2009) has become a de facto standard for force field calculation when coupled to B3LYP calculations.
A charge distribution and the resulting atomic charges are obviously dependent upon the particular geometry of the molecular system. This is a subtle point that may not be apparent when dealing with rigid ionic partners such as 1,3-dimethyl-imidazolium where the conformational mobility is limited to the CH_3 rotation. There are however examples where the conformational mobility is very large and it turns out not to be obvious which gas-phase geometry is representative of the liquid phase. We will make more comments to assess this specific point with an example in the final section of this chapter. It might be not obvious, in a system with a high conformational mobility, to choose one particular geometry in order to calculate charges. A polarizable force field partially solves and accounts for this problem, but it still relies upon a certain specific geometry to provide some important quantities. The example reviewed below represents an extreme case of conformational mobility. The geometry of the isolated cationic partner is completely different from that it assumes when in contact with the anion. The resulting charge distribution is even more diverse.
Another puzzling question which has to do with the determination of partial atomic charges is whether the ionic couple has to be considered as a real entity that survives in the ionic liquid or not. Having a net entire charge on the two ionic partners is obviously a great advantage if the main objective is that of creating a force field as transferable as possible to similar compounds. It has, however, been shown that fractional charges obtained by an optimization of the gas-phase ionic pair, might improve the reliability of the following molecular dynamics runs especially in terms of diffusion coefficients. It is interesting to note that less charged partners lead to less viscous fluids and to higher diffusion coefficients. This fact has a very simple explanation: due to the relatively high density of the liquid phase of these compounds, the ionic partner are considerably in close contact during the liquid dynamics. It is also easy to predict that in an ionic pair there will always be a certain amount of charge transfer (especially in protic ionic liquids). Therefore it might be concluded that a simulation in which the two ionic partners do not have a net entire charge might be more realistic. See (Borodin & Smith, 2006) for an example where a non integer charge on the cation and the anion were matched to an ab-initio simulation.

6.2 The role of dispersion energy

Given the strong electrostatic nature of the interaction energy it is clear that the contribution of dispersion forces is of secondary importance and limited to short range interactions. There is however a secondary, but still relevant possible contribution. The presence of π electron clouds on aromatic rings on the cation, and of long alkyl chains can make the importance of London interactions more important than one would initially consider. For example in ref. (Zahn & Kirchner, 2008; Zahn et al., 2008), the authors assess the importance of London forces on the structure of selected gas phase structures of typical ionic liquids pairs. They conclude that dispersion forces can play a role in determining the structure of ionic liquids. They also make a comparison of the performance of various GGA, meta-GGA and dispersion corrected density functionals concluding that in order to obtain reliable dissociation energies one has to use dispersion corrected functionals.

6.3 Molecular dynamics

A recent edition of Account of Chemical Research (see (Maginn, 2007)) contains several reviews focussing on the theoretical problem of modeling ionic liquids. Maginn's chapter in Reviews in Computational Chemistry is also a good starting point for a general discussion of molecular modeling of ionic liquids (Maginn, 2009).

Once a force field has been chosen, a sampling of the liquid phase space has to be performed. This can be achieved both using Monte Carlo and Molecular Dynamics. In the former a series of configurations of the system are generated along a Markov Chain and a transition probability from an element to the other has to be defined in order to allow the best exploration of the phase space. Statistical averages can then be run over each of the Markov Chain elements. In a molecular dynamics trajectory calculation, one solves the Newton equations using a finite step approximation (usually employing highly efficient simplectic integrators such as Verlet, velocity Verlet or Beeman). A snapshot of the system geometry, velocity and forces is then saved at a given time-step (much larger than the integrator time-step) creating what is generally called the "trajectory". The sequence of snapshots (trajectory) is then used to calculate statistical averages.

Here follows a series of examples for various properties that can be computed from the analysis of the trajectory.

- Density: the easiest to calculate quantity and the more commonly experimentally available one. It is always a good starting point for validating a force field even if it is well known that an incorrect force field can still give rise to a correct density. So in general it is possible to accurately reproduce density while other properties are only poorly reproduced. In general OPLS based force fields underestimate slightly the experimental densities.
- Liquid structure: a very strict test of the quality of the force field used in a simulation is represented by the comparison between the structural results obtained from the trajectory and X-ray or neutron diffraction experimental data. When a good agreement between theory and experiment is found, one can be sure that the molecular dynamics simulation provides a reliable and correct description of the structural properties of the system under investigation.

  In particular, one can calculate the static structure factor either of neutrons or of X-ray (Keen, 2000) from the trajectory of the liquid phase simulation and directly compare it with the experimental one. For X-ray scattering data the radial distribution function between particle type i and j can be used, which are defined as:

$$g_{ij}(r) = \frac{n_{ij}(r)}{4\pi r^2 dr \rho_j} \quad (3)$$

  where ρ_j is the number density of particles j, and $n_{ij}(r)$ is the number of particles of type j between distance r and $r + dr$ from a particle of type i. Given c_j the molar fraction of particles j, $\rho_j = c_j \cdot \rho_0$ where ρ_0 is the total number density of the sample. The total scattering for X-ray can be then written as:

$$F(Q) = \rho_0 \sum_{i=1}^{n} c_i c_j f_{ij}(Q) \int_0^\infty 4\pi r^2 [g_{ij}(r) - 1] \frac{\sin(Qr)}{Qr} \quad (4)$$

where Q id the scattering wave-vector modulus and f_i are the atomic Q-dependent X-ray scattering factors and where

$$f_{ij}(Q) = \frac{f_i(Q)f_j(Q)}{[\sum_{i=1}^{n} c_i f_i(Q)]^2} \tag{5}$$

- Melting point: this is particularly difficult to calculate. It is almost impossible to heat the solid cell and to obtain the liquid without introducing substantial overheating due to the high nucleation barriers that have to be overcome. One possible way to perform a calculation is that of finding the temperature at which the liquid and the solid phase have the same free energy. These methods are very demanding in terms of computational power.
- Heat capacity: this is fairly easy to estimate by running a simulation at two different adjacent temperatures and by applying the finite difference formula for enthalpy:

$$C_p \sim \left[\frac{H(T+\Delta T) - H(T)}{\Delta T}\right]_P \tag{6}$$

- Enthalpy of vaporization: this is counterintuitive for ionic liquids, but they can be distilled and have a small but measurable vapor pressure. It seems that ionic liquids vaporize by forming neutral ionic pairs which leave the bulk fluid. Protic ionic liquids may also form two different molecules by acid-base reactions e.g.: $RNH_3^+ + NO_3^- \rightarrow RNH_3 + HNO_3$. Vaporization enthalpies can be calculated by extracting the enthalpy of the liquid and by subtracting a perfect gas calculation of the enthalpy of an equal number of gas phase ion pairs.
- Diffusion: self diffusion is a simple molecular property that provides quantitative information on the motion of the ions in the liquid and can be easily computed over a long enough simulation time. In particular, the diffusion coefficient, D_i, can be calculated from the mean square displacements using the Einstein relation:

$$D_i = \frac{1}{6}\lim_{t\to\infty} \frac{\left\langle \|\mathbf{r}_i(t) - \mathbf{r}_i(0)\|^2 \right\rangle}{t} \tag{7}$$

where $\mathbf{r}_i(t)$ is the position of the ion at time t while $\mathbf{r}_i(0)$ is the ion initial position. Being a dynamic property, the diffusion coefficient can only be computed by molecular dynamics methods and not by Monte Carlo techniques. As we shall see, the difficult part for an ionic liquid is that the long enough simulation time can be of the order of 10-100 ns which is hardly attainable for current hardware and sophisticated molecular entities. Diffusive regimes where

$$\beta = \frac{\Delta log(\Delta r^2)}{\Delta log(t)} \sim 1 \tag{8}$$

can be obtained with some ionic liquids only after 10 ns of simulation time. In this case only very long trajectories can provide reliable information on the diffusion coefficients of the system. As we have mentioned above, non polarizable force fields, or those force fields which imply an integer charge on the ionic pairs constituents are often under-predicting diffusion coefficients to some extent. Polarizable force fields or charge scaling procedure yielded much better results.

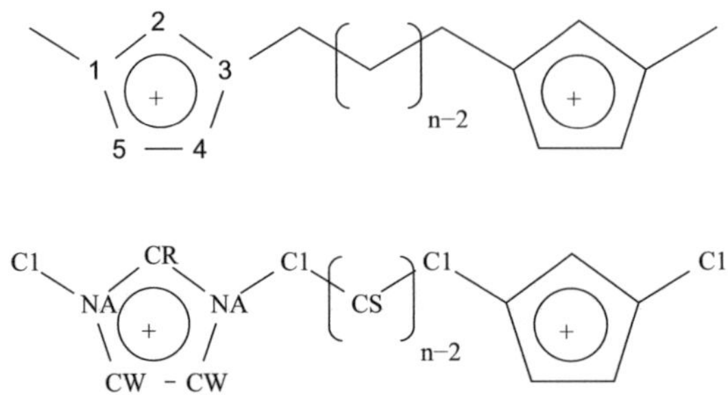

Fig. 2. Schematic structure of the $[C_n dmim][NTf_2]$ cations: n can be 3, 6, 9 and 12.

7. An example: dicationic imidazolium

A special class of ionic liquids have recently been synthetized using geminal imidazolium dications (Anderson et al., 2005; Payagala et al., 2007; Sun et al., 2009). This class of compounds represents an interesting variation of the cationic partner and may present several advantages over the traditional monocationic ionic liquid in their application as lubricants (Jin et al., 2006), solvents (Han & Armstrong, 2005; Xiao & Shreeve, 2005) and separation media (Anderson et al., 2005).
We have recently performed a theoretical study of the gas phase structure of these ionic liquids (Bodo & Caminiti, 2010) because it provides:

- a series of model structures that may be present also in the liquid phase. The ionic interactions in these complexes are often strong enough to preserve the gas-phase structure also in the liquid phase(Logotheti et al., 2009) or at least make it one of the dominant conformers;
- the electronic density and the molecular electrostatic potential which in turn can be fed into the force fields for the molecular dynamics simulations of the liquid phase through a fitting procedure generating partial atomic charges;
- the charge distribution which is of great importance for the description the structural properties of the liquid phase (see for example refs. (Li & Kobrak, 2009; Spohr & Patey, 2009)).

A schematic view of the molecular structures that we have examined is reported in figure 2: we have a linkage chain (whose length can be 3, 6, 9 or 12 for the compounds analyzed) that connects two imidazolium rings with a net positive charge on them and that are substituted with a methyl group. Various ionic liquids have been obtained by mixing these dications with Br^-, Tf_2N^-, BF_4^- and PF_6^-. These ionic liquids have thermal stabilities which are larger than their mono-cationic counterpart, their density decreases when increasing the linkage chain length (Anderson et al., 2005) while the viscosity increases. Except for those with the Tf_2N anions, most of these kind of dicationic ionic liquids (especially those with short linkage chains) are solid at room temperature and melt at temperature higher than 100°C (Anderson et al., 2005). Increasing the linkage chain length decreases the melting temperature and the C_{12} is a liquid at RT with all anion counterparts.

7.1 The minimum energy structure problem

When the number of atoms in a certain complex molecule or molecular system increases, the number of local minima of the potential energy surface (PES) quickly becomes so large that a systematic search is impossible. The study of stable geometries of large molecular systems characterized by a high conformational mobility, has to locate possible good candidates

for the time-consuming, *ab-initio* optimizations. One possible approach, in relatively small systems, is that of generating randomly (or by chemical intuition) many different starting structures and optimizing them separately. However, this route is impracticable in the present compounds because of the large number of possible initial realistic geometric guesses which are determined by the relative position of the chain and of the two anions.
We have employed a well tested force field for imidazolium based ionic liquids (Lopes et al., 2004) and we have used the *tinker* (Ren & Ponder, 2002) package in order to locate various minima on the PES. Our recipe is the following: we start with an isolated ionic complex made by the di-cation and two Tf_2N anions in close contact but in a random orientation. We run a 100 ns molecular dynamics trajectory at 150-200 K with 1 fs timestep saving a snapshot of the system each 1000 steps. We then minimize each snapshot and locate the various minimum energy structures. We select two or three structures among the lowest energy ones and we repeat the dynamics for each of these structures but with a lower temperature. After a few cycles, we are generally able to locate a recurring structure that, very likely, is the global minima for the adopted force field.
The global minima determined above has then been optimized at the CAM-B3LYP/6-31G(d) (Peach et al., 2006; Yanai et al., 2004) level using the Gaussian program (Frisch et al., 2009). Obviously, given the high number of local minima, we cannot be sure whether we are describing the global minima over the potential energy surface of these complexes or a local one. However, the molecular dynamics-based annealing procedure described above should be certainly able to produce reasonable structures so that we can consider the ab-initio analysis to describe realistically at least the key structural feature of the complexes. One should however always keep in mind that small energetic variations are certainly possible when considering the various position that the linkage chain can assume especially when it is long enough to acquire a certain conformational mobility as for the C_9 and C_{12} compounds.

7.2 The gas phase structures

Possible lowest energy structures for the linkage chains with 3, 6, 9 and 12 carbon atoms are reported in figure 3. The structures show clearly that the linkage chain "entangles" in order to maximize the interactions of the two imidazolium rings with the two anions. Crystals of $[C_2dmim][PF_6]$ have been obtained (Lee et al., 2010) and the reported structure showed a parallel arrangement of the imidazolium rings. As far as we know no solid state structures have been reported for longer linkage chains.
In general, in all the four structures, there are at least two close contacts between the "acidic" C(2)-H hydrogens and the oxygens of the Tf_2N anion (Kempter & Kirchner, 2010). For the C_3 and for the C_9 the distances we have found are 2.16 and 2.19 Å for the former and 2.08 and 2.16 Å for the latter. In the case of the C_6 and of the C_{12} instead, we have 2.32, 2.77 Å and 2.22, 2.88 Å. Moreover in the C_3 and C_9 we have that the two imidazolium rings are parallel while this is not the case for the other two compounds. Whether these geometric variations are relevant or not in determining the local structure in the liquid phase is difficult to say (see below).
Another important indicator that can help in the interpretation of the liquid phase properties is the charge distribution in the ionic complex. In the entangled configurations the distribution is rather different to what is found, for example, in other mono-imidazolium based liquids. In those systems one finds a highly polar head (the ring) connected through electrostatic interactions (and possibly through a weak, special hydrogen bond) to the anion and a non polar alkyl chain which is free to move creating an apolar tail with high rotational mobility. Here we have a rather spherical and compact structure with no or very low conformational

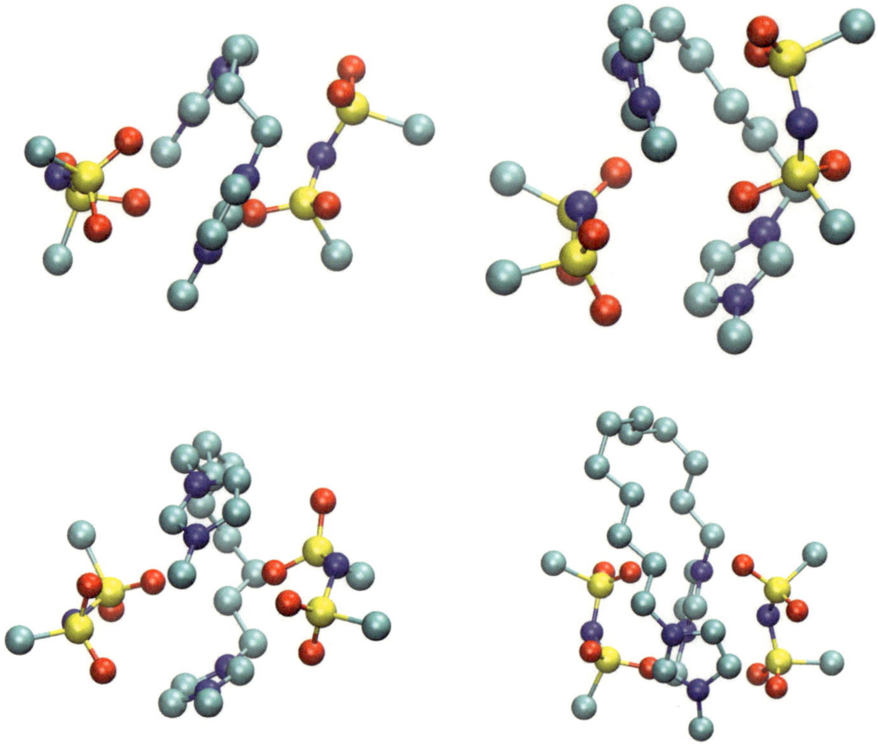

Fig. 3. Pictorial representation of the lowest energy structures associated with the 4 compounds. From the upper left corner, we have: C_3, C_6, C_9 and C_{12}. Hydrogens and Fluorine atoms have been omitted for clarity

mobility in which the charge is distributed in a quadrupolar way: two negatively charged region are counterbalanced by the two imidazolium rings and a rather apolar domain is created by the bent alkyl chain. We present the maps of the electrostatic potential for the C_9 compounds in 4. The positive charge is localized on the rings and the negative one on the oxygens of the Tf_2N.

We have run molecular dynamics liquid phase simulations using the above force field. This has been done in order to check if the compact structure survives in the liquid phase. What we have found is that the structure actually survives for the C_3 linkage chain, but is almost lost for the C_{12} one. The other two compounds present a somewhat mixed behavior. Anyway, given the lack of experimental structural studies, we have no validation of the performed dynamics and we cannot say if the adopted force field (especially the atomic charges) would lead to a realistic local liquid structure. Here we report just few of the results we have gathered on such systems. The molecular dynamics protocol we have followed is very similar to that used in our previous publication (Bodo et al., 2010). All-atom molecular dynamics simulations have been carried out on pure [$C_n$dmim][NTf_2] ionic liquids with n =3, 6, 9 or 12 using the DLPOLY2

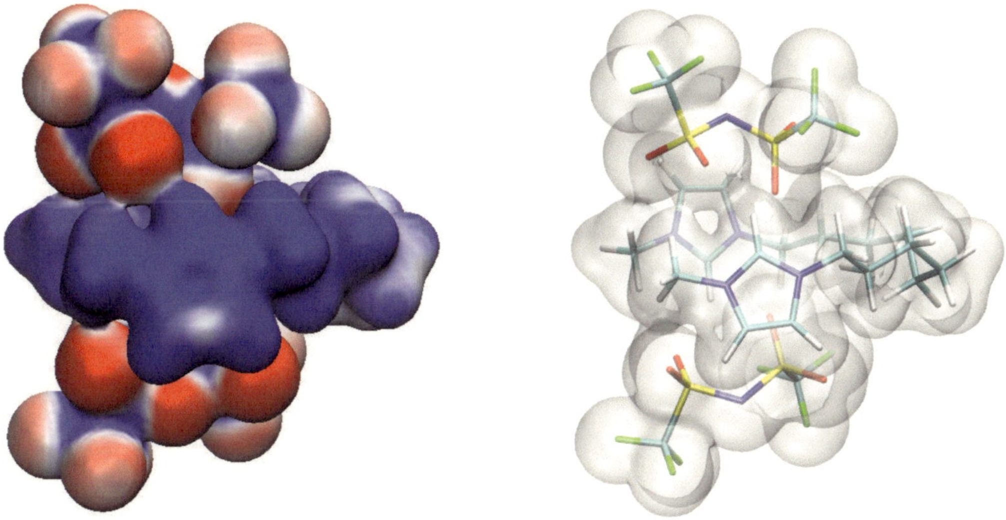

Fig. 4. Charge distribution mapped on the electronic density for the structure depicted on the right for the $C_9dmim(Tf_2N)_2$ ionic complex. Red regions are charged negatively, blue ones are charged positively.

(Smith et al., 2002) package. We have employed the force field from Canongia-Lopes and Pádua (Canongia Lopes et al., 2005; Lopes et al., 2004; Lopes & Pádua, 2004).
The initial configurations have been generated by randomly distributing 300 cations and 600 anions in very large simulation cells. A short (100 ps) isobaric-isothermal (NPT) run at 500 K and with 10^3 atmospheres pressure was carried out in order to provide densely-packed, mixed initial configurations. These initial configurations were left to relax during another NPT run of 300-500 ps at 1 atm and 300 K up to the point in which an approximately constant density was obtained. The densities we have obtained are within few percent from the experimental ones (Anderson et al., 2005). The volume was then fixed and an additional mixing was provided by a short (300 ps) NVT run at 300 K where the electrostatic interactions were turned off. The resulting configurations were finally left to evolve in a constant volume NVT production runs of about 5 ns with T = 300 K after an equilibration period.
Apart from the densities the other experimental quantities available are solid liquid transition temperatures and viscosities which are extremely expensive from the computational point of view and therefore we have not computed them.
However the agreement on density allows us to trace a few geometric considerations. One important feature of the liquid state is obviously represented by the position of the linkage chain: we have calculated the $g_{ij}(r)$ for the two lateral "C1" carbon atoms and we report it in figure 5 for the $[C_3dmim][Tf_2N]$ and $[C_9dmim][Tf_2N]$ compounds. We report in the same figure also the distances between the two lateral C1 carbon atoms in the ionic complex with two anions (smaller value) and in the isolated cation (larger value). As can be seen in both moieties the linkage in the liquid phase (at least in our simulations) seems to present a somewhat bent configuration which is not as compact as in the isolated ionic complex, nor extended as in the isolated cation. In any case the most abundant geometric conformation is the one in which the linkage chain is almost extended. Polarization effect might play an

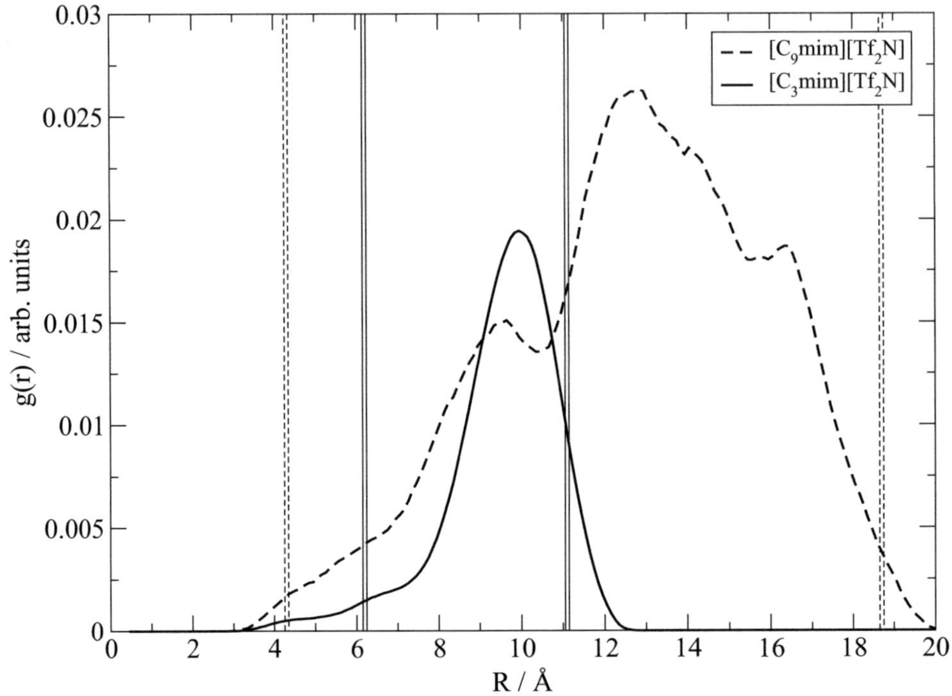

Fig. 5. Lateral C1 self radial distribution for the [C_3dmim][Tf_2N] and [C_9dmim][Tf_2N] compounds. The vertical lines identify the C1-C1 distances attained in the gas by the ionic complex with two Tf_2N or in the isolated cation.

important role in determining a different degree of entanglement of the complexes, but we were unable to verify it.

Another interesting feature of these ionic liquids is the relative position of the two imidazolium rings of the same cation. By looking at the gas phase structures above, one can see that some of the complexes show a parallel arrangement of the two imidazolium rings. We have calculated the angular distribution between imidazolium rings of the same molecule and we report it in Figure 6 for the [C_3dmim][Tf_2N] and [C_9dmim][Tf_2N] compounds. For the C_3 compound we find that the two imidazolium rings prefer a non parallel configuration in which the angle between them is around 50 degrees. We see, however, that a sizable fraction of molecules presents an angle near to zero, i.e. a parallel arrangement of the two rings.

8. Conclusions

In this Chapter we have reviewed the typical theoretical approaches that are commonly used to describe ionic liquids using *in silico* simulations. The study of these materials requires a combined approach made by ab-initio calculations and MD simulations. These techniques require a special attention in the application to ionic liquids because they of the intrinsic difficulties due to the extremely slow dynamics in the real systems. We have concluded our chapter by reporting few highlights from a recent work that we are carrying out on a special class of these compounds and we have pointed out a few interesting features we have discovered in order to provide an example of how powerful theoretical predictions can be for molecular systems which are still relatively unexplored experimentally.

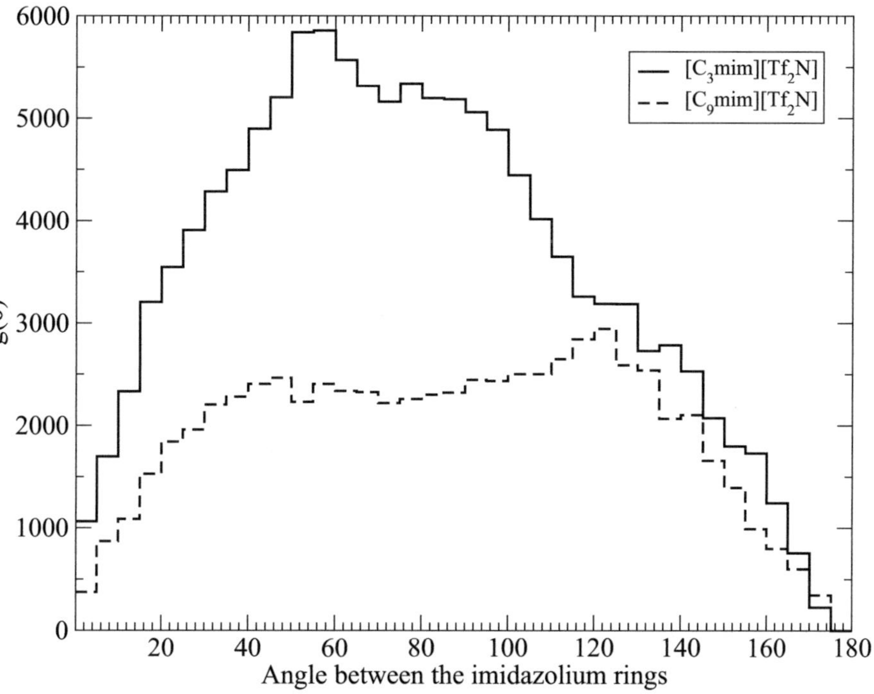

Fig. 6. Angle distribution for the two imidazolium rings of the same cation for the $[C_3dmim][Tf_2N]$ and $[C_9dmim][Tf_2N]$ compounds

9. References

Anderson, J. L., Ding, R., Ellern, A. & Armstrong, D. W. (2005). Structure and properties of high stability geminal dicationic ionic liquids, *J. Am. Chem. Soc.* 127: 593.

Bodo, E. & Caminiti, R. (2010). *J. Phys. Chem. A* 114: 12506.

Bodo, E., Gontrani, L., Caminiti, R., Plechkova, N. V., Seddon, K. R. & Triolo, A. (2010). Structural properties of 1-alkyl-3-methylimidazolium bis(trifluoromethyl)sulfonylamide ionic liquids: X-ray diffraction data and molecular dynamics simulations, *J. Phys. Chem. B* 114(49): 16398–16407.

Borodin, O. (2009). Polarizable force field development and molecular dynamics simulations of for iionic liquids, *J. Phys. Chem. B* 113: 11463.

Borodin, O. & Smith, G. D. (2006). Structure and dynamics of n-methyl-n-propylpyrrolidinium bis(trifluoromethanesulfonyl)imide ionic liquid from molecular dynamics simulations., *J. Phys. Chem. B* 110: 11481–11490.

Canongia Lopes, J. N. A. & Pádua, A. A. H. (2006a). Using spectroscopic data on imidazolium cation conformations to test a molecular force field for ionic liquids, *J. Phys. Chem. B* 110: 7485–7489.

Canongia Lopes, J. N., Deschamps, J. & Pádua, A. A. H. (2005). Modeling ionic liquids of the 1-alkyl-3-methylimidazolium family using an all-atom force field., *ACS Symp. Ser.* 901: 134–149.

Canongia Lopes, J. N. & Pádua, A. A. H. (2006b). Molecular force field for ionic liquids iii: Imidazolium, pyridinium, and phosphonium cations; chloride, bromide, and dicyanamide anions., *J. Phys. Chem. B* 110: 19586–19592.

Fei, Z., Geldbach, T. J., Zhao, D. & Dyson, P. J. (2006). From dysfunction to bis-function: On the design and applications of functionalised ionic liquids, *Chemistry Eur. J.* 12: 2122–2130.

Fischer, T., Sethi, A., Welton, T. & Woolf, J. (1999). Diels-alder reactions in room-temperature ionic liquids., *Tetrahedron Lett.* 40: 793–796.

Forsyth, S. A., Pringle, J. M. & MacFarlane, D. R. (2004). Ionic liquids-an overview., *Aust. J. Chem.* 57: 113–119.

Freemantle, M. (1998). *Chem . Eng. News.* 76: 32.

Freemantle, M. (2009). *An Introduction to Ionic Liquids*, RSC Publishing.

Frisch, M. J., Trucks, G. W., Schlegel, H. B., Scuseria, G. E., Robb, M. A., Cheeseman, J. R., Scalmani, G., Barone, V., Mennucci, B., Petersson, G. A., Nakatsuji, H., Caricato, M., Li, X., Hratchian, H. P., Izmaylov, A. F., Bloino, J., Zheng, G., Sonnenberg, J. L., Hada, M., Ehara, M., Toyota, K., Fukuda, R., Hasegawa, J., Ishida, M., Nakajima, T., Honda, Y., Kitao, O., Nakai, H., Vreven, T., Montgomery, Jr., J. A., Peralta, J. E., Ogliaro, F., Bearpark, M., Heyd, J. J., Brothers, E., Kudin, K. N., Staroverov, V. N., Kobayashi, R., Normand, J., Raghavachari, K., Rendell, A., Burant, J. C., Iyengar, S. S., Tomasi, J., Cossi, M., Rega, N., Millam, J. M., Klene, M., Knox, J. E., Cross, J. B., Bakken, V., Adamo, C., Jaramillo, J., Gomperts, R., Stratmann, R. E., Yazyev, O., Austin, A. J., Cammi, R., Pomelli, C., Ochterski, J. W., Martin, R. L., Morokuma, K., Zakrzewski, V. G., Voth, G. A., Salvador, P., Dannenberg, J. J., Dapprich, S., Daniels, A. D., Farkas, O., Foresman, J. B., Ortiz, J. V., Cioslowski, J. & Fox, D. J. Gaussian 09 Revision A.1. Gaussian Inc. Wallingford CT 2009.

Fujita, K., Nakamura, N., Igarashi, K., Samejima, M. & Ohno, H. (2009). Biocatalytic oxidation of cellobiose in an hydrated ionic liquid, *Green Chem.* 11: 351–354.

Galinski, M., Lewandowski, A. & Stepniak, I. (2006). Ionic liquids as electrolytes., *Electrochim. Acta* 51: 5567–5580.

Garcia, B., Lavallee, S., Perron, G., Michot, C. & Armand, M. (2004). Room temperature molten salts as lithium battery electrolyte., *Electrochim. Acta* 49: 4583–4588.

Gaune-Escard, M. & Seddon, K. R. (eds) (2010). *Molten Salts and Ionic Liquids: Never the Twain?*, Wiley.

Han, X. & Armstrong, D. W. (2005). Using geminal dicationic ionic liquids as solvents for high-temperature organic reactions, *Org. Lett.* 7: 4205.

Hanke, C., Price, S. & Lynden-Bell, R. (2001). *Mol. Phys.* 99: 801.

Hough, W. L. & Rogers, R. D. (2007). Ionic liquids then and now: from solvents to materials to active pharmaceutical ingredients., *Bull. Chem. Soc. Jpn.* 80: 2262–2269.

Hough, W. L., Smiglak, M., Rodriguez, H., Swatloski, R. P., Spear, S. K., Daly, D. T., Pernak, J., Grisel, J. E., Carliss, R. D., Soutullo, M. D., Davis, Jr., J. H. & Rogers, R. D. (2007). The third evolution of ionic liquids: Active pharmaceutical ingredients., *New J. Chem.* 31: 1429–1436.

Hu, Z. & Margulis, C. J. (2007). Room-temperature ionic liquids: Slow dynamics, viscosity, and the red edge effect., *Acc. Chem. Res.* 40: 1097–1105.

Jin, C.-M., Ye, C., Phillips, B. S., Zabinski, J. S., Liu, X., Liu, W. & Shreeve, J. M. (2006). Polyethylene glycol functionalized dicationic ionic liquids with alkyl or polyfluoroalkyl substituents as high temperature lubricants, *J. Mater. Chem.* 16: 1529.

Jorgensen, W. L., Maxwell, D. S. & Tirado-Rives, J. (1996). Development and Testing of the OPLS All-Atom Force Field on Conformational Energetics and Properties of Organic Liquids, *J. Am. Chem. Soc.* 118: 11225–11236.

Keen, D. A. (2000). A comparison of various commonly used correlation functions for describing total scattering, *Appl. Cryst.* 34: 172.

Kempter, V. & Kirchner, B. (2010). The role of hydrogen atoms in interactions involving imidazolium-based ionic liquids, *J. Mol. Struct.* 972: 22.

Ködderman, T., Paschek, D. & Ludwig, R. (2007). Molecular dynamic simulations of ionic liquids: A reliable description of structure, thermodynamics and dynamics, *ChemPhysChem* 8: 2464.

Lee, M., Niu, Z., Slabodnick, C. & Gibson, H. W. (2010). Structure and properties of n,n-alkilene bis(n′-alkylimidazolium) salts, *J. Phys. Chem. B* 114: 7312.

Li, H. & Kobrak, M. N. (2009). A molecular dynamics study of the influence of ionic charge distribution on the dynamics of a molten salt, *J. Chem. Phys.* 131: 194507.

Lipkowitz, K. B., Larter, R. & Cundari, T. (eds) (2009). *Atomistic simulations of ionic liquids*, Hoboken, NJ: Wiley.

Liu, Y., Shi, L., Wang, M., Li, Z., Liu, H. & Li, J. (2005). A novel room temperature ionic liquid sol-gel matrix for amperometric biosensor application., *Green Chem.* 7: 655–658.

Logotheti, G.-E., Ramos, J. & Economou, I. G. (2009). Molecular modeling of imidazolium based [Tf_2N^-] ionic liquids: Microscopic structure thermodynamic and dynamic properties and segmental dynamics, *J. Phys. Chem. B* 113: 7211.

Lopes, J. N. C., Deschamps, J. & Pádua, A. A. H. (2004). Modeling ionic liquids using a systematic all-atom force field, *J. Phys. Chem. B* 108: 2038.

Lopes, J. N. C. & Pádua, A. A. H. (2004). Molecular force field for ionic liquids composed of triflate or bistriflylimide anions, *J. Phys. Chem. B* 108: 16893.

Lynden-Bell, R. M., Atamass, N. A., Vasilyuk, A. & Hanke, C. G. (2002). Chemical potentials of water and organic solutes in imidazolium ionic liquids: a simulation study, *Mol. Phys.* 100: 3225–3229.

MacFarlane, D. R., Forsyth, M., Howlett, P. C., Pringle, J. M., Sun, J., Annat, G., Neil, W. & Izgorodina, E. I. (2007). Ionic liquids in electrochemical devices and processes: Managing interfacial electrochemistry, *Accounts of Chemical Research* 40(11): 1165–1173.

Maginn, E. J. (2007). Atomistic simulation of the thermodynamic and transport properties of ionic liquids., *Acc. Chem. Res.* 40: 1200–1207.

Maginn, E. J. (2009). Molecular simulations of ionic liquids: current ststus and future opportunities, *J. Phys. Cond. Matt.* 21: 373101.

Morrow, T. & Maginn, E. (2002). *J. Phys. Chem. B* 106: 12807.

Payagala, T., Huang, J., Breitbach, Z. S., Sharma, P. S. & Armstrong, D. W. (2007). Unsymmetrical dicationic ionic liquids: Manipulation of physicochemical propoerties using pecific structural architectures, *Chem. Mater.* 19: 5848.

Peach, M. J. G., Helgaker, T., Sałek, P., Keal, T. W., Lutnæs, O. B., Tozer, D. J. & Handy, N. C. (2006). Assessment of a coulomb-attenuated exchange–correlation energy functional, *Phys. Chem. Chem. Phys.* 8: 558.

Plechkova, N. & Seddon, K. (2008a). Applications of ionic liquids in the chemical industry, *Chem. Soc. Rev.* 37: 123.

Plechkova, N. V. & Seddon, K. R. (2008b). Applications of ionic liquids in the chemical industry, *Chem. Soc. Rev.* 37: 123–150.

Polshettiwar, V. & Varma, R. S. (2008). Microwave-assisted organic synthesis and transformations using benign reaction media, *Accounts of Chemical Research* 41: 629–639.

Ren, P. & Ponder, J. W. (2002). *J. Comput. Chem.* 23: 1487.

Rogers, R. D., Plechkova, N. V. & Seddon, K. R. (eds) (2009). *Ionic Liquids: From Knowledge to Application*, Vol. 1030 of *ACS Symp. Ser.*, American Chemical Society, Washington D.C.

Rogers, R. D. & Seddon, K. R. (2003). Ionic liquids–solvents of the future?, *Science* 302(5646): 792–793.

Rogers, R. & Seddon, K. (eds) (2005a). *Ionic Liquids IIIA: Fundamentals, Progress, Challenges, and Opportunities: Properties and Structure*, Vol. 901 of *ACS Symp. Ser.*, American Chemical Society, Washington D.C.

Rogers, R. & Seddon, K. (eds) (2005b). *Ionic Liquids IIIB: Fundamentals, Progress, Challenges, and Opportunities: Transformations and Processes*, Vol. 902 of *ACS Symp. Ser.*, American Chemical Society, Washington D.C.

Smith, W., Yong, C. W. & Rodger, P. M. (2002). Dl poly: Application to molecular simulation, *Molecular Simulation* 28: 892–7022.

Spohr, H. V. & Patey, G. N. (2009). Structural and dynamical properties of ionic liquids: The influence of charge location, *J. Chem. Phys.* 130: 104506.

Sun, H., Zhang, D., Liu, C. & Zhang, C. (2009). Geometrical and electronic structures of the dication and ion pair in the geminal dicationic ionic liquid 13-bis[3-methylimidazolium]propane bromide, *Theochem* 900: 37.

Walden, P. (1914a). *Bull. Acad. Imp. Sci. St.-Petersbourg* 8: 405.

Walden., P. (1914b). *Chem. Zentralbl.* p. 1800.

Wang, P., Zakeeruddin, S. M., Exnar, I. & Gratzel, M. (2002). High efficiency dye-sensitized nanocrystalline solar cells based on ionic liquid polymer gel electrolyte, *Chem. Commun.* pp. 2972–2973.

Wang, Y. & Voth, G. A. (2005). Unique spatial heterogeneity in ionic liquids., *J. Am. Chem. Soc.* 127: 12192–12193.

Wang, Y. & Voth, G. A. (2006). Tail aggregation and domain diffusion in ionic liquids., *J. Phys. Chem. B* 110: 18601–18608.

Welton, T. & Wasserscheid, P. (eds) (2008). *Ionic Liq. Synth.*, 2nd edn, Wiley VCH, Weinheim.

Wilkes, J. S. (2002). A short history of ionic liquids-from molten salts to neoteric solvents, *Green Chem.* 4: 73–80.

Wilkes, J. S. & Zaworotko, M. J. (1992). Air and water stable 1-ethyl-3-methylimidazolium based ionic liquids, *J. Chem. Soc., Chem. Commun.* pp. 965–967.

Wolfbauer, G., Bond, A., Eklund, J. C. & Macfarlane, D. (2001). *Sol. Energy Mater. Sol. Cells* 70: 85–101.

Xiao, J.-C. & Shreeve, J. M. (2005). Synthesis of 2,2′-biimidazolium-based ionic liquids: Use as a new reaction medium and ligand for palladium-catalyzed suzuki cross-coupling reactions, *J. Org. Chem.* 70: 3072.

Xu, W. & Angell, C. A. (2003). Solvent-free electrolytes with aqueous solution-like conductivities, *Science* 302(5644): 422–425.

Yanai, T., Tew, D. & Handy, N. (2004). A new hybrid exchange-correlation functional using the coulomb-attenuating method (cam-b3lyp), *Chem. Phys. Lett.* 393: 51.

Zahn, S. & Kirchner, B. (2008). Validation of dispersion corrected functional theory for ionic lquid systems, *J. Phys. Chem. A* 112: 8430.

Zahn, S., Uhlig, F., Thar, J., Spickermann, C. & Kirchner, B. (2008). Intermolecular forces in an ionic liquid ([mmim][cl]) versus those in a typical salt (nacl), *Angewandte Chemie International Edition* 47(19): 3639–3641.

6

Classical Density Functional Theory of Ionic Liquids

Jan Forsman[1], Ryan Szparaga[1], Sture Nordholm[2],
Clifford E.Woodward[3] and Robert Penfold[4]
[1]*Lund University*
[2]*Gothenburg University*
[3]*University of New South Wales, ADFA*
[4]*Institute of Food Research*
[1,2]*Sweden*
[3]*Australia*
[4]*UK*

1. Introduction

The current technological revolution based on ionic liquids (ILs) is driven by their unique properties. Though ILs have been known to scientists for close to a century (Walden , 1914), they are now the focus of intense activity mainly because of their promise for "environmentally friendly" applications. Some actual and potential uses of ILs include: specific solvents for heterogeneous and homogeneous catalysis; selective solvents for removal of heavy metal contaminants; electrolytes in various electrochemical processes and devices; and as dispersive agents for stabilization of nanoparticles. In all of these applications, the structural properties of ILs and their mixtures in the bulk phase and at interfaces are crucial to their performance (Maier et al. , 2010). In electrowinning processes, such as aluminium refining, metal ion speciation is a crucial consideration in determining the population of electroactive components (Rocher et al. , 2009) and consequently influences the pertinent dynamical processes that lead to oxidation and reduction. These populations are ultimately governed by the free energy of the IL solution, which in turn depends upon the mutual interactions between the IL molecules and the added components.

It is now clear that fundamental theoretical studies of structure-property relationships in ILs will provide important new insights, which will assist this kind of research. Not only in the choice and design of ILs for a specific application, but also for scenario based modelling of IL processes. For example, theoretical studies of the structure of ILs at charged surfaces are important for understanding the properties that they impart to electrodeposition processes, including factors such as ionic speciation and mass transport of electroactive species (MacFarlane et al. , 2010). As emphasized by Kornyshev in a recent review article (Kornyshev , 2007), applications of ILs at electrified interfaces to energy-storage systems, electrowetting devices or nanojunction gating media will be strongly promoted by a deeper understanding of the structure and properties of the interfacial double layer.

The importance of theoretical investigations of ILs is also manifested in the rapid accumulation of such studies in the scientific literature. This includes a large number of fully atomistic simulations of ILs, which first appeared some ten years ago (Hanke et al. , 2001; Lynden-Bell , 2010; Lynden-Bell et al. , 2007). These reports have provided IL researchers with valuable physical and chemical information with which to rationalize many experimental observations. By deploying numerically intensive Monte Carlo or Molecular Dynamics algorithms, computer simulations circumvent the technical difficulties inherent to dense liquids with strong electrostatic coupling that have hampered the development of a fundamental theory of ILs based upon statistical mechanical formalism. On the other hand, a fundamental theory of this type does exist for traditional aqueous electrolytes wherein ionic concentrations are generally low and electrostatic coupling is much weaker than in ILs. In particular, a simple mean-field theory such as the Poisson-Boltzmann Approximation (PBA) (Gouy , 1910) can be successfully applied. Calculations using the PBA are computationally rapid and reasonably accurate for traditional electrolytes. Even though seminal work by Jönsson and coworkers (Gulbrand et al. , 1984) has shown that the PBA can fail significantly for divalent ions, the PBA nevertheless remains a useful cornerstone theory for aqueous electrolytes and has provided much of the conceptual understanding in those systems.

Conditions that are found in typical ILs would normally cause a simple electrolyte to crystallize. Ionic liquids remain fluid, however, due to steric hindrance accompanying their larger and more complex molecular shapes. Thus, in ILs the steric length scale becomes relevant and there is strong interplay between electrostatic and steric interactions. Under these conditions the PBA fails. This has left something of a conceptual gap in the theoretical treatment of ILs which has been largely filled with detailed all-atom simulations. Unfortunately, these simulations require large computational resources and are not practical for rapid, exploratory calculations. Moreover, the simulation results may be difficult to interpret since important mechanisms may be hidden in the complex numerical calculations. In this chapter we describe an alternative theoretical approach that aims to play a similar role for ILs as the PBA does for traditional electrolytes. As with the PBA, this theory will be mathematically concise (albeit approximate) but contain sufficient corrections to the PBA to render it applicable for ILs. In order to make analytical progress, it will be necessary to use simpler coarse-grained models of ILs than those currently implemented in all-atom simulations. Though this will lead to an inevitable loss of detailed information, we anticipate that this will be outweighed by the value of an approximate but mechanistically more explicit and numerically less demanding theory for ILs. The advantages of such a theory to researchers in the field are both practical and technical:

- the ability to do rapid calculations will facilitate more ambitious large-scale scenario modelling, which allows us to reasonably cheaply explore a wide range of systems, inaccessible to all-atom simulations;
- mathematically concise, closed expressions for entropy related quantities (which are difficult to obtain in simulations) will allow identification of general and powerful mechanistic principles, that can be used to describe and organise large classes of phenomena in ILs.

2. Other approaches

ILs are fascinating from a theoretical standpoint. In order to treat ILs, a theory must account for both electrostatic and steric correlations. Attempts to develop such a theory have appeared in the literature. These include the lattice based model by Kornyshev and co-workers (Fedorov et al. , 2008; Kornyshev , 2007). In another lattice approach by Lauw and coworkers (Lauw et al. , 2009), both cations and anions were modelled by a star-like architecture, with each monomer subject to an independently specified dielectric response. Under certain circumstances, these models were able to qualitatively reproduce the so-called "camel-shaped" differential capacitance, C_D, profile observed in imidazolium ILs at carbon electrodes (Islam et al. , 2009; Lockett et al. , 2008a). This behaviour is an indication of subtle structuring effects at charged surfaces and is a stern test for theory. Although camel-shaped C_D profiles are predicted, qualitative discrepancy against experimental results remains. This is not surprising since these crude theories lack a proper description of electrostatic correlations despite accounting for steric contributions. Charge-charge correlations have been shown to be important in interfacial phenomena of highly coupled electrostatic fluids. As with dispersion forces, ionic correlations are always attractive and lower the total energy. Thus, we expect that they will also play a role in establishing the camel shape of the C_D. That this is the case, is seen from both simulations (Lamperski & Zydor , 2007; Vatamanu et al. , 2010) and modified Poisson-Boltzmann treatments (Lamperski et al. , 2009) of a restricted primitive model (RPM) electrolyte. These investigations show that the RPM displays a camel-shaped C_D at low temperatures and densities. However, electrostatic correlations do not tell the whole story for ILs. That is, for the RPM the C_D becomes bell-shaped and exhibits a single maximum at densities and temperatures typical for ILs. This suggests that the simple RPM is an inadequate model for ILs and other factors, such as dispersion forces and molecular structure (Fedorov et al. , 2010; Trulsson et al. , 2010), play an important role in determining their properties. In this work, we will use a model for ILs wherein the molecular ions interact via dispersion, as well as electrostatic, forces and the cation has an oligomeric structure consistent with imidazolium ions. This model has also been adopted by us in previous recent publications (Forsman et al. , 2011; Trulsson et al. , 2010), and employs the methods of polymer free energy density functional theory (DFT) (Woodward , 1991). The result will be a simple, albeit approximate DFT for ILs. We have already introduced the salient features of the theory in a previous publication (Forsman et al. , 2011). Here, we will present a much more thorough treatment, emphasizing the relation to theories of simpler fluids. We will also extend the scope of the study with some emphasis on temperature dependence and the effects of ion-specific adsorption. The IL-DFT that we will discuss here is an off-lattice theory, whose accuracy can be straightforwardly tested against computer simulations of the same molecular model. We will begin with a discussion of relevant earlier work, in order to introduce the essential mechanistic components that will be combined in our density functional theory for ionic liquids.

3. Classical density functional theory (DFT) of simple fluids

The ideal gas law is known and used by most physicists and chemists but a more complete description of a fluid including its ability to condense into a liquid requires a treatment of the interactions between its particles. The simplest realistic model of a molecular fluid assumes pairwise spherically symmetric interactions by a pair-potential $\phi(r)$ which has a repulsive (positive) inner region characterised by a length scale σ and an attractive tail that defines an energy scale ϵ. Such models are referred to as simple fluids. The most popular form of simple fluid pair-potential is the Lennard-Jones potential

$$\phi(r) = 4\epsilon\left(\left(\frac{\sigma}{r}\right)^{12} - \left(\frac{\sigma}{r}\right)^{6}\right) . \tag{1}$$

We shall let this familiar pair-potential illustrate our simple fluid in the discussion below. Even for a simple fluid, the the analysis quickly becomes quite complicated if we attempt to describe configurational states by explicitly locating all the particles as in simulation methods, or if we try to resolve the coupled hierarchy of correlation functions as in rigorous liquid state theory. Here we shall instead consider an approach based on deliberate and efficient approximations seeking to represent the essential features while hiding the less important details of the fluid properties. The most noteworthy early attempt in this direction was taken by J. van der Waals (van der Waals , 1873) who 140 years ago realized that the interactions in molecular fluids introduced two mechanisms: i) an excluded volume effect due to the size of the molecules and ii) a binding energy due to the attraction between molecules. His equation of state relating pressure P to volume per particle v and absolute temperature T reads

$$P = \frac{k_B T}{v - v_0} - \frac{a}{v^2} . \tag{2}$$

It incorporates an excluded volume v_0 per particle and a binding energy per particle $e_b = a/v$ due to the attractive interactions. Following his reasoning in the simplest way for the Lennard-Jones potential one would assume that particles do not approach closer than $r = \sigma$ and therefore exclude on the average the volume $v_0 = \frac{1}{2} \times 4\pi\sigma^3/3 = 2\pi\sigma^3/3$ per particle as evaluated in the low density limit. The attractive part of the potential is in van der Waals (vdW) theory assumed to be sampled without inducing any correlations, i.e. in the mean-field approximation so that the binding energy parameter becomes:

$$a = \frac{1}{2}\int_{\sigma}^{\infty} dr\, 4\pi r^2 \phi(r) = \frac{16\pi}{9}\epsilon\sigma^3 , \tag{3}$$

for the Lennard-Jones interaction. The corresponding configurational free energy per particle can be taken to be $f_c(v) = -k_B T \ln(v - v_0) - e_b = -k_B T \ln v_{\text{free}} - e_b$, where v_{free} is the free volume available to each particle after the excluded volume effect has been accounted for, i.e. $v_{\text{free}} = V/N - v_0$.

The origin of density functional theory can be traced back to the famous failure of the vdW equation of state in its prediction of unphysical "wiggles" in $P(v, T)$ where the pressure could increase with increasing volume and even become negative. The Maxwell reconstruction (Maxwell , 1875) eliminates these wiggles and is based on interpreting the vdW equation of state as the description of a uniform fluid with density $n(\mathbf{r}) = n_b = 1/v$ everywhere in the available volume. The total configurational free energy can then be obtained as

$$\begin{aligned}\mathcal{F}_c[n(\mathbf{r})] = \mathcal{F}_c(n_b) &= \int_V d\mathbf{r}\, n(\mathbf{r}) f_c(\mathbf{r}) = \int_V d\mathbf{r}\, n(\mathbf{r})\left(k_B T \ln\left(\frac{n(\mathbf{r})}{1 - n(\mathbf{r})v_0}\right) - a n(\mathbf{r})\right) \\ &= N\left(k_B T \ln\left(\frac{n_b}{1 - n_b v_0}\right) - a n_b\right) .\end{aligned} \tag{4}$$

Now the Maxwell reconstruction can be understood by recognising that the uniform fluid is not the stable state. Instead the fluid separates into a rarefied gas phase (of density $n_b^{(g)}$ and volume $V^{(g)}$) coexisting with a condensed liquid phase of density $n_b^{(l)} > n_b^{(g)}$ and volume $V^{(l)} < V^{(g)}$. Since the total volume and particle number are conserved we have the two relations: $V^{(g)} + V^{(l)} = V$ and $N^{(g)} + N^{(l)} = n_b^{(g)} V^{(g)} + n_b^{(l)} V^{(l)} = N$, which constrain the search for the most stable form of the fluid. At fixed T, minimizing the configurational free energy with respect to the choice of $n_b^{(g)}$ and $n_b^{(l)}$ will then produce the familiar Maxwell reconstructed isotherms with flat segments (tielines) replacing the wiggles.

3.1 The generalized van der Waals (GvdW) theory

The analysis of the traditional van der Waals theory above shows that:

i) equations of state can be understood as approximate local descriptions of energy and entropy in a fluid of given density and temperature;

ii) spatially variable densities allow the state of a fluid to be conveniently described.

In the treatment of bulk multiphase fluids as above one can assume that the variation of the density is slow on the length scale of either the repulsive or attractive range of interaction. This is certainly true well inside each phase but not neccessarily true in the interface between phases. Such interfaces are generally of minor extent ($N^{2/3}$) compared to the bulk phases (N) so that their effect can be neglected in the thermodynamic limit. We may also have slow density variations due to external fields. An example is our atmosphere which is held in place by the gravitational field. We could describe the atmosphere very well with a vdW functional as above parametrized by excluded volume and binding energy.

We now consider two improvements of the vdW functional. First we would like to improve the accuracy of the vdW equation of state for simple fluids over their full density range. In liquids, it turns out that the excluded volume per particle v_0 is severely overestimated by its low density limit value $2\pi\sigma^3/3$ as used above. Consideration of v_0 as an estimate of the volume per particle in the close packing limit shows that this volume would be just σ^3 in the simple cubic packing arrangement where each hard sphere of diameter σ is assigned a cubic volume in which it fits touching all sides. Thus, setting $v_0 = \sigma^3$ will improve the ability of the vdW equation of state to describe the full range of fluid densities.

Second, we shall restrict the assumption of local interactions to the short range repulsions (i.e., the excluded volume effect) but give the attractive interactions a nonlocal representation. The GvdW functional for our Lennard-Jones fluid becomes (Nordholm & Haymet , 1980):

$$\mathcal{F}_c[n(\mathbf{r})] = k_B T \int_V d\mathbf{r}\, n(\mathbf{r}) \ln\left(\frac{n(\mathbf{r})}{1 - n(\mathbf{r})\sigma^3}\right) + \int_V d\mathbf{r}\, n(\mathbf{r}) \left(\frac{1}{2}\int_V d\mathbf{r}'\, n(\mathbf{r}')\phi_{att}(\mathbf{r},\mathbf{r}') + V_{\text{ext}}(\mathbf{r})\right) . \tag{5}$$

Here the attractive part of the pair-potential is defined as

$$\phi_{att}(\mathbf{r},\mathbf{r}') = \begin{cases} 0 \, , & |\mathbf{r}-\mathbf{r}'| \leq \sigma \\ \phi(|\mathbf{r}-\mathbf{r}'|) \, , & |\mathbf{r}-\mathbf{r}'| > \sigma \end{cases} , \tag{6}$$

and we have added an energy due to an external field $V_{\text{ext}}(\mathbf{r})$. This functional can treat bulk phase equilibria as well as the main mechanisms of adsorption at a wall and gas-liquid surface tension in a simple fluid of this type (Johnson & Nordholm , 1981; Nordholm & Gibson , 1981).

3.1.1 The fine-grained GvdW theory

The key to many phenomena in nonuniform fluids is the nonlocality of interactions, that is the nonvanishing range of interaction on the length scale of density variations. From complete neglect of this nonlocality in the vdW analysis of phase separation we have proceeded above to account for the nonlocality of the pairwise attractive interaction leaving only the shorter range repulsive interactions to local approximation. There are, however, many applications where the nonlocality of the repulsive interactions is also important. We know, for example, that at higher densities hard spheres pack into distinct layers at hard walls. Such packing structures are not predicted by theories based on local representation of excluded volume effects. Moreover, it is clear that repulsive nonlocality will enter also into an accurate prediction of gas-liquid surface tension. For this reason fine-grained GvdW theory has been developed to allow both attractive and repulsive interactions to be treated in a nonlocal representation. The key step is to realize that the excluded volume effect is related to an average particle density over a spatial domain determined by the range of the repulsive interactions. If we, as above, treat the particles as interacting hard spheres of diameter σ then the simplest coarse-graining average reflecting the domain of the repulsive interactions is

$$\overline{n}(\mathbf{r}) = \frac{3}{4\pi\sigma^3} \int_{|\mathbf{r}-\mathbf{r}'|<\sigma} d\mathbf{r}' \, n(\mathbf{r}') \ . \tag{7}$$

Thus we assume that the excluded volume effect is related to $\overline{n}(\mathbf{r})$ while the density itself is $n(\mathbf{r})$ which may be considerably more finely structured. The free energy density functional is then

$$\mathcal{F}_c[n(\mathbf{r})] = k_B T \int_V d\mathbf{r} \, n(\mathbf{r}) \ln\left(\frac{n(\mathbf{r})}{1-\overline{n}(\mathbf{r})\sigma^3}\right) + \int_V d\mathbf{r} \, n(\mathbf{r}) \left(\frac{1}{2}\int_V d\mathbf{r}' \, n(\mathbf{r}')\phi_{att}(\mathbf{r},\mathbf{r}') + V_{\text{ext}}(\mathbf{r})\right) \ . \tag{8}$$

This functional can now describe hard sphere structuring at hard walls and related phenomena (Freasier et al. , 1989; Nordholm et al. , 1980).

Other equations of state, and coarse-graining approaches, have been suggested. For instance, Tarazona and Evans (Tarazona & Evans , 1984) based their free energy functional on the Carnahan-Starling (Carnahan & Starling , 1969) equation of state for hard spheres which yields a configurational free energy per particle

$$f_c = k_B T \left[\ln n_b - 1 + \frac{\eta(4-3\eta)}{(1-\eta)^3}\right] \ , \tag{9}$$

in terms of the volume fraction $\eta = \pi\sigma^3 n_b/6$. The "weighting" of the particle density was done by Tarazona (Tarazona , 1984) according to

$$\overline{n}(\mathbf{r}) = \int d\mathbf{r}' \, w(|\mathbf{r}-\mathbf{r}'|; n(\mathbf{r}))n(\mathbf{r}') \ , \tag{10}$$

$$w(r) = w_0(r) + w_1(r)n + w_2(r)n^2 \ , \tag{11}$$

where the weighting functions w_i are determined to fit the direct correlation function and the radial distribution function for a uniform hard sphere fluid as described by the Percus-Yevick integral equation theory.

4. Density functional theory (DFT) of simple ionic systems

The simple fluids considered above are all characterized by short range interactions where the second virial coefficient $B_2(T)$ is finite. In the case of ionic systems the Coulomb energy of two point charges q_i and q_j separated by a distance r reads

$$\phi_{ij}^{C}(r) = \frac{q_i q_j}{4\pi\varepsilon_0\varepsilon r} \ , \tag{12}$$

where ε_0 is the vacuum permeability and ε is the relative dielectric constant. Hence, the range of interaction is infinite and $B_2(T)$ diverges. This means that interactions have a more profound effect on these systems and the correlations between the ions play a more fundamental role than in the simple fluids discussed above.

4.1 The one-component plasma (OCP)

The simplest IL is the one-component plasma (OCP) where one of two components is treated as an inactive neutralizing background charge while the other is an active ionic component. An example could be a molten metal where the metal ions are taken to move in a background electron gas which is in its ground state. For active ions of magnitude q and density n_b the bulk charge density is qn_b while the neutralizing background charge density is $-qn_b$. In the absence of correlations, it follows that the system would behave as an ideal gas with zero potential energy on average. The active ions do repel each other, however, so that compact configurations are suppressed, the average potential energy is negative and the OCP is stable. The long range nature of the interaction means each ion is surrounded by a like-charge-depletion zone (a hole) with a shape that exactly neutralizes its charge q. Thus correlations play a fundamental role in the OCP. The GvdW free energy density functional for the ion density around a central ion can be written as

$$\mathcal{F}_c[n(\mathbf{r})] = k_B T \int_V d\mathbf{r}\, n(\mathbf{r}) \ln n(\mathbf{r}) + \frac{q^2}{4\pi\varepsilon_0\varepsilon} \int_V d\mathbf{r}\, \Delta n(\mathbf{r}) \left(\frac{1}{2} \int_V d\mathbf{r}' \frac{\Delta n(\mathbf{r}')}{|\mathbf{r}-\mathbf{r}'|} + \frac{1}{|\mathbf{r}|} \right) \ . \tag{13}$$

Here $\Delta n(\mathbf{r}) = n(\mathbf{r}) - n_b$ is the deviation of the ion density from its bulk value. On introducing the Lagrange multiplier α to accommodate the canonical constraint of fixed particle number, the variational derivative of this functional yields

$$\frac{\delta \mathcal{F}_c}{\delta n} = k_B T (\ln n(r) + 1) + \alpha + q\psi(r) \ , \tag{14}$$

$$\psi(r) = \frac{q}{4\pi\varepsilon_0\varepsilon} \left(\frac{1}{r} + \int_V d\mathbf{r}' \frac{\Delta n(\mathbf{r}')}{|\mathbf{r}-\mathbf{r}'|} \right) \ , \tag{15}$$

where we recognise the mean-field electrostatic potential $\psi(r)$. Setting this derivative to zero and writing $\beta = 1/(k_B T)$ yields the equation

$$n(r) = \exp(-1 - \beta\alpha - \beta q \psi(r)) = n_b \exp(-\beta q \psi(r)) \ . \tag{16}$$

This is the well known Poisson-Boltzmann equation: a mean-field approximation which can be solved by iteration. Linearizing this equation we obtain the Debye-Hückel equation

$$\Delta n(r) = \frac{-\kappa^2 \exp(-\kappa r)}{4\pi r} \ , \qquad \kappa^2 = \frac{n_b q^2}{\varepsilon_0 \varepsilon k_B T} \ . \tag{17}$$

Now, after defining the dimensionless coupling strength Γ by

$$\Gamma = \frac{q^2}{4\pi\varepsilon_0 \varepsilon k_B T a} , \tag{18}$$

where $a = (4\pi n_b/3)^{-1/3}$ is the ion-sphere radius, we get the internal potential energy per particle

$$u = -\frac{\sqrt{3}}{2} k_B T \, \Gamma^{3/2} . \tag{19}$$

The electrostatic contribution to the free energy is then

$$f_{\mathrm{el}} = \int_0^\Gamma \frac{d\Gamma'}{\Gamma'} u(\Gamma') = -\frac{1}{\sqrt{3}} k_B T \, \Gamma^{3/2} . \tag{20}$$

The Debye-Hückel theory is both a mean-field and a linear response theory. The radial distribution function is

$$g_{DH}(r) = 1 - \frac{\kappa^2}{4\pi n_b} \frac{1}{r} \exp(-\kappa r) . \tag{21}$$

Since physically $g(r)$ is greater or equal to zero the linear response cannot be applicable for small r where $g(r)$ becomes negative. The Debye-Hückel hole (DHH) theory (Nordholm , 1984a) rephrases the Debye-Hückel theory to assign $g(r)$ the value zero at these small r and then adds the exponential part continuously to give

$$g(r) = \begin{cases} 0 , & r < h \\ 1 - \frac{h}{r} \exp(-\kappa(r-h)) , & r \geq h \end{cases} . \tag{22}$$

Thus the κ-value is retained but the hole is reshaped to be physically realizable and consistent with the asymptotic DH screening of the central charge. At the density n_b the "hole" should contain one ion. Thus we have

$$n_b \int_0^\infty dr\, 4\pi r^2 (1 - g(r)) = 4\pi n_b \left[\frac{h^3}{3} + \int_h^\infty dr\, r h \exp(-\kappa(r-h)) \right] = 1 , \tag{23}$$

so that

$$h = \kappa^{-1} \left[\left(1 + (3\Gamma)^{3/2}\right)^{1/3} - 1 \right] . \tag{24}$$

In the DHH theory the charge screening distribution can be integrated over the Coulomb energy (12) to yield

$$u = -\frac{k_B T}{4} \left[\left(1 + (3\Gamma)^{3/2}\right)^{2/3} - 1 \right] . \tag{25}$$

As before, the electrostatic free energy can be obtained by a charging integration

$$f_{\mathrm{el}} = \frac{k_B T}{4} \left(1 + \frac{2\pi}{3\sqrt{3}} + \ln\left(\frac{\omega^2 + \omega + 1}{3}\right) - \omega^2 - \frac{2}{\sqrt{3}} \arctan\left(\frac{2\omega + 1}{\sqrt{3}}\right) \right) , \tag{26}$$

where

$$\omega = \left(1 + (3\Gamma)^{3/2}\right)^{1/3} . \tag{27}$$

The DHH theory (Nordholm , 1984a; Penfold et al. , 1991) produces thermodynamic properties in good agreement with simulation results for the full fluid range of coupling constants $1 < \Gamma < 160$. We note that when the ions are imbedded in hard spheres, as in the RPM of electrolytes, the DHH theory above should be applicable when the hard sphere diameter σ is smaller than the electrostatic hole size h above. The theory has been applied in the study of polyelectrolyte screening (Penfold et al. , 1990) and we shall find use for the DHH model of coion correlations in our treatment of ILs below. While the DHH provides an analytic solution to the a charged fluid problem, the more general DFT's discussed later will require numerical iterative solutions. With this in mind, it is useful to consider the convergence properties of the DHH theory, where an analytical solution is available, in order to elucidate some of the issues that may confront the numerical approaches.

4.1.1 An iterative scheme

By considering a simple statistical mechanical theory of bulk point plasmas, we aim to elucidate some general convergence characteristics of successive approximation schemes that are often deployed in free energy density functional calculations for Coulomb fluids. Although only a very crude level of theory is discussed here, we suggest that the conclusions are relevant for more sophisticated analyses of ILs.

Although the explicit DHH solution for the OCP is very valuable, the corresponding generalisation for arbitrary plasmas presents an ill-posed inverse problem (Penfold et al. , 2005) since the number of parameters grows faster with mixture components than does the number of constraint equations. Nevertheless, a simple iterative scheme can be formulated that resembles the successive approximation algorithms often used for more sophisticated free energy density functional calculations of Coulomb fluids. To study the convergence properties of this scheme in detail, it is useful to re-cast the procedure in terms of the DHH theory for the OCP. Recall the essential idea here is to "renormalise" the Debye-Hückel radial distribution function

$$g(r) = \begin{cases} 0 \ , & r < h \ , \\ 1 + \xi(h)\left(g_{DH}(r) - 1\right) \ , & r \geq h \ , \end{cases} \tag{28}$$

where g_{DH} is given by (21) and $\xi = \xi(h)$ is an asymptotic charge scaling factor fixed to ensure that g remains continuous and non-negative everywhere. By virtue of the previous analysis the appropriate DHH value is $\xi^{(\infty)} = h\exp(\kappa h)/(a\Gamma)$, but the following successive approximation scheme immediately suggests itself. Set the iteration counter $p = 0$ and specify an initial guess $\xi^{(0)}$. Now, invoke the continuity constraint to fix the current value for electrostatic exclusion hole size as the positive root of

$$l\exp(l) - \sqrt{3}\Gamma^{3/2}\,\xi^{(p)} = 0 \ , \tag{29}$$

where $l = \kappa h$. Next, appeal to the local electroneutrality condition (23) and obtain a new estimate for the renormalisation factor

$$\xi^{(p+1)} = \frac{1 - (3\Gamma)^{-3/2}\, l^3}{(1+l)\exp(-l)} \ . \tag{30}$$

Increment the counter $p \to p+1$ and repeat the nested cycle of equations (29) and (30) until convergence. This iterative procedure is related to the normal method of solving the nonlinear PB equation for screening charge densities. Indeed, the DHH analysis can be regarded as

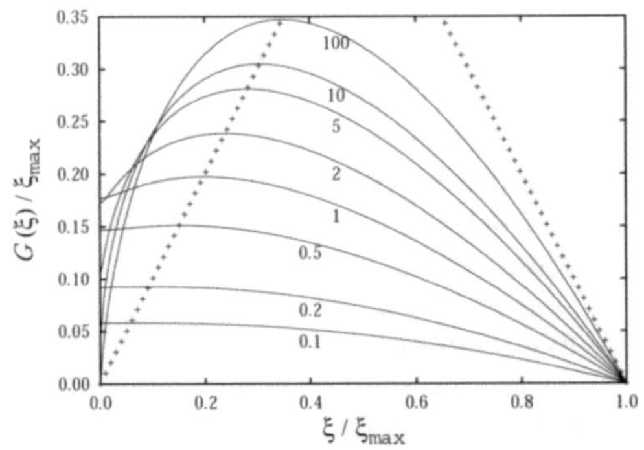

Fig. 1. A representative family of iteration function G for generating the sequence of asymptotic charge renormalisation factors. Coupling constant Γ values corresponding to each curve are indicated and span the range consistent with an OCP in the fluid state $(0 < \Gamma < 140)$. Lines of slope ± 1 are indicated by the symbols.

a very simplified PB theory. It will be convenient to express the iteration sequence by the recurrence relation

$$\xi^{(p+1)} = \frac{\left((3\Gamma)^{3/2} - l^3\right)}{3l(1+l)}\, \xi^{(p)} \equiv G\left(\xi^{(p)}\right) , \tag{31}$$

where $l = l\left(\xi^{(p)}\right)$ is given implicitly by (29) and the result (24) is clearly manifest as fixed points of the iteration function G.
Recall that, in general, the differentiable function $G = G(\xi)$ is a contraction mapping on the compact interval $[a,\, b]$ if:

i) $\xi \in [a,\, b] \Rightarrow G(\xi) \in [a,\, b]$

ii) there exists a (Lipschitz) constant $L < 1$ such that, $|G'(\xi)| \leq L$ for all $\xi \in [a,\, b]$.

For the OCP, charge reversal phenomena can be ignored so that ξ must remain positive. Consequently, by virtue of (29) and (31), the renormalisation factor is bounded

$$0 \leq \xi^{(p)} \leq \xi_{\max} = \frac{\exp\left(\sqrt{3\Gamma}\right)}{\Gamma} . \tag{32}$$

From Figure 1, it is evident that $\xi \in \left[0,\, \xi_{\max}\right]$ implies $G(\xi) \in \left[0,\, \xi_{\max}\right]$ and we are encouraged to expect that G will be a contraction map on a suitable sub-interval that also satisfies the gradient criterion (ii). Differentiation yields

$$\begin{aligned} G'(\xi) &= \frac{l}{3(1+l)^3}\left((3\Gamma)^{3/2} - ((1+l)^3 - 1)\right) \\ &> \frac{l}{3(1+l)^3}\left(l^3 - ((1+l)^3 - 1)\right) = -\left(\frac{l}{1+l}\right)^2 > -1 , \end{aligned} \tag{33}$$

where $\xi > 0$ demands $(3\Gamma)^{3/2} > l^3$ and the inequality is confirmed by Figure 1. The condition $G'(\xi) < 1$ is satisfied if and only if

$$L(l) \equiv \frac{1}{3}\left(\frac{3 + 9l + 12l^2 + 6l^3 + l^4}{l}\right)^{2/3} > \Gamma . \tag{34}$$

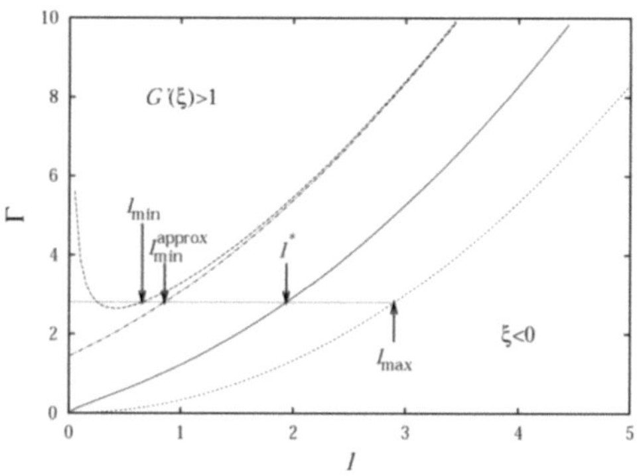

Fig. 2. Bounds on the Coulomb hole size l required for the iteration function G to be a contraction map are shown. The gradient condition (ii) leads to the result (34) where $L(l)$ is plotted here (broken line). For sufficiently large coupling parameters $\Gamma \geq \Gamma_{\text{cut}}$, the largest root l_{min} of $L = \Gamma$ sets a non-trival lower bound ξ_{lo} on the initial sequence term $\xi^{(0)}$. The corresponding solution l^* given by (24) is also indicated (solid line) as well as an upper bound (dotted line) presented by the non-negativity condition on the renormalisation factor ξ.

Figure 2 illustrates the behaviour of $L(l)$. Straightforward analysis establishes a single minimum value of

$$\Gamma_{\text{cut}} = \frac{(11+8\sqrt{2})^{2/3}}{3} \qquad \text{at} \qquad l_{\text{cut}} = \sqrt{2}-1 \ . \tag{35}$$

Hence, for $0 < \Gamma < \Gamma_{\text{cut}} \approx 2.64$ only the non-negatitivity constraint on ξ is active. It follows from the Banach Fixed Point Theorem (Sutherland , 2006) that the iteration sequence generated by $G(\xi)$ will converge to a unique solution from all initial values in the range

$$\xi_{\text{lo}} = 0 < \xi^{(0)} < \xi_{\text{hi}} = \xi_{\text{max}} \qquad (0 < \Gamma < \Gamma_{\text{cut}}) \ . \tag{36}$$

For $\Gamma \geq \Gamma_{\text{cut}}$, a positive lower bound on $\xi^{(0)}$ arises from the contraction map requirement (ii)

$$\xi_{\text{lo}} = \frac{l_{\text{min}} \exp(l_{\text{min}})}{\Gamma\sqrt{3\Gamma}} < \xi^{(0)} \quad \text{where} \quad L(l_{\text{min}}) = \Gamma \qquad (\Gamma_{\text{cut}} \leq \Gamma) \ . \tag{37}$$

Thus, the requirement (i) will also affect the upper bound on $\xi^{(0)}$ (see Figure 1) so that,

$$\xi^{(0)} < G^{-1}(\xi_{\text{lo}}) = \xi_{\text{hi}} \qquad (0 < \Gamma < \Gamma_{\text{cut}}) \ . \tag{38}$$

The results (36), (37) and (38) are plotted in Figure 3 to visualise the "basin" of attraction as a function of coupling stength. In accord with our intuition, it is evident that the success of the iteration scheme generated by G becomes increasingly more sensitive to the initial guess $\xi^{(0)}$ with increasing Γ. Since $G'(\xi^{(\infty)}) = 0$ (equation (33) and Figure 31), we also observe that the fixed point $\xi^{(\infty)}$ is super-attracting so that the rate of convergence is at least second order in the neighbourhood of the solution.

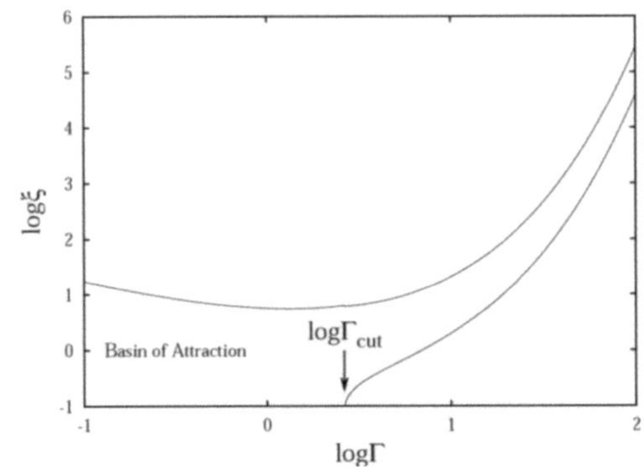

Fig. 3. The range of initial term values $\xi^{(0)}$ is shown as a function of couplping strength Γ that will ensure convergence of the iteration scheme defined by G to a unique fixed point.

4.2 Polymer density functional theory

The polymer density functional theory (DFT) is designed to treat flexible (Woodward , 1991) as well as semi-flexible (Forsman & Woodward , 2003) chains, and has the added advantage that solvent particles can be included explicitly. For this reason it is ideal for describing the underlying skeletal structure of the IL. In some cases, notably when solvent particles and monomers differ considerably in size (Forsman et al. , 2002; Forsman & Woodward , 2005) interesting phenomena arise. The polymer DFT has also been generalized to infinite, as well as polydisperse chains (Woodward & Forsman , 2008; 2009). The earliest expression of modern polymer density functional theory was described by Woodward (Woodward , 1991). That work built on the earlier formalism of Chandler and coworkers (Chandler et al. , 1986). Since then, most presentations of polymer density functional theory have used the formalism of Woodward, which separates the ideal and excess components of the free energy. Different versions of the polymer DFT generally amount to different approximations to the excess component. For a reasonably comprehensive review of poymer density functional methods we refer to the review article by Wu and Zhidong (Wu & Zhidong , 2007).

Here we will describe the polymer DFT for finite chains. We denote the position of the i-th monomer in a chain by $\mathbf{r}_i$. The complete polymer configuration of an r-mer can then be written as $\mathbf{R} = (\mathbf{r}_1, ..., \mathbf{r}_r)$ with the corresponding volume element $d\mathbf{R} = d\mathbf{r}_1 ... d\mathbf{r}_r$. Neighbouring monomers along a chain are connected by a bond of a fixed length σ. The Boltzmann factor containing the bonding potential is thus, $e^{-\beta V_B(\mathbf{R})} \propto \prod_{i=1}^{r-1} \delta(|\mathbf{r}_{i+1} - \mathbf{r}_i| - \sigma)$, where $V_B(\mathbf{R})$ is the bond potential. The r-point density distribution, $N(\mathbf{R})$ is defined such that $N(\mathbf{R})d\mathbf{R}$ is the number of polymer molecules having configurations between $\mathbf{R}$ and $\mathbf{R} + d\mathbf{R}$. We can then extract the monomer density as $n(\mathbf{r}) = \int \sum_{i=1}^{r} \delta(\mathbf{r} - \mathbf{r}_i) N(\mathbf{R}) d\mathbf{R}$. The free energy functional is decomposed into ideal and excess parts so that

$$\mathcal{F}[N(\mathbf{R})] = \mathcal{F}_{\text{id}}[N(\mathbf{R})] + \mathcal{F}_{\text{ex}}[N(\mathbf{R})] \ . \tag{39}$$

The ideal chain term is given by

$$\mathcal{F}_{\text{id}}[N(\mathbf{R})] = k_B T \int d\mathbf{R}\, N(\mathbf{R}) \left(\ln[N(\mathbf{R})] - 1\right) + \int d\mathbf{R}\, N(\mathbf{R}) (V_B(\mathbf{R}) + V_{\text{ext}}(\mathbf{R})) \ , \tag{40}$$

where $V_{\text{ext}}(\mathbf{R})$ is the externally applied potential. A similar expression is used for the solvent molecules. Note that for ideal chains in an ideal solvent, the theory is exact. The excess part

has to be approximated. Following Woodward (Woodward , 1991), we formulate it in terms of the monomer and solvent densities, so that $\mathcal{F}_{\rm ex}[n_m(\mathbf{r}); n_s(\mathbf{r}); \bar{n}_m(\mathbf{r}),; \bar{n}_s(\mathbf{r})]$, where $\bar{n}(\mathbf{r})$ is a weighted density (Nordholm et al. , 1980) as defined by (7). An explicit expression can be obtained by integrating an appropriate equation of state for polymer chains (Forsman et al. , 2002; Wichert et al. , 1996; Woodward & Yethiraj , 1994).
The grand potential is obtained as

$$\Omega[N(\mathbf{R}),\, n_s(\mathbf{r})] = \mathcal{F}_{\rm id}[N(\mathbf{R})] + \mathcal{F}_{\rm ex}[N(\mathbf{R})] - \mu_p \int d\mathbf{R}\; N(\mathbf{R}) - \mu_s \int d\mathbf{r}\; n_s(\mathbf{r}) \;, \qquad (41)$$

where μ_p and μ_s are the chemical potentials of the polymer and solvent fluids in the bulk. By minimising Ω we obtain the corresponding equilibrium density distributions for the polymer and solvent.

5. Model and theory of ionic liquids (ILs)

5.1 Coarse-grained models and dielectric response

In the last few years there has been a move away from full atomistic modelling, toward more coarse-grained descriptions (Fedorov et al. , 2008; Kornyshev , 2007; Lauw et al. , 2009). Coarse-grained models allow us to make some progress in our theoretical analysis. The models we will use must be simple enough to allow this analytical treatment, but sufficiently detailed to at least discriminate between broad classes of ILs and reproduce observed interfacial structures and behaviours. Steric interactions alone will usually give rise to interfacial structures on the length scale of the molecular size in a dense liquid. However, electrostatic and dispersion forces are also important. Indeed, monolayer enrichment of alkyl chains in imidazolium based ILs has been observed in angle-resolved x-ray photoelectron studies of the vapour/liquid interface (Lockett et al. , 2008b; Maier et al. , 2010), indicating the important role of attractive forces at the interface. Related to this are the observations of small-angle X-ray scattering experiments that appear to indicate structure on the mesoscale due to the incompatibility of the ionic and non-polar parts of cationic molecules, with sufficiently long carbon chains (Qiu & Texter , 2008). It has been conjectured that these structures are responsible for the unique solvent properties of some ILs, and to their ability to impart stability to dispersions of metallic nanoparticles. When mesoscopic structuring is present, added ions will most likely concentrate in the polar regions of the IL and affect their structure and stability. Other types of molecules added to ILs will also affect structure. Polymers and nanoparticles may stabilize existing mesostructures, or else may act as templates for others. More specific interactions can also occur in ILs, especially those based on dialkyl-imidazolium cations, which are known to form hydrogen-bonded and/or $\pi - \pi$ stacked supramolecular structures (Mele et al. , 2003). These structures are possibly responsible for the anomalous temperature dependence in DC measurements of imidazolium ILs (Lockett et al. , 2008a).
The model template for ILs that we will use in our work embodies the essential features of ILs and possesses a number of adjustable parameters to allow us the flexibility of representing different classes of ILs. It is schematically represented in Fig 4 for the case of imidazolium ILs and has the following general features:

- the cation is represented as a charged sphere with attached non-polar oligomeric chains (neutral spheres). This is adequate to model a wide variety of ILs including imidazoliums, pyridiums and ammoniums. Further discrimination between these cationic types can

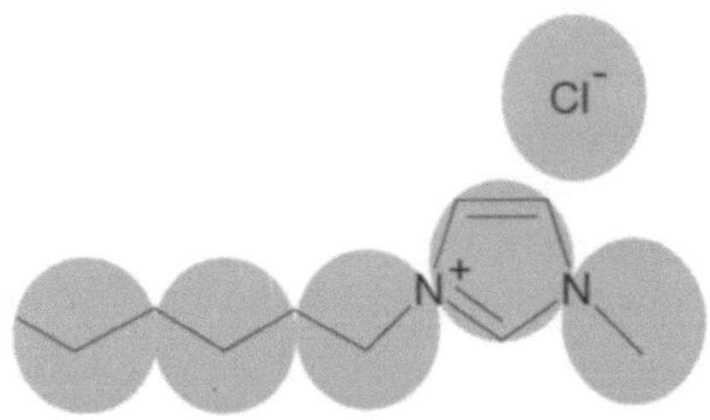

Fig. 4. A schematic picture, illustrating our coarse-grained IL model. Reproduced from reference (Trulsson:09), with permission from the American Chemical Society.

be obtained by using a partial charge distribution, rather than a single point charge. Partial charge models are widely available in the current literature. Further variability can be introduced via dispersion and steric interactions, which will be modelled by Lennard-Jones potentials centred on each sphere.

- the anion in many ILs is usually simpler than the cation and can often be represented as a negatively charged sphere. More complex anionic species can be modelled using similar approaches to that used for the cations.
- electrostatic interactions will be scaled by a relative dielectric constant, which mimics the low frequency polarizability of the molecular constituents.

Within the framework of this general model, other refinements are possible. For example, bond stiffness in the cation can be introduced via next nearest-neighbour interactions. More specific interactions can also be modelled. For example, some imidazoliums are believed to undergo $\pi - \pi$ stacking interactions.

One area of concern in modelling ILs, especially using approximate theory, is the appropriate choice of dielectric constant. Typical experimental values for the low frequency dielectric constant of ILs range between 10 and 20. As our model is coarse-grained, dielectric effects due to electronic, orientational and distortional polarization are neglected. The electronic contribution to ϵ_r is around 2, while the latter two contributions are due to charge distribution and molecular flexibility, respectively. In our model the molecules possess no permanent dipoles, as the unit charges are placed at a single center. Therefore, compared with "real" ILs, the electrostatic coupling may be overestimated. Recent work by Izgorodina and coworkers (Izgorodina et al. , 2009) shows that the orientational contribution to ϵ_r, can range between approximately 1 and 20, depending on the size of the molecular dipole moments, while the distortion contribution remains small (approximately unity).

The appropriate value for ϵ_r to be used in computer modeling is the subject of considerable debate in the literature (Izgorodina et al. , 2009; Kobrak & Li , 2010; Krossing et al. , 2006). This is because the simulations in principle capture a molecular low-frequency response, but not the electronic response. This leads to some uncertainty regarding the appropriate value with which to scale down the Coulomb interactions in a simulation. Paradoxically, the more approximate DFT approach is less ambiguous in this respect. The electrostatic correlation functional that we will use in this work is based on a one-component plasma (OCP) model. Long-ranged molecular responses of the kind just discussed, are not captured by the theory. This warrants the use of typical experimental values for the DFT calculations, so the value of ϵ_r can be chosen in a relatively unambiguous manner.

5.2 Density functional theory (DFT)

Now we have all the tools we need to construct our DFT for ILs, given our coarse-grained model. We start with a polymer DFT of a neutral oligomer in a solvent where the molar ratio in the bulk is fixed at 0.5 (commensurate with our subsequent electroneutrality criterion). Simple particles are considered comprised of monomers and solvent spheres with equal radii that interact via hard core plus Lennard-Jones potentials, and taking proper account of excluded volume as described above. All these spherical particles interact non-electrostatically with the electrode surfaces through a laterally integrated Lennard-Jones potential

$$\phi_w^\alpha(z) = \frac{2\pi\epsilon_w}{45}\left(2\left(\frac{\sigma_w}{z}\right)^9 - 15a_w^\alpha\left(\frac{\sigma_w}{z}\right)^3\right) , \tag{42}$$

where z is the distance from the plane of charge. In most cases, we have set $\sigma_w = \sigma$, but we also investigated the effect of varying σ_w/σ. The adsorption parameter a_w^α, will in this study vary between 0 (non-adsorbing electrodes) to 1 (adsorbing electrodes). The index α is introduced to facilitate ion-specific adsorption, in which case charged species carry an index "c". For example, $[a_w^c = 1; a_w^m = 0.5]$ signifies that the neutral monomers are weakly (non-electrostatically) attracted to the electrodes, while the charged monomer as well as the simple ions, are relatively strongly adsorbed, even at neutral electrodes.

At this stage, we can use polymer DFT to describe the Lennard-Jones oligomer plus solvent fluid, as described in preceding sections (see also (Forsman et al. , 2011)). Let us denote the corresponding grand potential Ω_{L-J}. Switching on charges, an additional contribution, Ω_{el}, must be added to the grand potential. At the simplest mean-field (Poisson-Boltzmann) level, this is given by

$$\Omega_{el}^{IL-PB} = \sum_\alpha \int_V d\mathbf{r}\, n_\alpha(\mathbf{r}) \left(\frac{1}{2}\sum_\lambda \int_V d\mathbf{r}' n_\lambda(\mathbf{r}')\phi_{\alpha\lambda}^C(|\mathbf{r}-\mathbf{r}'|) + V_{el}(\mathbf{r})\right) . \tag{43}$$

Here $n_\alpha(\mathbf{r})$ are ionic densities, where the indexes α and λ are summed over all the charged species. We shall by $\rho_m(\mathbf{r})$ denote the collective density of *all* monomers, neutral as well as charged. Note that only the second monomer along the cationic chain is charged, with a density denoted $n_m(\mathbf{r})$. The first term in eq.(43) is just the mean-field electrostatic interaction between ions. The second term describes the interactions between the ions and the uniform surface charge. We note that Ω_{el}^{IL-PB} is zero for the bulk phase IL.

A significant improvement results if we introduce DHH ideas, and add a correlation term to Ω_{el}^{IL-PB}. The resulting functional will simply be denoted DFT, and has the following form

$$\Omega_{el}^{DFT} = \Omega_{el}^{IL-PB} + \sum_\alpha \int_V d\mathbf{r}\, n_\alpha(\mathbf{r})\left(f_{ex}\left[\hat{n}_\alpha(\mathbf{r})\right] - \mu_c^\alpha\right) , \tag{44}$$

where μ_c^α is obtained by differentiating f_c at the bulk density of component α. We borrow results from DHH theory to determine the correlation free energy per particle, i.e. $f_{ex}[\hat{n}_\alpha(\mathbf{r})]$ is given by (26), with the bulk density being replaced by a coarse-grained density, $\hat{n}_\alpha(\mathbf{r})$, which is evaluated as a ion density average:

$$\hat{n}_\alpha(\mathbf{r}) = \frac{3}{4\pi h^3}\int_{|\mathbf{r}-\mathbf{r}'|<h} d\mathbf{r}' n_\alpha(\mathbf{r}') . \tag{45}$$

Notice that while the coarse-grained density is local, the range of integration h is conveniently approximated by the DHH hole in the bulk using (24). As discussed earlier, the correlation

correction used here accounts only for short-ranged effects, expected to dominate in ILs, and do not account for the ionic polarization contribution to the dielectric screening of electrostatic interactions. Thus the mean-field terms, that is the first two terms in (44), remain unaffected by the correlational term and the ionic polarization contribution must be included in the value of ϵ_r.
The total grand potential, $\Omega_{\text{tot}}^{\text{DFT}}$ is obtained as

$$\Omega_{\text{tot}}^{\text{DFT}} = \Omega_{\text{L-J}} + \Omega_{\text{el}}^{\text{DFT}} \ . \tag{46}$$

The functionals can be simplified by integrating out the (x, y) dimensions, parallel with the "electrode walls". Minimization of the chosen functional, with respect to the fluid particle densities then follows in the usual manner, wherein electroneutrality is ensured via a Donnan potential.

6. Results

6.1 Performance of the DFT for the full IL model

In our previous publication (Forsman et al. , 2011), we demonstrated that our DFT is able to reproduce simulated differential capacitance curves qualitatively, and with a semi-quantitative accuracy. Furthermore, the level of the curves, and their overall shape, agrees well with experimental results (Islam et al. , 2009; Lockett et al. , 2008a), which tends to support our simple model. Still, since we are armed with a theory and a microscopic model, we can go further and make scrutinizing analyses of the microscopic structure at the surfaces. These are not easily accessible by experimental methods, but are nevertheless of fundamental interest. One of the major strengths of a theoretical approach is the ability to correlate microscopic detail to macroscopic observations. We start by analyzing the ability of our DFT to predict density distributions at the electrodes. Comparisons with simulations are given in Figure 5. Details of the simulation parameters and methods are provided elsewhere (Forsman et al. , 2011). We note that the DFT tends to underestimate the height of the primary peak; presumably reflecting a slight underestimation of ion correlation effects. At a negative surface, the DFT predicts a too long-ranged oscillatory structure, and with a period that also seems a bit overestimated. Still, overall accuracy is quite reasonable. At a positive surface, we do observe some larger discrepancies, as seen in Figure 5 (b). In particular, the simulated data display a distinct primary peak of the charged monomers, outside the first anion "layer". A corresponding peak is detectable also with DFT (see inset), but it is much less pronounced. We believe this primarily results from our neglect of Coulomb truncation at short range, due to the hard cores of the particles. We are working on an improved DFT to remedy this problem. The total monomer density (of charged as well as neutral monomers) is given in Figure 5 (c). We believe the agreement is satisfactory, although the DFT in this case slightly *over*estimates the primary adsorption peak.
When making these comparisons, we should keep in mind the problem of ambiguity regarding the choice of ϵ_r in the simulations. We aim to describe an IL with $\epsilon_r = 12$ (used in DFT), but since the simulations capture a *part* of this (as discussed earlier), we have set $\epsilon_r = 7.6$ for the MC runs. This is probably a reasonable value, but there is still some degree of ambiguity.

6.2 A model system with only dispersion

Having demonstrated that our DFT is reasonably accurate, albeit not perfect, it might be a good idea to investigate the performance of the DFT that would result if we remove all

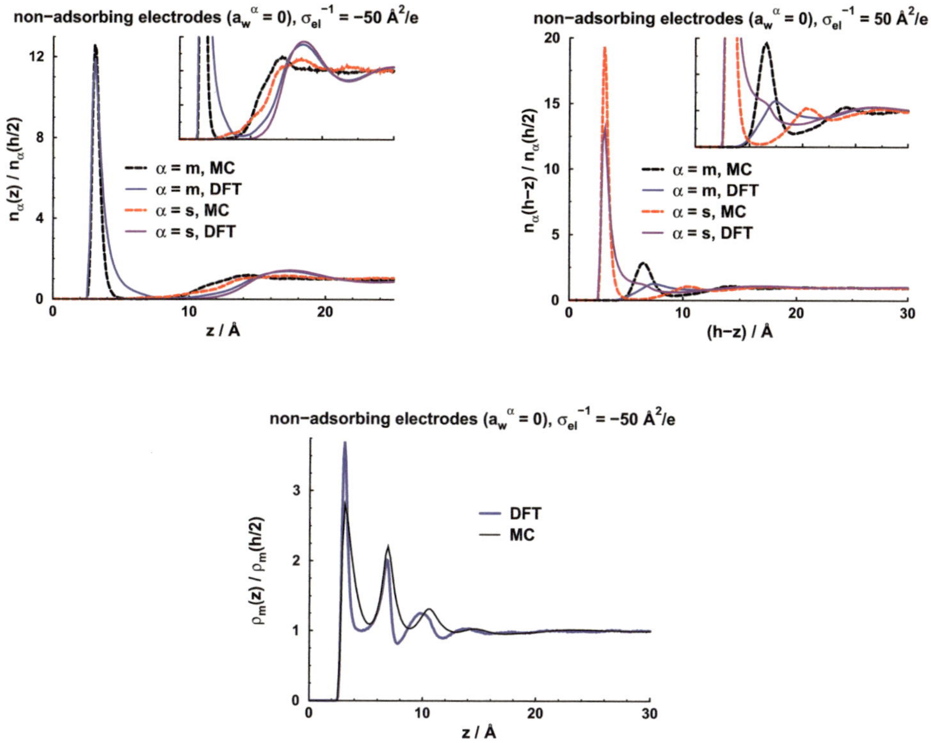

Fig. 5. Density profiles at "electrodes", as predicted by DFT, and simulated ("MC"), respectively. The "underlying" soft electrode-particle interaction is non-adsorbing, i.e. $a_w^\alpha = 0$ (with $\alpha =$ m,c)
(a) Simple anions ("s") and (positively) charged monomers ("m") at a strongly negative surface, $\sigma_{\mathrm{el}}^{-1} = -50$ Å$^2/e$. The densities are normalized by their value at the mid plane. The inset is a zoom-in.
(b) Simple anions and charged monomers, at a strongly positive surface, $\sigma_{\mathrm{el}}^{-1} = 50$ Å$^2/e$. The inset is a zoom-in.
(c) Total monomer density, outside at a strongly negative surface, $\sigma_{\mathrm{el}}^{-1} = -50$ Å$^2/e$.

charges, on ions as well as on the surfaces. We then end up with a binary Lennard-Jones mixture, where one component has a uniform oligomeric structure (a "pearl-necklace" of L-J monomers). We use the same densities as used for the fully charged system (in simulations and DFT, respectively). One difference from an ordinary binary L-J mixture is that we retain a "Donnan potential", ensuring that the number of simple particles (anions in the IL) in the slit region equals the number of oligomers. This approach, we believe, ensures that we make the system as similar as possible to the original IL, but without the charges. This should allow us to establish which of our approximations in the full DFT that are likely to be most problematic, i.e., where we should focus our efforts to improve the DFT. Structural comparisons are given in Figure 6, for adsorbing ($a_w^\alpha = 1$) as well as non-adsorbing ($a_w^\alpha = 0$) surfaces. All the qualitative features (humps and oscillations) are well reproduced by the DFT, although it does tend to overestimate the primary solvent peak somewhat. Given this satisfactory

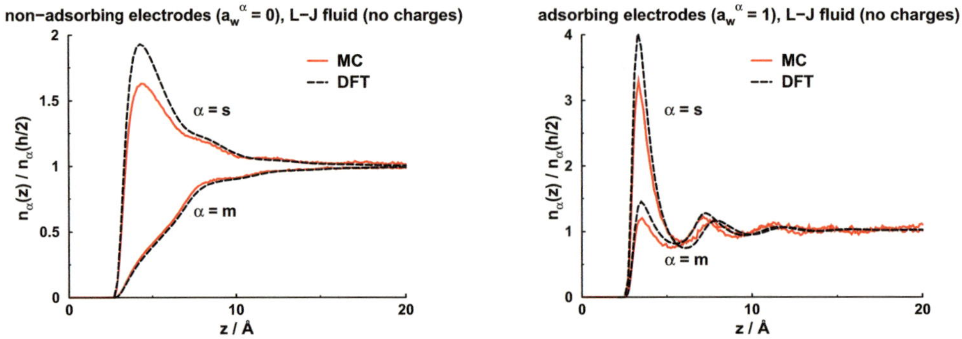

Fig. 6. Calculated and simulated density profiles, for our model fluids, but with all charges being removed.
(a) Non-adsorbing surfaces, $a_w^\alpha = 0$
(b) Adsorbing surfaces, $a_w^\alpha = 1$

performance for neutral species, we can almost with certainty state that the discrepancies found for ILs are primarily due to approximations of the Coulomb interactions; specifically the way in which these are altered by correlations in the system. As already mentioned, these problems are currently being addressed. After having evaluated the performance of the DFT (see also (Forsman et al. , 2011)), with the conclusion that it faithfully reproduces the true behaviour of the model, we will henceforth display DFT predictions without the corresponding time-consuming simulations.

6.2.1 Effects of the range of the soft surface repulsion

Let us now consider the role of the width of the depletion region in the case of non-adsorbing

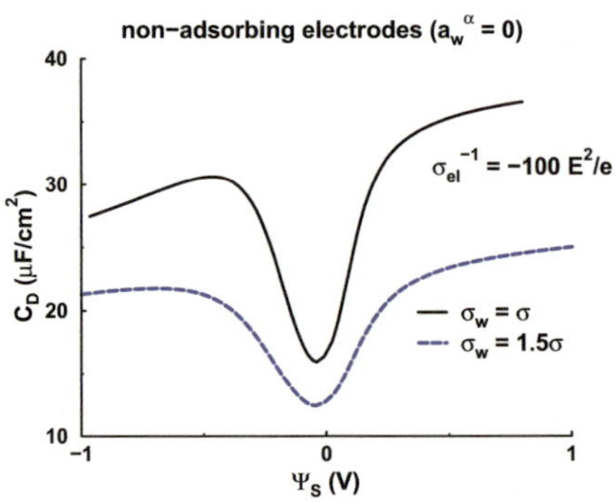

Fig. 7. DFT predictions of $C_D(\Psi_S)$ for non-adsorbing surfaces, as obtained with two different choices of $\sigma_w = \sigma$ or $\sigma_w = 1.5\sigma$ ($\sigma = 4$ Å, as usual).

surfaces. In Figure 7, we have compared differential capacitance curves with $\sigma = 4$Å and $a_w^\alpha = 0$, but with two different choices of σ_w. In our previous simulations/calculations, we have set $\sigma_w = \sigma$, but here we also consider the case $\sigma_w = 1.5\sigma$. The differential capacitance curve responds rather dramatically to a change in the distance of closest approach of the fluid! Within the exclusion region the potential varies strongly and linearly, giving rise to the so called "compact layer" contribution to the capacitance (Fedorov et al. , 2008). The

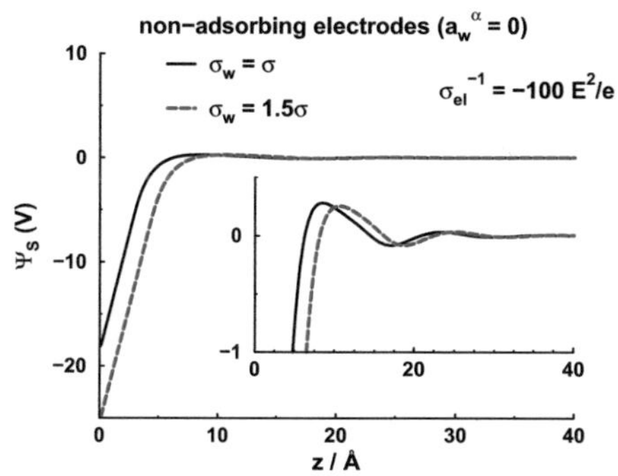

Fig. 8. The variation of the mean electrostatic potential, Ψ, with distance, z, from a negatively charged surface. The surfaces are non-adsorbing, with two different choices of σ_w, as in Figure 7.

"linear potential effect" is illustrated in Figure 8, where the corresponding potential profiles are shown. We see how these profiles are very similar, but "z-shifted" relative to the plane of charge. For the surface with a more long-ranged soft repulsion, $\sigma_w = 1.5\sigma$, the linear regime (where there are no responding ions) extends further out, which leads to a larger potential difference relative to the bulk value. While the "exclusion effect" in principle is trivial, we believe its relevance has been underestimated in the literature.

6.2.2 Effects of temperature and dielectric constant

In a recent experimental work, Lockett and coworkers (Lockett et al. , 2008b) claimed to establish the temperature dependence of the differential capacitance for imidazolium chloride ILs at around 393 K. We are not experimentalists, but have learned that it is quite difficult to ascertain the amplitude of C_D. Still, we will henceforth proceed under the assumption that the reported findings are correct. Specifically, they found that the C_D curve *increases* with temperature. Can we capture such a behaviour with our simple model, and if so, under what circumstances?

Let us first consider the process of merely increasing the temperature, at a constant pressure. The corresponding DFT predictions, for non-adsorbing electrodes, are collected in Figure 9. We see that the differential capacitance in this case *drops* with temperature, except at low surface potentials. This is in contrast with the experimental findings (Lockett et al. , 2008b). But what if the dielectric constant of the IL also changes with temperature?

Specifically, let us assume that the low-frequency dielectric constant increases with temperature. In Figure 10, we highlight the consequences to the response of C_D to an increased temperature. Under these circumstances, the differential capacitance is indeed predicted to *increase* with temperature, in accordance with the observations (Lockett et al. , 2008b)! Establishing the temperature dependence of ϵ_r for ILs is not an easy experimental task, but Hunger and coworkers (Hunger et al. , 2009) have recently obtained some estimates. It turns out that the static dielectric constant in most cases seems to descrease slightly with temperature. Still, there are exceptions, and at least one case with a positive value of $\partial\epsilon_r/\partial T$

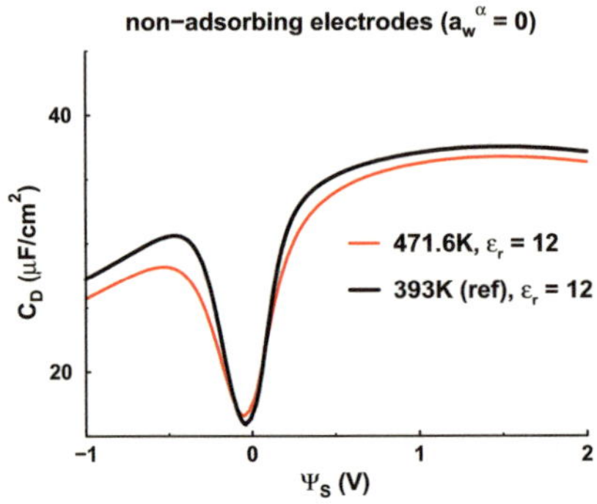

Fig. 9. Differential capacitance curves, with non-adsorbing electrodes, at two different temperatures. The dielectric constant is set to be temperature independent.

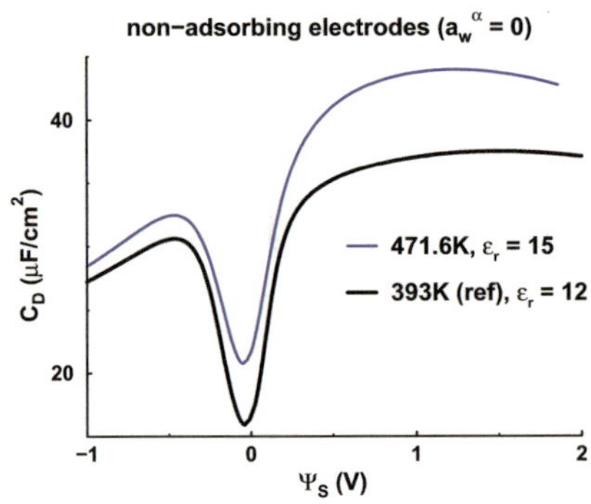

Fig. 10. The same scenario as in the previous Figure, although the dielectric constant is here assumed to increase with temperature.

was detected. We refrain from a discussion about the reliability of the various experimental measurements. Instead, we simply state that our simple theoretical model will predict an overall increase of C_D with temperature, provided that ϵ_r also increases.

6.2.3 Electrodes with adsorption or specific adsorption

Let us furthermore consider the case of adsorbing, or partially adsorbing electrodes. We recall the wall potential defined by (42) that allows us to treat the case of specific adsorption of the charged species, which presumably are more polarizable than CH_2 groups. Furthermore, there may be important contributions from image charge interactions, as discussed below. As we have reported in previous publications (Forsman et al. , 2011; Trulsson et al. , 2010), adsorption has a dramatic influence on the differential capacitance curve. Here we also consider the case of specific adsorption, with results presented in Figure 11. As might be anticipated, ion specific adsorption is quite efficient when it comes to "reducing the camel hump". Recall that the "hump" is the result of ion depletion at low surface potentials. This depletion is of course less pronounced when there is a non-electrostatic attraction to the electrodes. Note that these theoretical findings are commensurate with experimental

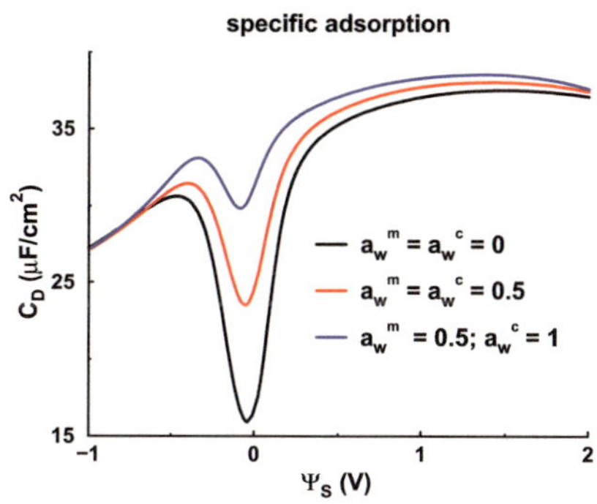

Fig. 11. Effects of non-electrostatic electrode adsorption. The top differential capacitance curve illustrates effects from ion specific adsorption. In this case, neutral monomers adsorb weakly ($a_w^m = 0.5$), while anions (simple) and cations (charged monomers) display a strong adsorption ($a_w^c = 1$).

differential capacitance data on metallic and non-metallic electrodes (Islam et al. , 2009). Specifically, one will in general find a bell-shaped C_D curve with metal electrodes (no camel-hump), and U-shaped C_D with glassy carbon electrodes. The latter scenario implies a camel-hump shaped complete C_D curve, although experimental limitations (redox-reactions etc.) may not reach the saturation limit. These findings can be rationalized if we consider that metallic electrodes are likely to be more strongly adsorbing at low surface potentials, on account of the expected attractive image charge forces, and strong dispersion interactions. As we have seen here, adsorption will tend to diminish "camel-hump" tendencies, ultimately resulting in a purely bell-shaped differential capacitance curve. Such adsorbing interactions are likely to be weaker with glassy carbon electrodes, which naturally will promote the occurrence of a camel-shaped C_D curve.

Finally, we investigate the temperature and dielectric response dependence for an "ion-specific adsorption" electrode, with $a_w^m = 0.5$; $a_w^c = 1$. The DFT predictions are provided in Figure 12. Here we quite clearly see how the differential capacitance will drop upon a

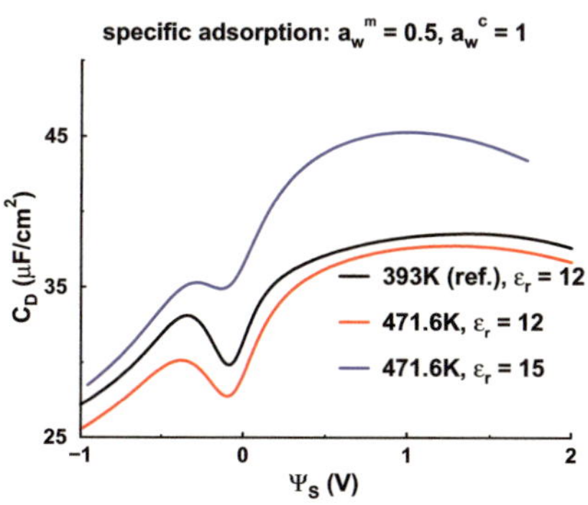

Fig. 12. Temperature response, for the "ion-specific adsorption" system in Figure 11. Cases of a constant as well as temperature-increasing low frequency dielectric response are given.

temperature increase, if ϵ_r remains unchanged. An increasing dielectric constant may on the other hand facilitate a C_D curve that increases with temperature.

7. Conclusions

We have tried to illustrate how classical DFT applied to simple coarse-grained models, can be used to gain an improved understanding of fundamental physical mechanisms underlying macroscopic behaviours of ILs. Furthermore, the DFT calculations are noise-free and exceptionally rapid, permitting more elaborate models to be studied. Examples of simple extensions/improvements are an intrinsic stiffness of the oligomer chain (Forsman & Woodward , 2003), distributed charges and/or a structurally more detailed model. Such extensions can easily be implemented.

8. References

Carnahan, N. F. & Starling, K. E. (1969). Equation of state for nonattracting rigid spheres. *J. Chem. Phys.*, Vol. 51, 635–636

Chandler, D.; McCoy, J. D.; Singer, S. J. (1986) Density functional theory of nonuniform polyatomic systems. I. General Formulation. *J. Chem. Phys.*, Vol. 85, 5971–5976

Izgorodina, E. I., Forsyth, M., MacFarlane, D. R., (2009) On the components of the dielectric constants of ionic liquids: ionic polarization? *Phys. Chem. Chem. Phys.*, Vol. 11, 2452–2458

Fedorov, M. V. and Kornyshev, A. A., (2008) Towards understanding the structure and capacitance of electrical double layer in ionic liquids. *Electroch. Acta*, Vol. 53, 6835–6840

Fedorov, M. V.; Georgi, N.; Kornyshev, A. A. (2010) Double layer in ionic liquids: The nature of the camel shape of capacitance. *Electrochem. Comm.*, Vol. 12, 296–299

Forsman, J.; Woodward, C. E.; Freasier, B. C. (2002). Density functional study of surface forces in athermal polymer solutions with additive hard sphere interactions: Solvent effects, capillary condensation, and capillary-induced surface transitions. *J. Chem. Phys.*, Vol. 117, 1915–1926

Forsman, J. & Woodward, C. E. (2003). An improved density functional theory of hard sphere polymer fluids at low density. *J. Chem. Phys.*, Vol. 119, 1889–1892

Forsman, J. & Woodward, C. E. (2005). Prewetting and layering in athermal polymer solutions. *Phys. Rev. Letts.*, Vol. 94, 118301

Forsman, J.; Woodward, C. E.; Trulsson, M. (2011). A classical density functional theory of ionic liquids. *J. Phys. Chem. B*, Vol. 115, 4606–4612

Freasier, B. C.; Woodward, C. E.; S. Nordholm, S. (1989). Generalized van der Waals theory of hard sphere oscillatory structure. *J. Chem. Phys.*, Vol. 90, 5657–5663

Gouy, G. (1910). Sur la constitution de la charge electrique ala surface d'un electrolyte, *J. Phys.*, Vol. 9, 457–468

Hanke, C. G.; Price, S. L.; Lynden-Bell, R. M. (2001). Intermolecular potentials for simulations of liquid imidazolium salts. *Molec. Phys.*, Vol. 99, 801–809

Hunger, J.; Stolla, A.; Schrödle, S.; Hefter, G.; Buchner, R. (2009). Temperature dependence of the dielectric properties and dynamics of ionic liquids *Chem. Phys. Chem.*, Vol. 10, 723–733

Islam, M. M.; Alam, M. T.; Okajima, T.; Oshaka, T. (2009). Electrical double layer structure in ionic liquids: an understanding of the unusual capacitance-potential curve at a nonmetallic electrode. *J. Phys. Chem. C*, Vol. 113, 3386–3389

Johnson, M. & Nordholm, S. (1981). Generalized van der Waals theory. VI. Application to adsorption. *J. Chem. Phys.*, Vol. 75, 1953–1957

Guldbrand, L.; Jönsson, B.; Wennerström, H.; Linse, P. (1984). Electrical double layer forces. A Monte Carlo study. *J. Chem. Phys.* , Vol. 84, 2221–2228

Kobrak, M. N. & Li, H. (2010). Electrostatic interactions in ionic liquids: the dangers of dipole and dielectric descriptions. *Phys. Chem. Chem. Phys.*, Vol. 12, 1922–1932

Kornyshev, A. A. (2007). Double-layer in ionic liquids: paradigm change? *J. Phys. Chem. B.*, Vol. 111, 5545–5557

Krossing, I.; Slattery, C.; Daguenet, J. M.; Dyson, P. J.; Oleinikova, A.; Weingartner, H. J. (2006). Dielectric response of imidazolium-based room-temperature ionic liquids. *J. Phys. Chem. B*, Vol. 110, 12682–12688

Lamperski, S. & Zydor, A. (2007). Monte Carlo study of the electrode/solvent primitive electrolyte interface. *Electrochim. Acta*, Vol. 52, 2429–2436

Lamperski, S.; Outhwaite, C.; Bhuiyan, L. (2009). The electric double-layer differential capacitance at and near zero surface charge for a restricted primitive model electrolyte. *J. Phys. Chem. B*, Vol. 113, 8925–8929

Lauw, Y.; Horne, M.; Rodopolous, T.; Leermakers, F. A. M. (2009). Room-temperature ionic liquids: Excluded volume and ion polarizability effects in the electrical double-layer structure and capacitance. *Phys. Rev. Lett.*, Vol. 103, 117801

Lockett, V.; Sedev, R.; Ralston, J.; Horne, M.; Rodopoulos, T. (2008a). Differential capacitance of the electrical double layer in imidazolium-based ionic liquids: Influence of potential, cation size and temperature. *J. Phys. Chem. C*, Vol. 112, 7486–7495

Lockett, V.; Sedev, R; Bassell, C.; Ralston, J. (2008b). Angle-resolved X-ray photoelectron spectroscopy of the surface of imidazolium ionic liquids. *Phys. Chem. Chem. Phys.*, Vol. 10, 1330–1335

Lynden-Bell, R. M. (2010). Screening of pairs of ions dissolved in ionic liquids. *Phys. Chem. Chem. Phys.*, Vol. 12, 1733–1740

Lynden-Bell, R. M.; Del Popolo, M. G.; Youngs, T. G. A.; Kohanoff, J.; Hanke, C. G.; Harper, J. B.; Pinilla, C. C. (2007). Simulations of ionic liquids, solutions and surfaces. *Acc. Chem. Res.*, Vol. 40, 1138–1145

MacFarlane, D. R.; Pringle, J. M.; Howlett, P. C.; Forsyth, M. (2010). Ionic liquids and reactions at the electrochemical interface. *Phys. Chem. Chem. Phys.*, Vol. 12, 1659–1669

Maier, F.; Cremer, T.; Kolbeck, C.; Lovelock, K. R. J.; Paape, N.; Schulz, P. S.; Wasserscheid, P.; Steinruck, H.-P. (2010). Insights into the surface composition and enrichment effects of ionic liquids and ionic liquid mixtures. *Phys. Chem. Chem. Phys.*, Vol. 12, 1905–1915

Maxwell, J. C. (1960). The scientific papers of James Clerk Maxwell, p. 424, Dover

Mele, A.; Tran, C.; De Paoli Lacerda, S. H. (2003). The structure of a room-temperature ionic liquid with and without trace amounts of water: The role of C–H–O and C–H..F interactions in 1-n-butyl-3-methylimidazolium tetrafluoroborate. *Angewandte Chemie*, Vol. 42, 4364–4366

Nordholm, S. (1984a). Simple analysis of the thermodynamic properties of the one-component plasma. *Chem. Phys. Lett.*, Vol. 105, 302–307

Nordholm, S. & Haymet, A. D. J. (1980). Generalized van der Waals theory. I. Basic formulation and application to uniform fluids. *Aust. J. Chem.*, Vol. 33, 2013–2027

Nordholm, S.; Johnson, M.; Freasier, B. C. (1980). Generalized van der Waals theory. III. The prediction of hard sphere structure. *Aust. J. Chem.*, Vol. 33, 2139–2150

Nordholm S. & Gibson J. (1981). Generalized van der Waals theory. VII. Interface profiles and surface tension. *Aust. J. Chem.*, Vol. 34, 2263–2282

Penfold, R.; Nordholm, S.; Jönsson, B.; Woodward, C. E. (1990). A simple analysis of ion-ion correlation in polyelectrolyte solutions. *J. Chem. Phys.*, Vol. 92, 1915–1922

Penfold, R.; Nordholm, S.; Jönsson, B.; Woodward, C. E. (1991). A simple analysis of the classical hard sphere one component plasma. I. Hole corrected Debye-Hückel theory. *J. Chem. Phys.*, Vol. 95, 2048–2056

Penfold, R.; Nordholm, S.; Nichols, N. (2005). Simple analysis of the thermodynamic properties for classical plasmas. *J. Stat. Mech.: Theory Expt.*, P06009, June, ISSN 1742-5468.

Qiu, Z. & Texter, J. (2008). Ionic liquids in microemulsions. *Curr. Op. Coll. Int. Sci.*, Vol. 13, 252–262

Rocher, N. M.; Izgorodina, E. I.; Ruther, T.; Forsyth, M.; MacFarlane, D. R.; Rodopolous, T.; Horne, M. D.; Bond, A. (2009). Aluminium speciation in 1-butyl-1-methylpyrrolidinium bis(triflouromethylsulfonyl)amide/$AlCl_3$ mixtures. *Chem. Euro. J.*, Vol. 15, 3435–3447

Sutherland, W. A. (2006). *Introduction to Metric and Topological Spaces*, Oxford University Press, Oxford.

Tarazona, P. (1984). A density functional theory of melting. *Mol. Phys.*, Vol. 52, 81–96

Tarazona, P. & Evans, R. (1984). A simple density functional theory for inhomogeneous liquids. *Mol. Phys.*, Vol. 52, 847–857

Trulsson, M.; Algotsson, J.; Forsman, J.; Woodward, C. E. (2010). Differential capacitance in room temperature ionic liquids: the role of dispersion forces. *J. Phys. Chem. Lett.*, Vol. 1, 1191–1195

Vatamanu, J.; Borodin, O.; Smith, G. D. (2010). Molecular insights into the potential and temperature dependencies of the differential capacitance of a room-temperature ionic liquid at graphite electrodes. *J. Am. Chem. Soc.*, Vol. 132, 14825–14833

Walden, P. (1914). Über die Molekulargrösse und elektrische Leitfähigkeit einiger geschmolzenen salze. *Bull. Acad. Imper. Sci. St. Petersburg*, Vol. 8, 405–422

van der Waals, J. D. (1873). Over de Continuïteit van den Gas - en Vloeistoftoestand, thesis, Leiden.

Wichert, J. M.; Gulati, H. S.; Hall, C. K. J. (1996). Binary hard chain mixtures. I. Generalized Flory equations of state. *J. Chem. Phys.*, Vol. 105, 7669–7682

Woodward, C. E. & Forsman, J. (2008). Density functional theory for polymer fluids with molecular weight polydispersity. *Phys. Rev. Lett.*, Vol. 100, 098301

Woodward, C. E. (1991). A density functional theory for polymers: Application to hard chain-hard sphere mixtures in slitlike pores. *J. Chem. Phys.*, Vol. 94, 3183–3191

Woodward, C. E. & Forsman, J. (2009). Interactions between surfaces in polydisperse semiflexible polymer solutions. *Macromol.*, Vol. 42, 7563–7570

Woodward, C. E. & Yethiraj A. (1994). Density functional theory for inhomogeneous polymer solutions. *J. Chem. Phys.*, Vol. 100, 3181–3186

Wu, J. & Zhidong, L. (2007). Density functional theory of complex fluids. *Ann. Rev. Phys. Chem.*, Vol. 58, 85–112

Part 3

Physical Properties

7

Interactions and Transitions in Imidazolium Cation Based Ionic Liquids

Madhulata Shukla, Nitin Srivastava and Satyen Saha
Department of Chemistry, Faculty of Science
Banaras Hindu University Varanasi
India

1. Introduction

Ionic liquids (ILs) raise considerable research interest not only as promising new solvents for the replacement of conventional solvents in synthesis, but also as new liquid materials.[1-2] It appears from the recent breakthroughs that the novel properties of various ILs are of more interest than mere application as 'green solvent' in traditional organic chemistry. ILs are now rediscovered to be whole new materials with many wonderful properties, much of it are yet to be discovered. The defining characteristic of ILs is of their constitutions - molecular ions as their building blocks as opposed to molecules in the traditional solvents. In other words, ILs or molten salts in general are defined as liquids composed of ions only, either at room temperature or at elevated temperatures (below 100°C). ILs are rather unique in the sense that in addition to ionic and covalent interactions, there are relatively weaker interactions such as H-bondings, and π-stacking, which are not commonly found in conventional solvents.[3-4] The nature of the forces in different ILs may however differ from one another and mainly control their physical properties. As the properties of any material depends on the structure of molecules in different phases, it is very important to understand the structural features of ILs in depth. Researchers have paid considerable attention towards pyridinium, imidazolium and pyrrolidinium based ILs in addition to ammonium and phosphonium ILs. Pyridinium based ILs are heterocyclic aromatic compounds proven to have great potential in organic synthesis and biocatalyst. Compared with the imidazolium based ILs, few studies have examined the biodegradability of pyridinium based ones.[5] Due to biodegradable property of these pyridinium and pyrrolidinium ILs, it has been extensively studied.[6-12] On the other hand, pyrrolidinium ILs are mainly involved in dye sensitized solar cell and batteries.[11] Since structure plays the vital role for any application, recently detailed x-ray scattering studies have been reported on the pyrrolidinium cations based ILs while varying the length of the alkyl chain attached with the ring.[12] Interestingly, diffraction pattern shows signature of intermediate range ordering similar to that of in imidazolium based ILs. Among all ionic liquids, imidazolium cation based ILs are the most extensively studied ILs, and therefore our discussion will mainly be confined with imidazolium cation based ILs.

As of true 'designer solvent', it has been observed that a small variation in imidazolium cation (such as increase or decrease in alkyl chain length) alters their physical properties (melting point, viscosity, conductivity etc.) drastically.[13-14] For instance, density of IL decreases as the length of the alkyl chain on the cation increases up to a certain length. For a given cation, the density increases as the molecular weight of the anion increases. Many of ILs tend to subcool easily, forming glasses at very low temperature rather than exhibiting crystallization or melting transitions. The thermal stability increases with increasing anion size, and heat capacities increases with temperature and increasing number of atoms in IL.[15] The work reported by Anderson et. al. [16] uses a linear free energy approach to characterize different imidazolium and pyridinium based Room temperature ionic liquids (RTILs) on the basis of their distinct multiple solvation interactions with probe solute molecules. This model provides data that can be used to identify the interactions and properties that are important for specific chemical applications. It has also been shown that the anion has greater effect on hydrogen bond basisity of RTIL while the effect of the cation was generally found to be small. RTILs found to exhibit multiple behaviour such that it acts as polar solvents in organic reactions containing polar molecules and acts as less polar solvents in the presence of less polar molecules.[16] Hence study of interactions present between cation and anion as well as with probe molecules in ILs are very important. To get more insight of interactions presents in ILs, theoretical calculations are found to be of great help.[17-20] Density Functional Theory (DFT) is a quantum mechanical theory used in physics and chemistry to investigate the electronic structure (mainly in ground state) and sometime used to calculate various thermodynamic properties of a molecule. With this theory, the property of many-electron system can be determined by using functional i.e. function of another function (electron density). At present DFT is the most popular and a versatile method available in computational chemistry.[21-23] Our work with DFT is mainly on the structural investigation of ILs as most of them are liquids at room temperature, so their structural investigations are very difficult by conventional techniques like single crystal x-ray diffraction (SCXRD). Hence DFT calculation found to be of great useful in predicting the molecular structure, as well as interactions present in a given molecule.[4, 21] Magnetic moment, dipole moment and many other physical properties of a molecule can also be calculated by the same. In addition, the calculated vibrational frequency (both IR and Raman) of the molecule gives us a strong base to analyze the experimental spectra and also the effect of interaction causing shifting in IR bands. Chang et. al have described using DFT that how the cation-anion interaction lead to shifting in vibrational frequencies.[24] Hence theoretical calculation found to be of great valuable in explaining the interactions present in a given IL.

Here we have addressed the following very specific issues related to important class of imidazolium cation based ILs. It has been found that with variation of anion, cation being the same, physical state of ILs changes drastically.[25] **1**) For example, while bmimCl and bmimBr found to be solid at room temperature [4,21] whereas bmimI is liquid at room temperature [bmim:1-butyl-3-methylimidazolium].[26] We have studied the structural features of these ILs by DFT calculation which found to predict the ILs structures quite well. Further it has been found that in bmimBF_4, bmimPF_6 and bmimNTf_2 multiple H-bondings were found between cation and anion at a specific orientation [18,27], whereas in bmimCl, bmimBr and bmimI, only one predominant H-bonding was found to exist. It is proved that H-bonding found to play crucial role in describing the physical state of different ILs.[21-22,24] In this chapter we have also addressed the colour of ionic liquids which is due to the transition in the visible wavelength range. While the chloride and bromide analogues (i.e., bmimCl and bmimBr) are colourless, the corresponding iodide salt is found to be coloured.[23,26] The colour issue is important since during the synthesis of these derivatives (e.g., halogen and PF_6, BF_4 anion containing alkylimidazolium ILs), often impurity imparts colour making it difficult to understand whether the colour is inherent to ILs concerned or is due to the colour impurity.[28-29] Especially, from spectroscopic application point of view it is very important to have optically transparent windows of the ILs.[28-30] TD-DFT theoretical calculation found to play a crucial role to explain, whether colour of given IL is intrinsic or due to impurity present in it.

2. Experimental

2.1 Synthesis of 1-butyl-3-methylimidazolium halide (bmimX, where X^-= Cl, Br, I, NTf_2, PF_6 and BF_4)

Synthesis of all bmimX have been done following the reported method [4,23,31] with some modifications to get the product as much pure as possible. The major modification is mainly of the temperature at which reactions were done. Since we have used considerable less temperature than generally reported, the time required for the reactions are also much higher. Nevertheless, this low temperature reaction has been found to provide much pure ILs.

2.1.1 Synthesis of bmimCl

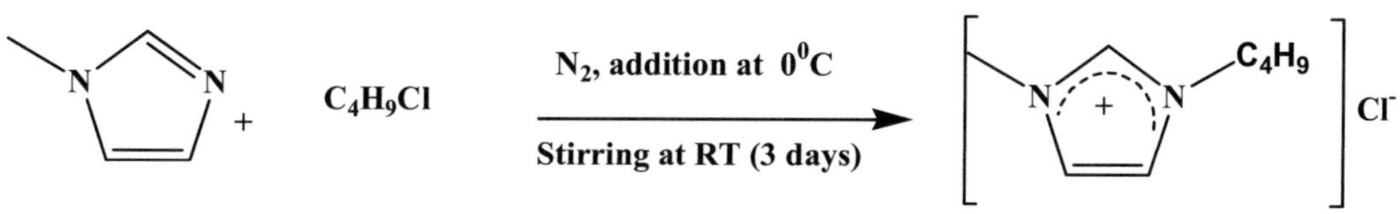

N-methylimidazole was dissolved in required amount of dry acetonitrile followed by slow addition of butyl halide at 0°C with constant stirring under N_2 atmosphere in dark. The solution was left at 0°C for 2h and then stirred for 3 days while keeping the temperature below 40°C. The progress of the reaction was monitored by checking the TLC (removal of methylimidazole). Acetonitrile (ACN) was then evaporated at reduced pressure at the same temperature. Pale yellow colour viscous liquid was obtained (yield: 70%) which was then washed with dry ethyl acetate and then followed by dry diethyl ether. During washing bmimCl was solidified. bmimCl was further purified with activated charcoal to remove coloured impurities. Finally, it was kept under reduced pressure (10^{-3} bar) for 5 h at 40°C. Precautions were taken to eliminate the presence of water or organic solvents in the purified IL. The product was confirmed by IR: 754 cm^{-1}, 1168 cm^{-1}, 1463 cm^{-1}, 1571 cm^{-1}, 1628 cm^{-1}, 2870 cm^{-1}, 2960 cm^{-1}, 3098 $cm^{-1}$and 3152 cm^{-1} and ^1H NMR, (in $CDCl_3$, ppm); 0.93(t), 1.36(sextet), 1.84(q), 4.11(s), 4.31(t), 7.50(s), 7.65(s), 10.38(s).

2.1.2 Synthesis of bmimBr

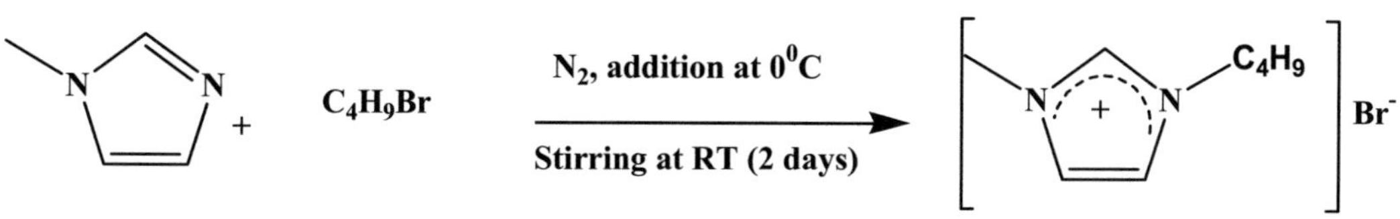

Following the above mentioned procedure, bmimBr was synthesized. Here stirring was carried out for 2 days. The pale yellow coloured solid was obtained which was further treated with activated charcoal to remove coloured impurities. Thus obtained bmimBr was kept under reduced pressure (10^{-3} bar) for 5 h at 40^{0}C. Precautions were taken to eliminate the presence of water or organic solvents in the purified IL. The product was confirmed by IR: 794 cm^{-1}, 1168 cm^{-1}, 1463 cm^{-1}, 1570 cm^{-1}, 1629 cm^{-1}, 2871 cm^{-1}, 2960 cm^{-1}, 3086 cm^{-1}and 3155 cm^{-1} and ^{1}H NMR, (in $CDCl_3$, ppm); 0.94(t), 1.35(sextet), 1.86(q), 4.13(s), 4.31(t), 7.39(s), 7.49(s), 10.35(s).

2.1.3 Synthesis of bmimI

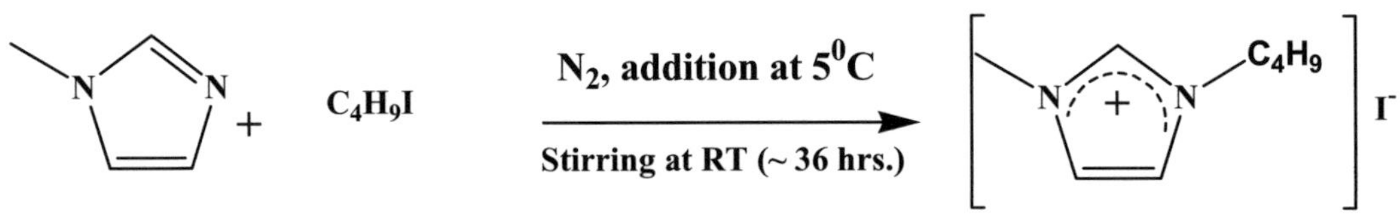

Above reported procedure was also followed to synthesize the bmimI. Pale yellow colour viscous liquid was obtained (yield: 94%) which was then washed with dry ethyl acetate and dry diethyl ether. The pale yellow coloured IL thus obtained was kept under reduced pressure (10^{-3} bar) for 5 h at 40 ^{0}C. Precautions were taken to eliminate the presence of water or organic solvents in the purified IL. The purity of product was confirmed by IR: 752 cm^{-1}, 1167 cm^{-1}, 1461 cm^{-1}, 1568 cm^{-1}, 1628 cm^{-1}, 2870 cm^{-1}, 2958 cm^{-1}, 3098 cm^{-1}and 3138 cm^{-1} and ^{1}H NMR, (in $CDCl_3$, ppm); 0.95(t), 1.39(sextet), 1.90(q), 4.13(s), 4.35(t), 7.53(s), 7.61(s), 9.95(s).

2.1.4 Synthesis of bmimNTf$_2$

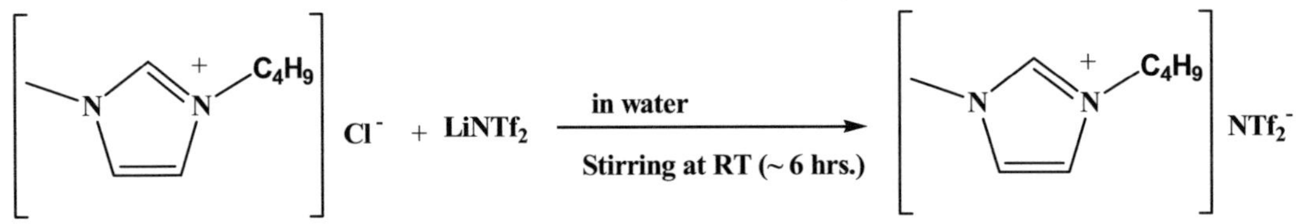

bmimNTf$_2$ was synthesised from bmimCl (7g, 43mmol) which was dissolved in 5 mL of triple distilled water followed by the slow addition of LiNTf$_2$ solution (13g, 47mmol in 13 mL triple distilled water) with stirring. After addition of LiNTf$_2$ solution in bmimCl solution, colour of solution changes to milky white and two layers were separated and the reaction mixture was stirred for further 6 hrs. Whole reaction mixture was then transferred in to a separating funnel and bmimNTf$_2$ was extracted with DCM. Then DCM layer was washed with cold water for removal of chloride ion which was monitored with the help of aqueous acidified solution of $AgNO_3$. DCM solution was dried over anhydrous $MgSO_4$ and then the dried DCM was evaporated providing colourless liquid which was kept under vacuum for removal of solvent and other volatile impurities. (yield: 85%) This was further treated with activated charcoal and then passed through column containing alumina and celite to remove coloured impurities. Precautions were taken to eliminate the presence of water or organic solvents in the purified IL. The product was confirmed by IR: 744 cm^{-1}, 845 cm^{-1}, 1057cm^{-1}, 1140 cm^{-1}, 1194 cm^{-1}, 1350cm^{-1}, 1464 cm^{-1}, 1573 cm^{-1}, 2937 cm^{-1}, 2964 cm^{-1}, and 3161 cm^{-1} and ^{1}H NMR, (in $CDCl_3$, ppm); 0.93(t), 1.32(sextet), 1.79(q), 3.93(s), 4.14(t), 7.26(s), 7.31(s), 8.73(s).

2.1.5 Synthesis of bmimPF_6

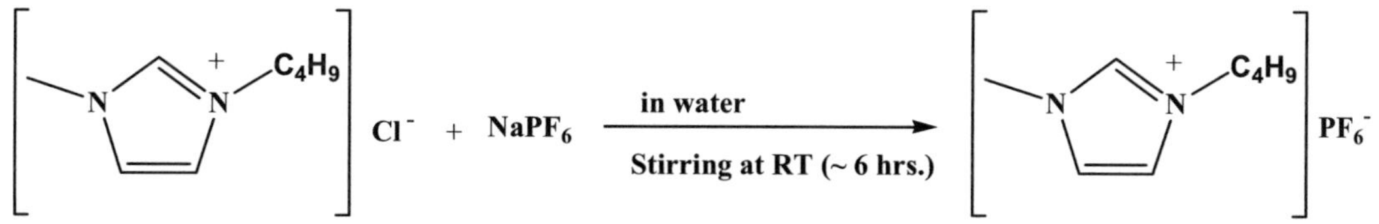

Above procedure reported for synthesis of bmimNTf_2 was followed for synthesizing bmimPF_6 also. Here instead of LiNTf_2, NaPF_6 was added to bmimCl solution. Yield was found to be 86%. The product was confirmed by IR: 748 cm^{-1}, 839 cm^{-1}(P-F stretching), 1112 cm^{-1}, 1164 cm^{-1}, 1464 cm^{-1}, 1573 cm^{-1}, , 2876 cm^{-1}, 2965 cm^{-1} and 3169 cm^{-1} and ^{1}H NMR, (in $CDCl_3$, ppm); 0.94(t), 1.23(sextet), 1.75(q), 3.94(s), 4.05(t), 7.26(s), 7.29(s), 8.33(s).

2.1.6 Synthesis of bmimBF_4

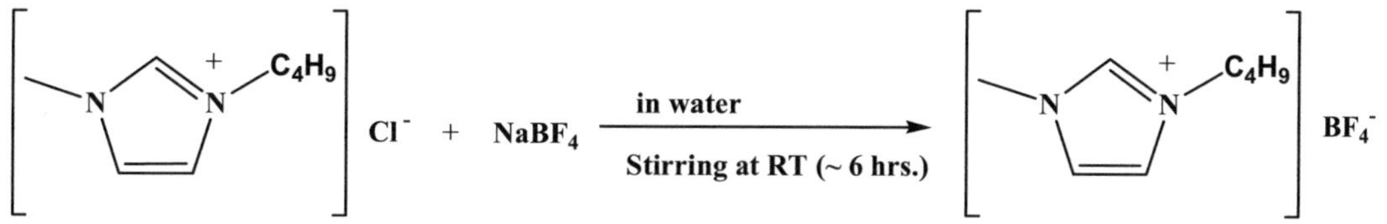

For synthesis of bmimBF_4 above mentioned procedure was followed, except that NaBF_4 was added instead of LiNTf_2. Yield of the product was found to be 80% Precautions were taken to eliminate the presence of water or organic solvents in the purified IL. The product was confirmed by IR: 757 cm^{-1}, 849 cm^{-1}, 1053cm^{-1} (B-F stretching), 1169 cm^{-1}, 1464 cm^{-1}, 1573 cm^{-1}, 1625 cm^{-1}, 2876 cm^{-1}, 2963 cm^{-1}, 3121 cm^{-1} and 3162 cm^{-1} and ^{1}H NMR, (in $CDCl_3$, ppm); 0.85(t), 1.29(sextet), 1.81(q), 3.89(s), 4.15(t), 7.33(s), 7.41(s), 8.82(s).

3. Computational method

The Gaussian 03 program [32] was used for the Density Functional Theory (DFT) calculation of different ILs. The basis sets already implemented in the program were used for the different types of calculations. The geometries of isolated ion pairs of ILs were optimized at the Becke's three parameter hybrid method with LYP correlation (i.e. B3LYP).[33] While 6-31G++(d,p) basis set was used for C, H, N, O, S, F and P, DGDZVP basis set was used for I atom.[34] Due to large atomic number of iodine atom, 6-31 G++(d,p) basis set is found to be incapable of predicting accurate structure and vibrational spectrum of bmimI. Optimisation, frequency calculation and Time-Dependent Density functional Theory (TD-DFT) were also performed at the B3LYP level of calculation. The absence of imaginary vibrational frequencies in vibrational spectrum ensures the presence of a true minimum.

4. Results and discussion

4.1 Studies of interactions by single crystal x-ray diffraction and Raman spectroscopic techniques

Among all imidazolium cation based ILs mentioned above, bmimCl and bmimBr are the two prototype ILs that have been studied extensively. [4, 21, 31] The interesting aspects which make these two monoatomic anion based ILs well studied is their existence both in solid and super cooled liquid phases. Therefore these two ILs give the opportunity to study the structures and interactions present in both solid and liquid state with a possibility to make a correlation. Crystal structures of bmimCl and bmimBr at room temperature have been well studied.[4,21,31] It has been discovered simultaneously and independently by,

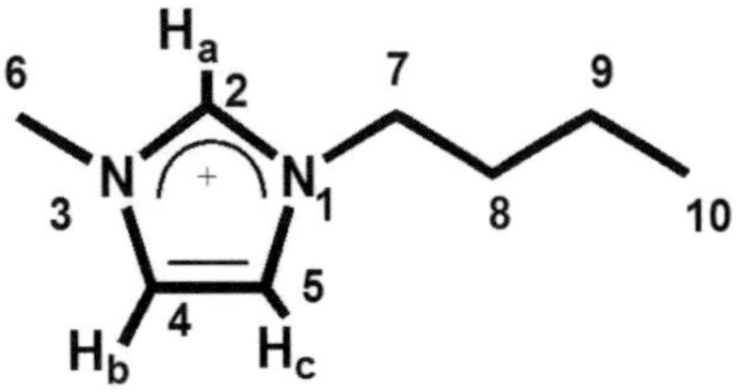

Fig. 1. Schematic representation of bmim catiom [bmim: 1-butyl-3-methylimidazolium]

Saha et. al. [4] and Holbrey et. al.[31] that bmimCl crystal shows two crystal polymorphism. Schematic diagram of bmim cation has been presented in Figure 1. In crystal 1 (Figure 2a), the *n*-butyl group takes *trans-trans* (**TT**) conformation with respect to C_7-C_8 and C_8-C_9 bonds, while in crystal 2 (Figure 2b), n-butyl group takes *gauche-trans* (**GT**) conformation with respect to C_7-C_8 and C_8-C_9 bonds similar to that of bmimBr conformation.[4, 25] Powdered x-ray diffraction patterns of the two polymorphs of bmimCl and bmimBr was recorded and found that the pattern for crystal 2 and bmimBr somewhat resembles each other, while that of bmimCl crystal 1 is distinct from the other two.[4,21]

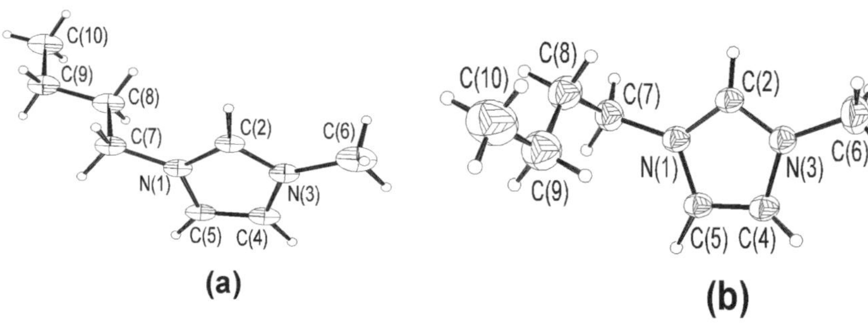

Fig. 2. The structure of bmim cation in bmimCl. (a) n-butyl chain has *trans-trans* conformation with respect to C7-C8 and C8-C9 bonds and therefore described as Crystal 1 or TT in text. (b) The crystal structure of bmim cation in bmimBr. In this case, n-butyl chain has *gauche-trans* conformation with respect to C7-C8 and C8-C9 bond and therefore termed as GT or Crystal 2 in the text.

Presence of polymorphs is found to be due to the different interactions in the crystal systems. In crystal 1, a couple of bmim cations form a pair through a *hydrophobic interaction* among stretched n-butyl group. Those pairs stack together and form a column in which all imidazolium ring planes are parallel with one another. Two types of cation columns with different orientations exist. Hamaguchi and co-workers have been studied these ILs extensively by Raman spectroscopic techniques.[21] It was found that Raman spectra of two polymorphs are markedly different from each other. In the wavenumber region 600-700 cm^{-1}, where ring deformation bands are expected, two bands appear at 730 cm^{-1} and 625 cm^{-1} in bmimCl crystal 1 (i.e., **TT** conformation), while another couple of bands appear at 701 and 603

cm^{-1} in bmimCl crystal 2 (**GT**) and bmimBr but not in bmimCl crystal 1. In order to clarify the origin of these bands, Ozawa et. al. has done extensive DFT calculation on these Ils.[22] It shows that the 625 cm^{-1} band of bmimCl crystal 1 and the 603 cm^{-1} band of bmimBr originates from the same ring but have different magnitude of couplings with the CH_2 rocking vibration of the C_7 carbon. The coupling occurs more effectively for the *gauche* conformation around the C_7-C_8 bond, resulting in a lower frequency in the GT form (603 cm^{-1}) than in the TT form (625 cm^{-1}). It was noted that the coupling with the CH_2 rocking mode having a higher frequency pushes down the frequency of the ring deformation vibration. The same coupling scheme holds for another ring deformation mode and the GT form has a lower frequency (701 cm^{-1}) than the TT form (730 cm^{-1}). It is therefore elucidated from the DFT calculation that the 625 and 730 cm^{-1} bands are the characteristics of the *trans* conformation around the C_7-C_8 bond, while the 603 and 701 cm^{-1} bands are the characteristics of the *gauche* conformation. In other words, one can use these bands as key bands to probe the conformation around the C_7-C_8 bond of the bmim cation. The 500 cm^{-1} band of the bmimBr is ascribed to the C_7-C_8-C_9 deformation vibration of the *gauche* conformation around the C_7-C_8 bond. The Raman spectra of all *liquid* bmimX (where X= Cl, Br, I, BF_4, PF_6) found to be similar, suggesting that bmim cation is similar for all these liquids. Both the two sets of marker bands, 625 and 730 cm^{-1} for the *trans* conformation and the 603 and 701 cm^{-1} band for the *gauche* conformation appears in all the liquid spectra. Therefore, at least two rotational isomers, one having a *trans* conformation and other having *gauche* conformation around the C_7-C_8 bond co-exist in liquid bmimX. The physical properties of these two polymorphs are found to be different, e.g., the melting point of crystal 1 is 41^{0}C while that of crystal 2 is 66^{0}C. Different melting points are due to different types of interactions present in the system which is further due to the different kind of molecular conformations (e.g., TT, GT etc.).

4.2 Studies of interaction through ^{1}HNMR

NMR data clearly explain the strength of interaction between cation and anion. Figure 3 shows ^{1}HNMR spectra of bmimX. C_2-H proton shiftings were recorded to be 10.38ppm,

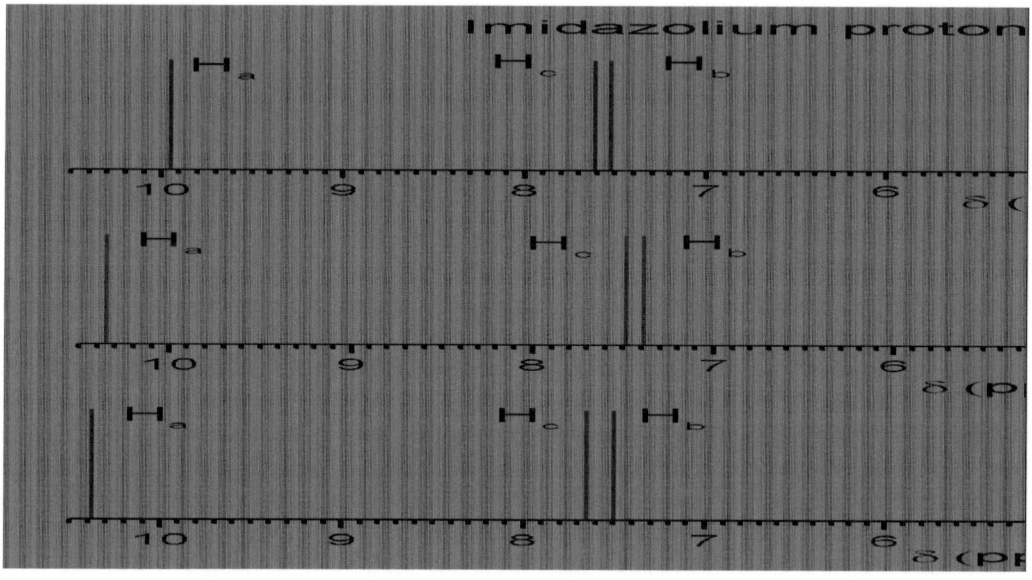

Fig. 3. ^{1}HNMR shift of 1-butyl-3-methylimidazolium (bmim) cation with various halide anions.

10.35ppm, 9.95ppm in bmimCl, bmimBr and bmimI respectively, explaining that H-X interaction is strongest in bmimCl and then in bmimBr and least in bmimI. When these values compared with the bmimNT_2, bmimBF_4 and bmimPF_6, this shift was found to be 8.73 ppm, 8.82 ppm and 8.33 ppm respectively, explaining that interaction between C_2-H and anion in these ILs is weak when compared with bmimCl, bmimBr and bmimI. Variation of IR spectra which is discussed later in this chapter have been found to be nicely correlated with the NMR studies of these ILs.

4.3 Study of interactions in molecular structure of different ILs using DFT calculations

The optimized molecular structure of bmimX (where X= Cl, Br, I, PF_6, BF_4 and NTf_2) determined at B3LYP/6-31G++(d,p) level of computation (except DGDZVP basis set used for I atom) are shown in **Figure 4(a-f)**

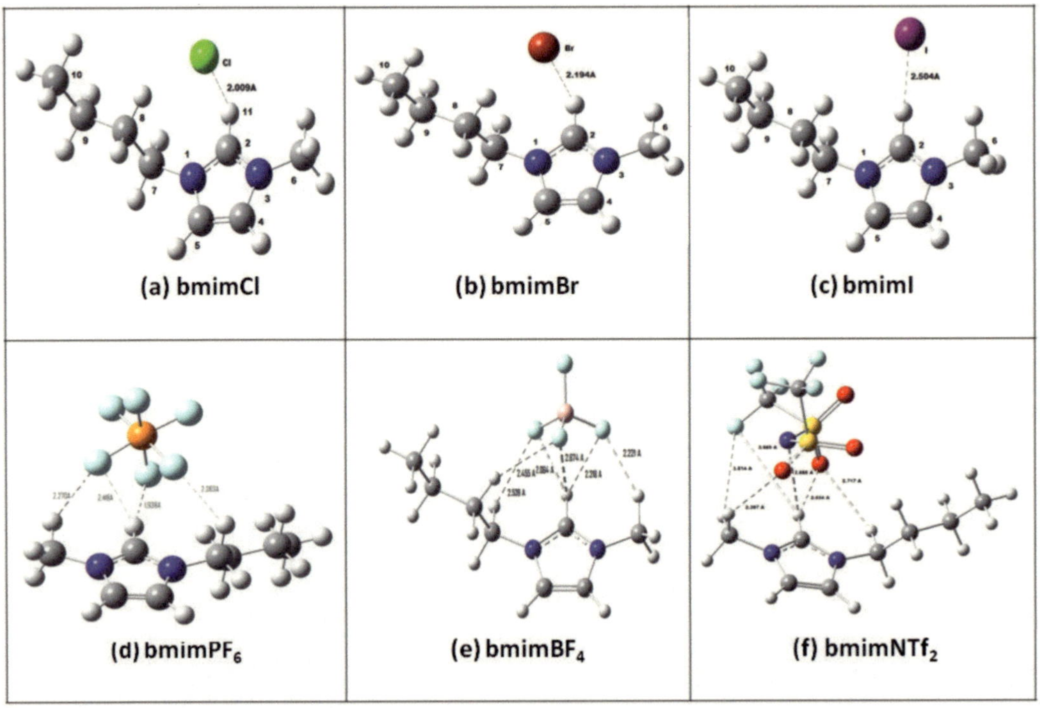

Fig. 4. DFT optimized structure of different imidazolium based Ils in gas Phase

The optimized structures reveal that the imidazolium ring is a planar pentagon as expected. The average bond length of 1.34 Å for the N1-C2 and C2-N3 in all the three ILs indicates the nature of conjugated double bond character. Further N1-C5 and N3-C4 bond lengths are (1.38 Å) shorter than a pure C-N single bond (1.47 Å). Bond length of C4-C5 is in the range of normal C=C bond (1.33 Å). It can be seen from the **Table 1** that dihedral angles (~180^0) of the butyl group explain the *trans* conformations in all 1-butyl-3-methyl-imidazolium halide ILs studied. Similar observation was observed for bmimPF_6, bmimBF_4 and bmimNTf_2 ILs. While the C^2-H-----X distance increases gradually from Cl^- to I^- following the trend of electronegativity (Cl^-=3.0, Br^-=2.8, I^-=2.5), the bond angle also decreases gradually from bmimCl to bmimI. The C_2-H---Cl interaction is thus found to be the strongest among all and the order was found to be C_2 -H----Cl > C_2- H----Br > C_2-H---I. Further the position of anion was found to be crucial in determining whether an IL will be liquid at ambient temperature or not. As mentioned earlier that bmimI exist in liquid state (m.p.-72°C) whereas its corresponding chloride, (bmimCl) and bromide (bmimBr) ILs are solids at 20°C. As mentioned earlier, melting point of bmimCl are 41°C and 66°C depending on the type of crystal polymorph, while that of bmimBr is 79°C.[21] From our calculations it is observed that while Cl^- and Br^- ions are present in the plane of the imidazolium ring, position of iodide anion is found to be below or above the plane.[23,34] The position of Cl^- and Br^- are satisfactorily matching with what was found in single crystal x-ray diffraction data. [21] This led us to conclude that because of the small size of the Cl^- and Br^-, there is compact packing of ions in the crystal lattice leading to higher melting points. Due to the large size of iodide anion, it does not fit well in the lattice. This makes it difficult to have close packing leading to liquid state at ambient temperature. It is important to mention here that in addition to ion size, interionic interactions also play major role in determining the state of an IL. Further, various dihedral angles of type <N_1-C_2-H-X (see **Table 1**) also indicates that the Cl^- and Br^- anions are in the plane of the imidazolium ring whereas I^- ion is out of plane.[17]

Parameters	bmimCl	bmimBr	bmimI
C_2-H	1.12 Å	1.11 Å	1.09 Å
C^2H---X	2.01 Å	2.19 Å	2.50 Å
C_2-N_1	1.34 Å	1.34 Å	1.34 Å
C_2-N_3	1.34 Å	1.34 Å	1.34 Å
N_1-C_5	1.38 Å	1.38 Å	1.38 Å
N_3-C_4	1.38 Å	1.38 Å	1.38 Å
C_4-C_5	1.36 Å	1.36 Å	1.36 Å
N_3-C_6	1.46 Å	1.46 Å	1.46 Å
<N-C-N	108.4^0	108.4^0	108.4^0
<C-H-X	158.8^0	153.7^0	143.4^0
<C_7-C_8-C_9-C_{10}	178.9^0	177.8^0	179.9^0

Table 1. Selected bond lengths, bond angles and dihedral angles as obtained from DFT calculation.

From **Figure 4(a-f)** it is clear that in bmimCl, bmimBr and bmimI, only one H-bonding was found with C_2-H, in a specific orientation whereas in case of BF_4, PF_6 and NTf_2 based imidazolium ILs multiple H-hydrogen bonding are seen. While in bmimCl, bmimBr and bmimI, C_2-H---X distances were found to be 2.01 Å, 2.19 Å and 2.50 Å respectively, in case of bmimPF_6, H----F distance was found to be 2.418 Å, 1.93 Å, 2.27 Å and 2.36 Å, in bmimBF_4, H----F distance found to be 2.53 Å, 2.45 Å, 2.21 Å and 2.08 Å and in bmimNTf_2, in addition to H---F distance (3.51 Å and 3.66 Å) O-H distance found to be 2.03 Å, 2.39 Å and 2.71 Å. Further in addition to the C2-H hydrogen bonding, C4-H and C5-H also participate in H-bonding interaction with halide anion but stabilization due to H-bonding interaction between cation and anion is found to be more in case of interaction involving C2-H compared to that of involving C4 and C5-H. [35] It is because, while C2 is positive (due to the electron deficient π bond formation of C=N bond), the C4 and C5 are essentially neutral (due to π bond formation between two carbon sharing equally the available electrons). [36]

4.4 Calculated vibrational spectra, analysis and its correlation with experiment

Vibrational spectra of all the ILs has been calculated at same level of calculation and using the same basis set in gas phase. The calculated vibrational spectra for bmimCl, bmimBr and bmimI have been shown in **Figure 5** and the spectral bands with the corresponding assignments are presented in **Table 2**. Vibrational frequencies corresponding to different vibrational motions such as stretching, in plane bending (*ip*) out of plane bending (*op*), wagging, rocking etc were assigned based on the predominant motion of the atoms and magnitude of the displacement vector associated with the atoms involved in concerned vibration. Most of the assignments were found to be matching with the literature values of similar compounds.[27] It is interesting to note that wavenumber for most of the motions for bmimCl and bmimBr are nearly same, but differs considerably for the same motion in bmimI. Among all, the most prominent and interesting observation is the shift of C_2-H stretching band in different ILs studied. While the band appears at 2652 cm^{-1} for bmimCl, 126 cm^{-1} blue shift is observed for the same in bmimBr. A further blue shift of 212 cm^{-1} is observed in bmimI (i.e., appears at 2990 cm^{-1}).

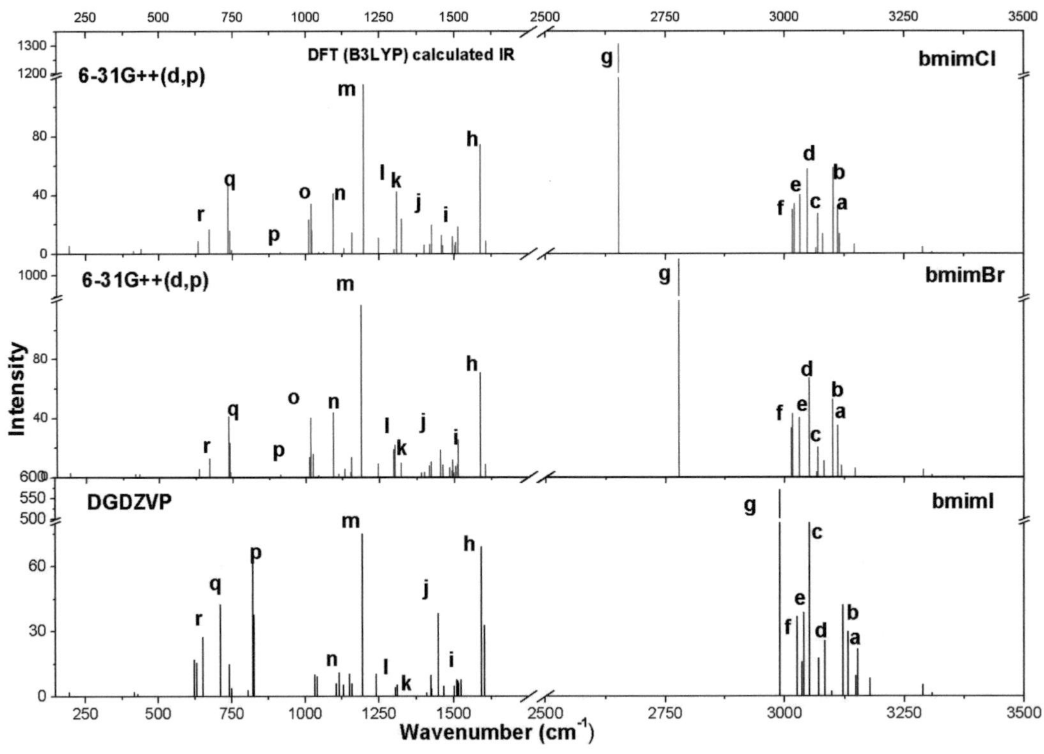

Fig. 5. DFT calculated IR spectra of bmimX where X=Cl^-, Br^-, I^- indicating the blue shift of the C_2-H stretching band (g) with different anions.

This can be explained considering the fact that this hydrogen participates in hydrogen bond formation with the halide anion. Since the strength of hydrogen bond depends on the type of anion (mainly on its electronegativity), shift in frequencies were observed.

Bands	Wavenumber /cm^{-1}			Assignment of bands
	bmimCl	bmimBr	bmimI	
a	3110	3110	3132	Asym. stretching of terminal CH_3 group
b	3101	3100	3121	Asym. stretching of C-H bond in butyl group
c	3069	3070	3052	Sym. stretching of C-H in N-CH_3
d	3048	3051	3084	Sym. stretching of C-H in N-CH_2
e	3032	3031	3040	Sym. stretching of C-H in terminal CH_3 group
f	3016	3016	3027	Sym. stretching of H-C-H in butyl group
g	2652	2778	2990	stretching of C_2-H bond
h	1588	1587	1593	*ip* bending of im. ring
i	1514	1513	1517	Scissoring of H-C-H in butyl group
j	1424	1423	1424	Wagging of H-C-H in butyl group
k	1323	1322	1327	Rocking of H-C-H in butyl group
l	1307	1300	1304	*ip* bend of im ring
m	1198	1190	1198	*ip* bend of im ring, CCCC str
n	1095	1094	1118	stretching of C-N(CH_3) & bending in N-CH_3
o	1019	1017	-	*op*. bend of C_2-H
p	841	844	827	Rocking in H-C=C-H ring
q	734	739	710	Wagging of HC=CH
r	669	671	651	*op* bending of im. ring

Where sym- symmetric, asym- asymmetric, ip- inplane, op- out of plane

Table 2. Selected DFT calculated harmonic vibrational frequencies for bmimX ILs in gas phase and their band assignment.

In other word, this shift in C_2-H stretching frequency is an indicator of the strength of the hydrogen bond which is formed between C_2-H and halide anion. These data suggests that the hydrogen bonding is strongest in case of bmimCl and weakest in bmimI.[23] This observation is exactly matching with the finding of ^{1}H NMR results asdiscussed earlier in this chapter. Further the single crystal X-ray crystallographic data that are available for bmimCl and bmimBr also corroborate with our results discussed here.[4,17,25] In addition, as expected, the position of the band arising due to 'out of plane' bending motion of C_2-H is also sensitive to the type of anion. Interestingly the I^- is found to be present out of plane of imidazolium ring which restricts the out of plane bending of C_2-H bond, making it absent in the IR spectrum whereas band at 1019 cm^{-1} in bmimCl and 1017 cm^{-1} in bmimBr are due to out of plane bending of C_2-H bond. Hence in bmimCl and bmimBr, Cl^- and Br^- found to exist in plane of imidazole ring and hence provide strong hydrogen bonding with C_2-H and therefore found to exist in solid state. Whereas in case of bmimI, I^- is present out of plane thus providing less interaction with C_2-H and therefore found to exist in liquid state.[23] Infrared (IR) spectrum of synthesised bmimI is compared with the calculated one and is presented in **Figure 6**. The overall correlation of band positions in calculated vibrational frequencies agrees reasonably well with that obtained experimentally. Hence, it appears thatB3LYP level of calculation gives reasonably good correlation between the experimental

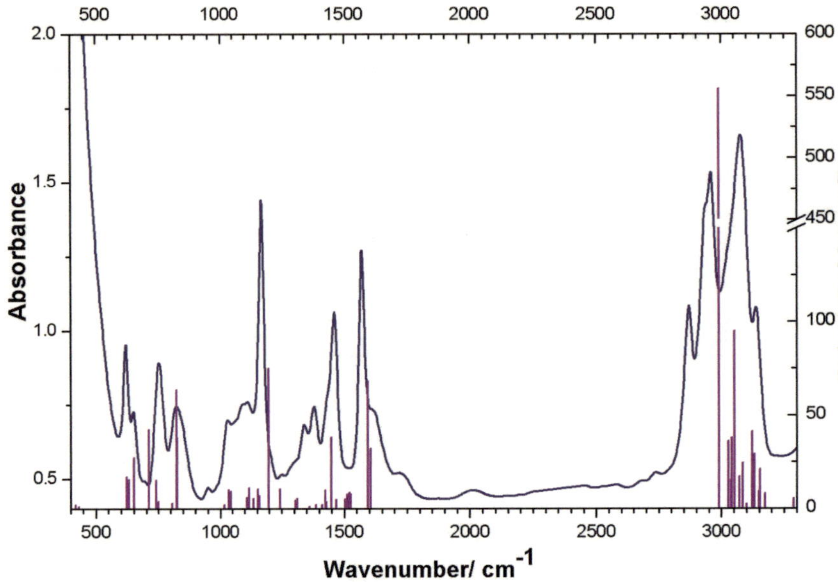

Fig. 6. Correlation of experimental and DFT calculated IR spectra of bmimI

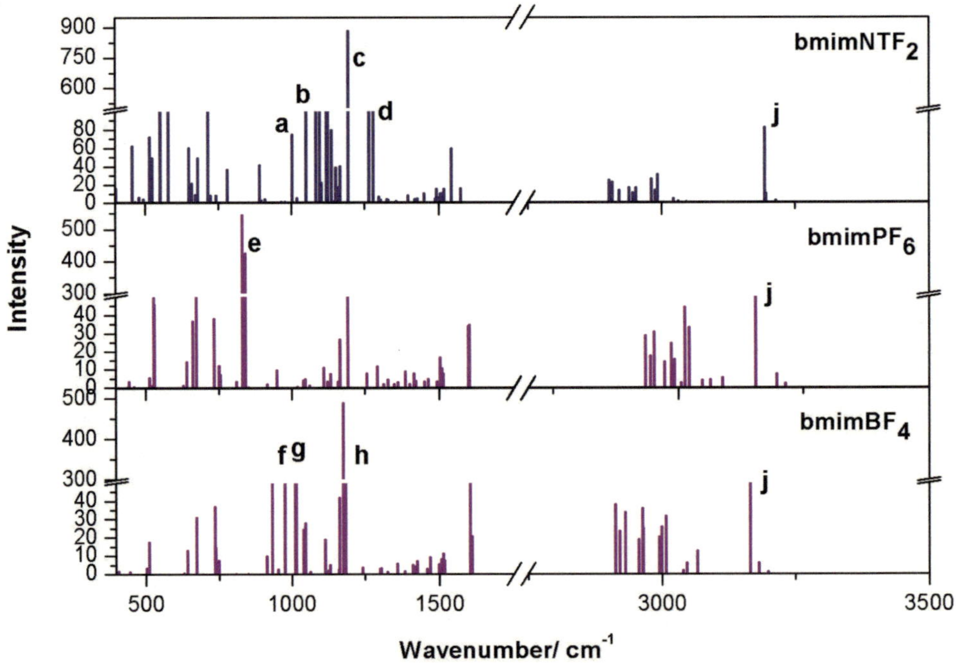

Fig. 7. DFT calculated IR spectra of bmimX where X=PF_6, BF_4 and NTf_2. Important bands have been marked. a, b, c, d bands are for C- S , C-F, N-S, S=O stretching respectively in $bmimNTf_2$. e band is for P-F stretching in PF_6 and f, g, h indicate in plane B-F stretching, and vertical B-F stretching in BF_4. Band marked as j indicates C_2-H stretch in all three ILs.

vibrational stretching frequency which is anharmonic in nature and the DFT calculated vibrational frequency which is harmonic in nature. Frequency of low wavenumber is found to match excellently while that of higher wavenumber deviated from theoretical values which are due to the increase in anharmonicity at higher wavenumber.

Hence a scaling factor of 0.9889 was used for all calculated values above 3000 cm^{-1}. The experimental results can be interpreted and rationalized in the light of theoretical investigations. Table 2 presents the assignments of different bands based on calculation. DFT calculated IR spectra for $bmimNTf_2$, $bmimPF_6$ and $bmimBF_4$ at B3LYP/6-31G++(d,p) level of calculation are shown in **Figure 7.** In $bmimNTf_2$ spectra S=O, N-S, C-F and C-S bond found to be at 1281 cm^{-1}, 1195 cm^{-1}, 1122 cm^{-1} and 1046cm^{-1}.[37] In $bmimPF_6$, P-F symmetric and asymmetric stretching found at 827 cm^{-1} and 841 cm^{-1} respectively. In $bmimBF_4$, B-F stretching (vertical) found at 1174 cm^{-1} and B-F stretching (in plane) found at 1009 cm^{-1} and 975 cm^{-1}. In higher wavenumber region intense peak found at 3373 cm^{-1}, 3233 cm^{-1} and 3278 cm^{-1} in $bmimNTf_2$, $bmimPF_6$ and $bmimBF_4$ respectively which is due to C_2-H stretching in imidazolium cation. Hence with variation of anion, C_2-H stretching frequency shift is observed. Experimental IR spectra of the $bmimNTf_2$, $bmimPF_6$ and $bmimBF_4$ ILs which we have synthesized have been correlated with the DFT calculated data and are shown in **Figure 8(a), 8(b) and 8(c)** respectively. To correlate well a scaling factor of 0.943, 0.975 and 0.965 has been applied in higher wavenumber region (above 3000cm^{-1}) for $bmimNTf_2$, $bmimPF_6$ and $bmimBF_4$ respectively. Hence the DFT calculated results found to be matching well with the experimental data.

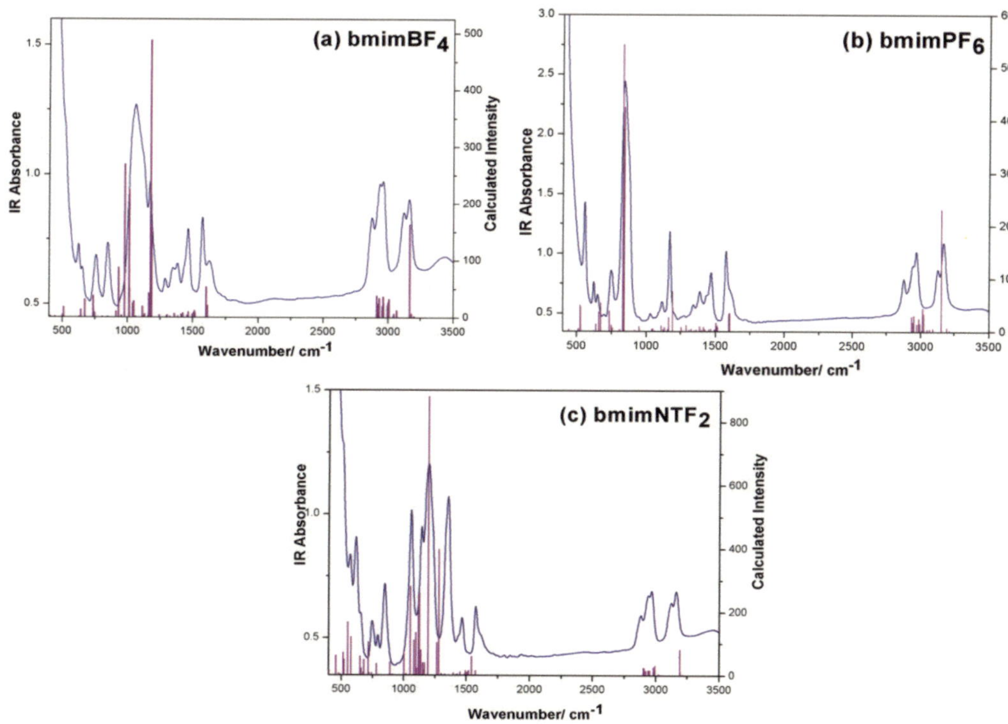

Fig. 8. (a). Experimental and DFT calculated IR spectra of (a) $bmimBF_4$. (b) $bmimPF_6$. (c) $bmimNTf_2$

As can be seen from the above figures, DFT can predict very well the vibrational bands of various ILs as discussed above. Therefore, assignment of experimental bands becomes a easy task, thereby helping us to understand completely the structure and interactions present in the ILs.

4.5 Transitions in bmimX: UV-Vis spectra and TD-DFT calculations

The UV-Visible spectrum of neat bmimCl and bmimBr could not recorded in absorbance mode, as they are solid at ambient temperature. The UV-Visible spectrum of pale yellow coloured bmimI was recorded and is presented in **Figure 9**.

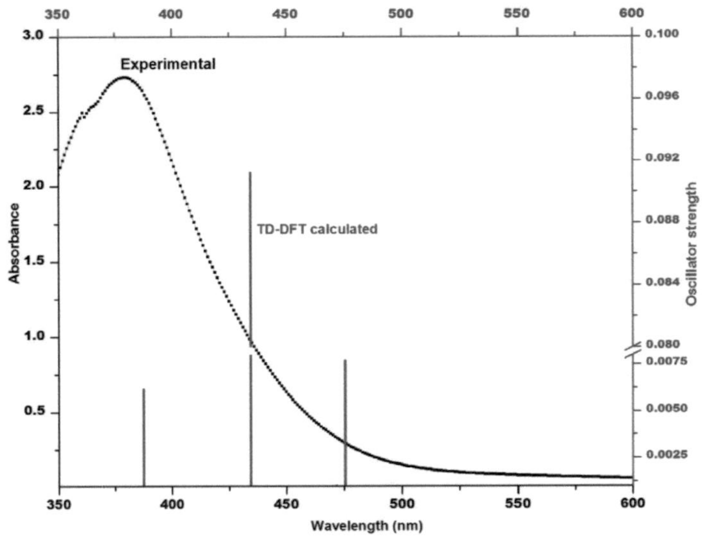

Fig. 9. UV-Vis spectrum (smooth line) of neat bmimI RTIL recorded using quartz cuvette having 5mm path length (Vertical lines correspond to calculated transitions using TD-DFT calculation))

While the peak of the spectrum appears at about 380 nm, long absorption tail is extended to about 525 nm. The interesting aspect is that the bmimI has pale yellow colour while the analogous derivatives of pure chloride and bromide are colourless. It is well known fact that during the process of synthesising these ILs, often colour impurity makes the resultant ILs coloured. In most of the cases improperly purified or unpurified ILs shows pale colour whereas on persistent and proper purifications, the final IL comes out to be colourless.[26, 38] So, to understand whether the bmimI is inherently pale yellow coloured or the colour is due to presence of impurity, time dependent DFT (B3LYP) calculations were performed on bmimCl, bmimBr and bmimI. For bmimCl and bmimBr, transitions found to occur below 356 nm (only in UV region), whereas in case of bmimI, transition found to occur in visible region also. The wavelengths of calculated transitions for bmimI, along with their oscillator strengths are presented in Figure 10. Even though the calculated oscillator strengths of various possible transitions do not exactly corresponds with that of experimentally observed spectrum (vide Figure 10), it does strongly indicate couple of transitions (though weak, at 371 and 380nm) in visible range.

The important point here to note is that a couple of transitions also occur at much longer wavelengths 434 and 475 nm having reasonably good oscillator strengths. Thus it is confirmed that the pale yellow colour of the bmimI is inherently due to the transition in bmimI itself.

Transition Wavelength (nm)	Transition occurring from one MO to other MO	Oscillator strength	Energy (eV)
475	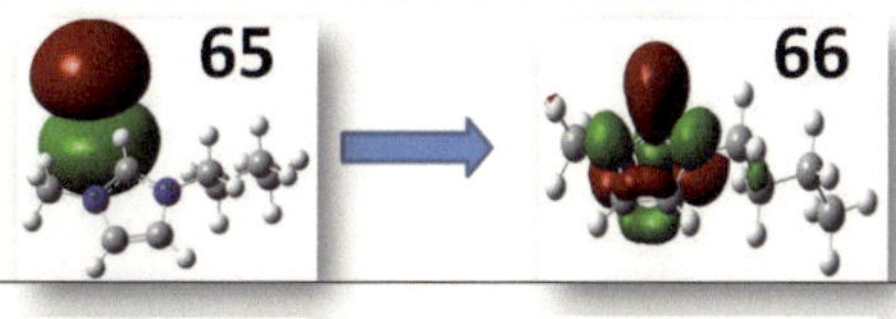	0.0027	2.6079
471	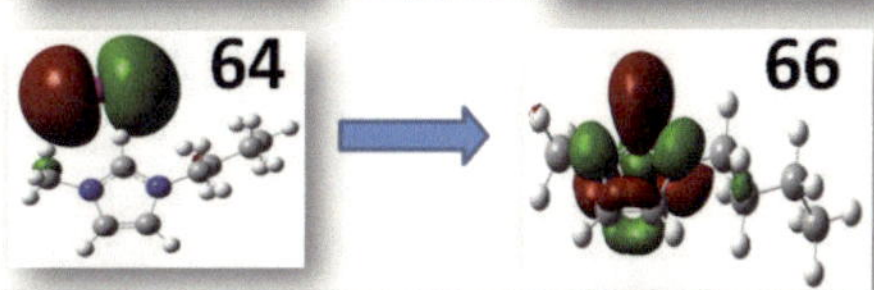	0.0002	2.6293
434	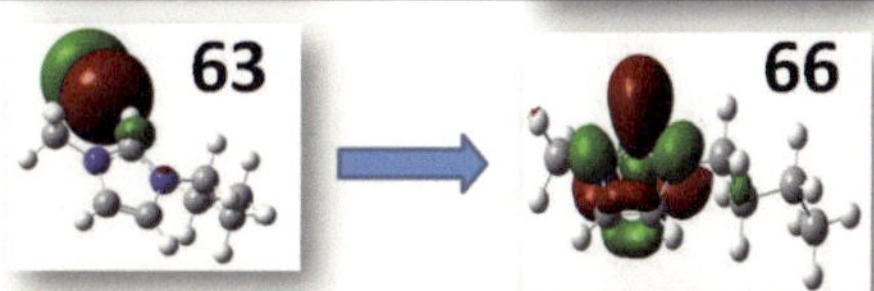	0.0912	2.8549
387	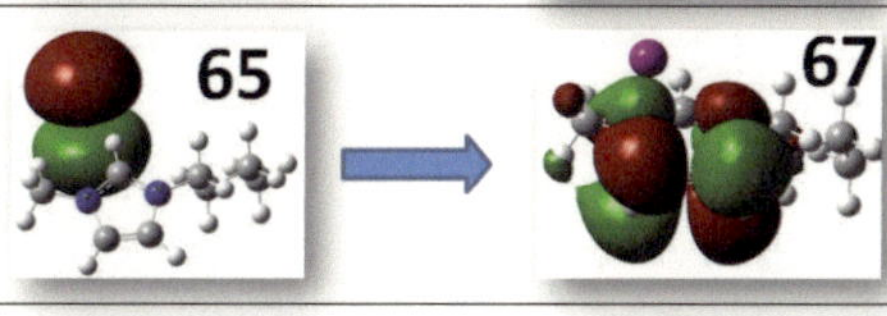	0.0017	3.2024
380	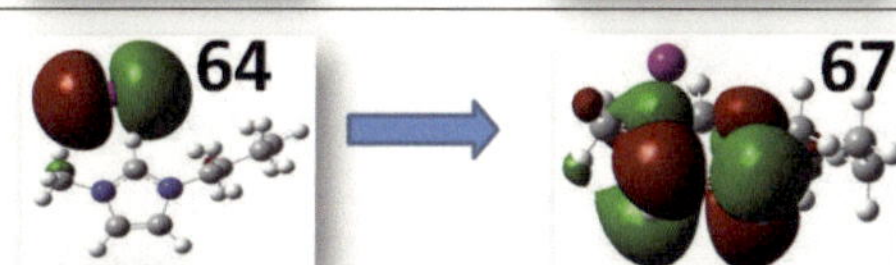	0.0011	3.2593
371	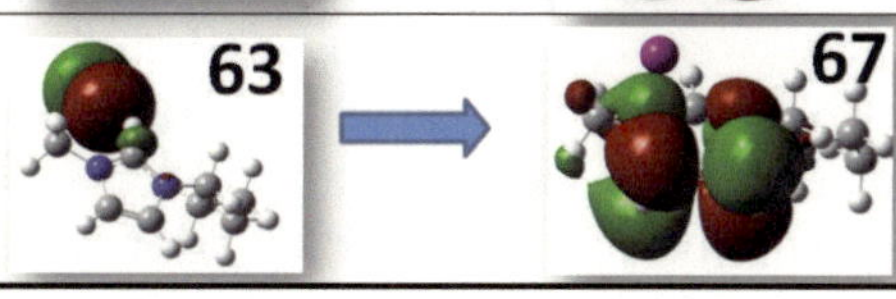	0.0008	3.3356

Fig. 10. Some important Frontier Molecular Orbitals of bmimI ion pair are shown which is obtained by calculation at B3LYP/DGDZVP level.

Being confirmed that the observed colour of bmimI is inherent to this IL, we have investigated the involvement of orbitals through Molecular Orbital (MO) analysis. **Figure 10** depicts the orbitals responsible for various low energy transitions. From this frontier molecular orbital pictures it is clear that HOMO is the non-bonding orbital localized on iodide ion (due to presence of lone pair electron) while LUMO is mainly localized on imidazolium ring and is antibonding in nature (π*). Thus the transition which is responsible for pale yellow colour of bmimI is due to charge transfer transition occurring from HOMO on iodide anion to LUMO on imidazolium cation. The strongest transition in the visible range (434nm) was found to be from MO 63 to MO 66 (shown in Figure 10) having oscillator strength of 0.0912.

Hence the pale yellow colour of bmimI is found to be intrinsic of this room temperature IL and is due to the charge transfer from iodide anion to imidazolium ring (n → π*). UV-visible spectra of bmimNTf2, bmimPF6 and bmimBF4 were also recorded and these ILs were found to be colourless liquids. Similar TD-DFT calculations for these RTILs are also done and supports the experimental observation. In all of these RTILs, as shown in Figure 11 (a-c) transitions found to occur in UV region only, that is why, they are transparent in visible region.

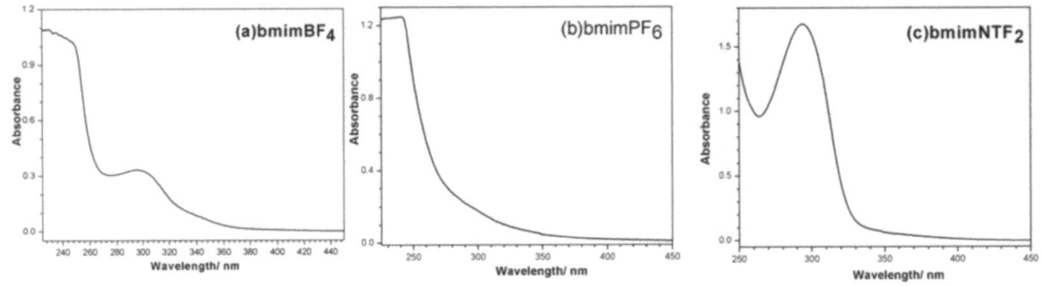

Fig. 11. UV-Visible spectra of (a) $bmimBF_4$,(b) $bmimPF_6$ and (c) $bmimNTf_2$. Quartz cuvette with 1cm path length was used for all the measurements.

Because of their clear optical window, these RTILs have extensively been used as a medium for photophysical studies of various probe molecules and solvation dynamics in recent years. [30, 39-41]

5. Conclusion

Modified synthesis of different imidazolium based ILs having simple mono atomic to complex anions have been described here. In addition to usual characterizations, ^{1}HNMR spectra are used to extract information on existence and strength of H-bondings present in the ILs having the mono-atomic anion. IR spectroscopy has extensively been used to analyse the interaction present in the ILs. A through DFT calculation at B3LYP level have been performed on the all the ILs to get fully optimized geometry and obtained complete vibrational spectra. These calculations help us to understand the structure and existence of various interactions including H-bonding interactions. Monoatomic anion based ILs (bmimCl, bmimBr and bmimI) found to have only one predominant H-bonding between cation and anion through C_2-H while multiple H-bonding interactions were found to exist in $bmimNTf_2$, $bmimPF_6$ and $bmimBF_4$ ILs. TD-DFT calculations were also presented to understand and interpret the experimental UV-Visible spectra of various ILs. That iodide ILs have pale yellow color while corresponding chloride and bromide ILs are colorless can well be explained with the TD-DFT calculations.

6. Acknowledgement

Financial assistance from CSIR, India (01(2210)/08/EMR-II) and DST, India (SR/FTP/CS-70/2006) are gratefully acknowledged. MLS and NS thanks UGC and CSIR respectively for fellowships. SS thanks Banaras Hindu University for providing the infrastructure and facilities.

7. References

[1] Plechkova, N. V. & Seddon, K. R., (2008) *Chem. Soc. Rev. Vol. 37*, pp. 123.
[2] Samanta, A., (2010). *J. Phys. Chem. Lett., Vol. 1*, pp. 1557.
[3] Saha, S. & Hamaguchi, H., (2006) *J. Phys. Chem. B. Vol. 110*, pp. 2777.
[4] Saha, S., Hayashi, S., Kobayashi, A. & Hamaguchi, H., (2003) *Chem. Lett. Vol. 32*, pp. 740.
[5] Zhao, H. & Malhotra, S. V., (2002) *Aldrichimica Acta Vol. 35*, pp. 75.
[6] Docherty, K. M., Dixon, J. K. & Kulpa, C. F. Jr (2007) *Biodegradation 18*, pp. 481.
[7] Zhang, C., Wang, H., Malhotra, S. V., Dodge, C. J. & Francis, A. J., (2010) *Green Chem., Vol. 12*, pp. 851.
[8] Domańska, U., (2010) *Int. J. Mol. Sci., Vol. 11*, pp. 1825
[9] Cadena, C., Zhao, Q., Snurr, R. Q., & Maginn, E. J., (2006) *J. Phys. Chem. B, Vol. 110,* pp. 2821.
[10] Kunze, M., Jeong, S., Paillard, E., Winter, M., & Passerini, S., (2010) *J. Phys. Chem. C, Vol. 114,* pp. 12364.
[11] Johansson, P., Fast, L. E., Matic, A., Appetecchi, G. B. & Passerini, S., (2010) *J. Power Sources Vol. 195,* pp. 2074 .
[12] Santos, C. S., Murthy, N. S., Baker, G. A., & Castner, E. W., Jr. (2011) *J. Chem. Phys. Vol.* 134, pp.121101-1
[13] Erdmenger, T., Vitz, J., Wiesbrock, F. & Schubert, U. S., (2008) *J. Mater. Chem., Vol. 18,* pp. 5267.
[14] Emelyanenko, V. N., Verevkin, S. P. & Heintz, A., (2009) *J. Phys. Chem. B, Vol. 113.* pp. 9871
[15] Fredlake, C. P., Crosthwaite, J. M., Hert, D. G., Aki, S. N. V. K. & Brennecke, J. F., (2004) *J. Chem. Eng. Data, Vol. 49*, pp 954.
[16] Anderson, J. L., Ding, J., Welton, T. & Armstrong, D. W., (2002) *J. Am. Chem. Soc., Vol. 124*, pp 14247.
[17] Logotheti, G. E., Ramos, J., & Economou, I. G., (2009) *J. Phys. Chem. B Vol. 113*, pp. 7211.
[18] Talaty, E. R., Raja, S., Storhaug, V. J., Dolle, A., & Carper, W. R. (2004) *J. Phys. Chem. B Vol. 108*, pp. 13177.
[19] Dong, K., Zhang, S., Wang, D. & Yao, X. (2006) *J. Phys. Chem. A, Vol. 110,* pp. 9775.
[20] Danten, Y., Cabaco, M. I. & Besnard, M., (2009) J. *Phys. Chem. A, Vol. 113,* pp. 2873
[21] Ozawa, R., Hayashi, S., Saha, S., Kobayashi, A., & Hamaguchi, H., (2003) *Chem. Lett. Vol.* 32, pp. 948.
[22] Hamaguchi, H. & Ozawa, R., (2005) *in Adv. Chem. Phys. Vol.* 131, pp. 85.
[23] Shukla, M., Srivastava, N. & Saha, S., (2010) *J. Mol. Struc. Vol.* 975, pp. 349.
[24] Chang, H. C., Jiang, J. C., Chang, C. Y., Su, J. C., Hung, C. H., Liou, Y. C. & Lin, S. H., (2008) *J. Phys. Chem. B, Vol. 112*, pp 4351.
[25] Seddon, K. R., (2003) *Nature mater., Vol. 2*, pp 363.
[26] Srivastava, N., Shukla, M. & Saha, S., (2010) *Ind. J. Chem. Vol. 49A*, pp. 757.
[27] Turner, E. A., Pye, C. C. & Singer, R. D., (2003) *J. Phys. Chem. A, Vol. 107*, pp 2277.
[28] Paul, A., Mandal, P. K., & Samanta, A., (2005) *J. Phy. Chem. B Vol. 109*, pp. 9148.
[29] Samanta, A., (2006) *J. Phys. Chem. B Vol.* 110, pp. 13704.
[30] Saha, S., Mandal, P. K., & Samanta, A., (2004) *Phys. Chem. Chem. Phys. Vol.* 6, pp. 3106.
[31] Holbrey, J. D., Reichert, W. M., Nieuwenhyuzen, M., Johnston, S., Seddon, K. R., Rogers, R. D., (2003) Chem. Commun., pp.1636,

[32] Frisch, M. J.; Trucks, G. W.; Schlegel, H. B.; Scuseria, G. E.; Robb, M. A.; Cheeseman, J. R.; Montgomery Jr., J. A.; Vreven, T.; Kudin, K. N.; Burant, J. C.; Millam, J. M.; Iyengar, S. S.; Tomasi, J.; Barone, V.; Mennucci, B.; Cossi, M.; Scalmani, G.; Rega, N.; Petersson, G. A.; Nakatsuji, H.; Hada, M.; Ehara, M.; Toyota, K.; Fukuda, R.; Hasegawa, J.; Ishida, M.; Nakajima, T.; Honda, Y.; Kitao, O.; Nakai, H.; Klene, M.; Li, X.; Knox, J. E.; Hratchian, H. P.; Cross, J. B.; Bakken, V.; Adamo, C.; Jaramillo, J.; Gomperts, R.; Stratmann, R. E.; Yazyev, O.; Austin, A. J.; Cammi, R.; Pomelli, C.; Ochterski, J. W.; Ayala, P. Y.; Morokuma, K.; Voth, G. A.; Salvador, P.; Dannenberg, J. J.; Zakrzewski, V. G.; Dapprich, S.; Daniels, A. D.; Strain, M. C.; Farkas, O.; Malick, D. K.; Rabuck, A. D.; Raghavachari, K.; Foresman, J. B.; Ortiz, J. V.; Cui, Q.; Baboul, A. G.; Clifford, S.; Cioslowski, J.; Stefanov, B. B.; Liu, G.; Liashenko, A.; Piskorz, P.; Komaromi, I.; Martin, R. L.; Fox, D. J.; Keith, T.; Al-Laham, M. A.; Peng, C. Y.; Nanayakkara, A.; Challacombe, M.; Gill, P. M. W.; Johnson, B.; Chen, W.; Wong, M. W.; Gonzalez, C.; & Pople, J. A. *Gaussian, Inc.:*, Wallingford, CT, 2004.

[33] Becke, A. D., (1993) *J. Chem. Phys. Vol.* 98, pp. 5648.

[34] Tsuzuki, S., Katoh, R., & Mikami, M., (2008) *Mol. Phys. Vol. 106*, pp. 1621.

[35] S. Tsuzuki, A. A. Arai, and K. Nishikawa, (2008) *J. Phys. Chem. B, Vol. 112*, 7739.

[36] V. Kempter, B. Kirchner, (2010) *J. Mol. Struc.*, doi:10.1016/j.molstruc.2010.02.003.

[37] Fujii, K., Seki, S., Fukuda, S., Kanzaki, R., Takamuku, T., Umebayashi, Y. & Ishiguro, S. (2007) *J. Phys. Chem. B Vol. 111*, pp. 12829.

[38] Stark, A., Behrend, P., Braun, o., Muller, A., Ranke, J., Ondruschka, B. & Jastorff, B. (2008) Green Chem. *Vol. 10*, pp 1152.

[39] Kanaparthi, R.K., Sarkar, M. & Samanta, A., (2009) *J. Phys. Chem. B Vol. 113*, pp. 15189.

[40] Santhosh, K. & Samanta, A., (2010) *J. Phys. Chem. B Vol. 114*, pp. 9195.

[41] Samanta, A., (2010) *J. Phys. Chem. Lett. Vol. 1*, pp. 1557.

8

High Pressure Phase Behavior of Two Imidazolium-Based Ionic Liquids, [bmim][BF_4] and [bmim][PF_6]

Yukihiro Yoshimura[1], Takekiyo Takekiyo[1], Yusuke Imai[2] and Hiroshi Abe[2]
[1]*Department of Applied Chemistry, National Defense Academy, Yokosuka*
[2]*Department of Materials Science and Engineering, National Defense Academy, Yokosuka*
Japan

1. Introduction

Room temperature ionic liquids (RTILs) consisting of organic cations and inorganic anions remain in the liquid state at room temperature (Welton, 1999). Due to their dual nature as both salts and fluids, vast interests in investigating the properties and applications of RTILs have been conducted. Because of the low melting point, almost zero vapor pressure, wide electrochemical window, and high recyclibility, RTILs are very attractive for a wide range of applications such as catalysis (Welton, 1999), synthesis (Itoh et al., 2004), and electrochemical devices (Galinski et al, 2006; Sato et al., 2004). Generally, RTILs are classified into several groups depending on the cationic series, imidazolium, pyridinium, and alphatic quaternary ammonium series.

Most of studies so far concern with the results at atmospheric pressure, but recently an increasing number of works using high pressure is reported (Chang et al., 2007, 2008a, 2008b, 2008c; Su et al., 2009, 2010; Umebayashi et al., 2009a, 2009b). A merit of the use of high pressure as a varialbe is summarized as follows. We can change, in a cotrolled way, the intermolecular interactions without encountering major pertubations, but an alteration in temperature of a chemical system at atmospheric pressure produces a simultaneous change in the thermal energy and the volume. To separate the thermal and volume effects, we need to execute high pressure experiments.

RTILs typically exhibit different bonding interactions, but unlike in other liquids, they are characterized by the presence of Coulomb interactions among the constituent ions. Thus, many RTILs are easily supercooled, and may introduce a rich phenomenology in the dynamics and also the phase transition. A bulky, asymmetric organic cation prevents ions from packing and the solidification. Therefore, it is most likely that the liquid structure of RTILs is determined by a balance between long-range electrostatic forces and local geometric factors. Application of high pressure is the ideal tool to tune and/or alter the liquid structure along with the bonding properties of the materials. If the RTIL system is compressed, the reaction moves to favor the components with smaller volume. In this

situation, an intriguing question is raised: if we compress RTILs using pressure as an external factor, what happens to the phase behavior; the crystallization occurs or the liquid state holds up to a certain pressure? The aim of the present manuscript is to show the phase stabilities of imidazolium-based (*i.e.*, known as prototype) RTILs at high pressures over 1 GPa. From the results, we may shed light on the relative importance of the hydrophilic vs. hydrophobic part of molecules in the collapse and aggregation (crystallization) processes.
Additionally, it is now well established that the 1-butyl-3-methylimidazolium ([bmim]) cation, has a *trans-trans* (*tt*)—*gauche-trans* (*gt*) equilibrium of the NCCC and CCCC angles of the butyl group as shown in Figure 1 (Hamaguchi & Ozawa, 2005). We wonder that the conformational analysis of [bmim] cation under high pressure is necessary to obtain the structural information of the high pressure phase behavior of the RTIL in relation to the packing efficiency. Actually, the solution structures including the conformational behavior of RTILs have been studied to reveal the physical and chemical properties such as conductivity, viscosity, and melting point (Chang et al., 2007, 2008c; Nishikawa et al., 2008, 2009; Rivera-Calzada et al., 2008; Seki et al., 2009).

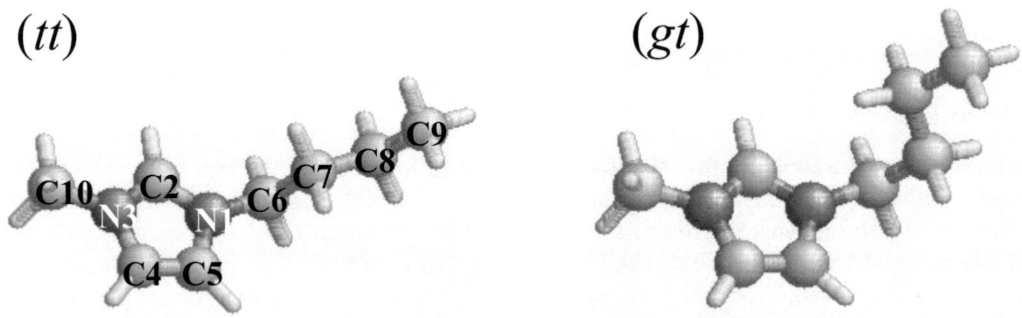

Fig. 1. Optimized structures of *trans-trans* (*tt*) and *gauche-trans* (*gt*) conformers of 1-butyl-3-methylimidazolium ([bmim]) cation by B3LYP/6-311G+(d) level.

The remainder of the paper is organized as follows. In section 2, we summarize the experimental methods. Section 3 presents the results of phase stabilities of [bmim][BF_4] and [bmim][PF_6] under high pressure. We discuss the results in the context of (1) the comparison with the phase behavior of [bmim][BF_4] at low temperatures, and (2) the conformational preferences of [bmim] cation in [bmim][BF_4] and [bmim][PF_6] along with the relation to the phase stabilities. Additionally, (3) the origin of the phase behavior by an interpretation from geometric volume calculation is proposed in this section. The conclusions of the present study are summarized in section 4.

2. Experimental methology

2.1 Samples

As samples, we used 1-butyl-3-methylimidazolium tetrafluoroborate ([bmim][BF_4], Cl < 0.005 %, Br < 0.005 %, F < 0.01 %, Na < 0.002 %, Li < 0.002 %, H_2O < 0.02 %, Kanto Chemical Co.) and 1-butyl-3-methylimidazolium hexafluorophosphate ([bmim][PF_6], Cl < 0.005 %, Br < 0.005 %, F < 0.01 %, Na < 0.002 %, Li < 0.002 %, H_2O < 0.02 %, Kanto Chemical Co.), because these are prototype RTILs and the most widely studied. The as-received sample may contain a small amount of water, so we doubly checked the water concentrations to be

116 ppm for [bmim][BF_4] and 98 ppm for [bmim][PF_6] using a Karl-Fischer titration method. Generally, RTILs are easily contaminated by vacuum drying to reduce H_2O, thus we used the samples without further purification. Special care with handling the samples was taken in a dry box not to contaminate chemicals further from atmosphere in a dry box with handling the samples. The chemical structure of [bmim] cation along with the numbering of each carbon is shown schematically in Figure 2.

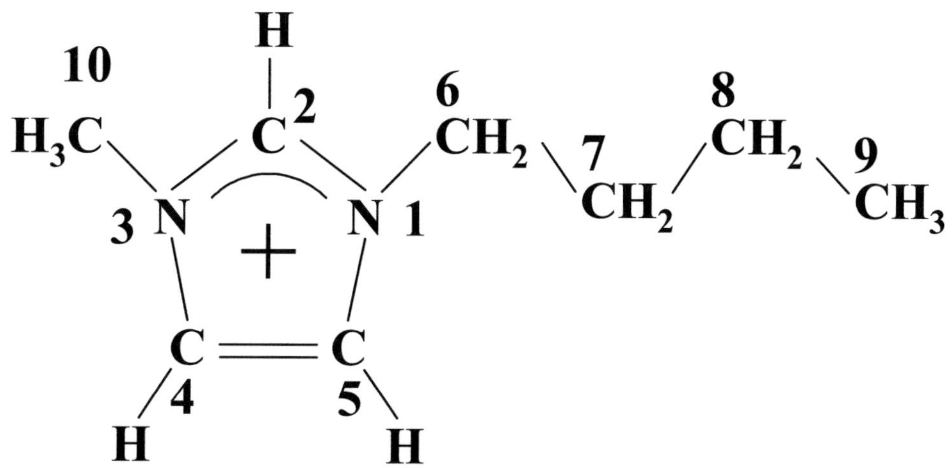

Fig. 2. Chemical structure of [bmim] cation along with numbering of the each carbon and nitrogen atoms.

2.2 High pressure measurements

For high pressure experiments, we used a diamond anvil cell (DAC, Kyowa Co. Ltd., SR-DAC-KYO3-3d) as shown in Figure 3. The sample solution was placed together with a small amount of powdered ruby chip for a pressure marker in a SUS301 gasket mounted in the DAC (Figure 4). The spectral shift of the R_1 fluorescence line of ruby chip was used for a pressure calibrant (Mao et al., 1978). The error of the pressure was ±0.05 GPa. The DAC was mounted under the microscope of a JASCO NR-1800 spectrophotometer (Figure 5).

2.3 Raman measurements

Raman spectroscopy is often used to explore bonding structures of liquids, because it provides information on the local structure in the liquid state. The observation of the C-H stretching vibration is one of the keys to characterize the imidazolium-based cations and can serve as a useful probe to reflect the phase transition behavior. Raman spectra were measured by a JASCO NR-1800 Raman spectrophotometer equipped with a single monochromator and a charge coupled device (CCD) detector. The 514.5 nm line from a Lexel Ar^+ laser was typically used as an excitation source with a power 250 mW.

For complimentary measurements of temperature-induced Raman spectral changes at a normal pressure, we controlled the sample temperature by LINKAM THMS-600 (Japan hitech Co.). The cooling rate was 5 K/min. For the experiments on conformational variation with temperature, temperature was decreased from 298 K to 113 K in 10 K decrements. The obtained Raman spectra were fitted with Gaussian-Lorentzian mixing function using the GRAMS/386 software (Galactic Ind. Co. Ltd.).

Fig. 3. Schematic picture of diamond anvil cell (DAC).

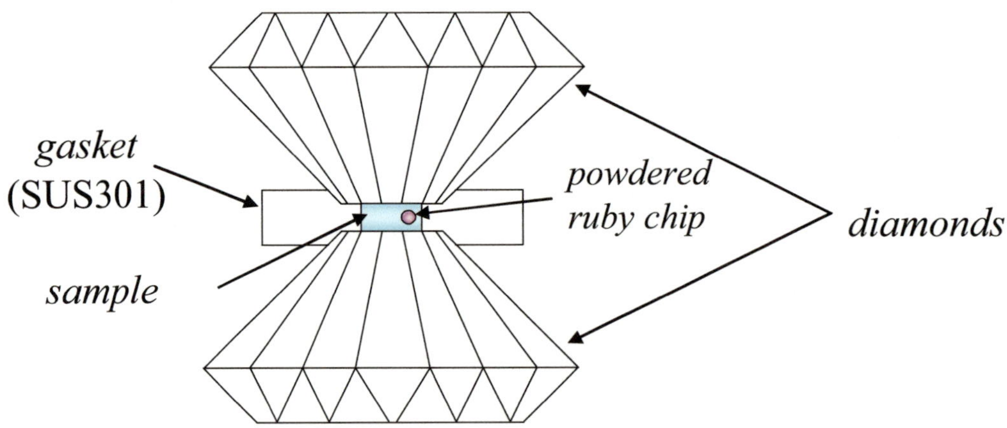

Fig. 4. Schematic diagram of diamond anvils with a gasket and a powdered ruby chip.

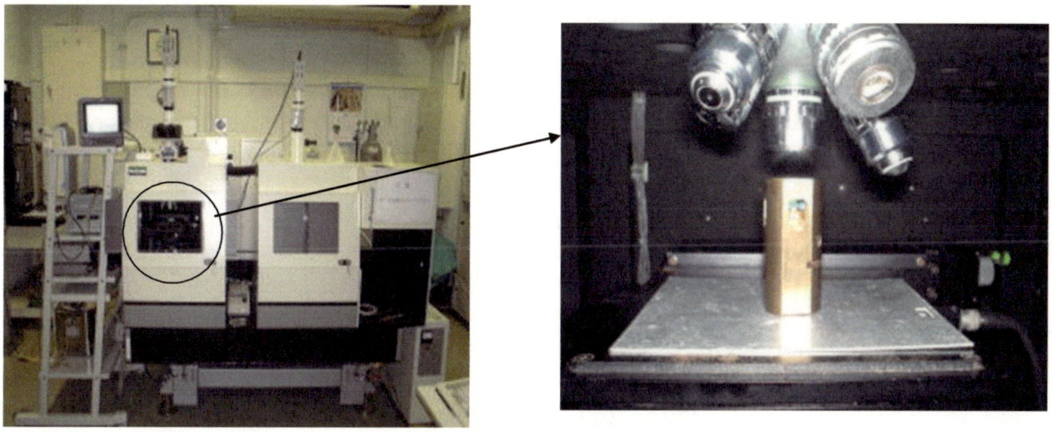

Fig. 5. Schematic picture of diamond anvil cell set upped with a Raman spectrophotometer (JASCO NR-1800).

2.4 Geometric volume calculations

Density functional theory (DFT) calculations have been used as an investigation methodology for the conformational analysis (Holomb et al., 2008; Umebayashi et al., 2005). We performed the DFT calculations of geometries of the *tt* ([bmim])−anion and the *gt* ([bmim])−anion complexes. All DFT calculations were carried out using GAUSSIAN03 program (Frish et al., 2003). For the present calculation, we used the Becke's three-parameter (B3) exchange function (Becke, 1998). The B3 exchange function was combined with Lee-Yang-Peer correlation function (B3LYP) (Lee et al., 1998). The 6-311+G(d) basis set was used. Geometric volume calculations have been conducted as an analysis methodology for obtaining the information on a relationship between the partial molar volume (PMV) and the structure of a molecule (Edelsbrunner et al., 1995; Imai et al., 2005; Ling et al., 1998; Takekiyo et al., 2006a, 2006b). Geometric component of [bmim] cation in the RTILs enables the calculation of the solvent accessibility of the molecular surface by assuming that RTILs are hard-sphere probes. The details of geometric volume calculations were described in the previous papers (Edelsbrunner et al., 1995; Ling et al., 1998; Takekiyo et al., 2006a, 2006b). We have calculated the van der Waals (V_w) and the molecular volumes (V_M). V_M is composed of the V_w and the void volume (V_v). V_v is structural void within the solvent inaccessible core of the solute molecule. Hence, the V_v can be obtained by the subtraction of V_w from V_M ($V_v = V_M - V_w$). Figure 6 shows the definition of respective volume contributions schematically. For the radius of the probe, the average radius of each optimized [bmim] conformer-anion using B3LYP/6-311+G(d) level as a hard-sphere probe was used. The radius of the solvent probe was estimated to be 5.93 Å for [bmim][PF_6].

3. Experimental results and discussion

3.1 Phase stabilities of 1-butyl-3-methylimidazolium tetrafluoroborate ([bmim][BF_4]) and hexafluorophosphate ([bmim][PF_6]) at high pressures

Firstly we provide the results of the Raman spectral changes of [bmim][BF_4] as a function of pressure in Figure 7. We follow the detailed peak assignments of the CH stretching mode (ν_{CH}) reported previously (Heimer et al., 2006). The wavenumber region from 2750 to 3075 cm^{-1} is due from the v_{CH} of the alkyl chain of [bmim] cation and that from 3075 to 3200 cm^{-1} is from the v_{CH} mode of the imidazolium ring, respectively. We can see that the v_{CH} band shifts to a higher frequency with accompanying a broadening of the band upon compression. This frequency shift might arise from the contraction of C-H bonds and the overlap repulsion effect by applied pressure (Chang et al., 2008a). We note that at 1.4 GPa a larger change in the spectrum of [bmim][BF_4] was obtained. The spectral resolution of each peak in the v_{CH} of the alkyl chain becomes unclear; the peak intensities at 2873 cm^{-1} and 2910 cm^{-1} due from the symmetric v_{CH} ($v_{CH\ ss}$) of propyl (C6, C7 and C8) group and from the $v_{CH\ ass}$ of butyl (C6, C7, C8 and C9) group apparently decrease along with a pronounced peak broadening, whereas the peak at 3168 cm^{-1} assigned to the ring $v_{CH\ ss}$ of C(4)H and C(5)H groups can be still clearly observable. Importantly, spectral changes due from the alkyl chain are larger than those from the imidazolium ring, indicating that the environment around the butyl chain is larger perturbed than that around the imidazolium ring with applied pressure. This pressure-enhanced change in the v_{CH} mode might be a key to provide the stability of the material under high pressure conditions.

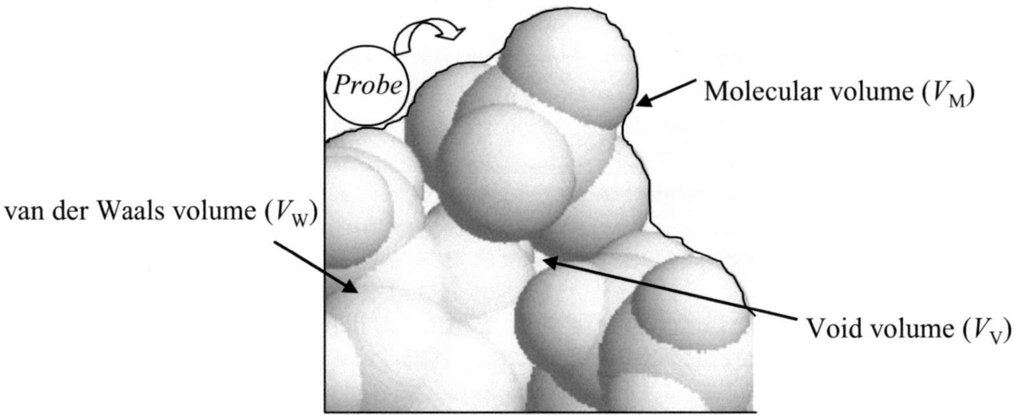

Fig. 6. Definition of geometric volume contributions

Specifically, the wavenumber shift of the most intense peak at around 2965 cm^{-1} assigned to the C(9)H asymmetric stretching mode ($v_{C(9)H\ ass}$) of [bmim] cation as a function of pressure is shown in Figure 8. The $v_{C(9)H\ ass}$ linearly shifts to a higher frequency with increasing pressure without any discontinuous jump under 1.4 GPa. A distinct phase change such as crystallization could not be recognized and [bmim][BF_4] is still stable at this pressure range.
Additionally, unlike common molecular liquids, RTILs show a structural feature on the nanoscopic level, which is one of the characteristic features appear in RTILs. Based on the simulation (Raabe & Koehler, 2008) and experimental (Xiao et al., 2007) studies, imidazolium-based RTILs show the nanoscale spatial heterogeneity (nano-phase separation) where an anion and a positively charged imidazolium ring pair up, and the pairs aggregate and form polar domains. Alkyl side chains of imidazolium cations also aggregate and form nonpolar domains. This domain structure due from the separation of each polar/nonpolar domain is a cause of the heterogeneity. The nano-phase separations exist when an imidazolium cation has over 4 (*i.e.*, longer than butyl chain) on the alkyl side chain (Lopes & Padua, 2006; Wang & Voth, 2005). The "nano-heterogeneity" may have an influence on the non-equilibrium phase behavior.

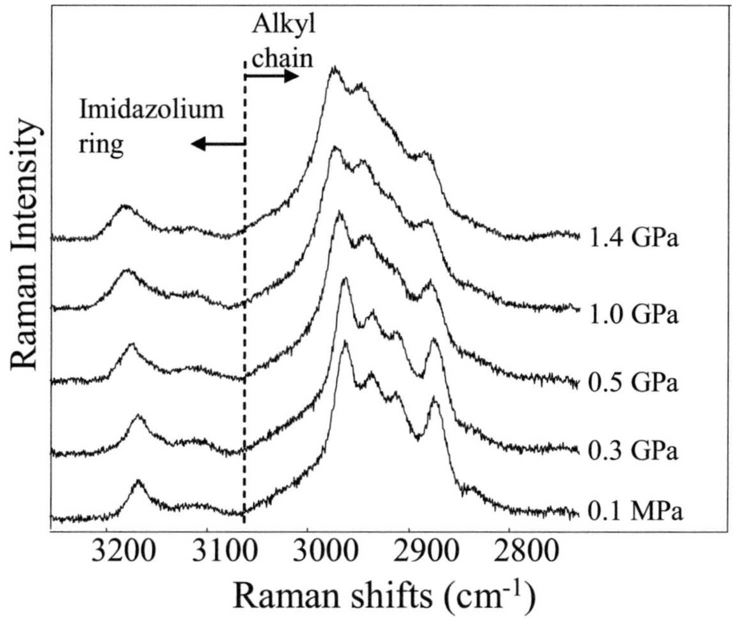

Fig. 7. Raman CH stretching spectral changes of [bmim][BF_4] as a function of pressure.

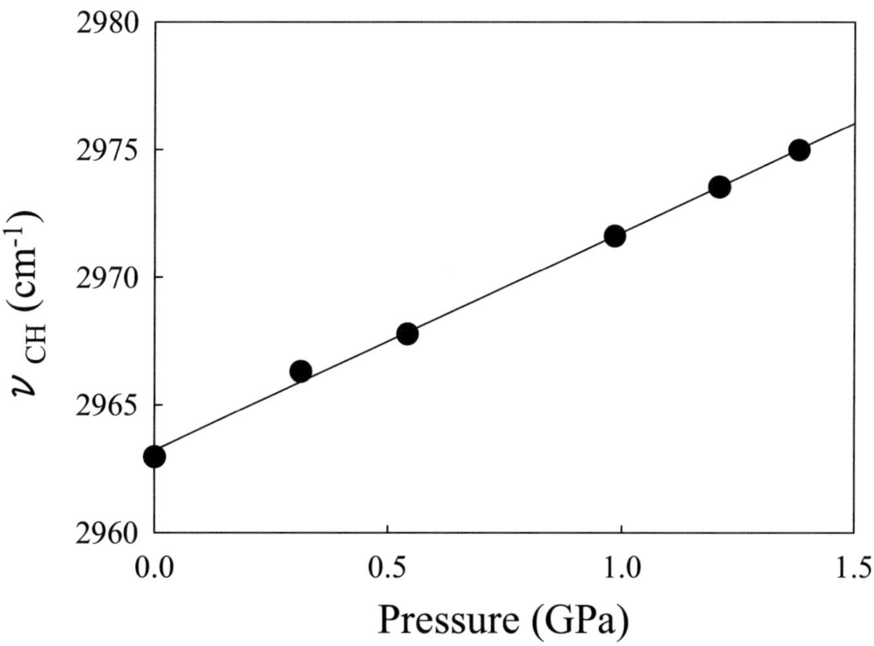

Fig. 8. Wavenumber shift of the CH asymmetric stretching mode of C(9)H_3 group of [bmim] cation as a function of pressure.

On the other hand, Figure 9 shows the pressure-induced Raman ν_{CH} spectral changes of [bmim][PF_6] having the hydrophobic anion, as compared to the BF_4^- anion. We can see that the fine structures are appeared in the Raman ν_{CH} spectrum at 0.2 GPa. Previous high pressure-differential thermal analysis (HP-DTA) study of [bmim][PF_6] showed that the phase transition occurs at ~0.1 GPa (Su et al., 2009). Therefore, this change corresponds to the (liquid-crystalline) phase transition. Further compression above 0.2 GPa, all the observed Raman ν_{CH} frequencies shift to higher frequencies as shown in Figure 10. Interestingly, the Raman ν_{CH} band at around 2916 cm^{-1} due from the butyl chain clearly separates into two peaks (▲: 2923 cm^{-1} and △: 2935 cm^{-1}) at 1.1 GPa. Additionally, a new peak at 2985 cm^{-1} (◇) appeared at above 1.7 GPa. These spectral changes might imply the existence of other high pressure solid-solid phase transitions at around 1.1 and 1.7 GPa other than reported one at 0.2 GPa. In previous differential scanning calorimetry (DSC) and Raman studies in the heating process (scanning rate =5.0 mK/s), Endo *et al.* (Endo et al., 2010) found that [bmim][PF_6] has three low-temperature crystalline phases, which are α (226.5 K), β (250.3 K), and γ-phases (276 K). Based on our results on the pressure-induced Raman ν_{CH} spectral changes, [bmim][PF_6] might also have three high-pressure crystalline phases under 2.0 GPa.

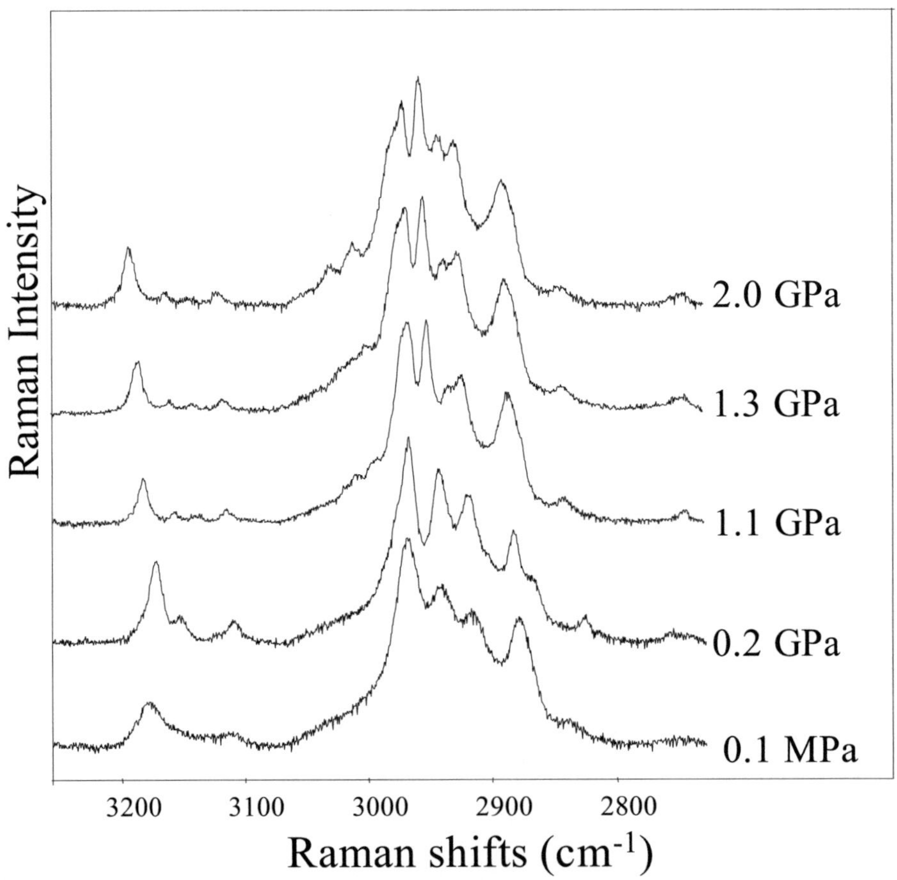

Fig. 9. Raman CH stretching spectral changes of [bmim][PF_6] as a function of pressure.

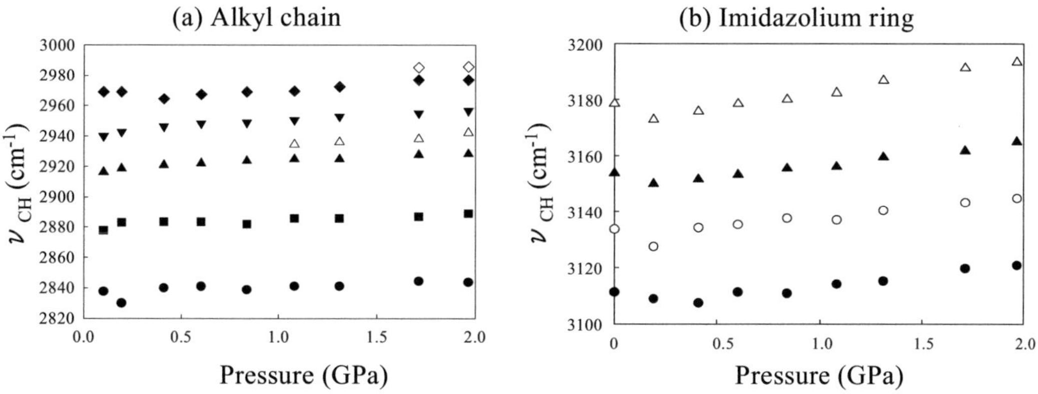

Fig. 10. Changes in the CH stretching frequencies of (a) alkyl chain and (b) imidazolium-ring of [bmim][PF_6] as a function of pressure.

3.2 Comparison with the phase behavior of [bmim][BF_4] at low temperatures

Previously, we have investigated the phase behavior of a quaternary ammonium-type RTIL, *N*,*N*-diethyl-*N*-methyl-*N*-(2-methoxyethyl) ammonium tetrafluoroborate (designated as [DEME][BF_4]), under high pressure (Imai et al., 2010). The Raman spectra of [DEME][BF_4] changed at around 1.0 GPa, but the Raman spectrum at high pressure is totally different from that obtained by slow cooling at a normal pressure. In the same way, we show the temperature-induced Raman spectral changes of [bmim][BF_4].

Figure 11 shows the Raman spectrum of [bmim][BF_4] at 113K. For a comparison, the spectra of 0.1 MPa and 1.4 GPa at 298 K is shown in the same figure. We can see that the spectral feature at 113 K basically remains unchanged from the liquid state at 298 K and 0.1 MPa, though the peak intensity at around 2910 cm^{-1} of the [bmim] cation slightly increases at 113 K. Consulting with the literatures (Katayanagi et al., 2004; Ozawa et al., 2003), [bmim][BF_4] easily transforms to the glassy state at low temperatures. The glass transition temperature (T_g) was reported to be around 180 K. Therefore, the Raman spectrum at 113 K should correspond to the glassy state of [bmim][BF_4].

As mentioned above, imidazolium [C_n-mim] cations having longer alkyl chains of $n \geq 4$ tend to aggregate within the polar or non-polar regions and the nanoscale spatial heterogeneity occurs (Lopes & Padua, 2006; Wang & Voth, 2005). We suppose that there may be a shortening of the size of the micelle-like domain structure in [bmim][BF_4] upon compression. As another important feature of [bmim] cation in the liquid state, we point out that two stable conformers, the *tt* (*trans-trans*) form and the *gt* (*gauche-trans*) form (Hamaguchi & Ozawa, 2005; Ozawa et al., 2003) due to the rotational isomerism of butyl group exist, which will be described in detail below. These may be a cause for the difference in the phase behavior in response to the external functions (pressure and temperature). As stated briefly above, Raman spectral changes in the low temperature phases of [bmim][PF_6] were already reported by Endo *et al.* (Endo et al., 2010). We will discuss the relation to the conformational changes of [bmim] cation in the next section.

3.3 Conformational preferences of [bmim][BF_4] and [bmim][PF_6,] with the relation to the phase stabilities

Thermodynamic study concerning the conformational equilibrium of RTILs is useful to clarify the relationship between the conformation and complicated phase transition behavior of RTILs (Endo et al., 2010; Hamaguchi & Ozawa, 2005; Imanari et al., 2008; Katayanagi et al., 2004).

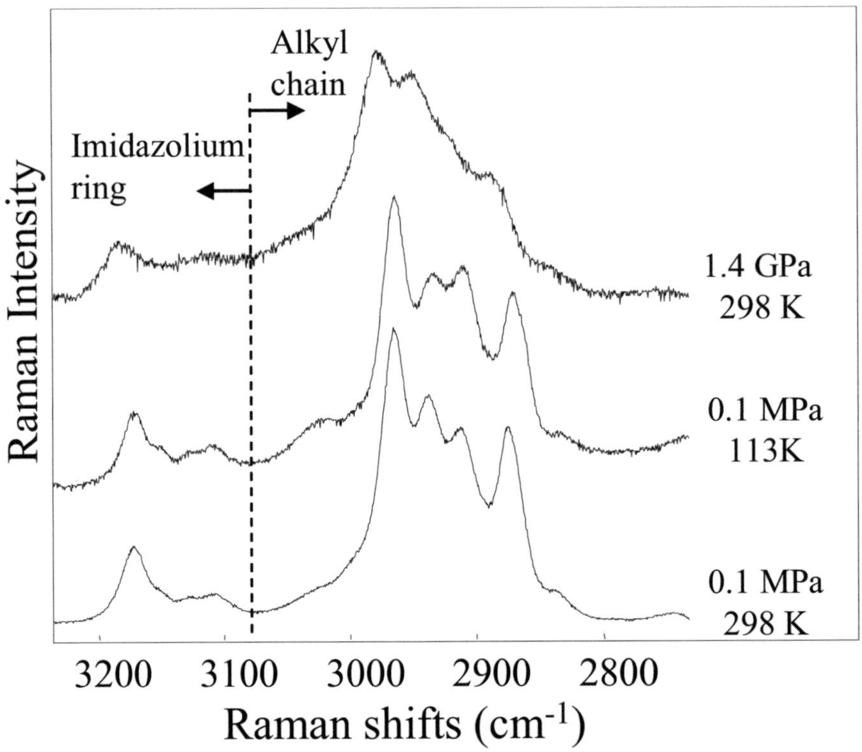

Fig. 11. Comparison of Raman CH stretching spectra of [bmim][BF_4] at various *p*-*T*s.

Now it is well established that [bmim] cation in the liquid state has a *tt*−*gt* equilibrium of the NCCC and CCCC angles of the butyl group as already shown in Fig. 1 (Hamaguchi & Ozawa, 2005; Ozawa et al., 2003). Recently, Endo *et al.* (Endo et al., 2010) reported that three low-temperature crystalline phases of [bmim][PF_6] closely correlate with the conformational changes of [bmim] cation. Subsequently, Su *et al.*(Su et al., 2009) reported the phase transitions of [bmim][PF_6] using HP-DTA up to 1.0 GPa. The transition from the liquid to crystalline phases was reported to occur at ~0.1 GPa. However, detailed information on the transformed crystalline phase of [bmim][PF_6] still requires further study. We suppose that the conformational analysis of [bmim] cation in its relation to the phase transition behavior is helpful to obtain the structural information of the high pressure phases.

Firstly, we provide the results of the conformational changes of [bmim] cation associated with the high pressure phase transitions of [bmim][PF_6] as shown in Figure 12. Two peaks at 580 and 620 cm^{-1} assigned to the CH_2 rocking mode of NCCC moiety are due to the *gt* and *tt* conformers of [bmim] cation, respectively (Hamaguchi & Ozawa, 2005; Ozawa et al., 2003). We found that the Raman intensity due to the *gt* conformer drastically increases at 0.2 GPa, and the intensity of the *tt* conformer decreases. The increase of *gt* conformer of [bmim] cation of [bmim][PF_6] at high pressure is consistent with the conformational change of [bmim][Cl] upon crystallization under high pressure (Chang et al., 2007). To discuss the observed results further, the pressure dependence of the intensity ratio (I_{gt}/I_{tt})

between the conformers of [bmim] cation is shown in Figure 13. The value of I_{gt}/I_{tt} drastically increases up to 0.2 GPa, but after high pressure phase transition (crystallization) occurs, the I_{gt}/I_{tt} shows a constant value up to 1.0 GPa. From these results, we can conclude that the high pressure phase of [bmim][PF_6] prefers the *gt* conformer. We note that further compression up to 2.0 GPa induces two more changes in the I_{gt}/I_{tt} ratio, implying the existence of other high-pressure crystalline phase transitions in agreement with the results as shown in section 3.1.

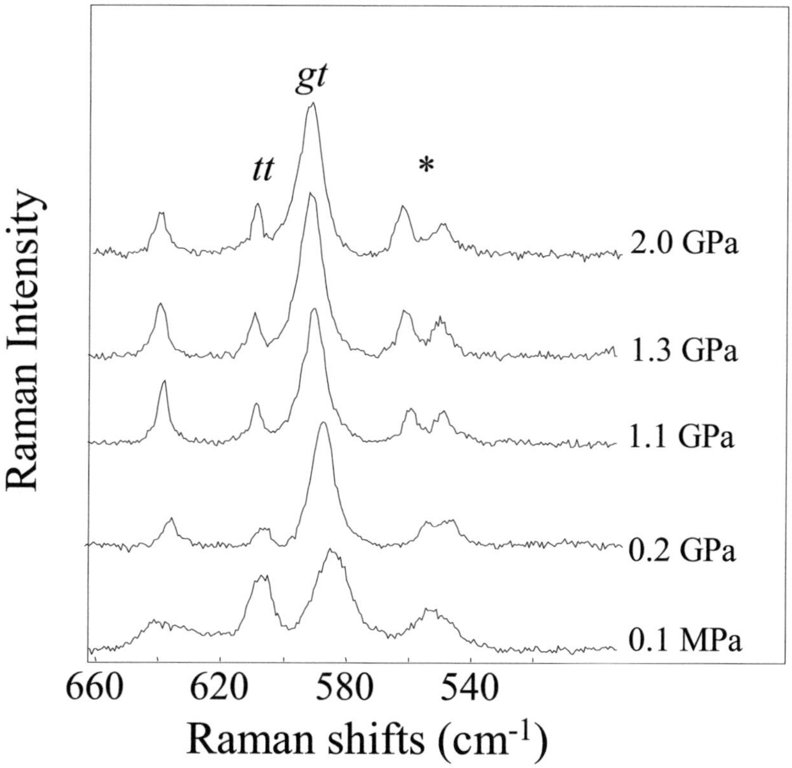

Fig. 12. Raman spectral changes in the lower frequency region of [bmim][PF_6] as a function of pressure. The asterisk represents a vibrational mode of the PF_6^- anion. (in figure caption of Figure 12)

On the other hand, as to the [bmim][BF_4], the value of I_{gt}/I_{tt} slightly increases with applied pressure as shown in Figure 13, *i.e.*, the *gt* conformer is also a dominant specie and stabilized under high pressure (Takekiyo et al., 2011). Therefore, the predominance of the *gt* conformer under high pressure seems to be less dependent on the anionic species such as PF_6^-, BF_4^- (Takekiyo et al., 2011), and Cl^- (Change et al., 2007). In relation to this, previous Raman and IR studies of simple organic liquids such as dihaloethane (Taniguchi et al., 1981), halopropane (Kato & Taniguchi, 1990), and *n*-alkane in neat solution (Kato & Taniguchi, 1991; Schoen et al., 1979) show that the *gauche* conformer is more stable than the *trans* conformer under high pressure. Thus, the higher stability of *gt* conformer of [bmim][PF_6] under high pressure is agreement with the pressure-induced conformational change of the simple organic liquids (Kato & Taniguchi, 1990, 1991; Schoen et al., 1979; Taniguchi et al., 1981).

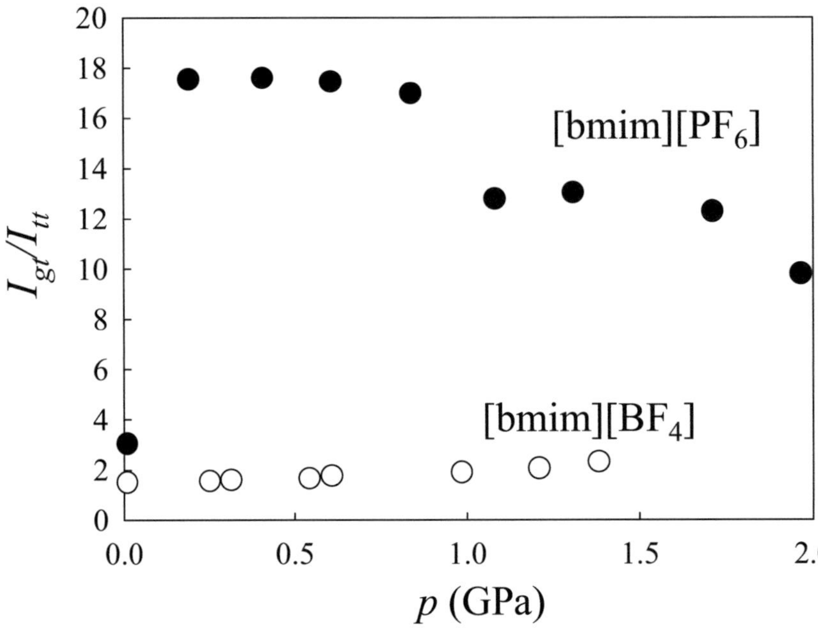

Fig. 13. Pressure dependence of the intensity ratio between the *gt* and *tt* conformers of [bmim] cation for [bmim][PF$_6$] and [bmim][BF$_4$]. The closed and open circles represent the I_{gt}/I_{tt} values for [bmim][PF$_6$] (●) and [bmim][BF$_4$] (○), respectively.

3.4 The origin of the phase behavior: Interpretation from geometric volume calculations

In this section, we discuss why the *gt* conformer of [bmim] cation has a higher stability at high pressure. Here, to clarify the main factor of the conformational stability of [bmim] cation under high pressure, we calculated the optimized structures of the *tt* ([bmim])—counter anion (*e.g.*, PF_6^-) and the *gt* ([bmim])—anion complexes using B3LYP/6-311+G(d) level.

We expected that the pressure shifts the conformational equilibrium to the conformation with the smaller partial molar volume (PMV). From the pressure dependence of the I_{gt}/I_{tt} value in Figure 13, we can predict the PMV difference between the conformers for [bmim] cations in [bmim][PF$_6$] and [bmim][BF$_4$]. We found that the PMV of the *gt* conformer of [bmim] cation is smaller than that of the *tt* conformer ($\Delta V^{tt\rightarrow gt}<0$). To discuss the main factor of $\Delta V^{tt\rightarrow gt}<0$ of [bmim] cation, we additionally performed the geometric molecular volume calculation of [bmim] cation by simple molecular volume calculation method (Imai et al., 2005; Takekiyo et al., 2006a, 2006b). Here, we assume that the molecular volume (V_M) composes of the van der Waals volume (V_W) and the structural void volume (V_v). Figure 14 shows the change of V_M of [bmim] cation as a function of the NCCC angle. Conformational change from the *trans* to the *gauche* conformers accompanies with the decrease of V_M value ($\Delta V_M^{tt\rightarrow gt}$ = -1.55 cm^3/mol). In this calculation the value of -0.13 cm^3/mol for $\Delta V_W^{tt\rightarrow gt}$ and -1.42 cm^3/mol for $\Delta V_v^{tt\rightarrow gt}$ are obtained. We found that $\Delta V_v^{tt\rightarrow gt}$ is the predominant factor in $\Delta V_M^{tt\rightarrow gt}$. This calculation results well explain the experimental results ($\Delta V^{tt\rightarrow gt}<0$). Therefore, the negative volume change of [bmim] cation is mainly ascribed to the contribution from the $\Delta V_v^{tt\rightarrow gt}$. We conclude that the preference of the *gt* conformer having the smaller PMV in the high-pressure phase of [bmim][PF$_6$] is largely due to the void volume contribution, which relates to the difference in the molecular packings between the conformers of [bmim] cation.

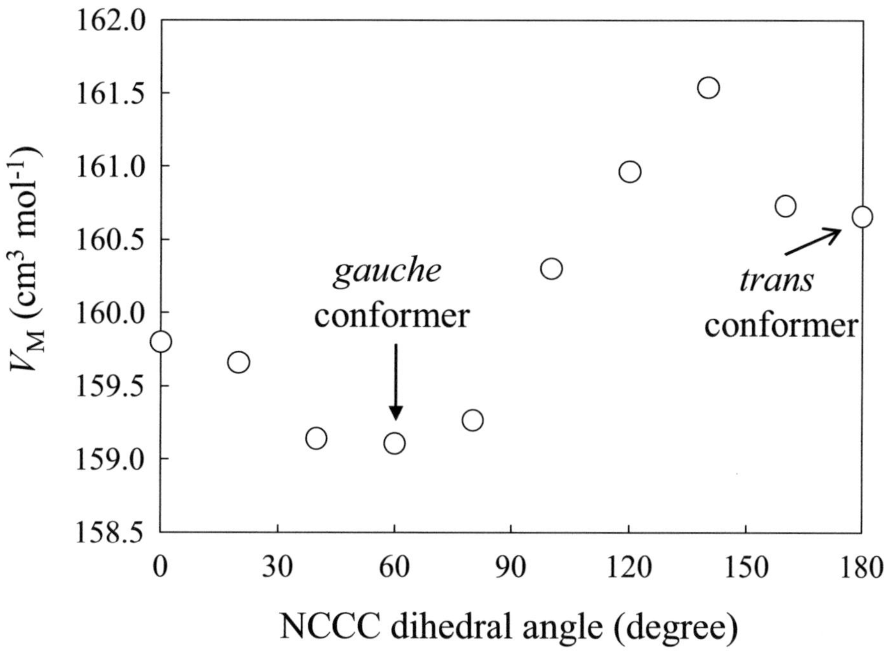

Fig. 14. Variation of the molecular volume (V_M) of [bmim] cation as a function of NCCC dihedral angle.

4. Concluding remarks and future prospects

In summary, we have demonstrated the phase stabilities of two prototype RTILs, [bmim][BF_4] and [bmim][PF_6] under high pressure. Interestingly, [bmim][PF_6] easily crystallizes upon compression, but [bmim][BF_4] maintains the liquid state up to over 1 GPa. The results indicate that a contribution of the anion is significant to the phase stability at high pressures. These clearly contrast with the results in the low-temperature phase behaviors of both RTILs.

Another important fact deduced from the present study is conformational change in [bmim] cation occurs concordant with the phase transition. In an effort to elucidate the origin of the phase behavior, interpretation from theoretical calculations were done. We showed the conclusion that a key to understand the conformational preference of *gt* conformer in the high-pressure [bmim][PF_6] phase is the void volume contribution of [bmim] cation .

RTILs have received significant attention for their use in multidisciplinary chemistry areas (Itoh et al., 2004; Galinski et al., 2006; Sato et al., 2004; Welton, 1999). Investigations on the high pressure behavior of the materials have important significance regardless of their basic knowledge or understanding of the potential application. As Sui *et al.* (Su et al., 2009) pointed out, one direct impact of the understanding of the high pressure phase behavior of RTILs can be recycling or purification of RTILs. High pressure crystallization may be an effective way to recycle RTILs at the room temperature, as crystallization temperature increases as the pressure rises. The green nature of RTILs as solvent is difficult to evaporate, which prevents the pollution on environment. In another point of view, however, exactly because of the merit in RTILs, distillation through evaporation, which is an important approach to recycle and purity common organic solvents, cannot be applied to RTILs (Chang et al., 2008b; Su et al., 2010). Moreover, RTILs are very hazardous to the environment if they are released. For instance, a hydrolysis of BF_4^- anion in the presence of water is sometimes known to occur (Freire et al., 2010).

On the other hand, as present results shown, pressure plays a critical role in regulating the structures and properties of materials. High pressure is unique tool to explore "new" chemistry by creating novel substances and altering the properties of existing compounds. Pressure may use as a way to explore the concealed properties of fluids and solids. One example is that the high-pressure development of functional materials with potential industrial applications, such as high-energy density materials, superhard materials and hydrogen storage materials, has become a vibrant area of high pressure research.

5. References

Becke, A. D. (1988) Density-functional Exchange-energy Approximation with Correct Asymptotic Behavior, *Phys. Rev. A* Vol. 38: 3098-3100.

Chang, H. -C., Jiang, J. -C. Su, J. -C., Chang, C. -Y., & Lin, S. -H. (2007) Evidence of Rotational Isomerism in 1-Butyl-3-methylimidazolium Halides: A Combined High-Pressure Infrared and Raman Spectroscopic Study, *J. Phys. Chem. A* Vol. 111: 9201-9206.

Chang, H. -C., Jiang, J. -C., Chang, C. -Y, Su, J.-C., Hung, C. -H., Liou, Y., & Lin, S. -H. (2008a) Structural Organization in Aqueous Solutions of 1-Butyl-3-methylimidazolium Halides: A High Pressure Infrared Spectroscopic Study on Ioinic Liquids, *J. Phys. Chem. B* Vol. 112: 4351-4356.

Chang, H. -C., Jiang, J. -C., Liou, Y. -C., Hung, C. -H., Lai, T. -Y., & Lin, S. -H. (2008b) Effects of Water and Methanol on the Molecular Organization of 1-Butyl-3-methylimidazolium Tetrafluoroborate as Functions of Pressure and Concentration, *J. Chem. Phys*. Vol. 129: 044506-6.

Chang, H. -C., Jiang, J. -C., Liou, Y. -C., Hung, C. -H., Lai, T. -Y., & Lin, S. -H. (2008c) Local Structures of Water in 1-Butyl-3-methylimidazolium Tetrafluoroborate Probed by High-Pressure Infrared Spectroscopy, *Ana. Sci.* Vol. 24: 1305-1309.

Edelsbrunner, H., Facello, M., Fu, P., & Liang, J. (1995) Measuring Proteins and Voids in Proteins, *Proceedings of the 28 th Annual Hawai International Conference on System Science*; IEEE Computer Society Oress: Los Alamos, CA, Vol. 5 256-264.

Endo, T., Kato, T., Tozaki, K., & Nishikawa, K. (2010) Phase Behaviors of Room Temperature Ionic Liquid Linked with Cation Conformational Changes: 1-Butyl-3-methylimidazolium Hexafluorophosphate, *J. Phys. Chem. B* Vol. 114: 407-411.

Freire, M. G., Neves, C. M. S. S., Marrucho, I. M., Coutinho, J. A. P., & Fernandes, A. M. (2010) Hydrolysis of Tetrafluoroborate and Hexafluorophosphate Counter Ions in Imidazolium-Based Ionic Liquids, *J. Phys. Chem. A*. Vol. 114: 3744-3749.

Frish, M. J., Trucks, G. W., Schlegel, H. B., Scuseria, G. E., Robb, M. A., Cheeseman, J. R., Zakrzewski, V. G., Montogomery, J. A., Daniels, A. D., Kudin, K. N., Strain, M. C., Farkas, O., Tomasi, J., Barone, V., Cossi, M., Cammi, R., Mennucci, B., Pomelli, C., Adamo, C., Clifford, S., Ochterski, J., Petersson, G. A., Ayala, P. Y., Cui, Q., Morokuma, K., Malick, D. K., Rabuck, A. D., Raghavechari, K., Foresman, J. B., Cioslowski, J., Ortiz, J. V., Baboul, A. G., Stefanov, B. B., Liu, G., Liashenko, A., Piskorz, P., Komaromi, I., Gomperts, R., Martin, R. L., Fox, D. J., Kieth, T., Al-Laham, M. A., Peng, C. Y., Nanayaakkara, A., Gonzalez, C., Challacombe, M. P., Gill, M. W., Johnson, B., Chen, W.,

Wong, M. W., Andres, J. L., Gonzalez, C., Head-Gordon, M., Replogle, E. S., & Pople, J. A. (2003) GAUSSIAN 03, Gaussian, Inc., Pittsburgh, PA.

Galinski, M., Lewandwski, A., & Stepniak, I. (2006) Ionic Liquids as Electrolytes, *Electrochim. Acta* Vol. 51: 5567-5580.

Hamaguchi, H. & Ozawa, R. (2005) Structure of Ionic Liquids and Ionic Liquid Compounds: Are Ionic Liquids Genuine Liquids in the Conventional Sense?, *Adv. Chem. Phys.* Vol. 131: 85-104.

Heimer, N. E., Sesto, R. E. D., Meng, Z., Wilkes, J. S., & Carper, W. R. (2006) Vibrational Spectra of Imidazolium Tetrafluoroborate Ionic Liquids, *J. Mol. Liquids* Vol. 124: 84-95.

Holomb, R., Martinelli, A., Albinsson, I., Lassègues, J. C., Jahansson, P., & Jacobsson, P. (2008) Ionic Liquid Structure: the Conformational Isomerism in 1-Butyl-3-methylimidazolium Tetrafluoroborate [bmim][BF_4], *J. Raman Spectrosco.*Vol. 39: 793-805.

Imai, T., Takekiyo, T., Kovalenko, A., Hirata, F., Kato, M., & Taniguchi, Y. (2005) Theoretical Study of Volume Changes Associated with the Helix-Coil Transition of an Alanine-Rich Peptide in Aqueous Solution, *Biopolymers* Vol. 79: 97-105.

Imai, Y., Abe, H., Goto, T., Miyashita, T., & Yoshimura, Y. (2010) Pressure-induced Change in the Raman Spectra of Ionic Liquid [DEME][BF_4]-H_2O Mixtures, *J. Phys. Conf. Ser.* Vol. 215: 12069.

Imanari, M., Nakakoshi, M. Seki, H., & Nishikawa, K. (2008) ^{1}H NMR Study on Reorientational Dynamics of an Ionic Liquid, 1-Butyl-3-methylimidazolium Bromide, Accompanied with Phase Transitions, *Chem. Phys. Lett.* Vol. 511: 241-246.

Itoh, H., Naka, K., & Chojo, Y. (2004) Synthesis of Gold Nanoparticles Modefied with Ionic Liquid Based on the Imidazolium Cation, *J. Am. Chem. Soc.* Vol. 126: 3026-3027.

Katayanagi, H., Hayashi, S., Hamaguchi, H., & Nishikawa, K. (2004) Structure of an Ionic Liquid, 1-*n*-butyl-3-methylimidazolium Iodide, Studied by Wide-angle X-ray Scattering and Raman Spectroscopy, *Chem. Phys. Lett.* Vol. 392: 460-464.

Kato, M. & Taniguchi, Y. (1990) Effect of Pressure on Conformational Equilibria of Liquid 1-Chloropropane and 1-Bromopropane, *J. Chem. Phys.* Vol. 93: 4345-4.

Kato, M. & Taniguchi, Y. (1991) High Pressure Study on Molecular Conformational Equilibira of n-Pentane, *J. Chem. Phys.* Vol. 94: 4440-6.

Lee, C., Yang, W., & Parr, R. G. (1988) Development of the Colle-Salvetti Correlation-Energy Formula into a Functional of the Electron Density, *Phys. Rev. B* Vol. 37: 785-789.

Liang, J., Edelsbrunner, H., Fu, P., Sudlhakar, P. V., & Subramaniam, S. (1998) Analytical Shape Computation of Macromolecules: I. Moleuclar Area and Volume Through Alpha Shape, *Proteins*, Vol. 33: 1-17.

Lopes, J. N. A. C. & Padua, A. A. H. (2006) Nanostructual Organization in Ionic Liquids, *J. Phys. Chem. B* Vol. 110: 3330-3335.

Mao, H. K., Bell, P. M., Shaner, J. W., & Steinberg, D. J. (1978) Specific Volume Measurments of Cu, Mo, Pd, and Ag and Calibration of the Ruby R_1 Fluorescence Pressure Gauge from 0.06 to 1 Mbar, *J. Appl. Phys.* Vol. 49: 3276-3284.

Nishikawa, K., Wang, S., & Tozaki, K. (2008) Rhythmic Melting and Crystallization of Ionic Liquid 1-Butyl-3-methylimidazolium Bromide, *Chem. Phys. Lett.* Vol. 458: 88-91.

Nishikawa, K., Wang, S., Endo, T., & Tozaki, K. (2009) Melting and Crystallization Behaviors of an Ionic Liquid, 1-Isopropyl-3-methylimidazolium Bromide, Studied by Using Nanowatt-Stabilized Differential Scanning Calorimetry, *Bull. Chem. Soc. Jpn.* Vol. 82: 806-812.

Ozawa, R., Hayashi, S., Saha, S., Kobayashi, A., & Hamaguchi, H (2003) Rotational Isomerism and Structure of the 1-Butyl-3-methylimidazolium Cation in the Ionic Liquid State, *Chem. Lett.* Vol. 32: 948-949.

Raabe, G. & Kőehler, J. (2008) Thermodynamical and Structural Properties of Imidazolim Based Ionic Liquids from Molecular Simulation, *J. Chem. Phys*. Vol. 128: 154509-7.

Rivera-Calzada, A., Kaminski, K., Leon, C., & Paluch, M. (2008) Ion Dynamics under Pressure in an Ionic Liquid, *J. Phys. Chem. B*. Vol. 112: 3110-3114.

Sato, T., Matsuda, G., & Takagi, K. (2004) Electrochimical Properties of Novel Ionic Liquids for Electric Double Layer Capacitor Applications, *Electrochim. Acta* Vol. 49: 3603-3611.

Seki, S., Hayamizu, K., Tsuzaki, S., Fujii, K., Umebayashi, Y., Mitsugi, T., Kobayashi, T., Ohno, Y., Kobayashi, Y., Mita, Y., Miyashiro, H., & Ishiguro, S. (2009) Relationship between Center Atom Species (N, P) and Ionic Conductivity, Viscosity, Density, Self-diffusion Coefficient of Quaternary Cation Room-Temperature Ionic Liquids, *Phys. Chem. Chem. Phys*. Vol. 11: 3509-3514.

Schoen, P. E., Proest, R. G., Sheridan, G. P., & Schnur, J. M. (1979) Pressure induced Changes in Liquid Alkane Chain Conformation, *J. Chem. Phys*. Vol. 71: 317-7.

Su, L., Li, L., Hu, Y., Yuan, C., Shao, C., & Hong, S. (2009) Phase Transition of [C_n-min][PF_6] under High Pressure up to 1.0 GPa, *J. Chem. Phys*. Vol. 130: 184503-4.

Su, L., Li, M., Zhu, X., Wang, Z., Chen, Z., Li, F., Zhou, Q., & Hong, S. (2010) In Situ Crystallization of Low-Melting Ionic Liquid [BMIM][PF_6] under High Pressure up to 2 GPa, *J. Phys. Chem. B* Vol. 114: 5061-5065.

Takekiyo, T., Imai, T., Kato, M., & Taniguchi, Y. (2006a) Understanding High Pressure Stability of Helical Conformation of Oligopeptides and Helix Bundle Protein -High Pressure FT-IR and RISM Theoretical Studies-, *Biochim. Biophys. Acta* Vol. 1764: 355-363.

Takekiyo, T. & Yoshimura, Y. (2006b) Raman Spectroscopic Study on the Hydration Structures of Tetraethylammonium Cation in Water, *J. Phys. Chem. A* Vol. 110: 10829-10833.

Takekiyo, T., Imai, Y., Hatano, N., Abe, H., & Yoshimura, Y. (2011) Conformational Preferences of Two Imidazolium-based Ionic Liquids at High Pressures, Chem. Phys. Lett. Vol. 511: 241-246.

Taniguchi, Y., Takaya, H., Wong, P. T. T., & Whalley, E. (1981) Effect of Pressure on Molecular Conformations. II. *Trans-Gauche* Equilibrum of 1,2-Dichloroethane and 1,2-Dibromoethane, *J. Chem. Phys*. Vol. 75: 4815-8.

Umebayashi, Y., Fujimori, T., Sukizaki, T., Asada, M., Fujii, K., Kanzaki, R., & Ishiguro, S. (2005) Evidence of Conformational Equilibrium of 1-Ethyl-3-methylimidazolium in Its Ionic Liquid Salts: Raman Spectroscopic Study and Quantum Chemical Calculations, *J. Phys. Chem. B* Vol. 109: 8976-8982.

Umebayashi, Y., Jiang, J.-C., Shan, Y. -L., Lin, K. -H., Fujii, K., Seki, S., Ishiguro, S., Lin, S. – H., & Chang, H. -C. (2009a) Structural Change of Ionic Association in Ionic Liquid/Water Mixtures: A High-Pressure Infrared Spectroscopic Study, *J. Chem. Phys*. Vol. 130: 124503-6.

Umebayashi, Y., Jiang, J. -C., Lin. K. -H., Shan, Y.-L., Fujii, K., Seki, S., Ishiguro, S., Lin, S. - H., & Chang, H. -C.(2009b) Solvation and Microscopic Properties of Ionic Liquid/Acetnitrile Mixtures Probed by High-Pressure Infrared Spectroscopy, *J. Chem. Phys*. Vol. 131: 234502-7.

Wang, Y. T. & Voth, G. A. (2005) Unique Spatial Heterogenity in Ionic Liquids, *J. Am. Chem. Soc*. Vol. 127: 12192-12193.

Welton, T. (1999) Room Temperature Ionic Liquids. Solvents for Synthesis and Catalysis, *Chem. Rev*. Vol. 99: 2071-2084.

Xiao, D., Rajian, J. R., Cady, A., Li, S. F., Bartsch, R. A., & Quitevis, E. L. (2007) Nanostructural Organization and Anion Effects on the Temperature Dependence of the Optical Kerr Spectra of Ionic Liquids, *J. Phys. Chem. B* Vol. 111: 4669-4677.

9

Dielectric Properties of Ionic Liquids Proposed to Be Used in Batteries

Cserjési Petra, Göllei Attila,
Bélafi-Bakó Katalin and Gubicza László
University of Pannonia
Hungary

1. Introduction

Ionic liquids (ILs) opened new technological possibilities (Marsh, 2004) at the end of the 1990s for many application fields (Mutelet & Jaubert, 2006) due to their special physical and chemical properties (Welton, 1999). They have several advantages compared to commercial organic solvents (Mutelet & Jaubert, 2006), or electrolyte liquids (Plechkova & Seddon, 2008) since they are liquid over a wide temperature range, easy to recycle, non-combustible (Matsumoto, 2005), non-flammable (Saruwatari, 2010), display wide electrochemical windows (Giroud, 2008), high inherent conductivities (Stracke, 2009), have negligible vapor pressure (Endres, 2008), high thermal (Wasserschied & Welton, 2007) and electrochemical stability (Arnold, 2004), tuneable physicochemical properties (Koel, 2008) and lack of reactivity in various electrochemical or industrial applications (Seddon, 2003).

Therefore ILs have been intensively studied recently as novel, much safer electrolyte materials (Sakaebe, 2007) for electrochemical devices (Ohno 2005) and energy storing devices, such as Li batteries for cellular phones (Xu, 2006), batteries for vehicles, fuel cells, supercapacitors (Sato, 2004), solar cells (Stathatos, 2005.), etc.

In order to decide whether an IL could be successfully used as electrolyte it is necessary to first gain enough information about the electrochemical properties, hence the dielectric behaviour of the selected liquid.

However, several methods have already been published about determination of the dielectric properties of ILs - generally characterized by the dielectric constant (ε') (Wakai, 2005) and by the dielectric loss factor (ε'') (Bright & Baker, 2006; Wakai, 2006), so far no measuring method was introduced. This can be explained by the fact that the dielectric properties of ionic liquids cannot be measured by classical methods due to their high conductivity (Krossing, 2006).

In this chapter we intend to describe a novel method and a self-designed microwave dielectrometric apparatus built to measure the (ε') and (ε'') values and the electrical conductivity (G) of several ionic liquids, which might be used as electrolyte in batteries, at 2.45 GHz and at different temperatures. Furthermore the connections between the structure of the investigated ionic liquids and the dielectric characteristics will be determined.

2. Ionic liquids used in batteries

The most investigated ILs for the use of batteries are composed of quaternary ammonium (Mastumoto, 2001), pyridinium (Xiao & Malhotra, 2005), pyrrolidinium (Forsyth, 2001), pyrazolium (Alarco, 2004), piperidinium (Sakaebe, 2005), imidazolium (Ito, 2000) cations and $[BF_4]^-$, $[PF_6]^-$ anions.

In 2000 Hagiwara and Ito (Hagiwara & Ito, 2000) reported a dramatic increase of activity in this field of research (Wilkes, 1982) and published several physical and chemical properties of ILs which could be used for electrolysis, electrochemical deposition and the production of batteries. They found that the investigated ILs possesssed many excellent characteristics as electrolytes.

Few years later Diaw and co-workers studied the temperature dependence of the electrochemical properties of ILs mixed with organic solvents used as electrolyte for lithium batteries. They concluded that this new system did not only permit the same amount of charging and discharging without significant loss in capacity as the commercial batteries, but it could also be used at higher temperature without any safety problems (Diaw, 2005).

Matsumoto and co-workers published the electrochemical properties of ILs built up of amide anions and aliphatic onium cations at the temperature of 25°C. They found that not only the physical but also the electrochemical properties of the ILs depend mainly on the type of anions and cations (Matsumoto, 2005). Also it was found that the introduction of alkoxy groups into the ILs might lower the viscosity hence increase the effectiveness of the battery based on such liquid (Matsumoto, 2000).

The electrochemical windows, feasibility of ILs and the cycle life of batteries prepared with ILs were studied by Yang and co-workers (Xu, 2006). It was concluded that the battery containing *N*-methyl-*N*-propylpiperidinium bis (trifluoromethanesulfonyl) imide as electrolyte base (Sakaebe & Matsumoto, 2003) had a wide potential-stable window (> 5.5 V) and that it showed stable efficiency of 90-92% after 10 cycles.

Sakaebe and co-workers investigated the advantages and drawbacks of ILs used in batteries. It was found that the application of ILs could lead to relatively high electrochemical window, excellent thermal stability and that this kind of cell design had to be customised for further development (Sakaebe, 2007).

Stracke and co-workers reported the successful application of ILs as electrolytes in a toxic compound free Leclanché batteries and found that higher ionic conductivities were related to higher Leclanché battery potentials (Stracke, 2009).

The electrochemical behaviour and ionic conductivity of IL electrolytes and the resistance of the battery cells were studied by Saruwatari and co-workers. Both the temperature and the chemical structure of ILs had an effect on the ionic conductivity data. The increase in resistance of electrolyte meant the decrease of the ionic conductivity of electrolyte. Furthermore the cell showed as large discharge capacity as $235 mAhg^{-1}$ at 20°C (Saruwatari, 2010).

Sutto and co-workers prepared new type of batteries containing IL electrolyte and determined their ionic conductivity between 20°C and 100°C and carried out tests on measuring their capacity. It was demonstrated that new, environmentally sound energy storage devices could be developed with discharge voltages ranging up to 1.8 V depending on the used anode and cathode (Sutto, 2011).

Ionic conductivity, dynamic viscosity and electrochemical and thermal stability of several ILs were reported by Giroud and co-workers. Thermogravimetric investigations showed

that the IL solution were damaged above 450°C hence they could be used even for electric vehicle applications. It was also stated that the IL electrolyte showed better cycling performances than organic electrolyte and that operating potentials for IL electrolyte batteries might be higher than that of conventional lithium ion batteries (Giroud, 2008).
Most of the authors having studied the electrochemical properties of ILs and their possible application in batteries as electrolyte concluded that ILs could be successfully used to create new types of rechargeable systems or to even replace the standard lead-acid batteries.

3. Dielectric properties

3.1 Permittivity

The permittivity (ε) is a characteristic of a material, which describes how any electric field affects and is affected by a material (a dielectric medium). In nonconducting materials (insulators or dielectrics) charges do not move freely only might be slightly displaced from their equilibrium position (Heaviside, 2007). Permittivity is determined by the ability of a material to polarize in response to the field, and reduce the total electric field inside the material, hence it can be calculated by Equation 1, which gives the electric field of a point charge (Q) at the distance r from Q.

$$E = \frac{Q}{4\pi \cdot \varepsilon \cdot r^2} \rightarrow \varepsilon = \frac{Q}{4\pi \cdot E \cdot r^2} \qquad (1)$$

Equation 2 gives the unit of ε, which is conventionally measured in farads per meter [F/m], or coulombs per volt-meter [C/Vm], as well.

$$[\varepsilon] = \frac{[Q]}{[E] \cdot [r^2]} = \frac{1As}{1\frac{V}{m} \cdot 1m^2} = 1\frac{As}{Vm} \qquad (2)$$

3.2 Dielectric constant

For a given system the ratio of the electric flux density (D) to the electric field strength (E) is a constant, known as the absolute permittivity (Equation 3), which can also be defined as the product of the permittivity of free space (or vacuum permittivity) (ε_0) and the relative permittivity (ε_r), or sometimes also called as dielectric constant (ε') as shown in Equation 4 (Housecroft & Sharpe, 2005).

$$\varepsilon = \frac{D}{E} \qquad (3)$$

$$\varepsilon = \varepsilon_0 \cdot \varepsilon_r \qquad (4)$$

where the value of vacuum permittivity is; $\varepsilon_0 = 8.8541878176 \times 10^{-12}$ As/Vm (Fleisch, 12008).
The dielectric constant, or relative permittivity, which is the permittivity of a material relative to that of free space and equals to 1 for vacuum by definition. Due to ability of any material to polarize more than vacuum the dielectric constant of a given material is always higher than 1. The ε' is dimensionless since it is just a ratio of two similar quantities.
The most common application of dielectric materials is to increase the capacitance and maximum operating voltage of capacitors. The capacitance (ability to store charge) of a

parallel-plate capacitor (*C*) is proportional to the area of the plates (*A*) and the distance between the plates (*d*) and is given by Equation 5 (Carrier & Soga, 1999).

$$C_0 = \varepsilon_0 \frac{A}{d} \quad (5)$$

The capacitance of a capacitor is increased by placing a dielectric between the capacitor plates and is given by Equation 6 (Benenson, 2002).

$$C = \varepsilon \frac{A}{d} = \varepsilon_0 \varepsilon_r \frac{A}{d} \quad (6)$$

The dielectric constant (ε') in reality is not always constant. In general, it can be different depending on the position in the medium, the frequency of the electric field, temperature, humidity, and other parameters.

3.3 Dielectric loss factor

When a capacitor of value *C* is connected to a generator of RMS voltage (*U*) at angular frequency (ω) will draw a current (*I*) according Equation 7.

$$I = jU\omega C \quad (7)$$

Supposing that the permittivity has an imaginary component and is written by Equation 8

$$\varepsilon_r = \varepsilon' - j\varepsilon'' \quad (8)$$

then substituting Equation 6 and Equation 8 into Equation 7 the current will be;

$$I = U\omega \frac{\varepsilon_0 A}{d} \left(j\varepsilon' + \varepsilon'' \right) \quad (9)$$

The first therm in parentheses shows the component of current in phase quadrature with the voltage as it would be in the case of lossless capacitors. The second term is a component in phase with the applied voltage and therefore it reperesents power dissipation (*P*), which is given by Equation 10.

$$P = U^2 \omega \frac{\varepsilon_0 A}{d} \varepsilon'' \quad (10)$$

where ε'' is called the dielectric loss factor of a dielectric and it can quantify the power dissipation in a dielectric filled capacitor.

The dielectric loss angle (δ) is defined as the angle by which the resultant current is different from the ideal 90° phase angle relative to the voltage and is given by Equation 11 (Meredith, 1998).

$$tg\delta = \frac{\varepsilon''}{\varepsilon'} \quad (11)$$

3.4 Electrical conductivity

The electrical conductivity is a measure of how easy it is for an electrical current to flow through a material or how easily electrical charge may be transported through it. High value corresponds to the materials that carry electrical currant easily and do not readily remain electrically charge. Due to this behaviour these material might be excellent candidates for electrolytes in certain batteries (Heiken, 1991). Electrical conductivity is defined by Ohm's law for direct current (Equation 12), measured in units of Siemens (S) and is given by the inverse of the resistance as shown in Equation 13 (Mohos, 2010).

$$U = IR \tag{12}$$

where U is voltage (V), I is current (A) and R is resistance (Ω).

$$G = \frac{1}{R} \tag{13}$$

4. Measurement methods for dielectric properties

Several different techniques have been developed for the determination of the dielectric constant and the dielectric loss factor of solid and liquid materials. The most commonly used ones are the following.

4.1 Perturbation technique

The perturbation technique is normally used in the case of homogeneous materials due to its simplicity, high temperature capability and easy data reduction (Tong, & Lentz, 1993). One of its greatest advantages is that it can be highly accurate in the determination of small loss factors. It is based on the measurement of either reflection coefficients or resonant frequencies (Sheen, 2009). In this later case the investigated materials are characterized to load a resonant cavity and the sample permittivity is evaluated from the shift of the resonant frequency value, compared to that of the empty cavity (Murthy & Raman, 1989). For the evaluation of dielectric properties of homogeneous dielectric materials a new, even more accurate perturbation technique has been developed where a special cavity has been designed with very small slot at the center of the broader side of the wave-guide in order to insert the sample (Kumar & Sharma, 2007).

4.2 Waveguide and coaxial transmission line method

The determination of dielectric constant and loss factor is based on the transmission line theory (Ebara, 2006). According to this theory the values of ε' and ε'' can be determined by measuring the phase and amplitude of reflected microwave signals both from a solid or liquid sample placed against the end of a short circuit transmission - waveguide or coaxial - line (Kozhevnikov, 2009). Charreyre and co-workers were among the first researchers who have described an apparatus designed to measure dielectric parameters by this technique (Charreyre, 1984).

4.3 Open-ended probe technique

In this case a coaxial line with a tip, which can sense the signal reflected from the investigated material, must be applied (Sheen & Woodhead, 1999). The tip is either in contact with the sample by touching the probe to a flat face of the solid material (Gardner & Anantaraman, 1995) or it is immersed into the liquid sample during the measurements (Guo, 2011). The dielectric properties are calculated from the phase and amplitude of the reflected signal at the end of the open-ended coaxial line. This technique is valid for materials with loss factor higher than 1, between the frequencies of 915 and 2450 MHz.

4.4 Time domain spectroscopy method

Although, this method is rather expensive it is an excellent tool to investigate the interaction of the electromagnetic energy and materials over a wide frequency range (Afsar, 1986). The reflection characteristics of the investigated material are used to compute the dielectric properties. The measurement is very rapid and the accuracy is very high, within a few percent error range (Lee, 2003).

4.5 Free-space transmission technique

It is considered to be a non-destructive and contact-less measuring method, which is particularly suitable for determining the dielectric properties of materials at high temperature and for inhomogeneous dielectrics. It can also be easily implemented in industrial applications for continuous monitoring and control (Kraszewski, 1980). The accuracy of measurement depends mainly on the performance of the measuring system and the validity of the equations used for the calculation (Håkansson, 2007).

The above described conventional techniques to measure the dielectric properties of ionic liquids or electrolyte solutions mostly tend to fail because the samples are largely short circuited by the high electrical conductance (Wakai, 2005). Hence it is quite difficult to measure the dielectric constant and loss factor of conducting materials. The main challenge is to develop an accurate method for determining the dielectric properties of such materials while the method is not only effected by the frequency of interest, the required accuracy, the geometry of the sample holder (Roumeliotis & Kokkorakis, 1994) and the adequate modeling of the circuit, but also by the physical and electrical characteristics of the studied liquid (Venkatesh & Raghavan, 2005).

Therefore we intended and managed to design and build a so-called microwave dielectrometric measurement device in which - based on the compensation of phase change due to the microwave energy absorption of the liquid sample – the dielectric properties of ionic liquids can be successfully measured in a continuous and automatic way.

5. Experimental

5.1 Materials

All ILs used in this work were synthesized at the Research Institute of Chemical and Process Engineering, University of Pannonia, Hungary. Their names, structural formula and water content can be seen in Table 1. The water content of the ILs was measured by Karl-Fischer titration and the values ranged between 0.08% and 5.11% (w/w). After a 24 h long vacuum treatment all ILs contained less than 0.01% (w/w) water.

5.2 Experimental set-up

The schematic representation of the self-designed experimental set-up applied for automatic and online measurement of dielectric properties of well conducting liquids in a definite temperature range is shown in Figure 1. It is composed of the following devices and instruments; cylindrical sample holder unit, thermostat, peristaltic pump, waveguide, temperature sensor, displaceable piston, stepper motor, magnetron, detectors, control unit, and a PC.

Ionic liquids			
Short name	Name	Structure	Water content(w/w%)
[DiEtMeIm][PF_6]	1,3-diethoxymethyl imidazolium hexafluorophosphate	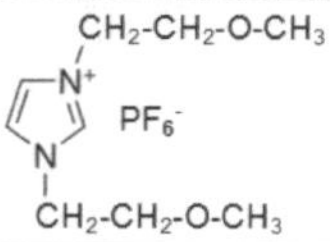	5.11
[DiEtMeIm][BF_4]	1,3-diethoxymethyl imidazolium tetrafluoroborate	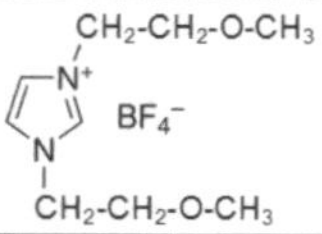	3.22
[DiEtEtIm][PF_6]	1,3-diethoxyethyl imidazolium hexafluorophosphate	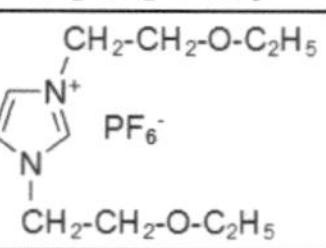	1.58
[DiEtEtIm][BF_4]	1,3-diethoxyethyl imidazolium tetrafluoroborate	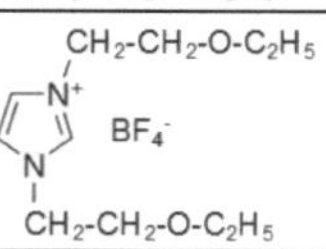	0.08
[DiEtEtIm][Br]	1,3-diethoxyethyl imidazolium bromide	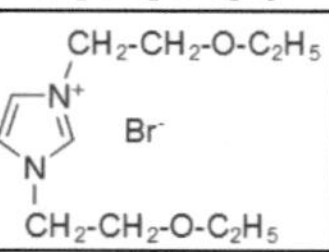	0.76
[DiEtEtIm][$N(CN)_2$]	1,3-diethoxyethyl imidazolium dicyanamide	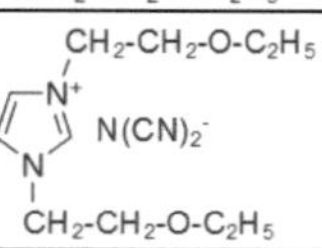	0.93
[DiEtEtIm][$N(SO_3CF_3)_2$]	1,3-diethoxyethyl imidazolium bis(trifluoromethyl sulfonyl) imide	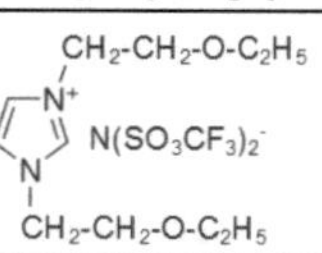	1.17
[DiEtEtIm][SCN]	1,3-diethoxyethyl imidazolium thiocyanate	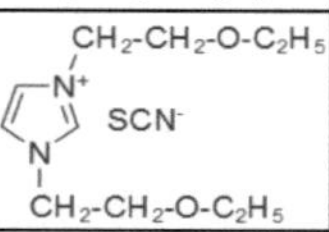	0.88
[DiEtPrIm][PF_6]	1,3-diethoxypropyl imidazolium hexafluorophosphate	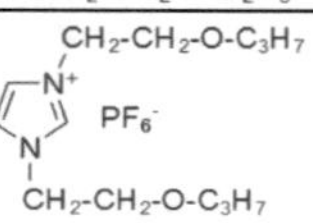	0.95
[DiEtPrIm][BF_4]	1,3-diethoxypropyl imidazolium tetrafluoroborate	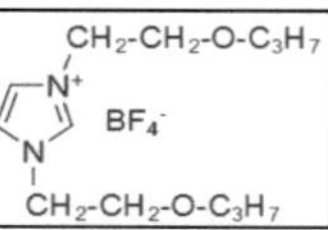	2.27
[DiEtBuIm][PF_6]	1,3-diethoxybutyl imidazolium hexafluorophosphate	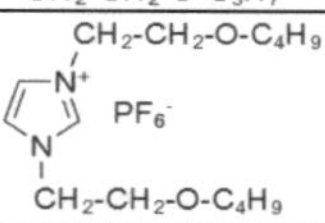	1.96
[DiEtBuIm][BF_4]	1,3-diethoxybutyl imidazolium tetrafluoroborate	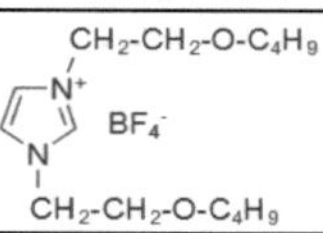	2.35

Table 1. Name, structural formula and water content of the studied ionic liquids

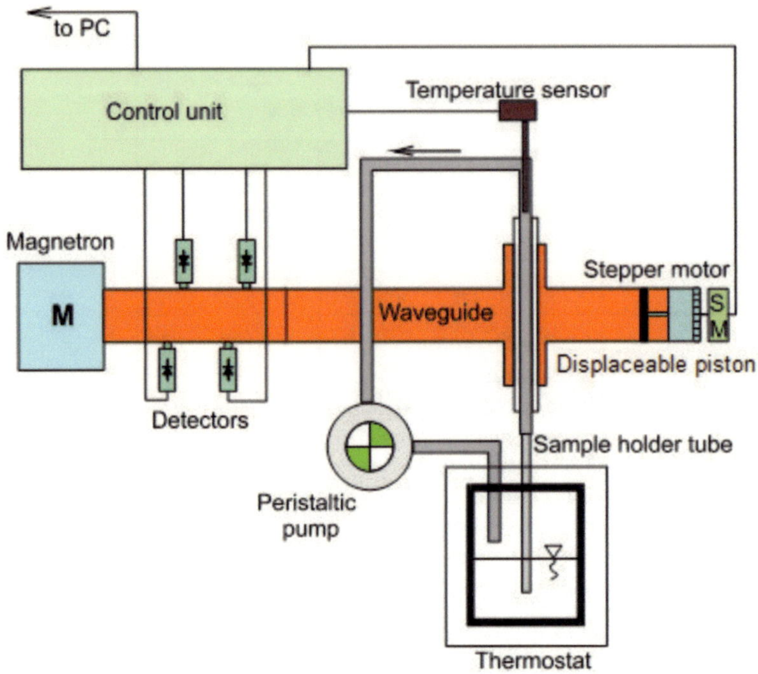

Fig. 1. Scheme of the experimental set-up

The IL sample, which is placed in the thermostat to keep it at the desired constant temperature, is flown across the waveguide having a length of about 3λ through the sample holder tube with the help of a peristaltic pump. The electric energy is transformed into microwave energy by the magnetron. The stepper motor is controlled by a microprocessor control unit, which contains an Intel 8-bit microcontroller, 12 bit A/D converters for receiving the four diode-detector signals, a stepper motor driver and a RS-232 serial interface to connect it to the PC. The control unit collects the detector signals and the temperature data determined by the temperature sensor and sends them to the PC. Furthermore it controls the position of the short circuit displaceable piston on the basis of algorithm software elaborated for this purpose.

5.3 Measurement method

The method is based on the compensation of phase change due to the microwave energy absorption of the liquid sample. The short circuit piston situated behind the sample must be actuated for compensation. The energy conditions created by the wave front in the waveguide are measured by four diode-detectors. Based on the diode detector signals the wave-front pictures are developed in the transmission-line as it is represented on Figure 2. As it shown, essentially three arrangements can be formed.

Figure 2 a) shows the case when the sample holder is empty and the piston is at the position x_0, then a field maximum is detected by diode D_3. The short circuit piston must be moved to a distance $3/2\lambda$ from the diode detector D_1, and the signal of D_1- being at a distance $\lambda/4$ from the detector D_3 - exhibits a minimum. As a result of suitable geometric and electric arrangements, the signals of the detector D_2 and D_4 - which are placed with $\lambda/4$ spacing - are identical. The incorrect position of the short circuit piston is indicated by the difference between D_2 and D_4. The differences between the signals determine the distance from the required position of the piston, while its sign refers to the direction where the short circuit piston must be moved. These latter detectors are applied to measure the equilibrium state, while D_1 and D_3 serve for the detection of the standing wave ratio.

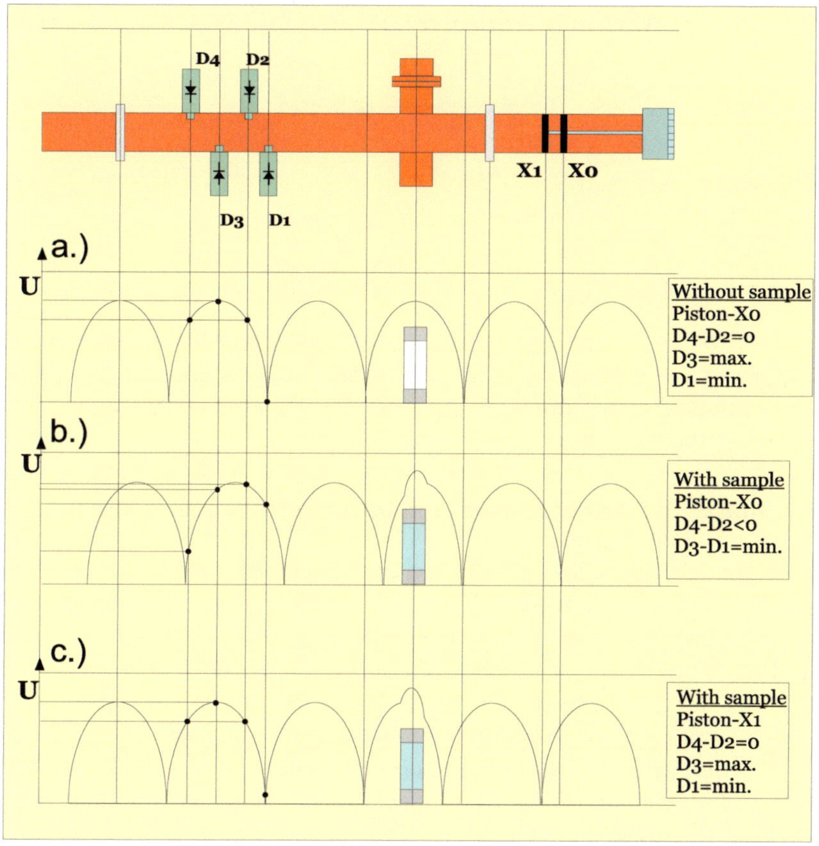

Fig. 2. a-c. Illustration of wave-fronts developed in the transmission line

Figure 2 b) shows the situation when the sample is introduced into the sample holder such a way, that the position of the piston remains unchanged. However, the reflection on the short circuit takes place in the same way as previously, the signals of the detectors are different from the previous state (demonstrated in Figure 2 a) when the shape of the wave-front formed in the transmission line is modified by the sample of dielectric properties. Due to the equilibrium shift between the D_2 and D_4 detectors, their signals become different and the quotient of the signals of D_1 and D_3 decreases because of the change of the traveling/reflected wave ratio, where the latter depends on the dielectric constant.

Figure 2 c) represents the state when the sample is still in the sample holder, but the equilibrium is restored by moving the short circuit piston. At the correct position of the piston the wave-front is restored to the state demonstrated in Figure 2 a). The position of the short circuit piston will be at a new x_1 position after the detector signals turn into the initial positions on the standing wave. The difference between position x_1 and x_0 (Δx) is used to calculate the dielectric constant ε' (Equation 14). The standing wave ratio is equivalent to the quotient of the maximum and minimum values of the electromagnetic waves developed in the transmission line, which are measured by the detectors marked with odd numbers. When the sample holder is empty the standing wave ratio is equal to 1 at equivalent state ($D_2 = D_4$). When the sample is inserted into the sample holder the standing wave ratio increases and the calculation of the dielectric loss factor ε'' is based on the extent of this increase of the standing wave ratio (Equation 15).

$$\varepsilon' = 1 + \frac{a\lambda^2}{\lambda_T \pi^2 d^2} tg(\beta \cdot \Delta x) \tag{14}$$

where $\beta = 2\pi / \lambda_T$ and $\Delta x = x_1 - x_0$

$$\varepsilon'' = \frac{a\lambda^2}{\lambda_T \pi^2 d^2} \cdot \frac{1}{r} \tag{15}$$

where $1/r = \sqrt{(U_1/U_3)}$, a is the large waveguide dimension, d is the inner diameter of the sample holder λ and λ_T are wavelength in vacuum and in the waveguide, respectively, U_1/U_3 is the so called standing wave ratio, arising from D_1 and D_3 detector signals and Δx is the displacement of short circuit piston (Charreyre, 1984).

5.4 Determination of the dielectric constant, dielectric loss factor and the electrical conductivity of ionic liquids

Dielectric constants, dielectric loss factors and the temperature dependence of the dielectric properties of ionic liquids intended to be used in batteries were determined by the above described self-designed microwave dielectrometric apparatus (Figure 3) at the frequency of 2.45 GHz and at different temperatures (30°C, 40°C, 50°C, 60°C, 70°C, 80°C, 90°C, 100°C, 110°C and 120°C). The speed of the change in temperature depends on the electrical field strength in the material (E), the absorbed microwave power, (P_v) density (ρ), the specific heat capacity (C_p) and the dielectric loss factor ε'' and can be given by Equation 16 (Göllei, 2009). The electrical conductivity values (G) were calculated by Equation 12 and Equation 13.

$$\frac{\Delta T}{\Delta t} = \frac{P_v}{\rho C_p} = \frac{jE^2 f \varepsilon''}{\rho C_p} \tag{16}$$

6. Results and discussion

6.1 Effect of anion on the dielectric properties

The dielectric constant (E_1), dielectric loss factor (E_2) and electric conductivity (G) values of ILs built up of the same [DiEtEtIm] cation and six different anions ($[PF_6]$, $[BF_4]$, [Br], $[N(CN)_2]$, $[N(SO_3CF_3)_2]$ and [SCN]), at different temperatures between 30°C and 120°C are shown in Figure 4, Figure 5 and Figure 6, respectively.

It can be seen that while at the temperature of 30°C dielectric constants are similar for all ILs, ranging from 6 to 8, a slight increase of E_1 values can be seen with the increase in temperature up to 80°C. Above this temperature ILs containing different anions behave in a slightly different way; the E_1 values for [DiEtEtIm]$[N(CN)_2]$ and [DiEtEtIm][SCN] decrease, tha values for [DiEtEtIm]$[PF_6]$ sharply increase and the values for [DiEtEtIm]$[BF_4]$,

[DiEtEtIm][Br] and [DiEtEtIm][N(SO_3CF_3)$_2$] remain more or less the same between 80°C and 120°C. At 120°C temperature the highest dielectric constant 18.2 belongs to [DiEtEtIm][SCN] and the lowes, 5.6 to [DiEtEtIm][N(CN)$_2$].

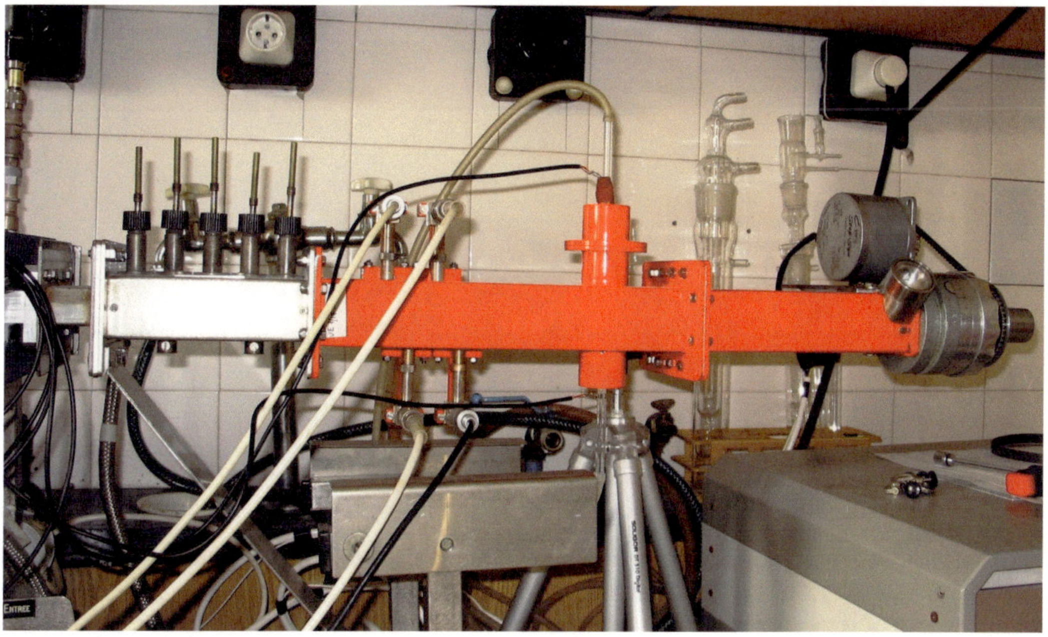

Fig. 3. Picture of the microwave dielectrometric apparatus

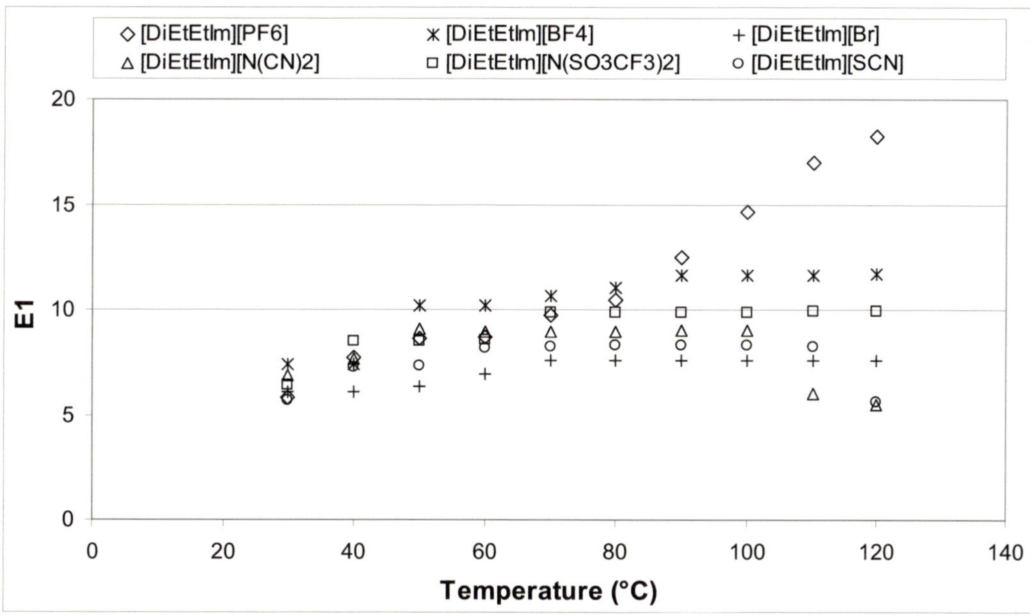

Fig. 4. Dielectric constant of ILs containing common [DiEtEtIm] cation and different anions

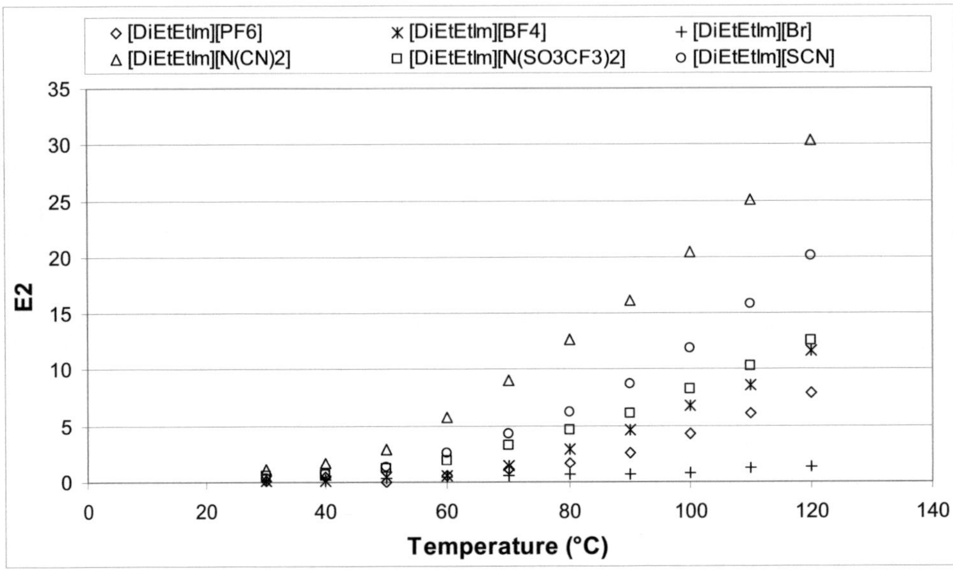

Fig. 5. Dielectric loss factor of ILs containing common [DiEtEtIm] cation and different anions

The dielectric loss values are almost the same and quite low for all six ILs at the temperature of 30°C. Except for the E_2 of [DiEtEtIm][Br], which reaches only the value of 2 at 120°C, the dielectric loss of the other five investigated ILs increase constantly with higher temperature. It is interesting to note that the highest E_2 value belongs to the [DiEtEtIm][N(CN)$_2$] IL, which has the lowest E_1 value.

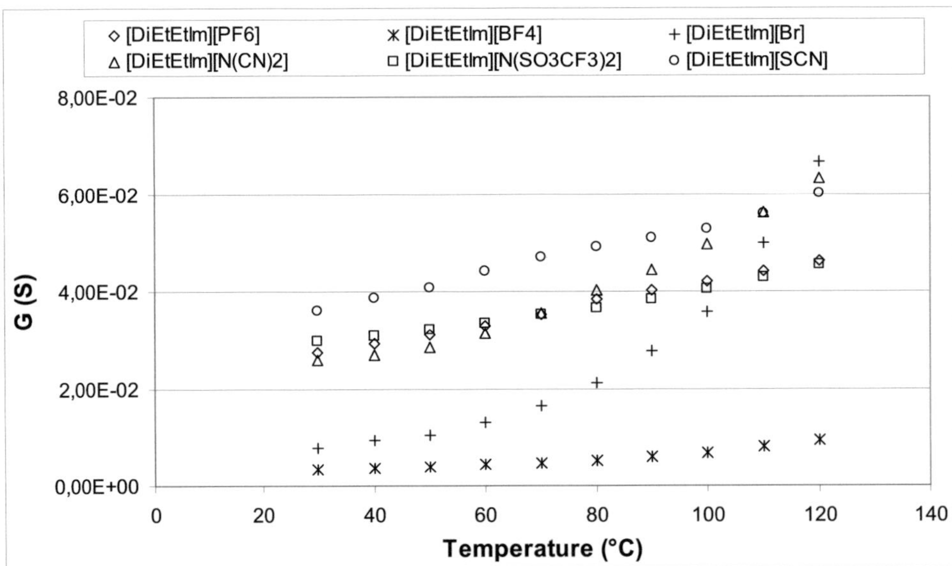

Fig. 6. Electrical conductivity of ILs containing common [DiEtEtIm] cation and different anions

At the initial 30°C temperature the lowest electric conductivity value of 3x 10^{-3} S is exhibited by [DiEtEtIm][BF_4], followed by [DiEtEtIm][Br] with the value of 8x10^{-3} S and then the other three ILs around 3x10^{-2} S and finally by the [DiEtEtIm][SCN] with 3.8x10^{-2} S. The conductivity of [DiEtEtIm][Br] and [DiEtEtIm][PF_6] show a steep increase, while that of the other ILs indicate only a slight increase with the increase in temperature between 30°C and 120°C. The *G* values of [DiEtEtIm][PF_6], [DiEtEtIm][$N(SO_3CF_3)_2$], [DiEtEtIm][$N(CN)_2$] and [DiEtEtIm][SCN] are more than four times higher than that of water at the initial temperature and further increase at elevated temperatures. The [DiEtEtIm][BF_4] has extremely low conductivities, at the temperature range between 30°C and 120°C they hardly reach that of water. However, the conductivity values for [DiEtEtIm][Br] exceed that of water at elevated temperature, at the general operation temperature of batteries it shows extremely low values. Therefore neither of these ILs should be considered as candidates for elctrolytes.

It is worth noticing that the ILs containing the same cation and different anions have quite different E_1 values, and that their E_2 values become more distinct with the increase in temperature between the temperature range of 80°C and 120°C.

It can be concluded that there might be a connection between the dielectric constant, dielectric loss factor and electrical conductivity of the investigated ILs and the structure of the ILs but the connection is not yet correctly defineable and further experiments should be carried out in this field.

6.2 Effect of cation on the dielectric properties

The dielectric constant (E_1), dielectric loss factor (E_2) and electrical conductivity (*G*) values of ILs built up of the same [PF_6] anion and six different cations containing alkyl chains with different lengths ([DiEtMeIm], [DiEtEtIm], [DiEtPrIm] and [DiEtBuIm]) at different temperatures between 30°C and 120°C are shown in Figure 7, Figure 8 and Figure 9, respectively.

The dielectric constants for [DiEtEtIm][PF_6], [DiEtPrIm][PF_6] and [DiEtBuIm][PF_6] exhibit more or less the same values around 5 and that of [DiEtMeIm][PF_6] is four times higher at 30°C. With the increase in temperature the E_1 values of [DiEtMeIm][PF_6] and [DiEtEtIm][PF_6] successively increase, while in the case of [DiEtPrIm][PF_6] and [DiEtBuIm][PF_6] a sudden break can be noticed in the constant values between 50°C and 70°C. This might be associated with some sudden changes in the structure of the ILs or in the physicochemical interactions between the ions.

Similarly to the dielecetric constant, the highest dielectric loss of 4.3 belongs to [DiEtMeIm][PF_6], while the other ILs have more than four times lower E_2 values. With increasing temperature between the range of 30°C and 120°C the dielectric loss value of [DiEtMeIm][PF_6] shows a sharp increase, the values for [DiEtEtIm][PF_6] and [DiEtBuIm][PF_6] moderately increase and the increase of [DiEtPrIm][PF_6] E_2 value exhibits such a sudden change in the values between 100°C and 110°C as it is shown in the case of E_1.

The highest 2.9x10^{-2} S electrical conductivity value belongs to [DiEtMeIm][PF_6] followed by the 2.8 x10^{-2} S of [DiEtBuIm][PF_6] then by the 4.4 x10^{-2} S value for [DiEtPrIm][PF6] and finally by the 3.8 x10^{-2} S for [DiEtBuIm][PF_6] at 30°C. The conductivity of the former two ILs sharply increase, while those of the latter two ILs show only a slight increase with increasing temperature.

It is worth noticing that the [DiEtMeIm][PF_6], which contains the shortest alkyl chain in its cation, exhibits the highest E_1, E_2 and *G* values, hence in the case of ILs containing the same

[PF_6] anion it can be stated that the cation has an effect on the dielectric constant and the dielectric loss factor of the investigated liquids. Furtehrmore it can be concluded that the [DiEtMeIm][PF_6] and [DiEtetIm][PF_6] ILs seem to be the most adequate for battery applications, however further investigation must be carried out.

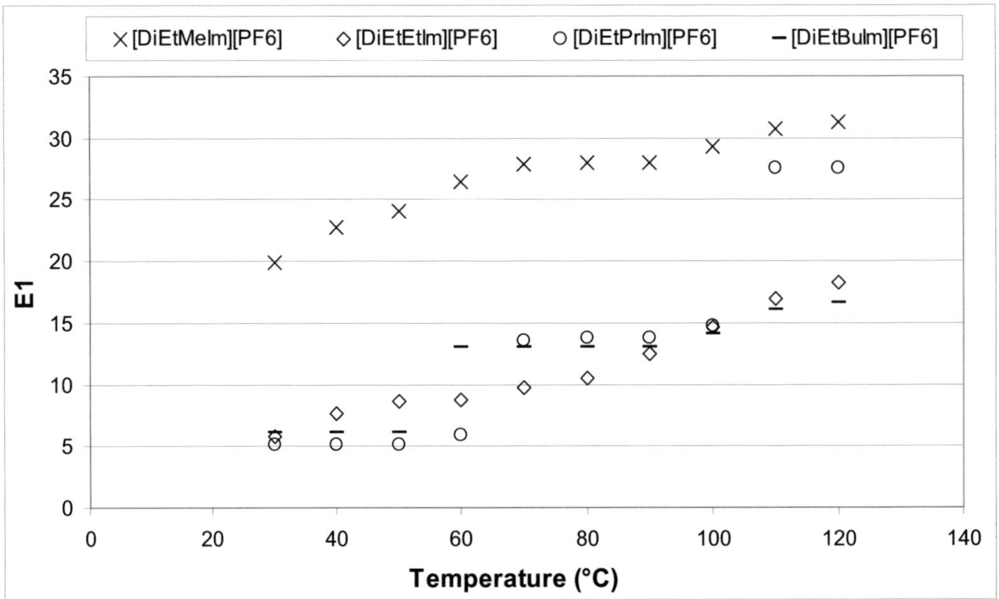

Fig. 7. Dielectric constant of ILs containing common [PF_6] anion and different cations

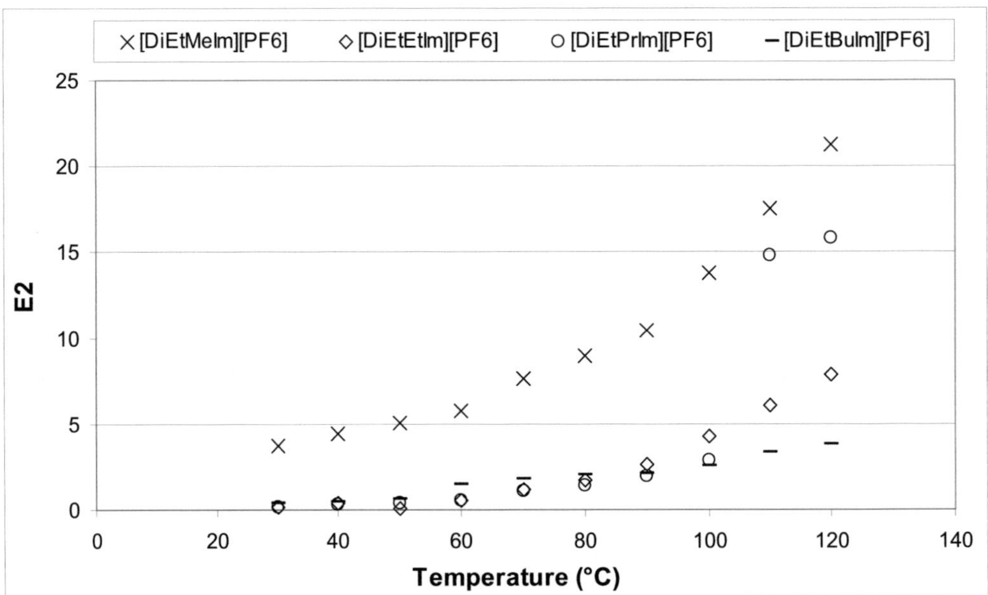

Fig. 8. Dielectric loss factor of ILs containing common [PF_6] anion and different cations

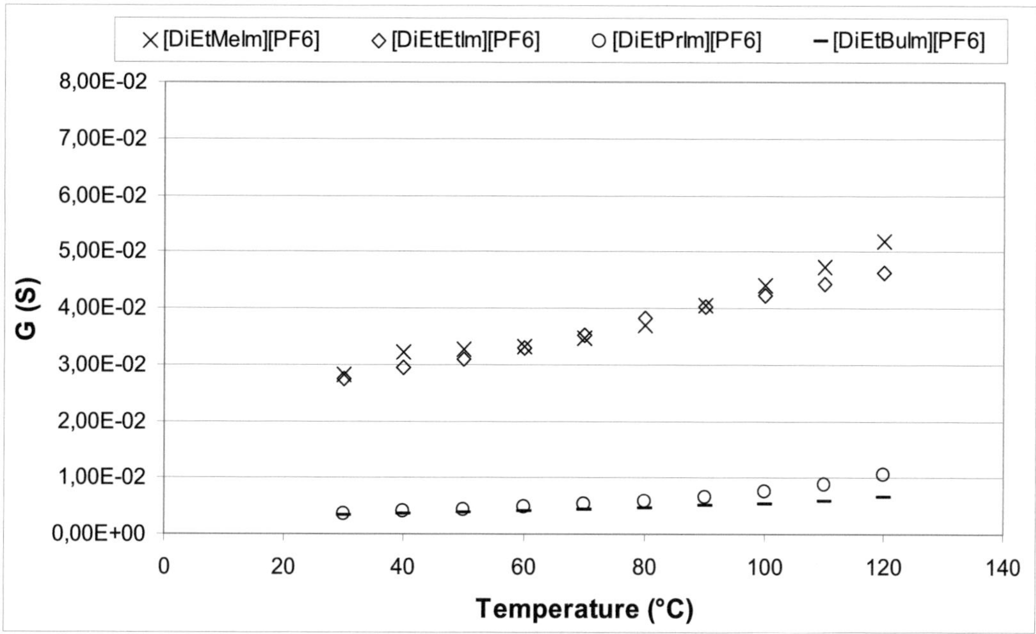

Fig. 9. Electrical conductivity of ILs containing common $[PF_6]$ anion and different cations

The dielectric constant (E_1), dielectric loss factor (E_2) and electrical conductivity (G) values of ILs built up of the same $[BF_4]$ anion and six previously described cations containing alkyl chains with different lengths ([DiEtMeIm], [DiEtEtIm], [DiEtPrIm] and [DiEtBuIm]) at different temperatures between 30°C and 120°C are shown in Figure 10, Figure 11 and Figure 12, respectively.

At the initial measuring temperature of 30°C all four of the studied ILs have similar dielectric constants around 7. With the increase in temperature the E_1 values for $[DiEtMeIm][BF_4]$, $[DiEtEtIm][BF_4]$ and $[DiEtPrIm][BF_4]$ slightly increase up to 13, while the dielectric constant for $[DiEtBuIm][BF_4]$ shows a sudden break at 90°C and at 120°C it reaches the value as high as 27. This could be explained with some sudden changes in the structure of the ILs or in the physicochemical interactions between the anion and the cation. Excluding the results for $[DiEtBuIm][BF_4]$ the values at elevated temperature show that the highest E_1 value belongs to $[DiEtMeIm][BF_4]$, followed by $[DiEtEtIm][BF_4]$ and then by the $[DiEtPrIm][BF_4]$, hence the dielectric constant increases with the decrease in the alkyl chain length of the cation.

It can be seen that all of the ILs have similar dielectric loss factor values at 30°C and sharply increase with the increase in temperature. Similar to E_1 the E_2 value of $[DiEtBuIm][BF_4]$ shows a sudden change between the temperature range of 80°C and 90°C. At higher temperatures the highest dielectric loss factor of 19 belongs to $[DiEtBuIm][BF_4]$, followed by $[DiEtMeIm][BF_4]$, $[DiEtEtIm][BF_4]$ and then by the $[DiEtPrIm][BF_4]$. Therefore it can be said that the longer the alkyl chain length in the IL cation, the highest the E_2 values are.

It is important to notice that the E_1 and E_2 values of ILs containing cation with longer alkyl chains tend to exhibit a sudden break and increase at elevated temperatures.

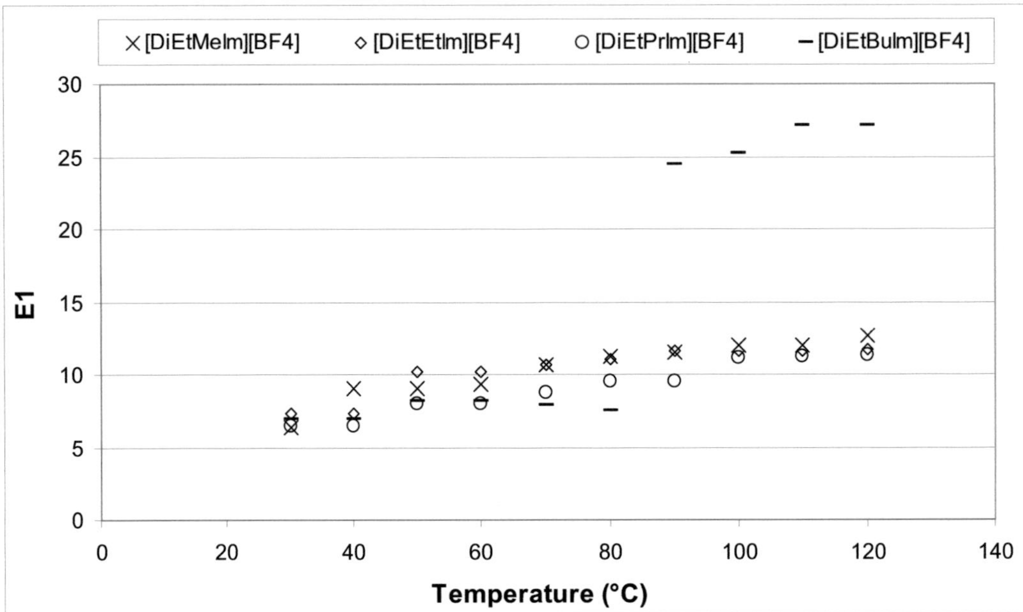

Fig. 10. Dielectric constant of ILs containing common [BF_4] anion and different cations

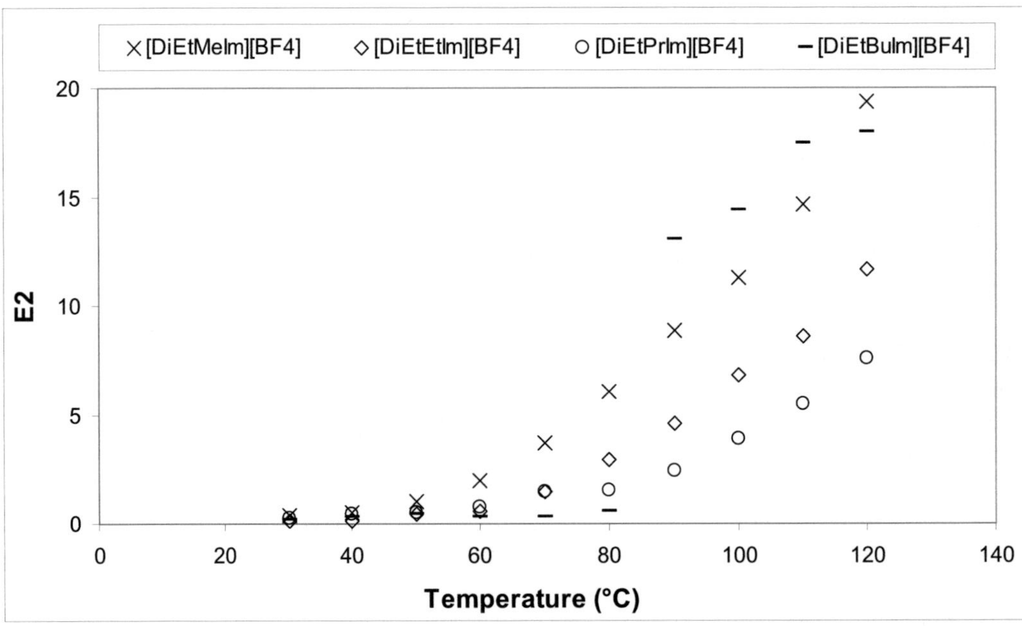

Fig. 11. Dielectric loss factor of ILs containing common [BF_4] anion and different cations

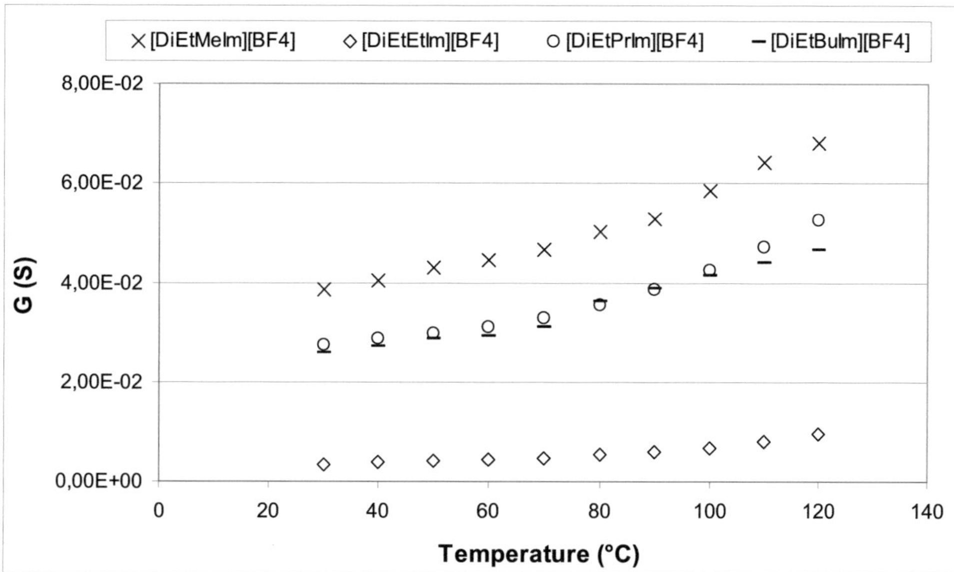

Fig. 12. Electrical conductivity of ILs containing common $[BF_4]$ anion and different cations

The highest electrical conductivity of 3.94x10^{-2} S belongs to $[DiEtMeIm][BF_4]$, the $[DiEtPrIm][BF_4]$ and $[DiEtBuIm][BF_4]$ ILs have the second highest G values of about 2.8x10^{-2} S and the lowest 3x10^{-3} S conductivity is exhibited by $[DiEtEtIm][BF_4]$ at the temperature of 30°C. With the increase in temperature the electrical conductivity values of $[DiEtEtIm][BF_4]$ moderately and those of the other three examined ILs sharply increase.

It is worth noticing that excluding the results of $[DiEtEtIm][BF_4]$, which has a surprisingly low conductivity compared to the other three studied ILs, the *G* values increase along with the decrease in the alkyl chain length of the IL's cation. Therefore it can be stated that there might be a similar connection between the chemical structure of the cation and the dielectric properties of the ILs containing common $[BF_4]$ anion as it has been shown for the ILs built up of the common $[PF_6]$ anion.

Taking into account the dielectric constant, dielectric loss and electrical conductivity results of all the investigated ILs it can be concluded that the alkyl chain length of the cation and the structure of the anion strongly influence the dielectric properties of the ILs and that the highest *G* value is exhibited by the $[DiEtMeIm][BF_4]$ IL at 30°C temperature, hence it is the most suitable candidate for battery applications. Furtermore it can be stated that implementing the knowledge about the connection between the IL structure and the dielectric properties another ILs should be studied in order to select the ones most adequate for electrolyte application.

7. Conclusion

Due to the special characteristics of ILs, such as wide electrochemical windows, high inherent conductivities, high thermal and electrochemical stability, tuneable physicochemical properties, etc., they might be excellent candidates for environmentally sound, green electrolites in batteries. In order to decide whether they could be successfully applied it is essential to gain information about their dielectric properties.

However, several techniques have already been published about determination of the dielectric properties of ILs, so far no measuring methods were introduced due to the fact that such measurements cannot be realized by classical methods because of the high conductivity of ILs.
In this chapter first a self-designed microwave dielectrometric apparatus and a novel method for the measurement of the dielectric properties of ILs were described, then the dielectric constant, dielectric loss factor and electrical conductivity of several ILs were measured. Furthermore the connections between the structure of the investigated ionic liquids and the dielectric characteristics were determined.
It was concluded that there was a certain connection between the structure of the investigated ILs and that this knowledge was worth implementing in studying further ILs that might be applied as novel, environmentally sound electrolytes.

8. Acknowledgement

The research work was supported by the research program "Livable environment and healthier people - Bioinnovation and Green Technology Research at the University of Pannonia TÁMOP-4.2.2-08/1/2008-0018"

9. References

Afsar, M.N., Birch, J.R., Clarke, R.N., Chantry, G.W. (1986). The measurement of properties of materials. *IEEE Transactions on Instrumentation and Measurement*, 74., 1., 183-199., 0018-9456

Alarco, P.J., Yaser, A.L., Ravet, N., Armand, M. (2004). Lithium conducting pyrazolium imides plastic crystals: a new solid state electrolyte matrix. *Solid State Ionics*, 172., 1-4., 53-56., 0167-2738

Arnold, C.B. Sutto, T.E. Kim, H.K. Pique, A. (2004). Direct-write laser processing for miniature electrochemical systems. *Laser Focus World*, 40., 5., 9-12., 1043-8092

Benenson, W., Harris, J.W., Stocker, H., Lutz, H. (2002). *Handbook of Physics*, Springer-Verlag New York, Inc., ISBN: 0-387-95269-1, New York, USA

Bright, F., Baker, G.A. (2006). Comment on "How Polar Are Ionic Liquids? Determination of the Static Dielectric Constant of an Imidazolium-based Ionic Liquid by Microwave Dielectric Spectroscopy". *The Journal of Physical Chemistry B*, 110., 11., 5822-5823., 1520-6106

Carrier, M., Soga, K. (1999). A four terminal measurement system for measuring the dielectric properties of clay at low frequencies. *Engineering Geology*, 53., 2., 115-123., 0013-7952

Charreyre, M., Thiebaut, J., Roussy, G. (1984). Permittivity measurement of materials during high-power microwave irradiation and processing. *Journal of Physics E: Scientific Instruments*, 17., 8., 678-682., 0022-3735

Diaw, M., Chagnes, A., Carré, B. Willmann, P., Lemordant, D. (2005). Mixed ionic liquid as electrolyte for lithium batteries. *Journal of Power Sources*, 146., 1-2., 682-684., 0378-7753

Ebara, H., Inoue, T., Hashimoto, O. (2006). Measurement method of complex permittivity and permeability for a powdered material using a waveguide in microwave band. *Science and Technology of Advanced Materials*, 7., 1., 77-83., 1468-6996

Endres, F., Abbott, A.P., MacFarlane, D.R. (2008). *Electrodeposition from Ionic Liquids*, Wiley-VCH Verlag GmbH & Co. KGaA, ISBN: 978-3-527-31565-9, Weinheim, Germany

Fleisch, D. (2008). *A Student's Guide to Maxwell's Equations*, Cambridge University Press, ISBN: 978-0-521-87761-9, Cambridge, UK

Forsyth, S., Golding, J., MacFarlane, D.R., Forsyth, M. (2001). N-methyl-N-alkylpyrrolidinium tetrafluoroborate salts: ionic solvents and solid electrolytes. *Electrochimica Acta*, 46., 10-11., 1753–1757., 0013-4686

Gardner, C.L., Anantaraman, A.V. (1995). Measurement of membrane conductivities using an open-ended coaxial probe. *Journal of Electroanalytical Chemistry*, 395., 1-2., 1995, 67-73., 1572-6657

Giroud, N.M., Rouault, H., Chainet, E., Poignet, J.C. (2008). Ionic liquids based electrolytes for lithium ion battery. *ECS Meeting Abstracts*, 802., 3044., 1091-8213

Göllei, A., Vass, A., Pallai, E., Gerzson, M., Ludányi, L., Mink, J. (2009). Apparatus and method to measure dielectric properties (ε′ and ε″) of ionic liquids. *Review of Scientific Instruments*, 80., 4., 0034-6748

Guo, W., Liu, Y., Zhu, X., Wang, S. (2011). Temperature-dependent dielectric properties of honey associated with dielectric heating. *Journal of Food Engineering*, 102., 3., 209-216., 0260-8774

Hagiwara, R., Ito, Y. (2000). Room temperature ionic liquids of alkylimidazolium cations and fluoroanions. *Journal of Fluorine Chemistry*, 105., 2., 221-227., 0022-1139

Håkansson, E, Amiet, A., Kaynak, A. (2007). Dielectric characterization of conducting textiles using free space transmission measurements: Accuracy and methods for improvement. *Synthetic Metals*, 157., 24., 1054-1063., 0379-6779

Heaviside, O. (2007). *Electromagnetic Theory, Volume I.*, Cosimo, Inc., ISBN: 978-1-60206-271-9, New York, USA

Heiken, G.H., Vaniman, D.T., French, B.M. (1991). *LUNAR sourcebook; a user's guide to the moon*, Press Syndicate of the University of Cambridge, ISBN: 0-521-33444-6, Melbourne, Australia

Housecroft, C.E., Sharpe, A.G. (2005) *Inorganic chemistry*, 2nd Edition, Pearson Education Limited, ISBN: 978-0-13-039913-7, Essex, England

Ito, K., Nishina, N., Ohno, H. (2000). Enhanced ion conduction in imidazolium-type molten salts. *Electrochimica Acta*, 45., 8-9., 1295–1298., 0013-4686

Koel, M. (2008). *Ionic liquids in Chemical Analysis*, CRC Press Taylor & Francis Group, ISBN: 1420046462, USA

Kozhevnikov, N.O. (2009). Applying the transmission line theory to study ungrounded horizontal loop self-transients. *Russian Geology and Geophysics*, 50., 3., 222-233., 1068-7971

Kraszewski, A.W. (1980). Microwave aquametry-A review. *Journal of Microwave Power and Electromagnetic Energy*, 15., 4., 209-220., 0832-7823

Krossing, I., Slattery, J.M., Daguenet, C., Dyson, P.J., Oleinikova, A., Weingartner, H. (2006). Why are ionic liquids liquid? A simple explanation based on lattice and solvation energies. *Journal of the American Chemical Society*, 128., 41., 13427-13434., 0002-7863

Kumar A., Sharma, S. (2007). Measurement of dielectric constant and loss factor of the dielectric material at microwave frequencies. *Progress In Electromagnetics Research*, 69., 47-54., 1559-8985

Lee, K.-S., Lu, T.-M., Zhang, X.-C. (2003). The measurement of the dielectric and optical propertiesof nano thin films by THz differential time-domain spectroscopy. *Microelectronics Journal*, 34., 1., 63-69., 0026-2692

Marsh, K.N., Boxall, J.A., Lichtenthaler, A. (2004). Room temperature ionic liquids and their mixtures-a review. *Fluid Phase Equilibria*, 219., 1., 93-98., ISSN: 03783812

Mastumoto, H., Kageyama, H., Miyazaki, Y. (2001). Room temperature molten salts based on tetraalkylammonium cations and bis(trifluoromethylsulfonyl)imide. *Chemistry Letters*, 30., 2., 182-183., 0366-7022

Matsumoto, H., Sakaebe, H., Tatsumi, K. (2005). Preparation of room temperature ionic liquids based on aliphatic onium cations and asymmetric amide anions and their electrochemical properties as a lithium battery electrolyte. *Journal of Power Sources*, 146., 1-2., 45-50., 0378-7753

Matsumoto, H., Yanagida, M., Tanimoto, K., Kojima, K., Tamiya, Y., Miyazaki, Y. (2000). Highly conductive room temperature molten salts based on small trimethylalkylammonium cations and bis(trifluoromethylsulfonyl)imide. *Chemistry Letters*, 29., 8., 922., 0366-7022

Meredith, R. (1998). *Engineers' handbook of industrial microwave heating*, The Institution of Electrical Engineering, ISBN: 0 85296 916 3, London, United Kingdom

Mohos, F. (2010). *Confectionery and Chocolate Engineering; Principles and Applications*, John Wiley & Sons Ltd, ISBN: 978-1-4051-9470-9, West Sussex, United Kingdom

Murthy, V.R.K., Raman, R. (1989). A method for the evaluation of microwave dielectric and magnetic parameters using rectangular cavity perturbation technique. *Solid State Communication*, 70., 8., 847–850., 0038-1098

Mutelet, F., Jaubert, J.N. (2006). Accurate measurements of thermodynamic properties of solutes in ionic liquids using inverse gas chromatography. *Journal of Chromatography A* 1102., 1-2., 256-267., 00219673

Ohno, H. (2005). *Electrochemical Aspects of Ionic Liquids*, 2nd Ed., John Wiley and Sons, Inc., ISBN: 978-0-471-64851-2, Hoboken, New Jersey, USA

Plechkova, N.V., Seddon, K.R. (2008). Applications of ionic liquids in the chemical industry. *Chemical Society Reviews*, 37., 1., 123-150., 0306-0012

Roumeliotis, J.A., Kokkorakis, G.C. (1994). Resonant frequencies in an electromagnetic cylindrical/spherical cavity with an internal off-axis small dielectric sphere. *Electromagnetics*, 14, 195-215, 0272-6343

Sakaebe, H. Matsumoto, H. (2003) N-Methyl-N-propylpiperidinium bis (trifluoromethane-sulfonyl)imide (PP 13-TFSI) - novel electrolyte base for Li battery. *Electrochemistry Communications*, 5., 7., 594–598., 1388-2481

Sakaebe, H. Matsumoto, H., Tatsumi, K. (2005). Discharge–charge properties of Li/$LiCoO_2$ cell using room temperature ionic liquids (RTILs) based on quaternary ammonium cation - Effect of the structure. *Journal of Power Sources*, 146., 1-2., 693-697., 0378-7753

Sakaebe, H., Matsumoto, H., Tatsumi, K. (2007). Application of room temperature ionic liquids to Li batteries. *Electrochimica Acta*, 53., 3., 1048-1054., 0013-4686

Saruwatari, H., Kuboki, T., Kishi, T., Mikoshiba, S., Takami, N. (2010). Imidazolium ionic liquids containing LiBOB electrolyte for lithium battery. *Journal of Power Sources*, 195., 5., 1495-1499., 0378-7753

Sato, T., Masuda, G., Takagi, K. (2004). Electrochemical properties of novel ionic liquids for electric double layer capacitor applications. *Electrochimica Acta*, 49., 21., 3603-3611., 0013-4686

Seddon, K.R. (2003). Ionic liquids: A taste of the future. *Nature Materials*, 2., 6., 363-365., 1476-1122

Sheen, J. (2009). Measurements of microwave dielectric properties by an amended cavity perturbation technique. *Measurement*, 42., 1., 57-61., 0263-2241

Sheen, N.I., Woodhead, I.M. (1999). An open-ended coaxial probe for broad-band permittivity measurement of agricultural products. *Journal of Agricultural Engineering Research*, 74., 2., 1999, 193-202. 0021-8634

Stathatos, E. Lianos, P. Jovanovski, V., Orel, B. (2005). Dye-sensitized photoelectrochemical solar cells based on nano-composite organic-inorganic materials. *Journal of Photochemistry and Photobiology A: Chemistry*, 169., 1., 57–61., 1010-6030

Stracke, M.P., Migliorini, M.V., Lissner, E., Schrekker, H.S., Dupont, J., Gonçalves, R.S. (2009). Imidazolium ionic liquids as electrolytes for manganese dioxide free Leclanché batteries. *Applied Energy*, 86., 9., 1512-1516., 0306-2619

Sutto, T.E., Duncan, T.T., Wong, T.C., McGrady, K. (2011). Ionic liquid batteries: Chemistry to replace alkaline/acid energy storage devices. *Electrochimica Acta*, 56., 9., 3375-3379., 0013-4686

Tong, C.H., Lentz, R.R. (1993). Dielectric properties of bentonite pastes as a function of temperature. *Journal Food Processing and Preservation*, 17., 2., 139-145., 0145-8892

Venkatesh, M.S., Raghavan, G.S.V. (2005). An overview of dielectric properties measuring techniques. *Canadian Biosystems Engineering*, 47., 7., 15-30., 1492-9058

Wakai, C., Oleinikova, A., Ott, M., Weingartner, H. (2005). How polar are ionic liquids? Determination of the static dielectric constant of imidazolium-based ionic liquid by microwave dielectric spectroscopy. *The Journal of Physical Chemistry B*, 109., 36., 17028-17030., 1520-6106

Wakai, C., Oleinikova, A., Weingartner, H. (2006). Reply to "Comment on "How polar are ionic liquids? Determination of the static dielectric constant of an imidazolium-based ionic liquid by microwave spectroscopy"". *The Journal of Physical Chemistry B*, 110., 11., 5824., 1520-6106

Wasserschied, P., Welton, T. (2007). *Ionic Liquids in Synthesis*, VCH–Wiley, ISBN: 978-3-527-31239-9, Weinheim, Germany

Welton, T. (1999). Room-temperature ionic liquids, solvents for synthesis and catalysis. *Chemical Reviews*, 99., 8., 2071-2083., 0009-2665

Wilkes, J.S., Levisky, J.A., Wilson, R.A., Hussey, C.L. (1982). Dialkylimidazolium chloroaluminate melts: a new class of room-temperature ionic liquids for electrochemistry, spectroscopy and synthesis. *Inorganic Chemistry*, 21., 3., 1263-1264., 0020-1669

Xiao, Y., Malhotra, S.V. (2005). Friedel-crafts acylation reactions in pyridinium based ionic liquids, *Journal of Organometallic Chemistry*, 690., 15., 3609-3613., 0022-328X

Xu, J., Yang, J., NuLi, Y., Wang, J., Zhang, Z. (2006). Additive-containing ionic liquid electrolytes for secondary lithium battery. *Journal of Power Sources* 160., 1., 621-626., 0378-7753

10

Translational and Rotational Motions for TFSA-Based Ionic Liquids Studied by NMR Spectroscopy

Kikuko Hayamizu
National Institute of Advanced Industrial Science and Technology
AIST Tsukuba Central 2, Tsukuba, Ibaraki
Japan

1. Introduction

Solution-state NMR spectroscopy is an important method to study chemical structures, intramolecular local motions and translational and rotational motions of a whole molecule. NMR parameters of the solution-state NMR are chemical shift, spin-spin coupling constant, spectral pattern including line width, spin-lattice relaxation time (T_1) and spin-spin relaxation time (T_2). The assignment of the spectral signals is the starting task for advanced studies. Various two-dimensional (2D) NMR techniques can help to make assignments for complex compounds. Additionally, 2D-NMR techniques afford information on the intermolecular interactions or molecular diffusion. Generally, chemical structures of ionic liquids (IL) are not complex and the NMR spectra can be assigned easily using general rules. Once the assignment is fixed, we can obtain the precise data for the individual species for ILs. Multinuclear NMR techniques can be applied to magnetic active nuclei, and the nuclei relating to ILs are ^{1}H, ^{2}H, ^{7}Li, ^{11}B, ^{13}C, ^{19}F, ^{31}P and others. In addition to general NMR spectra, self-diffusion coefficient (D) of each ion can be measured by the pulse-gradient spin-echo (PGSE) NMR method. The D's of the individual ions in ILs are important to understand translational motion of ions.

In this chapter, NMR studies for four ILs having a common anion bis(fluoromethylsulfonyl)amine (TFSA) will be described. The cations are N-ethyl-N-methylimidazolium (EMIm), N,2-dimethyl-N-propylimidazolium (DMPIm), N-methyl-N-propylpyrrolidinium (P13), and N,N-diethyl-N-methyl-N-(2-methoxylethyl)ammonium (DEME). These ILs are important candidates to use as electrolytes in various batteries or cells. The basic properties of molecular weight (MW), viscosity (η) and cation and TFSA diffusion coefficients (D_{cation} and D_{TFSA}) measured at 30°C are given in Table 1 for the four ILs. The experimental data have been published in our previous papers for individual ILs.[1-4] In the present chapter, the individual experimental data for the four ILs are plotted on a sheet to clarify the cation properties. In addition, newly derived plots of D_{cation} versus correlation times of molecular motion are included. The chemical structures of the ions are shown.

To study the translational and rotational motions of ILs by NMR spectroscopy, the D's of the individual ions and the T_1's of the individual protons are important parameters in the

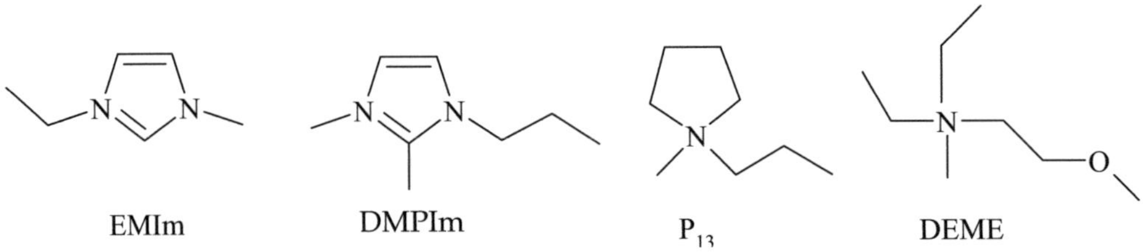

IL	MW	Viscosity η mPas	D_{cation} m^2s^{-1}	D_{TFSA} m^2s^{-1}
EMIm-TFSA	111.2 + 280.1	27.8	6.6×10^{-11}	3.9×10^{-11}
DMPIm-TFSA	139.2 + 280.1	74.8	2.2×10^{-11}	1.5×10^{-11}
P_{13}-TFSA	128.2 + 280.1	49.8	2.9×10^{-11}	2.2×10^{-11}
DEME-TFSA	146.3 + 280.1	56.6	2.3×10^{-11}	2.0×10^{-11}

Table 1. Molecular weight, viscosity and diffusion coefficients of the four ILs at 30 °C.

temperature range for which ILs are in liquid state. Solution-state NMR affords information on molecular motions for liquids or mobile species which behave like liquids. In this chapter, the properties of the ILs in the liquid states will be described, where the spectral lines are sharp enough to measure individual peaks in the solution-mode NMR. The temperature-dependent D values were measured for the cations (D_{cation}) by ^{1}H resonance and the anion of TFSA (D_{TFSA}) by ^{19}F resonance using the PGSE NMR method. Arrhenius plots of ^{1}H T_1 for the cations exhibited T_1 minimum in the temperature range in the liquid state, and the correlation time τ_c(cation) were evaluated for the same temperature range. The molecular motion of a whole molecule of the cations can be assigned to a phenomenon of the T_1 minimum and the timescale estimated was 1~500 ps depending on the temperature. The motion which gives the minimum in the ^{1}H T_1 Arrhenius plots is assumed to be a librational flip rather than an all over reorientational motion of a whole molecule. The τ_c(cation) is related to viscosity and D_{cation}. Since Arrhenius plots of the ^{19}F T_1 did not show a T_1 minimum, the correlation times including internal rotations of CF_3 could not be calculated but also related to the viscosity of the ILs.

In this chapter, complex dynamic properties of the four ILs will be discussed based on the experimental data such as D_{TFSA}, D_{cation}, τ_c(cation), bulk viscosity, ionic conductivity and density in the temperature range from 253 to 353 K or above freezing, while the data of the viscosity and density were measured between 283 and 353 K. Since ILs are in wide variety of structures and rich in properties and applications, the description in this chapter is limited for the four ILs.

2. Measurements of diffusion coefficients by NMR

It is very popular to obtain positional information using nuclear magnetic resonance, so called NMR imaging including medical MRI. In addition of a large coil to produce static magnetic field (presently super-conductive magnet (SCM)), small three-dimensional coils are placed in an NMR probe to produce pulse field gradient (PFG). The diffusion measurement for liquids is based on the same principle, where coils are wound on a NMR probe to produce PFG to the direction of the magnetic field of the SCM. In a special case, if three-dimensional coils were placed, it is possible to measure diffusion coefficients along x, y, z-directions in an anisotropic circumstance. Application of the PFG is controlled by a pulse sequence in Fig. 1 for (a) the modified Hahn echo (PGSE) and (b) stimulated echo (STE) pulse sequences, and the latter is valid to measure species having short T_2.[5] The two most commonly used pulse sequences in Fig. 1 are important to measure the D's of ILs. Generally, the D's of viscous ILs are smaller compared with solutions prepared by organic and aqueous solvents and thus require larger PFG's to obtain sufficient echo attenuation for accurate D data in ILs at least up to 15 T/m depending on temperature. Generally, a high PFG probe is necessary for the studies on translational diffusion of ILs.

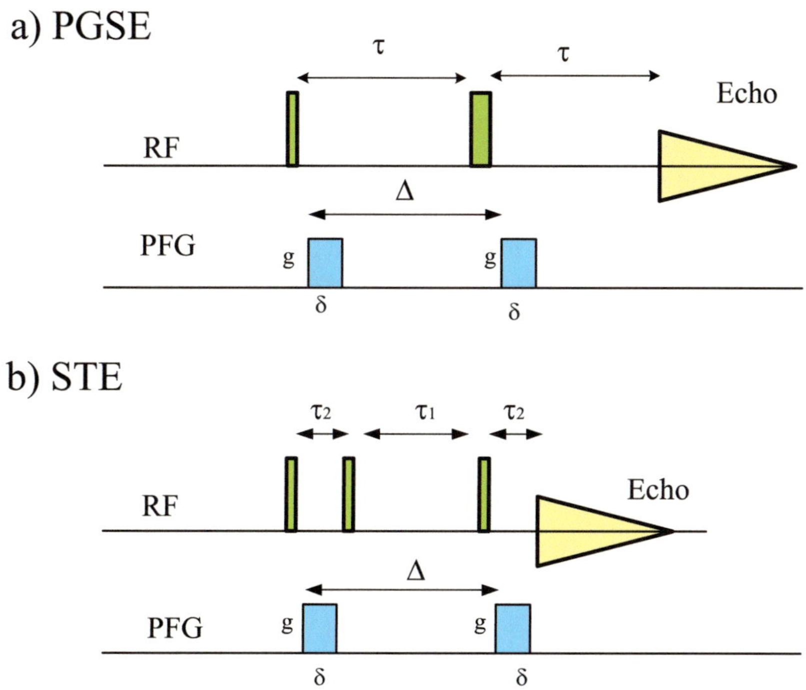

Fig. 1. (a) PGSE and (b) STE pulse sequences used to measure the self-diffusion coefficient (D). The τ's are the intervals of RF pulses, g is the strength of PFG for duration δ, and Δ is the interval between the leading edges of the gradient pulses.

Positional information is acquired in phases of the NMR signals, and basically two equivalent PFG's are applied to the sample. The first PFG pulse can put a mark in the initial position of a target molecule and the second PFG pulse detects the change of the position during the time interval Δ. Originally, Hahn echo sequence (90°-τ-180°-τ-Acq) was proposed to T_2 measurements. The modified Hahn spin-echo (SE) was proposed to incorporate two PFG's.[6] For nuclei having shorter T_2, the stimulated echo (STE) (90°-τ_2-90°-τ_1-90°-τ_2-Acq) was introduced, where τ_2 is related to T_2 process and τ_1 is governed by T_1 process.[7] PGSE data obtained using either pulse sequence can be analyzed by the Stejskal and Tanner equation (1).[6]

$$E = \exp(-\gamma^2 \delta^2 g^2 D(\Delta - \frac{\delta}{3})), \quad (1)$$

where γ is the gyromagnetic ratio, *g* is the strength of the PFG of duration *δ*, and Δ is the interval between the leading edges of the gradient pulses.

The basic assumptions in the setup of NMR diffusion measurements to be analyzed by the Stejskal and Tanner equation are: (1) all the species must be in a homogeneous external magnetic field to prevent signal attenuation from background magnetic field gradients. The magnitude of the background field gradient can be estimated by the line width of signals. (2) The sample must be placed in the constant region of the PFG, and (3) the RF field must be homogeneous to produce accurate 90° and 180° pulses over the whole sample. In particular, condition (2) is important to achieve accurate measurement for viscous ILs since the requisite large PFG's with constant profile must be generated over limited volume. If the sample conditions were not proper, the measurements may include artifacts. Usually, RF pulse width is in μs order and δ and Δ are in ms order. The PFG magnitude is determined by g × δ where g is larger than approximately 2 T/m for ILs and the largest δ is less than 4 ms to prevent the decay of NMR signal during the PFG application. Eq. (1) implies that the PFG pulse is homogeneous and the decay of the echo signal corresponds to the movement of molecules whose positions are marked by the first PFG pulse and detected by the second PFG pulse. Thus the sample volume must be confined in the PFG coil. Cares are necessary to the external perturbation like convection effect induced by thermal inhomogeneity within the sample, particularly in high temperatures. When the PGSE measuring conditions are adequately prepared to obtain an accurate self-diffusion coefficient, the value becomes a physical constant at a certain temperature. As an example, the attenuation of echo signals following Eq. (1) are shown in Fig. 2 for ^{19}F resonance of TFSA in EMImTFSA measured at 30 °C.

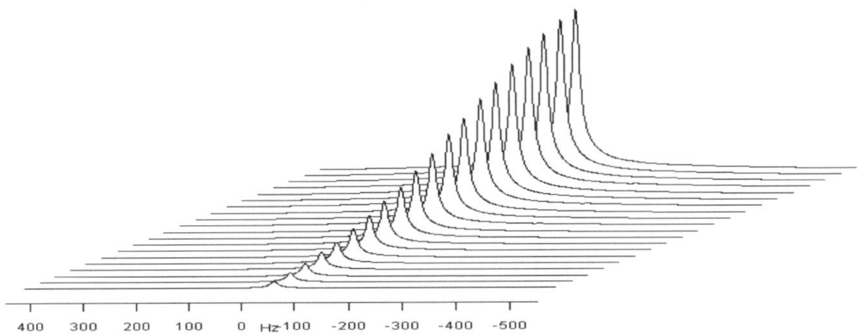

Fig. 2. The attenuation of ^{19}F echo signals of TFSA in EMImTFSA at 30 °C. Δ = 50 ms, *g* = 2.56 T/m and δ changed from 0.1 to 2 ms (20 points). The effective PFG is determined by *g* × δ. Following to Eq. (1), the gradient (*D*) obtained by the plot of ln(*E*) (*E*=echo amplitude) versus $-\gamma^2\delta^2 g^2(\Delta - \delta / 3)$ was 3.80 ± 0.02 × $10^{-11} m^2 s^{-1}$.

Since the PGSE method measures the movement of the center of gravity of a molecule under study, all the nuclei in a molecule should have the same diffusion coefficient, for example, ^{19}F and ^{11}B resonances of the anion BF_4 in EMIm-BF_4 give the same values of D_{BF4} under suitable measuring conditions. Also, in homogeneous liquids the *D* values should be independent of Δ, if the experimental conditions are set properly. However, as described later, many ILs having large viscosity behave differently from usual isotropic homogeneous solutions in the diffusion measurements.

3. Spin-lattice relaxation time T_1 for ILs

The ^1H and ^{19}F T_1 data for the present ILs are needed in order to set the PGSE measuring conditions and we found that the Arrhenius plots exhibit a T_1 minimum in the ^1H resonance. Generally, Arrhenius plots in ^1H T_1 for organic solutions are linear where the T_1 values become longer with the increase of temperature. On the contrary, in the solid-state NMR, organic materials often show ^1H T_1 minima. Many studies had been made for the ^1H T_1 in the solid-state and widely developed about 30 years ago before the attractive techniques of "high resolution solid-state NMR" were introduced. Recently, the studies on ^1H T_1 measurements are seldom reported even in the solid-state NMR. Although ILs are liquids, some protons in ILs show a ^1H T_1 minimum in the Arrhenius plots where the classical Bloembergen, Purcell and Pound (BPP) equation[8] can be applied to obtain a physical constant of the correlation time τ_c for a motion of ^1H-^1H dipoles within a molecule in the liquid state. The BPP equation is given in Eq. (2).

$$\frac{1}{T_1} = C\left(\frac{\tau_c}{1+\omega_0^2\tau_c^2} + \frac{4\tau_c}{1+4\omega_0^2\tau_c^2}\right), \tag{2}$$

where ω_0 is the observe frequency (rad s^{-1}), τ_c is the correlation time of the dipole-dipole motions. C in Eq. (2) is written as

$$C = \frac{3}{10}\gamma^4\hbar^2\sum_j \frac{1}{r_j^6}, \tag{3}$$

where r is atomic distance between protons and the summation index j is over all interacting dipoles. The T_1 value measured at a certain temperature is not a physical constant, and depends on the measuring frequency and relaxation mechanisms. Owing to the term in brackets in Eq. (2), the T_1 is minimum when $\omega_0\tau_c = 2\pi\nu_0\tau_c = 0.616$. Nicely, the $\sum \frac{1}{r_j^6}$ term in the constant C in Eq. (3) is not necessary to calculate in order to obtain τ_c. It is known that the internal rotations around the three symmetrical axis of CH_3 and segmental motions of long alkyl chains included in ILs are activated and averaged out in the condensed state. Since the cations of the ILs in this study exhibited the ^1H T_1 minima in their Arrhenius plots, the correlation time τ_c, a physical constant for molecular motion, can be evaluated at each temperature. Arrhenius plots of ^{19}F T_1 for the CF_3 of TFSA show that the T_1 values become linearly longer as the increase of the temperature. These phenomena are often observed for molecules in isotropic solution state and defined as the extreme narrowing condition, i.e. $\omega_0\tau_c << 1$ and $1/T_1 \propto \tau_c$ from Eq. (2).

The temperature dependence of ^{13}C T_1 have been studied for 1-butyl-3-methyl hexafluorophospate (BMIm-PF_6),[9-11] 1-methyl-3-nonylimidazolium hexafluorophospate (MNIm-PF_6),[12] 1-ethyl-3-methylimidazolium butanesulfonate (EMIm-BSO_3),[13] 1,2-dimethyl-3-propylimidazolium bis(trifluoromethylsulfonyl)amide (DMPIm-TFSA),[1] BMIm-Br[14] and 1-alkyl-3-methylimidazolium butanesulfonate,[15] and the ^{13}C T_1 minima have been observed. Usually, ^{13}C NMR measurements are performed under ^1H decoupling, and the transfer efficiency of the ^1H polarization (nuclear Overhauser effect, NOE) differs for individual carbons. Furthermore, the ^{13}C relaxation mechanisms include ^1H-^{13}C dipolar interaction, spin-rotation and ^{13}C chemical shift anisotropy effects and thus great efforts are required to evaluate the correlation time $\tau_c(^{13}C)$. Our Arrhenius plots of the ^{13}C T_1 for DMPIm-TFSA showed the minima, but it was difficult to obtain physical constants. Then in this chapter we will not describe the ^{13}C resonance further more.

4. Experimental procedure

The data in this chapter were collected from our published papers[1-4] and compiled for the discussion. Although the experimental procedures have been written in each paper, the essential parts are summarized here. For NMR diffusion measurements, the samples were placed into a 5-mm NMR microtube (BMS-005J, Shigemi, Tokyo) to a height of 5 mm and sealed with epoxide resin under argon atmosphere to exclude moisture. The small height of the sample is important to prevent convection effects and also to obtain the homogeneity of the PFG. All NMR spectra were measured on a Tecmag Apollo with a 6.35 T wide bore magnet using a multinuclear JEOL PFG probe controlled by a JEOL console. The measurements of various nuclei are possible by tuning the probe for resonance frequencies with keeping the temperature. The gradient strength was calibrated by using H_2O and D_2O (^2H NMR at the frequency 41.5 MHz). The maximum PFG strength is 20 T/m. The T_1 measurements were performed using the inversion recovery (180°-τ-90°-Acq.) sequence. The ^1H and ^{19}F NMR spectra were measured at 270.2 and 254.2 MHz, respectively. ^1H spectra of the four cations and the spectral assignments are shown in Fig. 3.

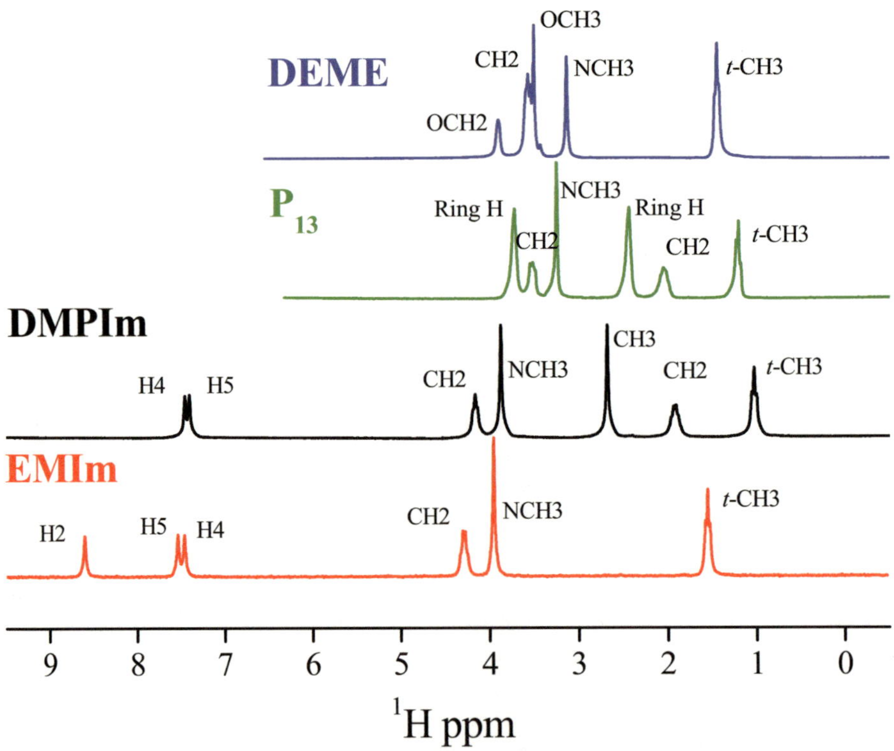

Fig. 3. ^1H spectral patterns of the four cations at 30 °C and the spectral assignments.

Although the line widths of ^1H spectra are not very sharp for ILs owing to the high viscosity, the ^1H spectra show splitting due to the ^1H-^1H spin coupling as shown in Fig. 3. The spin-spin coupling J influences the phases of the resonances in the echo signals depending on the J-values. For example, echo signals for resonances with J = 7 Hz are distorted significantly around τ = 23 (= $1/(2\pi J)$), 45 (= $2/(2\pi J)$), 68 (= $3/(2\pi J)$) ms. The J-coupling modulations produce deleterious effects in PGSE experiments such that depending on the choice of τ. There can be large loss in signal-to-noise and distortion of the line-shape. Nevertheless, providing that there is sufficient signal-to-noise, the correct diffusion coefficient can be obtained. For references, the effects of J-evolution on the EMIm echo signals obtained with the PGSE and STE pulse sequences are shown in Fig. 4(b) and 4(c).

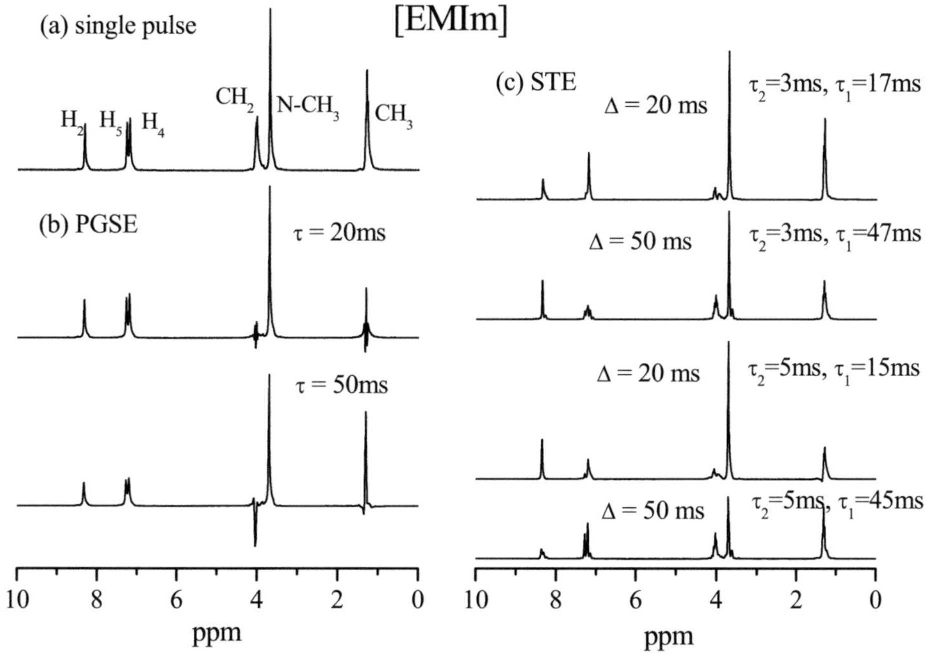

Fig. 4. ^1H spectra of EMIm measured with (a) single pulse, (b) Hahn (PGSE) and (c) stimulated echo (STE) sequences. The STE echo signals were measured with small and short PFGs.

Although the ethyl signals are significantly influenced by the J-evolution, the N-CH_3 and imidazolium ring proton signals with negligible spin-spin coupling are almost unchanged and in all cases the evolution effects are insensitive to the presence or absence of the PFGs. The signals obtained with the STE sequence for different delay combinations under short and small PFGs are shown in Fig 4(c). Without PFGs, the J-coupled signals were significantly distorted. For freely diffusing species the values of these three parameters are somewhat arbitrary, if the sufficient echo signal attenuation can be achieved to enable an accurate diffusion measurement as shown in Fig. 2.

The time interval Δ in the pulse sequences shown in Fig. 1 is an important parameter for the diffusion measurements. Practically, long Δ induces the reduction of S/N ratio due to the T_2 effect and T_1 effect in the STE pulse sequence. Also the convection effects may have a chance to enlarge the diffusion coefficients by the long interval due to the thermal inhomogeneity within a sample. A short Δ requires a large effective PFG. It is important to confirm that the two strong PFG's are equivalent for short Δ.[16] If the pulse duration, δ were too long, the magnetization may decay during the PFG application. The accuracy of short δ values is related to a rectangular shape of the PFG pulse. It is necessary to estimate the accuracy of D values measured by individual PFG probes. We found that the reliable limit of short Δ is 20 ms in our PFG probe.

The Δ-dependence of D for species in structurally heterogeneous systems like polymers and porous materials is referred to as anomalous diffusion and been reviewed by Metzler and Klafter.[17] ILs cannot be assumed to have such structures resulting in motional restrictions. Experimentally, the Δ-dependence *D's* for ILs have been observed in the lower temperature region especially in viscous ILs. As an example, the Δ-dependence observed for DEME-TFSA at 273 K are shown in Fig. 5. The diffusion plots following Eq. (1) are always linear with good precisions similar to an example shown in Fig. 2 for different protons of DEME in ^{1}H resonance and CF_3 in ^{19}F resonance.

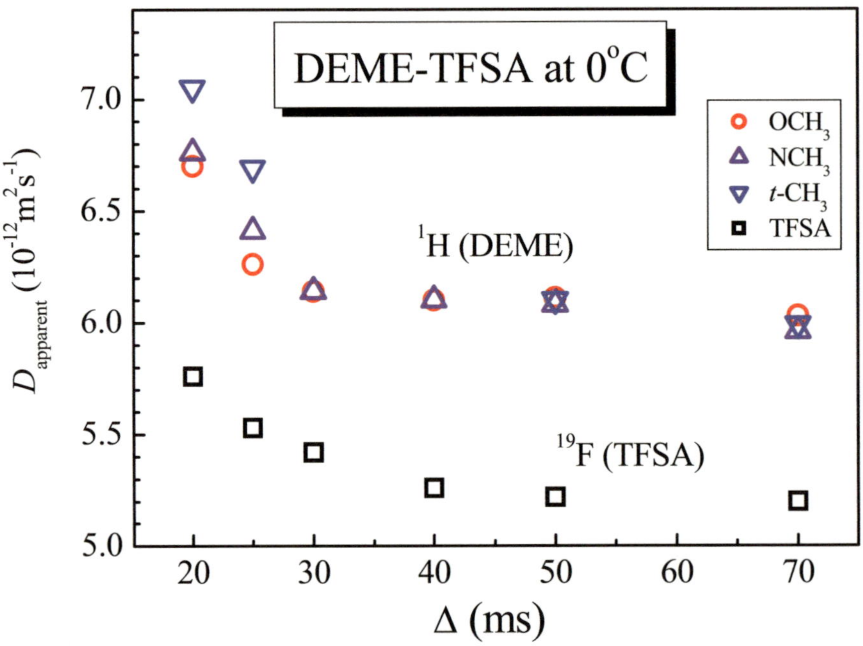

Fig. 5. The Δ-dependence of the apparent diffusion constant ($D_{apparent}$) for DEME (^{1}H) and TFSA (^{19}F) measured at 0 °C.

Clearly when Δ became shorter, apparent diffusion constant ($D_{apparent}$) became larger both in DEME and TFSA. At shorter Δ, the individual protons in DEME showed slightly different values. At the longer Δ, the $D_{apparent}$ values approached to equilibrated values for all proton signals and agreed within experimental errors. The similar experimental results were observed for ILs with high viscosity. Actually, DEME-TFSA does not show a Δ-dependence above 30 °C, while very viscous DEME-BF_4 showed significant Δ-dependent $D_{apparent}$ values at 30 °C. For the longer Δ values, each signal has the same values of $D_{apparent}$ and we adopt such value as a diffusion coefficient D. In our experiments, we always observe values of $D_{apparent}$ for several Δ values and determine D at each temperature. Generally, in high temperatures the values measured in short Δ are adopted, while the values at long Δ are adopted in low temperatures.

5. Results

5.1 Temperature dependent T_1

Arrhenius plots of ^1H T_1 for various protons of the four cations are shown in Fig. 6 for EMIm, DMPIm, P_{13} and DEME. Since not all the protons show a T_1 minimum in their plots, the curves having minima are shown in solid marks. As a general trend, the protons located near the center of the molecules have T_1 minimum. For example, t-CH_3 of the propyl group in P_{13} gave the longest T_1 and did not exhibit a T_1 minimum, indicating that rapid local motion around the CH_3 group is contributed to T_1. Similarly, t-CH_3 in the ethyl group of EMIm did not give the T_1 minimum, while the t-CH_3 in the propyl group in DMPIm showed the T_1 minimum. When the T_1 minimum is observed, it is possible to calculate τ_c using Eq. (2).

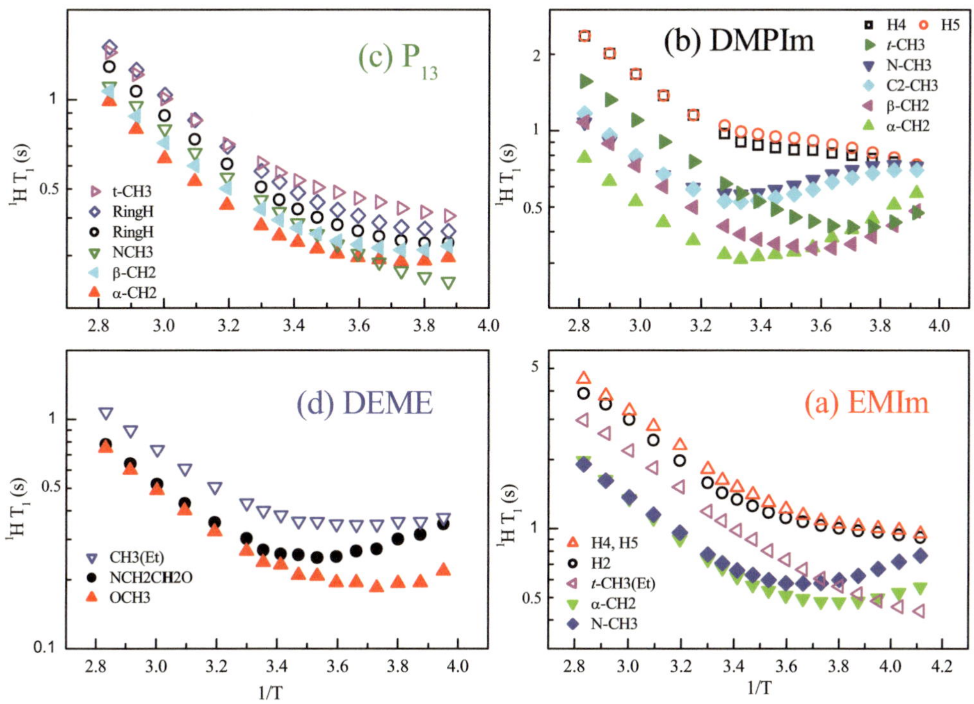

Fig. 6. Arrhenius plots of ^1H T_1 for the four cations, (a) EMIm, (b) DMPIm, (c) P_{13} and (d) DEME.

From Eq. (2), at the temperature of T_1 minimum, $\omega_0\tau_c = 2\pi\nu_0\tau_c = 0.616$ holds as described above. Since the present $\nu_0(H) = 270.17$ MHz, $\tau_c = 3.63 \times 10^{-10}$ s (363 ps) at the T_1 minimum temperatures. From Fig. 6, it is possible to calculate τ_c values for each proton within a molecule at every temperature. The dominant relaxation process showing [1]H T_1 minimum in the present temperature range can be assumed to be an isotropic motion of a whole molecule.

The calculated τ_c can be contributed by additional relaxation processes like fast local motions such as internal rotation around a three symmetrical axis in CH_3, segmental motions of alkyl chains, conformational exchange of imidazorium ring or ring inversional motion of pyrrolidinium ring as given in Eq. (4).

$$\frac{1}{\tau_c} = \frac{1}{\tau_o} + \frac{1}{\tau_s}, \quad (4)$$

where τ_o and τ_s are the correlation times for the isotropic reorientational and local intramolecular motions, respectively. Each proton having T_1 minimum afforded a little different τ_c values within a molecule. We adopted the longest τ_c values as the isotropic reorientational motion of a whole molecule, since the shorter τ_c values include the contribution of faster local motion. The Arrhenius plots of τ_c(cation) are shown in Fig. 7, and the activation energies are given in Table 2, where the τ_c(cation) values at 30°C are included. Since the Arrhenius plots are almost linear for the four ILs, the same mode of molecular motion is predominant to the T_1 process and the relaxation process is the isotropic reorientational motion. Clearly, the molecular rotation of DMPIm was the slowest in the whole temperature, and the other three cations have similar τ_c(cation) values at the high temperatures and in the low temperature range the isotropic motions are quicker in the order of P_{13}, EMIm, DEME and DMPIm.

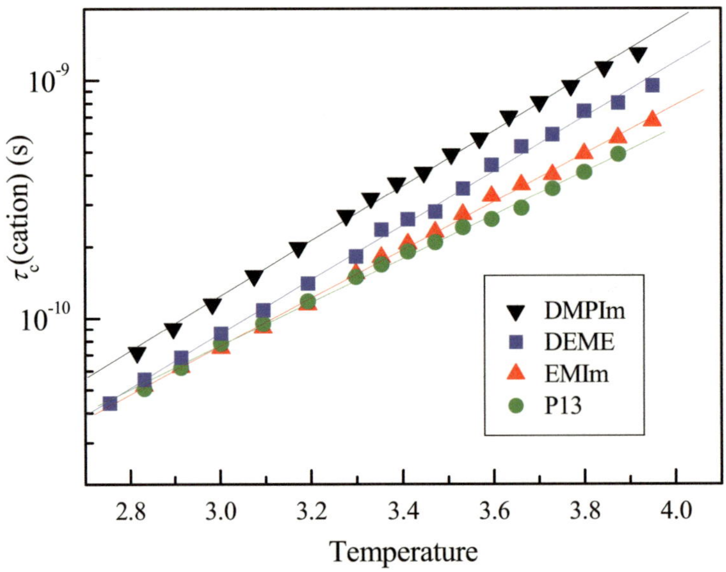

Fig. 7. Arrhenius plots of τ_c(cation) for the four ILs.

Cation	E_a for τ_c (kJ/mol)	τ_c at 30°C (s)	E_a for ^{19}F T_1 (kJ/mol)	
EMIm	19.4 ± 0.2	1.5 × 10^{-10} (15 ps)	14.4 ± 0.2 (above 40°C)	8.8 ± 0.2 (below 5°C)
DMPIm	22.1 ± 0.3	2.7 × 10^{-10} (27 ps)	15.4 ± 0.3 (above 40°C)	8.3 ± 0.2 (below 20°C)
P_{13}	17.5 ± 0.2	1.5 × 10^{-10} (15 ps)	14.1 ± 0.2 (above 30°C)	8.1 ± 0.2 (below 25°C)
DEME	21.8 ± 0.2	1.8 × 10^{-10} (18 ps)	10.2 ± 0.2 (above 40°C)	7.1 ± 0.4 (below 20°C)

Table 2. The activation energies E_a (kJ/mol) of the τ_c for the cation motion and the τ_c (s) at 30°C, and the E_a (kJ/mol) for ^{19}F T_1 of CF_3.

As shown in Fig. 6, the protons of the imidazolium rings of EMIm and DMPIm showed longer T_1 than those of the side chains and in the low temperature region the T_1 values did not increase. In other words T_1 minimum was not observed for the imidazolium ring protons. In the present discussion, the reorientational motion should affect significantly ^1H T_1 of the ring protons.

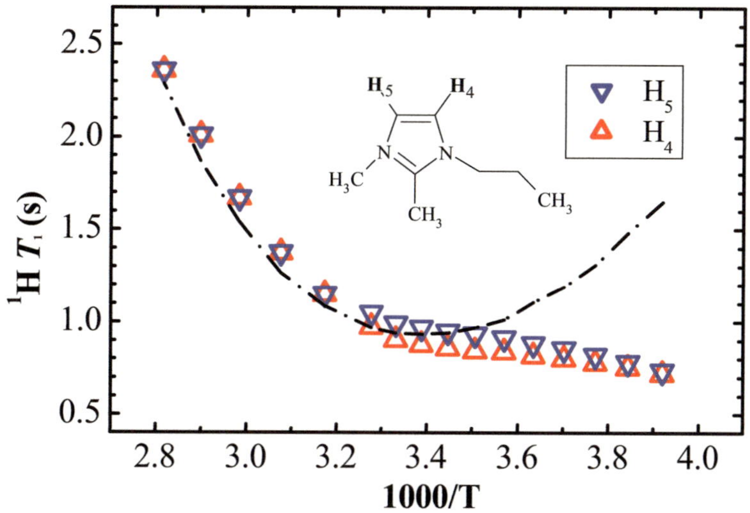

Fig. 8. The observed T_1 values of H_4 and H_5 of the imidazolium ring of DMPIm and the calculated T_1 (dotted line) for the H_5 using the τ_c(DMPIm) in Fig. 6 and Eqs. (2) and (3).

The trial was made to calculate the T_1 values for the H_5 of DMPIm using the τ_c(DMPIm) in Fig. 7. Here the H-H distances for the H_5 between H_4 and CH_3 were included. Clearly, above 303 K the calculated T_1 of H_5 agreed with the observed T_1's of the H_4 and H_5 of which the T_1 values were almost the same. As the temperature decreased, the observed T_1 became shorter since additional fast relaxation mechanisms become operative. The calculations were made for the ^1H T_1 for the H_4 and H_5 of the imidazolium protons in EMIm and the similar results were obtained.[4] In the low temperature, the rate of the isotropic reorientation of the whole molecule becomes slower and the long T_1 is reduced by local motions such as planar-

nonplanar conformational changes of the imidazolium ring changes having the activation energy of about 5 kJ/mol observed by Raman spectroscopy.[18]
Arrhenius plots of ^{19}F T_1 of CF_3 in TFSA are shown in Fig. 9 for the four ILs. The plots are curved and the activation energies were calculated for the higher and lower temperature regions as given in Table 2. Two relaxation processes are possible for the ^{19}F T_1, i. e. the internal rotation around the three-symmetrical axis and the isotropic reorientational molecular motion. Probably an overall molecular motion is the major relaxation process in the high temperature region with larger activation energies and the internal rotation of the CF_3 contribute predominantly to the ^{19}F T_1 in the lower temperature region with the smaller activation energies. Among the four ILs, the values of ^{19}F T_1 of the TFSA in EMIm-TFSA are much longer than those of the other three ILs, owing to faster motions of the fluorine atoms in the smallest viscosity.

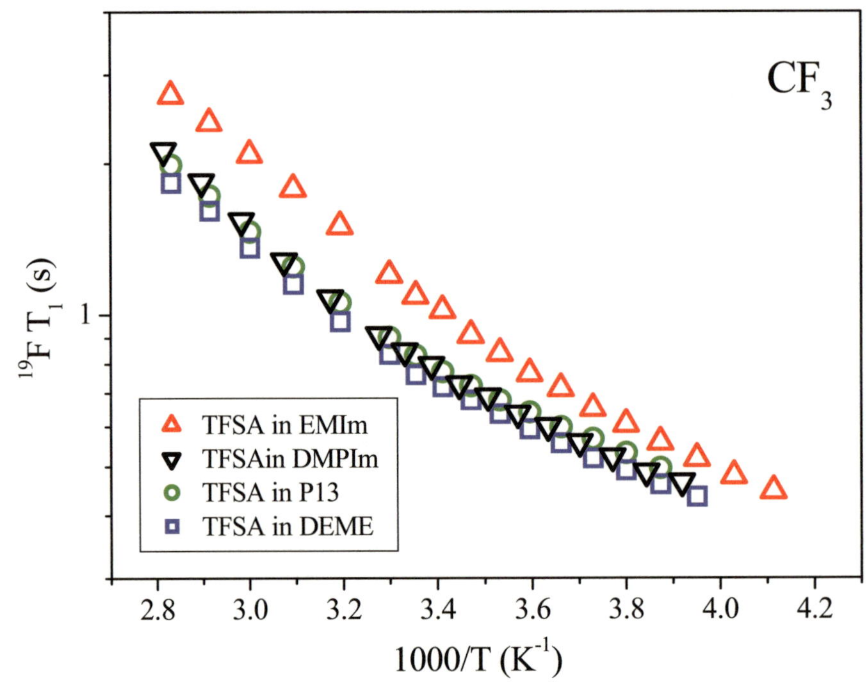

Fig. 9. Arrhenius plots of ^{19}F T_1 for CF_3 of the TFSA in the four ILs.

5.2 Translational diffusion

The diffusion coefficients are plotted versus temperature in Fig. 10 for (a) cations and (b) TFSA. The order of D_{cation} and D_{TFSA} is EMIm > P_{13} > DEME > DMPIm, which is consistent with the order of $1/\eta$ in Table 1. Within the same IL it always holds $D_{cation} > D_{TFSA}$ in the present four ILs.

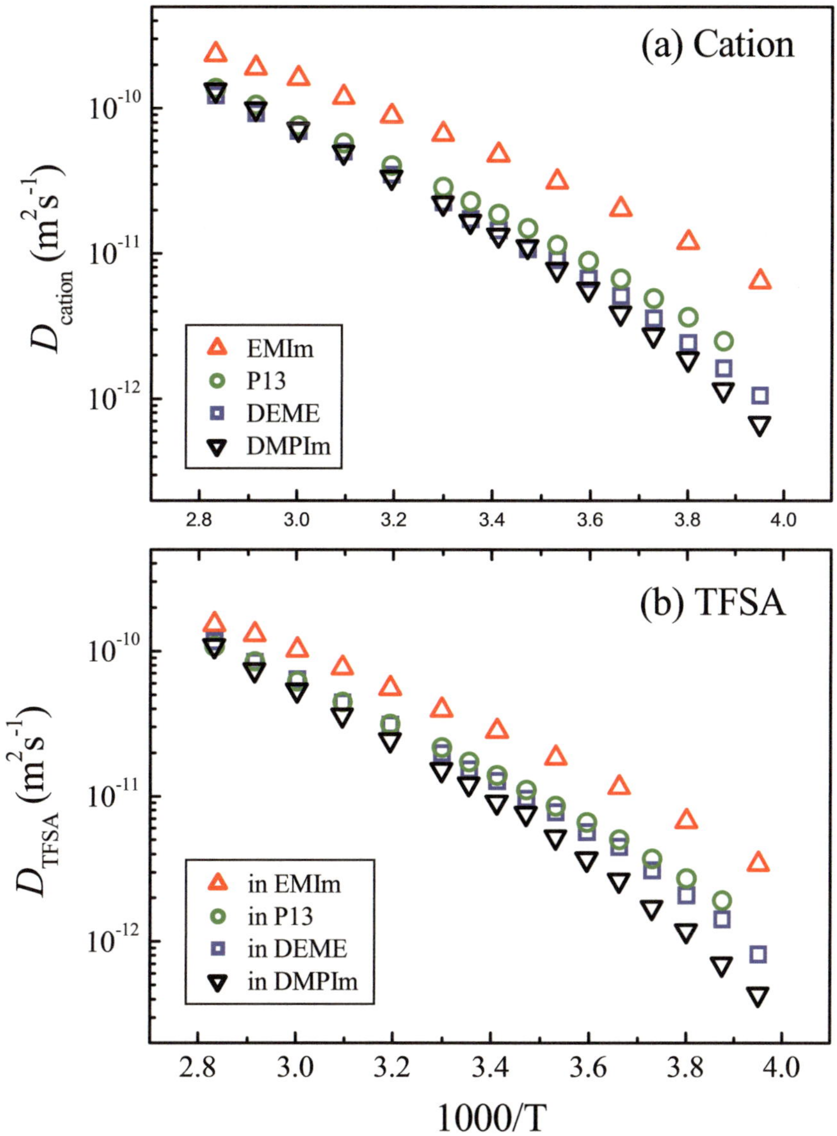

Fig. 10. (a) The D_{cation} and (b) the D_{TFSA} are plotted versus temperature.

5.3 Viscosity, ionic conductivity and density

The viscosity (η) is plotted versus temperature in Fig. 11(a) and the Arrhenius-type plots were made for 1/η for inverse of temperature in Fig. 11(b) for the four ILs.

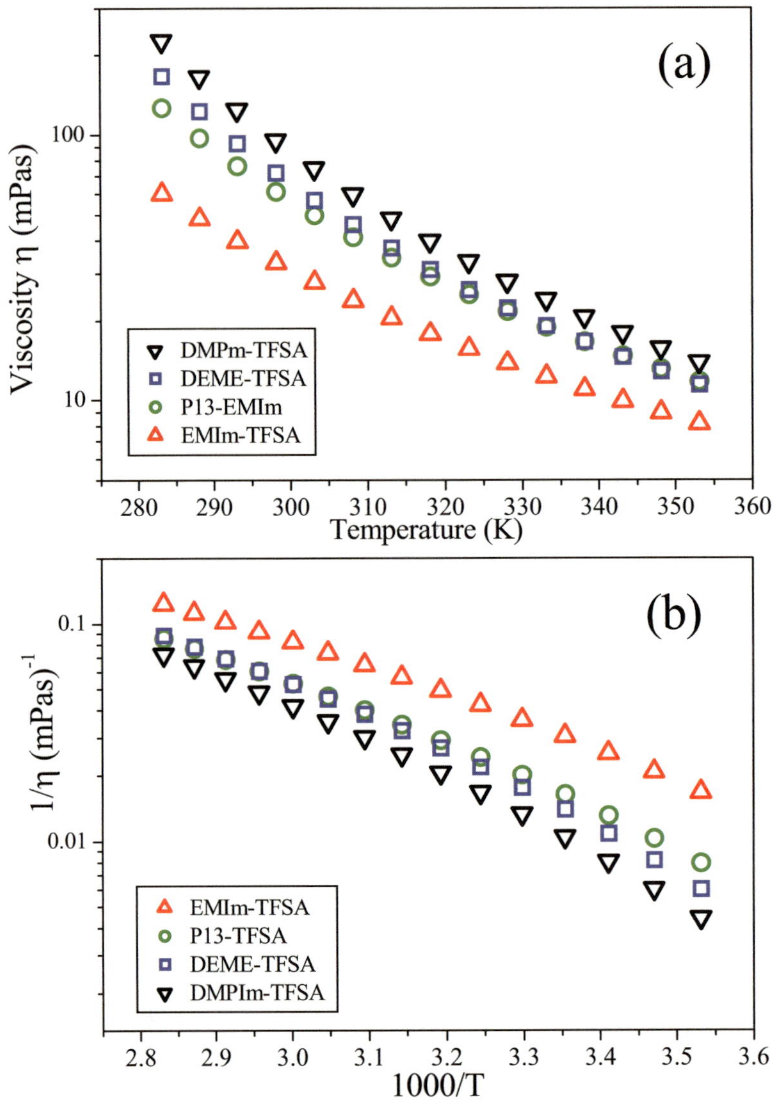

Fig. 11. Temperature dependent viscosity for the four ILs (a) η vs temperature and (b) Arrhenius plots of 1/η.

The viscosity differences among the ILs became small at the higher temperature. In the low temperature region the order of the viscosity is EMIm-TFSA << P_{13}-TFSA < DEME-TFSA <DMPIm-TFSA. The values of η at 30°C are given in Table 1.

The ionic conductivities (σ) of the four ILs are plotted in Fig. 12 and the order of the magnitudes is EMIm-TFSA >> P_{13}-TFSA ≥ DEME-TFSA ~DMPIm-TFSA. The largest σ for EMIm-TFSA was obtained for the smallest η.

The densities are plotted against temperature in Fig. 13. The order is EMIm-TFSA >> DMPIm-TFSA > P_{13}-TFSA > DEME-TFSA and for reference the order of the molecular weight given in Table 1 is EMIm-TFSA < P_{13}-TFSA < DMPIm-TFSA < DEME-TFSA.

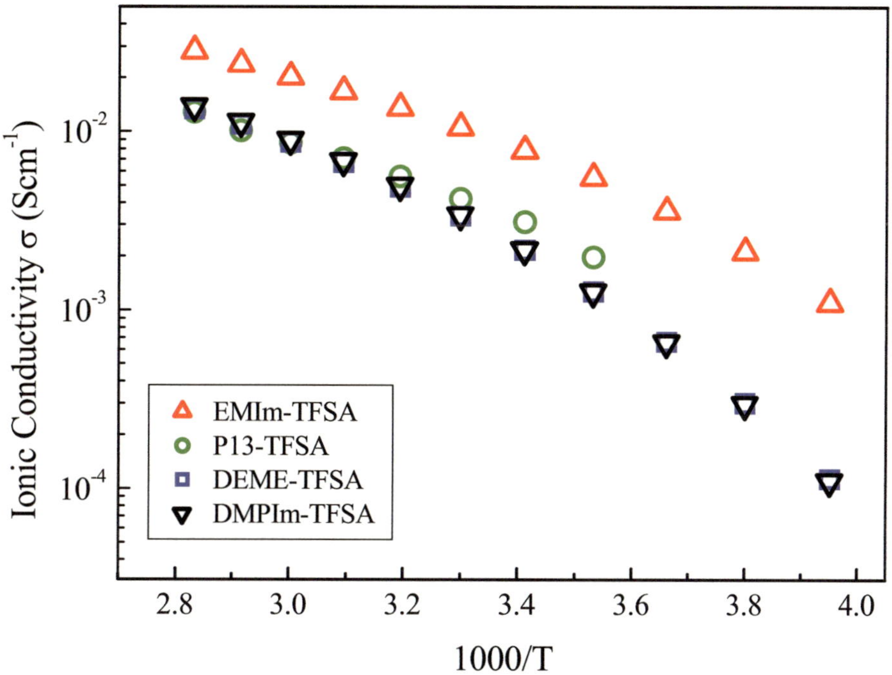

Fig. 12. The temperature dependence of ionic conductivity for the four ILs

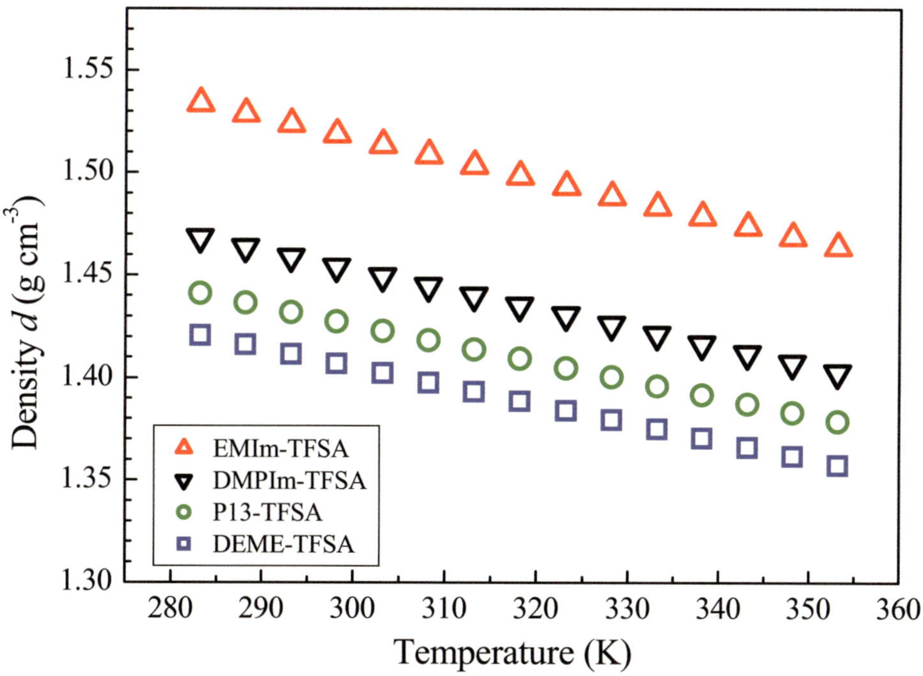

Fig. 13. The densities are plotted against temperature for the four ILs.

6. Discussion

6.1 Stokes-Einstein relation

By the classical Stokes-Einstein (SE) (or Einstein-Sutherland) relation we will describe our experimental results between the individual ion diffusion coefficients in Fig. 10 and the viscosity in Fig. 11. The SE equation is given as

$$D = \frac{kT}{c\pi\eta a} \tag{4}$$

where a is the radius of the diffusing species, k is the Boltzmann constant and the constant c theoretically ranges between 4 to 6 for slip and stick boundary conditions, respectively.[16,19] In the limit of the slip condition (c = 4), no interaction is assumed between a particle and surrounding molecules. When a particle interacts strongly with solvent molecules, the stick condition holds (c = 6). Here, we attempted to plot the experimentally determined D versus $1/\eta$ according to the classical SE by assuming that c is an empirical parameter. The D values are plotted against $kT/\pi\eta$ for (a) D_{TFSA} and (b) D_{cation} of the four samples in Fig 14. When we presume that the SE equation is valid, the gradient of each line corresponds to the inverse of $c \times a$. In Fig. 14 the plots are linear for the all ions. The experimental values of $c \times a$ are summarized in Table 3, where the errors estimated for the linear plots were less than 3 %. Clearly, the classical SE relationship (4), i.e., $D \propto 1/\eta$ can be used to obtain physical meaning between the translational ion motion and bulk viscosity in the present four ILs. Here we assume that the van der Waals radius calculated from the van der Waals volume can be used as the radius of diffusing ion as given in Table 3.

Ion	Counter ion	c × a (nm)	Van der Waals volume (nm³)	Van der Waals radius (nm)	c
EMIm	TFSA	0.833 ± 0.016	0.11756	0.304	2.7_5
DMPIm	TFSA	0.850 ± 0.061	0.15053	0.330	2.5_6
P_{13}	TFSA	0.982 ± 0.013	0.15025	0.330	2.9_8
DEME	TFSA	1.138 ± 0.007	0.17806	0.349	3.2_6
TFSA	EMIm	1.323 ± 0.019	0.14950	0.329	3.7_5
TFSA	DMPIm	1.162 ± 0.013[a)]	0.14950	0.329	3.5_3
TFSA	P_{13}	1.228 ± 0.013	0.14950	0.329	3.7_3
TFSA	DEME	1.204 ± 0.018	0.14950	0.329	3.6_6

[a)]Two points in DMPIm were not included for the calculation, which were largely deviated in the low temperature region

Table 3. SE relationships for each ion in the four ILs.

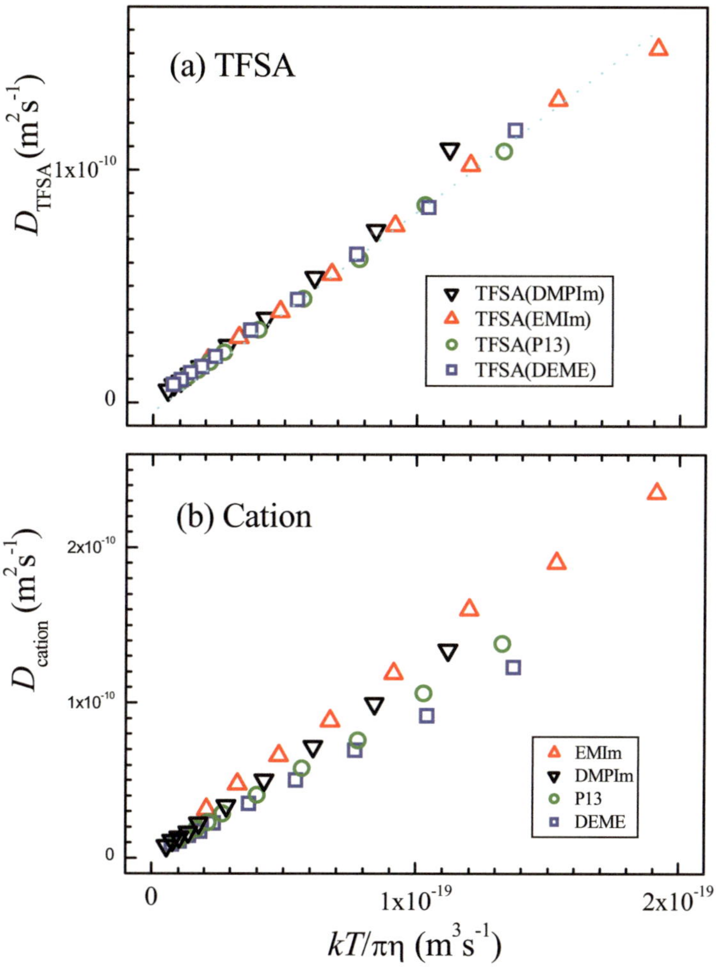

Fig. 14. Diffusion coefficients were plotted *vs* $kT/\pi\eta$ for (a) TFSA and (b) cations.

It is noted that the *c* values of TFSA range between 3.5 and 3.8 and the cations have the smaller *c* values compared to TFSA. Previously we have reported that the *c* values of bis(fluorosulfonyl)amide (FSO_2-N^--SO_2F, FSA, MW = 180.1 and van der Waals radius = 0.284 nm) are 2.8 and 3.1 for EMIm-FSA and P_{13}-FSA, respectively. Also, the *c* values of EMIm and P_{13} were 2.5 and 3.0 in the FSA-based ILs.[3,4] The *c* values are purely empirically derived parameters for individual ions in neat ILs based on the SE equation. As a general trend, ions in ILs have their individual *c* values. The empirical c values are smaller than 4, the theoretically predicted smallest limit. Experimental values were calculated as $c \times a$ for each ion. If we assume the size of the diffusing ions become larger owing to the ion association, the *c* values becomes smaller. The classical SE relation was derived for a diffusing rigid particle with radius *a* in a medium having viscosity η. Ionic liquids are quite different materials from the assumptions of the SE equation. Fractional SE relation (F-SE) has been proposed to explain the relation between *D* and η as $D \propto (T/\eta)^m$.[20-22] We attempted to apply the F-SE and found that even when the extra parameter *m* was introduced, the fitting in the linear plots ln(*D*) versus ln($kT/\pi\eta$) were not much improved.

6.2 Walden plot

Walden's rule in electrochemistry describes the relationship between ionic conductivity σ with viscosity η, that is, σ × η = constant.[23] The electrochemical molar conductivity (Λ_{imp}) can be calculated from ionic conductivity as $\Lambda_{imp} = \sigma M/\rho$, where M is the molecular weight and ρ is the density. The Walden plots of the molecular conductivity versus $1/\eta$ in the unit of 0.1 Pas (=poise) proposed by Angell and coworkers[24] are shown in Fig. 15 for the present four ILs. The Walden rule relates the molecular mobility ($1/\eta$) to the molar conductivity induced from the charged ions in solution electrolytes and characterizes ILs. The fully dissociated ions such as diluted aqueous KCl solution give a behavior shown as a line in Fig. 15. The deviation from the ideal plot, ΔW has been proposed to relate with the ion pairing of ILs.[25] In the present ILs, the ΔW for EMIm-TFSA exhibited the smallest value, followed by DMPIm-TFSA, DEME-TFSA and P_{13}-TFSA.

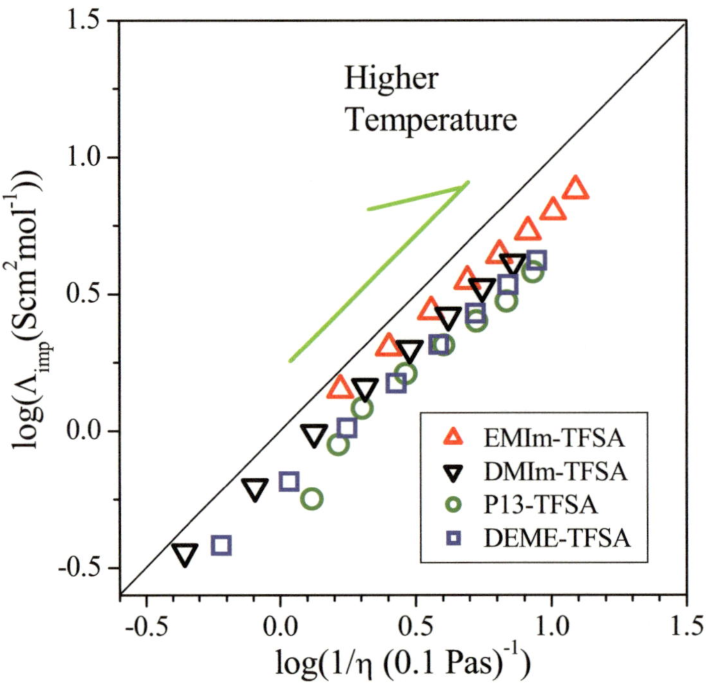

Fig. 15. Walden plots for the four ILs.

6.3 Nernst-Einstein relation

The relationship between the ion diffusion and ionic conductivity has been studied by the Nernst-Einstein (NE) equation (5) for molar conductivity Λ.

$$\Lambda = \frac{zF^2}{RT}(D_+ + D_-), \qquad (5)$$

where z is the ionic valency, F is Faraday's constant, R is gas constant, and D_+ and D_- are the self-diffusion coefficients of the cation and anion, respectively under the assumption of complete ion dissociation.[23] Studies of aqueous electrolytes have been made to verify the NE equation. It is known that when a salt concentration increases, the modification of the original NE equation is necessary to give Eq. (6)

$$\Lambda_{NE} = \frac{zF^2}{RT}(D_+ + D_-)(1-\xi), \tag{6}$$

where ξ is a parameter to relate with the ion association. Then 1-ξ means a degree of ion dissociation. In the early stage of our studies for PGSE NMR diffusion measurements of solution electrolytes, we verified that Eq. (6) is valid for electrolytes including lithium salts in organic solvents (propylene carbonate and γ-butyrolactone) near the infinitesimal concentration by measuring the ion diffusion coefficients and ionic conductivity. As the concentration of the salt increased, the values of 1-ξ became smaller to indicate the promotion of ion association. The diffusion coefficients of the solvents became smaller as the increase of the salt concentration owing to the increase of viscosity.[26] Since the PGSE self-diffusion measurements reflect the average D's for species in both free and associated forms, the value of the NMR molar conductivity (Λ_{NMR}) calculated using Eq. (5) ($\Lambda = \Lambda_{NMR}$) is always overestimated. The calculated Λ_{NMR} and Λ_{imp} are plotted versus temperature in Fig. 16 for the four ILs. Clearly the Λ_{NMR} values are always larger than Λ_{imp}, and precise watching indicates that the relationship between the Λ_{NMR} and Λ_{imp} are a little different dependent on temperature and ILs.

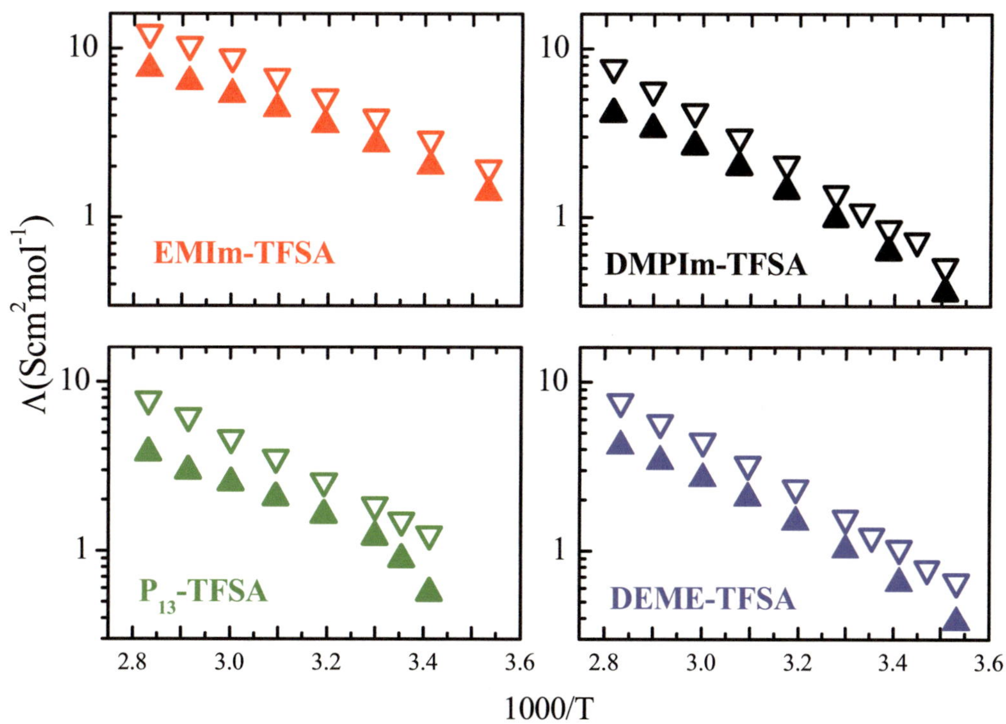

Fig. 16. The molar conductivities derived from ionic conductivity Λ_{imp}(solid) and NMR Λ_{NMR}(open) are plotted versus temperature for the four ILs.

The Λ_{imp} can be obtained from the electrochemical impedance method by measuring only the charged ions. The relationship between ionic conductivity σ and the mobility μ of ions is expressed as $\sigma = nq\mu$, where n is the number of carrier ions, q is the electric charge of ions and μ is the mobility of ions. The ion mobility can be assumed to be parallel to the *D*, although the question remains whether isolated and paired ions diffuse with the same diffusion coefficients. The actual values are averaged ones. The discrepancy between Λ_{imp} and Λ_{NMR} is considered to be related to the relative numbers of the charged and paired ions, and/or the lifetime of the paired ions. Then, the Haven ratio, $\Lambda_{imp}/\Lambda_{NMR}$, calculated as the ratio of Λ_{imp} divided by the Λ_{NMR} has been named as "ionicity" of ILs proposed by Tokuda et al[27-28] and is related to the ratio of the charge carrying ions. Previously, no temperature dependence of the ionicity was observed, but our recent measurements indicate a decrease in ionicity with increasing temperature.[1-4] The $\Lambda_{imp}/\Lambda_{NMR}$ values of the four ILs are plotted against temperature in Fig. 17 and it is shown that the temperature dependent behaviors of the $\Lambda_{imp}/\Lambda_{NMR}$ are a little different with each other for the individual ILs.

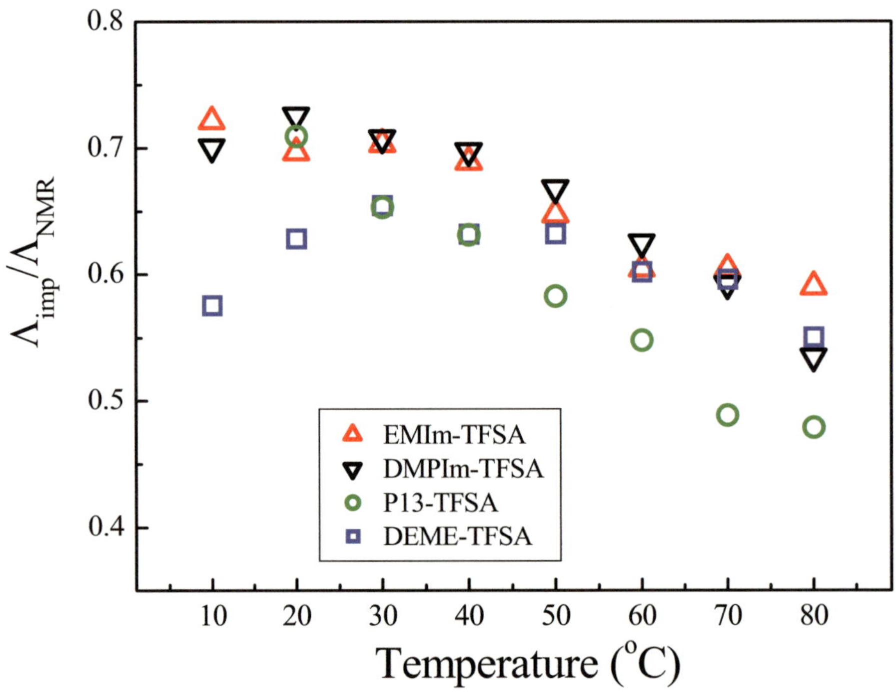

Fig. 17. The $\Lambda_{imp}/\Lambda_{NMR}$ (ionicity) are plotted *vs* temperature for the four ILs.

The experimental values of $\Lambda_{imp}/\Lambda_{NMR}$ were calculated from the empirical values of σ, ρ, D_{cation}, and D_{anion}, all of which were measured independently. Even if an error bar of each measurement is small, the scattering of the $\Lambda_{imp}/\Lambda_{NMR}$ can easily become larger. In the higher temperature region, the values of EMIm-TFSA, DMPIm-TFSA and DEME-TFSA are similar and that of the P_{13}-TFSA is smaller than the other three ILs. When the temperature decreased, the value of DEME-TFSA decreased largely. Generally, the $\Lambda_{imp}/\Lambda_{NMR}$ values became smaller as the higher temperature region.

From Eqs. (5) and (6), $\Lambda_{imp}/\Lambda_{NMR}$ corresponds to 1-ξ, which is defined as the degree of ion dissociation or ion dissociation parameter in aqueous and organic solution electrolytes. It has been established that (1) as the ion concentration increases, the η increases and the σ decreases owing to the formation of ion pairs, and (2) when the η increases, the *D* decreases without distinguishing isolated or paired ions. The degree of ion dissociation 1-ξ (=1 is the complete ion dissociation) has parallel relationships with 1/η, σ and *D* and increases with the increase of temperature for series of a solution electrolyte. The empirical behaviours of $\Lambda_{imp}/\Lambda_{NMR}$ for the ILs in Fig. 17 conflict with general concepts that ion dissociation becomes larger in higher temperature. Comparing the Arrhenius plots of ionic conductivity (Fig. 12), $1/\eta$ (Fig. 11 (b)) and diffusion coefficients (Fig.10), the curvatures in the ionic conductivities were larger than those of the other parameters in the high temperature region. Phenomenologically, the decreasing trends of $\Lambda_{imp}/\Lambda_{NMR}$ in the higher temperature region can be interpreted from the facts that Arrhenius plots of the ionic conductivity for most ILs are curved significantly at the high temperatures and the values do not increase as expected from viscosity and ion diffusion coefficients.

We must say that "ionicity" in ILs is not a static "degree of ion dissociation" as pointed out by Harris,[29] but the original definition of "ionicity" as the ratio of the charge carrying ions by Watanabe's group may be still valid. The experimental facts that the $\Lambda_{imp}/\Lambda_{NMR}$ increases with the increase of the viscosity is very strange and conflicts with the general properties of solution electrolytes. More precise studies are necessary to connect experimental "ionicity" ($\Lambda_{imp}/\Lambda_{NMR}$) to ion pairings in ILs including timescales of ionic interactions, such as dynamic ion associations and static ion-ion interactions.

6.4 Cation molecular motion

We assume that the ^{1}H correlation time τ_c for the cations in Fig. 7 corresponds to whole molecular motion rather than intramolecular local motions. The segmental motion of alkyl substituents and CH_3 rotational motion are accelerated and averaged out in the condensed state. The activation energies for the τ_c(cation) in Table 2 are 17 to 22 kJ/mol in the temperature range studied. The conformational change of the ethyl chain attached to imidazolium ring for EMIm has been reported to take place with activation energies about 5 kJ/mol.[18] Similarly, the activation energies of the pseudo-rotation of the pyrrolidinium ring was reported to less than 5 kJ/mol.[30-31] Then, the motion of a whole molecule can contribute to the temperature-depending ^{1}H T_1 having T_1 minima. The activation energies for translational diffusions (26, 30, 32 and 33 kJ/mol for EMIm, P13, DEME and DMIm, respectively for the linear plots in the higher temperature regions in Fig. 10 (a)) are much larger. In NMR relaxation theory, the translational correlation time can be defined as $\tau_c = 2a^2/D$ following Abragam,[32] where a is the radius of the diffusing particle. The temperature dependences of τ_t values for the four cations are shown in Fig. 18 together with the τ_c(cation) values obtained from the ^{1}H T_1.

The overall isotropic molecular correlation time τ_2 is related to the viscosity η by the Stokes-Einstein-Debye (SED) relation as

$$\tau_2 = \frac{V\eta}{kT} \tag{7}$$

where V is the effective molecular volume. The model was originally proposed for rotational time of a solute in a solvent of the shear viscosity η. Significant studies have been reported about the SE and SED relationships based on MD simulations, and the validities and limitations of the SE and SED have been discussed.[33,34] Experimental studies on the rotational motion of small molecules dissolved in ILs have also been reported.[35-39] We calculated τ_2 in Eq. 7 by using the measured viscosity data and assuming V of the van der Waals volumes of the cations in Table 3. The calculated τ_2 values are also shown in Fig. 18 together with the τ_c and τ_t values for the temperature range between 283 and 353 K.

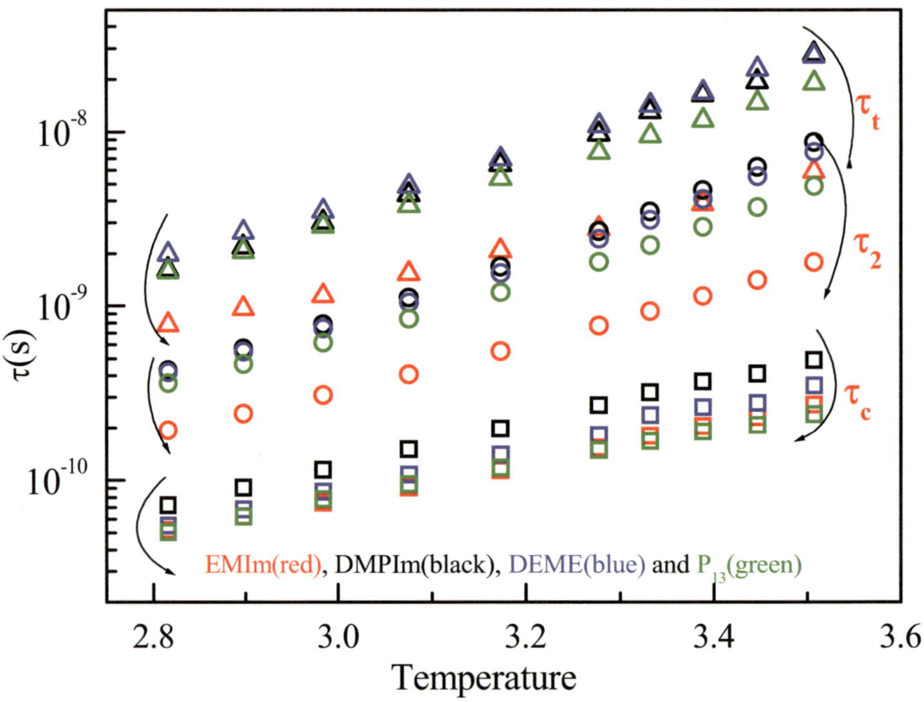

Fig. 18. Temperature dependences of the correlation times τ_c for librational motion, τ_2 for reorientational motion and τ_t for translational motion of the four cations.

The τ_c values evaluated from NMR relaxation times are always shorter than the values estimated using the SED model. The NMR relaxation takes place for one motion of a single dipole-dipole flip and can describe the time interval for the motion but not the amplitude. As proposed in our previous papers,[1-4] we assume that the τ_c process is the small-angle librational motion of the cation. Assuming the τ_2 value calculated from η correspond to a 360° rotation, the angles for the librational motions were calculated and plotted versus temperature in Fig. 19. The flip angle for EMIm is particularly larger than those for the other anions. The molecular flip amplitude becomes smaller at lower temperature.

In the present study, we propose experimental correlation times for the three different modes of motion with different times, i.e. τ_t, τ_2 and τ_c derived from the experimental data of D_{cation}, η (and the volume) and ^{1}H T_1 of the cations, for the translational, molecular overall rotational and librational motions of the cations, respectively. Clearly, the order was τ_c (librational) < τ_2 (overall molecular rotation) < τ_t(translation) for the all cations in the present temperature range. It is noted that although the values τ_c for EMIm enter the values of the other cations, the τ_t and τ_2 for EMIm are particularly short compared with those of the three cations probably due to the extremely small viscosity of EMIm-TFSA.

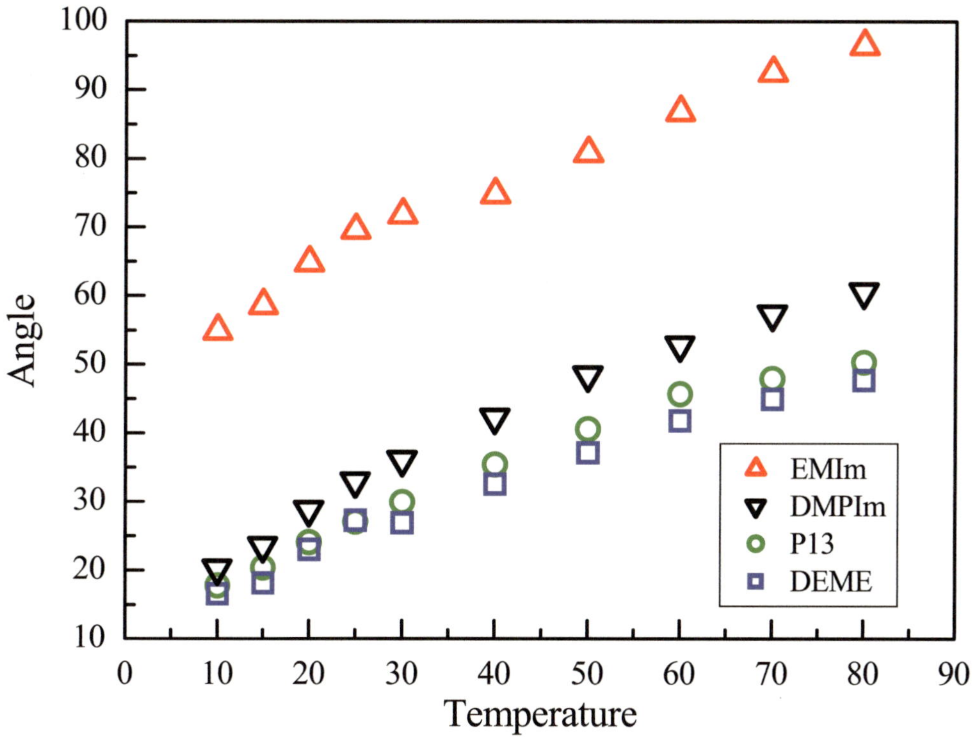

Fig. 19. The flip amplitude of the cations versus temperature for the four cations.

6.5 Relationships of τ_c(cation) with η and *D*

The τ_c(cation) is defined as a time of a motion for the librational flip of a whole molecule. An attempt was made to plot the τ_c(cation) versus viscosity as shown in Fig. 20. When the viscosity is small (about less than 50 mPas) above room temperature, the relations between the τ_c(cation) and the bulk η are linear, while in the lower temperatures, the plots were curved and the gradients became smaller. It is clearly shown that the viscosity became larger, the time interval for the librational flip of the cation became longer. The larger gradients for the relations τ_c(cation) versus η were obtained for EMIm and DMPIm having imidazolium rings and P_{13} exhibited the smallest gradient. The rate of the librational motion is proportional to the bulk viscosity.

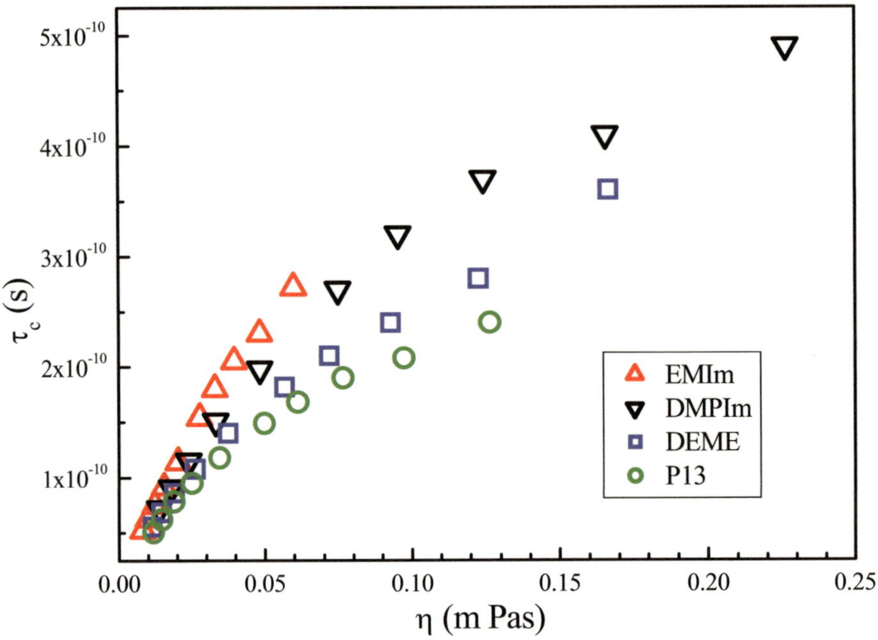

Fig. 20. The τ_c was plotted versus viscosity for the four cations.

Log-log plots of the D_{cation} versus the τ_c(cation) are shown for the four cations in Fig. 21. Although the plots are curved, the D_{cation} and the τ_c(cation) are clearly correlated.

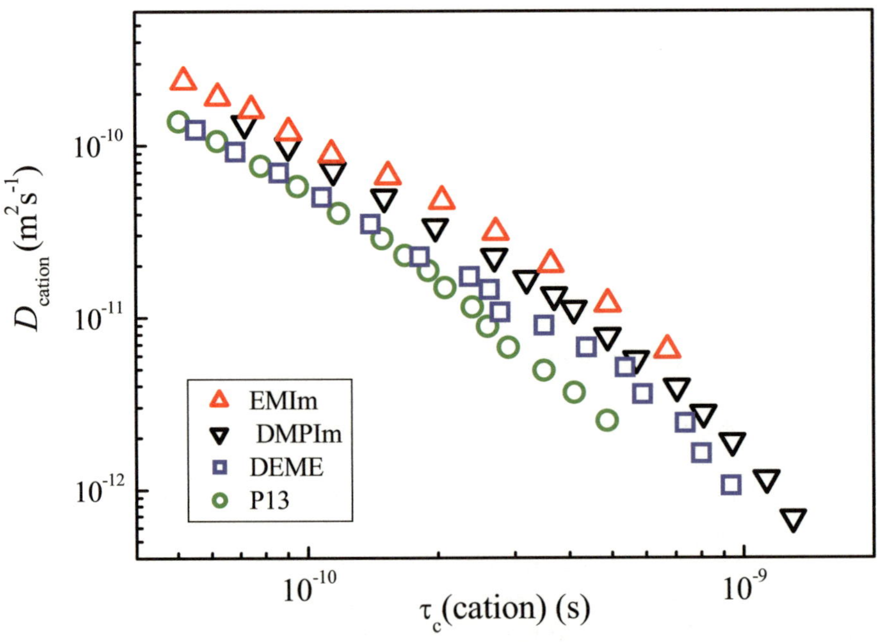

Fig. 21. The plots of D_{cation} versus τ_c(cation) for the four cations.

It is interesting to compare the Arrhenius plots of D_{cation} in Fig. 10 (a) and the plots of D_{cation} against τ_c(cation) in Fig. 21. Since the viscosity of EMIm-TFSA is much smaller compared with that of DMIm-TFSA, D_{EMIm} is much faster than D_{DMPIm} at the same temperature as shown in Fig. 10(a), while the D's are closer for the same τ_c(cation) in Fig 21, suggesting the correlation time of librational flip is related to translational diffusion.

Next, the values of the D_{cation} were plotted against the rates of librational motion ($1/\tau_c$(cation)) as shown in Fig. 22. Although the plots were bent, linear relations were obtained. EMIm exhibited the linear relation in the whole temperature range. The gradients were calculated for the plots D_{cation} versus $1/\tau_c$(cation) above and below 30°C. Since the unit of the gradients is m^2, the square root values are summarized in Table 4 as a diffusion distance during one-flip for each cation. Except for EMIm they became a little longer in the higher temperature region. It is noted that the distance is not sensitive to temperature. As a general trend, the distances are in the order for EMIm ~ DMPIm > P13 ~ DEME. Since the distance is much smaller than the van der Waals radius given in Table 3, translational movement is closely coupled with a molecular librational flip. In other words, the distance for translational movement of the cation during one flip is smaller than the molecular size. The librational flip and translational motion are strongly coupled, and the translation takes place by accompanying librational flip motions. The diffusion scheme for the cation in ILs is different from the rigid-body diffusion model which is the basic concept to drive the SE relation.

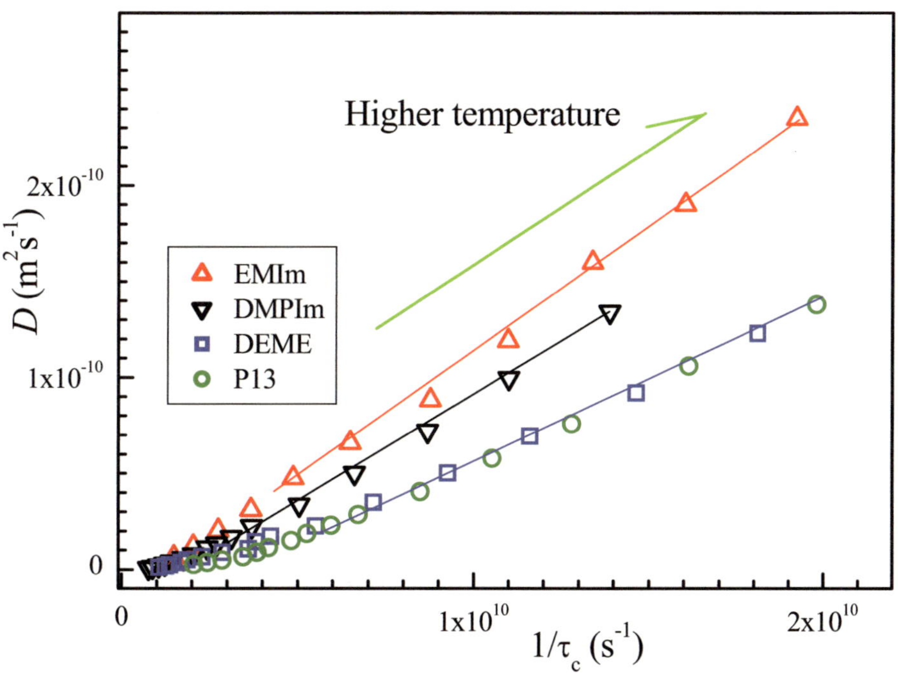

Fig. 22. The D_{cation} is plotted versus the rate of the librational flip.

	Above 30°C	Below 30°C
EMIm	$0.11_4 \pm 0.002$	
DMPIm	$0.12_2 \pm 0.002$	$0.08_3 \pm 0.02$
P13	0.09 ± 0.01	0.07 ± 0.02
DEME	0.09 ± 0.01	0.07 ± 0.02

Table 4. Translational distances during a librational flip of the cations (nm)

6.6 CF_3 motion of TFSA

The Arrhenius plots of ^{19}F T_1 for CF_3 of the TFSA in Fig. 8 were almost linear and the T_1(F) values became longer as the temperature increased, indicating the ^{19}F relaxation takes place under the extreme narrowing condition, i. e. $\omega_F \tau_c(F) << 1$ and thus, τ_c(F) is proportional to $1/T_1(^{19}F)$. In Fig. 23, the viscosity was plotted against $1/T_1(^{19}F)$. The tendency exists that the τ_c(F) is affected by the bulk viscosity and became longer as the viscosity increased.

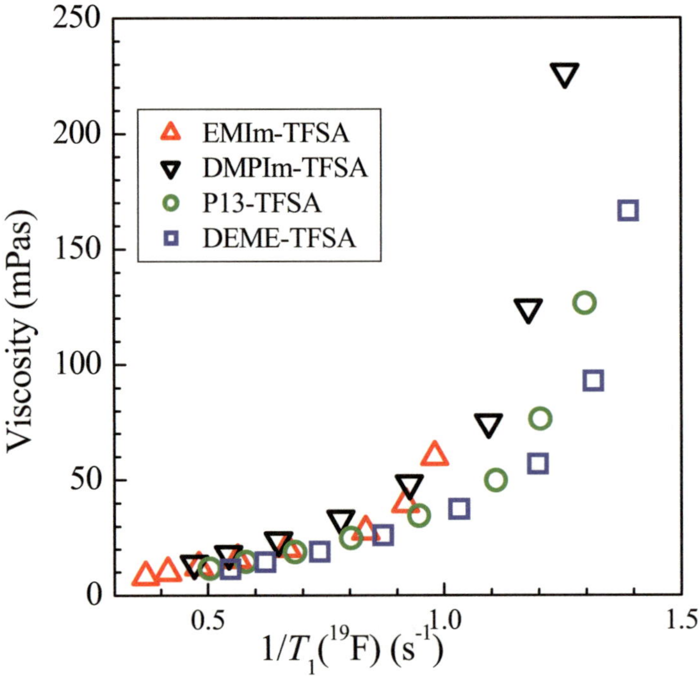

Fig. 23. The viscosity was plotted versus $1/T_1(^{19}F)$ of CF_3.

7. Conclusion

The motions of cations for the four TFSA-based ILs with EMIm, DMPIm, P_{13} and DEME were studied by ^{1}H NMR spectroscopy by measuring the self-diffusion coefficient (D) and

^{1}H spin-lattice relaxation time (T_1). The correlation time τ_c for the cation motion was derived from the Arrhenius plots of T_1 where the T_1 minimum was observed. The relationship of the D and τ_c suggests that the translational motion is coupled with a small-angle librational flip of the cation. In other words, the position change of a cation takes place by accompanied with a flip motion. The flip amplitude is related to the viscosity and the shape of cation. By using the physicochemical parameters such as ionic conductivity, density, and viscosity, the experimental approaches were discussed for the classical SE, NE, and SED relations. We propose an empirical constant of SE for individual ions in ILs. Since ILs have specific properties induced by ion-ion interactions, deviations from classical relations are clearly shown. In this chapter, the experimental data are described following to the classical relationships without any modification to indicate the varieties of ILs. Particularly, cation diffusion in ionic liquids is shown to couple with a small-angle librational flip. Strong ion-ion interactions must be effective in various time-scales and distances in space. Theoretical approaches are waited to interpret the analysis based on the experimental data for ionic liquids.

8. Acknowledgements

The author expresses hearty thanks to Dr. Shiro Seki for the precise data of ionic conductivity, density and viscosity. She is also indebted to him and Drs. S. Tsuzuki, and Y. Umebayashi for their exciting discussion on the complex properties of the ILs.

9. References

[1] K. Hayamizu, S. Tsuzuki and S. Seki, *J. Phys. Chem.* A 112, 12027 (2008).
[2] K. Hayamizu, S. Tsuzuki, S. Seki, Y. Ohno, H. Miyashiro and Y. Kobayashi, *J. Phys. Chem.* B 112, 1189 (2008).
[3] K. Hayamizu, S. Tsuzuki, S. Seki, K. Fujii, M. Suenaga and Y. Umebayashi, *J. Chem. Phys.* 113, 194505 (2010).
[4] K. Hayamizu, S. Tsuzuki and S. Seki, and Y. Umebayashi, *J. Chem. Phys.* 135, 84505 (2011).
[5] W. S. Price, *NMR Studies of Translational Motion*, Cambridge University Press, Cambridge (2009).
[6] E. O. Stejskal and J. E. Tanner, *J. Chem. Phys.* 42, 288 (1965).
[7] J. E. Tanner, *J. Chem. Phys.* 52, 2523 (1970).
[8] D. D. Traficante, *Relaxation: An Introduction, in Encyclopedia of Nuclear Magnetic Resonance*, ed. M. D. Grant and R. K. Harris (Wiley, New York, 1996). Vol. 6, p. 3988.
[9] H. Antony, D. Mertens, A. Dölle, P. Wasserscheid and W. R. Carper, *ChemPhysChem.* 4, 588 (2003).
[10] W. R. Carper, P. G. Wahlbeck, J. H. Antony, D. Mertens, A. Dölle and P. Wasserscheid, *Anal. Bioanal. Chem.* 378, 1548 (2004).
[11] W. R. Carper, P. G. Wahlbeck and J. Dölle, *J. Phys. Chem.* A 108, 6096 (2004).
[12] J. H. Antony, A. Dölle, D. Mertens, P. Wasserscheid, W. R. Carper and P. G. Wahlbeck, *J. Phys. Chem.* A 109, 6676 (2005).
[13] N. E. Heimer, J. S. Wilkes, P. G. Wahlbeck and W. R. Carper, *J. Phys. Chem.* A 110, 868 (2008).

[14] M.Imanari, H. Tsuchiya, H. Seki, K. Nakanishi and M. Tashiro, *Magn. Reson. Chem.* 47, 67 (2009)

[15] M.Imanari, K. Uchida, K. Miyano, H. Seki and K. Nishikawa, *Phys. Chem. Chem. Phys.* 12, 2959 (2010).

[16] W. S. Price, K. Hayamizu, H. Ide, Y. Arata, *J. Magnetic Resonance,* 139, 205-212 (1999)

[17] R. Metzler and J. Klafter, *J. Physics Reports*, 2000, *339*, 1

[18] Y. Umebayashi, T. Fujimori, T. Sukizaki, M. Asada, K. Fujii, R. Kanzaki and S. Ishiguro, *J. Phys. Chem.* A 109, 8976 (2005).

[19] H. J. V. Tyrrell and K. R. Harris, Diffusion in Liquids: A Theoretical and Experimental Study (Butterworths, London, 1984).

[20] S. R. Becker, P. H. Poole and F. W. Starr, *Phys. Rev. Lett.* 97, 55901 (2006).

[21] A. Voronel, E. Veliyulin, V. S. Machavariani, A. Kisliuk and D. Quitmann, *Phys. Rev. Lett.* 80, 2630 (1988).

[22] K. R. Harris, *J. Chem. Phys.* 131, 054503 (2009).

[23] J. O'M. Bockris and A. K. N. Reddy, *Modern Electrochemistry*, Vol. 1, Plenum Press, New York (1998).

[24] W. Xu, E. I. Cooper and C. A. Angell, *J. Phys. Chem.* B 107, 6170 (2003).

[25] K. J. Fraser, E. I. Izgorodina, M. Forsyth, J. L. Scott and D. R. MacFarlane, *Chem. Commun.* 37, 3817 (2007).

[26] Y. Aihara, K. Sugimoto, W. S. Price and K. Hayamizu, *J. Chem. Phys.* 113, 1981 (2000).

[27] H. Tokuda, S. Tsuzuki, M. A. B. H. Susan, K. Hayamizu and M. Watanabe, *J. Phys. Chem.* B 110, 19593 (2006).

[28] 28 K. Ueno, H. Tokuda and M. Watanabe, *Phys. Chem. Chem. Phys.* 12, 1649 (2010) and references therein.

[29] R. Harris, *J. Phys. Chem.* B 114, 9572 (2010)

[30] Y. Umebayashi, T. Mitsugi, K. Fujii, S. Seki, K. Chiba, H. Yamamoto, J. N. C. Lopes, A. A. H. Pádua, M. Takeuchi, R. Kanzaki and S. Ishiguro, *J. Phys. Chem.* B 113, 4338 (2009).

[31] T. Fujimori, K. Fujii, R. Kanzaki, K. Chiba, H. Yamamoto, Y. Umebayashi and S. Ishiguro, *J. Mol. Liq.* 131–132, 216 (2007).

[32] A. Abragam, *The Principles of Nuclear Magnetism*, page 302 (Oxford, 1961).

[33] M. G. Mazza, N. Giovambattista, H. E. Stanley and F. W. Starr, *Phys. Rev.* E 76, 031203 (2007) and references therein.

[34] T. Köddermann, R. Ludwig and D. Paschek, *ChemPhysChem* 9, 1851 (2008).

[35] J. A. Ingram, R. S. Moog, N. Ito, R. Biswas and M. Maroncelli, *J. Phys. Chem.* B 107, 5926 (2003).

[36] Y. Yasaka, C. Wakai, N. Matubayasi and M. Nakahara, *J. Chem. Phys.* 127, 104506 (2007).

[37] H. Cang, J. Li and M. D. Fayer, *J. Chem. Phys.* 119, 13017 (2003).

[38] K. Dahl, G. M. Sando, D. M. Fox, T. E. Sutto and J. C. Owrutsky, *J. Chem. Phys.* 123, 84504 (2005).

[39] K. S. Mali, G. B. Dutt and T. Mukherjee, *J. Chem. Phys.* 128, 54504 (2008).

[40] N. E. Heimer, J. S. Wilkes, P. W. Wahlbeck and W. J. Carper, *J. Phys. Chem.* A 110, 868 (2006).

[41] A. Wulf, R. Ludwig, P. Sasisanker and H. Weingärtner, *Chem. Phys. Lett.* 439, 323 (2007).

Part 4

Applications in Synthesis

11

Ionic Liquids Recycling for Reuse

Samir I. Abu-Eishah
United Arab Emirates University
United Arab Emirates

1. Introduction

Ionic liquids (ILs) are unique new materials that offer novel solutions to the chemical industry as well as to its customers. They are organic salts composed entirely of ions, typically large organic cations and small inorganic anions. They have various striking features such as being liquid salts at ambient temperatures, thermally stable liquids over a range of 300 K, have very low vapor pressure at room temperature, and have the potential to replace many volatile organic solvents. They have judicious variation of the alkane chains on the organic cation, and may generate an infinite set of "designer solvents" because their physical properties can be modified according to the nature of the desired reactions by modification of their cations and anions [Candeias et al., 2009]. Since ILs have no detectable vapor pressure and do not emit volatile organic compounds (VOCs), they are expected to provide a basis for clean manufacturing "green chemistry." [Anonymous, 2004].

In addition, many ILs have unusual solubility and miscibility properties, attractive electric conductivity, quite interesting polarity nucleophilicity for catalysis, and remarkable tribologic properties [Werner et al., 2010]. The thermodynamics and reaction kinetics of processes carried out in ILs media are different from those carried out in conventional media. This creates new opportunities for catalytic reactions, separations, electrochemistry, and combined reaction/separation processes [Anonymous, 2004].

There has been an explosion of interest in ILs in the last decade that has resulted in the discovery of a vast number of new ILs with a wide range of applications. However, ILs may be sensitive to contaminants and may require frequent rejuvenation or replacement. The degree of sensitivity to impurities and the rate of degradation may have a large impact on IL functionality in industrial processes. Process engineering studies are required to obtain sufficient data on ILs stability under long-term exposure to process conditions and exposure to air, moisture, heat, corrosion products, trace impurities (e.g., SO_x, NO_x, etc.) and other key industrial application components. The useful lifetime (life-cycle costs), impact of water content, sensitivity to contaminants, thermal stability, aging, IL losses, etc., over time, as well as recyclability data, are crucial in evaluating ILs viability in commercial applications [Anonymous, 2004].

Rakita (2003) raised similar issues that need to be addressed to achieve efficient commercial synthesis of ILs and make them available on industrial scale. For example, what will the ILs look like after they have been used in a synthetic procedure? Are there contaminants left behind in the IL solvents? What are they? Do they matter, and if so, how can they be removed? These questions need to be answered in the context of specific applications, not abstractly or in an absolute sense.

Despite the excitement of ILs, chemists are much too cautious about the propaganda of ILs' greenness. In the first major US meeting on ionic liquids, held in 2001 at the American Chemical Society meeting in San Diego, Robin Rogers (the Director of Center for Green Manufacturing at the University of Alabama in Tuscaloosa) said: "Before we can say that ionic liquids are green, we have to look at their entire life cycle. People are calling ionic liquids green because they are not volatile, but we have to look at how they are made all the way through to recycling and disposal" [Sen, 2006]. Rogers added: "Researchers also need to find better ways to recycle ionic liquids." Many processes for cleaning up ionic liquids involve washing with water or VOCs. "This needs more work. We don't want to create a secondary waste."

Although ILs are invariably described as highly stable green solvents, thorough investigations to quantify their environmental impact have lagged behind the pace of other research in the area [Scammells et al., 2005]. These issues are extremely important as well as the approaches for the recovery of the ILs (after their use) for further reuse (i.e., recycling). A very important challenge is to use the unique solvent properties of the ILs to develop efficient method for product separation and IL recycling [Galán Sánchez, 2008; Marsh et al., 2004; Gordon, 2001; Welton, 2004; Meindersma et al., 2005; Heintz, 2005; Olivier-Bourbigou & Magna, 2002; Wasserscheid & Keim, 2000; Wasserscheid & Welton, 2003; Wishart & Castner, 2007; Seddon et al., 2002; Holbrey & Seddon, 1999; Holbrey, 2004; Olivier, 1999].

Despite their extremely low vapor pressure that prevents IL emission to the atmosphere, they are, at least, partially miscible with water and will inevitably end up in the aqueous environment (as what happens in the application of ILs for the electro-deposition of metals). During industrial application, ILs will get mixed with other products that necessitate efficient separation and recycling of ILs for economical and ecological needs [Haerens et al., 2010].

Many researchers have pointed out the major concern in the use of ILs is their relative high cost, which makes their recycling an important issue for further study [Laali & Gettwert, 2001; Gonzalez et al., 2006; Verma et al., 2008]. Currently, ionic liquids are quite expensive in comparison with conventional molecular solvents, and large amounts of ILs are needed in the various applications, thus efficient recycling of ILs is important for their economic use, especially in large-scale applications [Wagner & Hilgers, 2008]. However, the cost of ILs is expected to decrease, due to their higher production levels, which is expected to render ILs economically competitive with organic solvents [Khodadoust et al., 2006].

To overcome the cost problem, some low cost and simple synthesis of the ILs are essential for their recycling and reuse. Attri et al. (2010), for example, recovered ILs from binary mixtures of ILs/N,N-dimethyl formamide (DMF) by the removal of the DMF component under vacuum. No appreciable change in the physical properties of the recovered ILs was observed and the recovered ILs were reused at least four times without loss of their purity.

Ionic liquids are now available from a number of commercial manufacturers and from many suppliers. However, IL preparation, work-up to separate/extract products and their recycling usually involve other solvents (e.g., VOCs) to extract unwanted residues, which are toxic persistent (to some extent). However, this diminishes the "green" aspect of their usage due to the cross-contamination arising from the use of organic solvents. Some ILs are certainly toxic and have been shown to have a detrimental effect on aquatic life that will tend to persist in local environment, on land/sea, etc. Thus there are worries about ILs end of life disposal, and materials incompatibility [Rayner, N.D.; Weatherley, N.D.]. One should not forget that ionic liquids are quite different products compared to traditional organic solvents, simply because they cannot be easily purified by distillation, due to their very low volatility. This, combined with the fact that small amounts of impurities influence the IL properties significantly, makes the quality of ILs a quite important consideration for many applications [Seddon et al., 2000].

Also ionic liquids have to be much superior to conventional organic solvents where activity, selectivity and stability of the catalyst are concerned besides being environmentally benign [Song et al., 2001]. However, we have to keep in mind that organic solvents can be recycled and purified by distillation, while it is difficult to recycle ILs because immiscible organic solvents are required to separate the products [Fioroni et al., 2003; Hassan et al., 2006].

For academic researchers it is possible and desirable to obtain perfectly pure ILs with negligible amounts of impurities. However, such efforts are simply not necessary for most applications and would significantly add to the cost of ILs when produced on a commercial scale. Thus, once a customer has decided in which way to use a certain IL in a specific application, the set of specification parameters (purity, by-products, halide content, water content, color) have to be defined altogether [Wagner & Hilgers, 2008]. The recycling protocols usually make use of the low solubility of the ILs in organic solvents. This allows products and residual organics to be extracted using an organic solvent while the salt by-products present in the water-immiscible ILs can be washed out with water. On the other hand, water-miscible ILs are more difficult to recycle bacause inorganic by-products cannot be easily removed. Furthermore, if reaction products are sufficiently volatile they can be distilled directly from the IL, since ionic liquids are inherently non-volatile [Wagner & Hilgers, 2008].

Kanel (2003) mentioned some of the methods that have been developed to recover and reuse ILs like heating or evaporation of volatiles under vacuum, extractions with VOC solvents (obviate some of the advantages of using ILs), supercritical CO_2 ($scCO_2$) extraction, and distillation/stripping of the solute from the ILs (for thermally stable ILs).

In their Minireview, Wu et al. (2009) raised the question: do we understand the recyclability of ILs in a real sense? They have discussed various methodologies for the effective recycling of ILs and focused on the methods needed for the ILs separation from their "working" environment. They proposed that the appropriate separation method should be selected according to the different systems. To better understand the separation of ILs, fundamental research on the existence forms (ions, ion pairs or supermolecule) of ILs in solvents is vitally important.

2. Separation of solutes from ionic liquids by distillation/stripping

Distillation can be used to recover ILs from compounds with low boiling points, because of the ILs negligible vapor pressure. However, direct vacuum distillation protocol is energy-consuming, particularly for non-volatile compound/IL systems. Moreover, if the IL has a tendency to undergo hydrolytic decomposition, such as those that contain $[PF_6]^-$ ions, direct heating should be avoided or at least minimized [Wu et al., 2009]. Distillation can also be used to recover thermo-stable compounds with low boiling points from ILs [Han & Armstrong, 2007; Kanel, 2003]. The recycling of organic solvents requires that they can be purified by distillation, while the recycling of ILs is difficult because immiscible organic solvents are required to separate the products [Fioroni et al., 2003;. Hassan et al., 2006]. Kralisch et al. (2007) carried out the work-up by distillation, as well as the recycling of IL solvents, using a rotary evaporator fitted with a water-bath. The pressure decay was adjusted to the boiling point of the solvent.

BASF (2011) pointed out that the recycling of ILs is easy if protonated cations (HY or RY) are used. In this case the ILs can be switched off by deprotonation (See Fig. 1). Imidazolium cations can be deprotonated by bases to form neutral carbene molecules. The resulting carbenes (amine or imidazole) are found surprisingly stable and can be distilled for recycling or purification purposes.

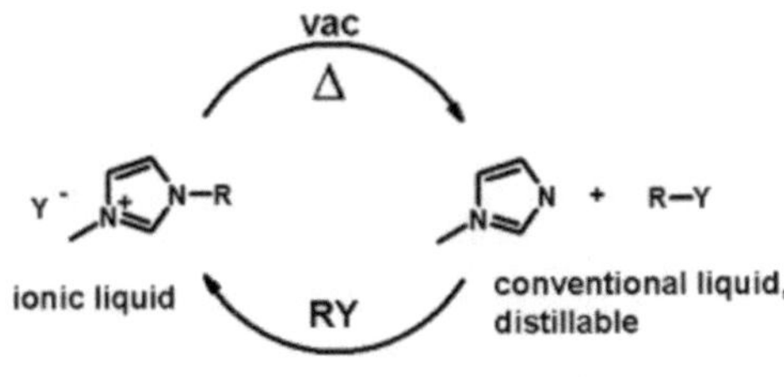

Fig. 1. ILs switching off by deprotonation

It is more difficult with alkylated cations. Apart from purification or recycling by liquid-liquid extraction, two principal "distillation" methods have been reported. The first is the formation of distillable carbenes [WO 01/77081; WO 05/019183] and the second is the back-alkylation of the anion (See Fig. 2) [WO 01/15175].

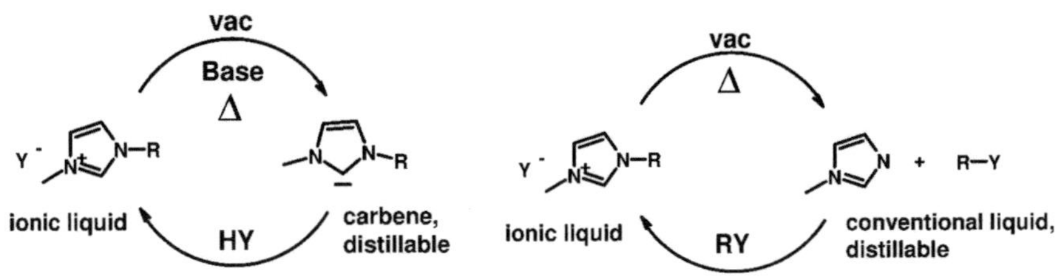

Fig. 2. Formation of distillable carbenes and back-alkylation of the anion

The IL can be recycled by further reaction of the carbene with an acid. The controlled decomposition reaction allows for a recycling or purification process of the IL [WO 01/15175]. In this case the IL is thermally cleaved. The neutral imidazole and alkylating agent are distilled, collected and re-reacted. However, in some cases it is even possible to distill the entire IL without decomposition. For example, the ionic liquid 1-Hexyl-3-methyl-imidazolium-chloride, [Hmim]Cl, can be distilled at 150 °C at 0.5 mbar [WO 2005/068404]. This has recently been demonstrated to be true also for fully alkylated cations [Earle et al., 2006]. These materials require higher temperatures of about 300 °C and 6 mbar vacuums.

Early before ionic liquids are adopted as solvents for the production of nitro-aromatics, Dal & Lancaster (2005) studied the nitration of aromatics using acetyl nitrate (HNO_3-Ac_2O) in two different ionic liquids, [bmpy][OTf] and [bmpy][$N(Tf)_2$], and developed a regime by which the solvent may be recovered and reused. The recycling of the IL was achieved by dissolving the post-reaction mixture into dichloromethane and adding water. It was then possible to extract any unreacted nitric acid, plus the acetic acid (HOAc) generated by the reaction with water. After removing the dichloromethane, the organics were removed from the IL by steam distillation, extracted from the water with dichloromethane and rotary evaporated to remove the solvent and substrate. The IL was recovered by extracting with

dichloromethane and then heating *in vacuo*. This process is illustrated in Fig. 3. The samples of the IL withdrawn after each run, and analyzed before being reused revealed that there was no difference between the recovered and the original IL [Dal & Lancaster, 2005].
The yield of chlorobenzene and the amount of IL [bmpy][N(Tf)$_2$] recovered after each reaction (after removal of the sample for NMR analysis) are shown in Table 1, which shows a steady loss of IL after each run. No degradation of the IL was observed during the reaction (as evidenced by the NMR spectra of the IL after each run and the fact that there was no discolouration). Therefore, Dal & Lancaster (2005) proposed that these losses are purely mechanical, arising from the washing procedure employed.

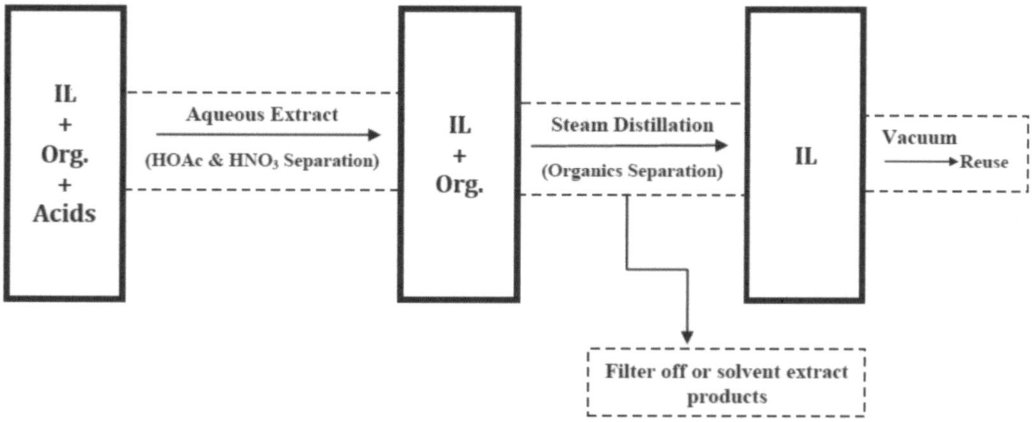

Fig. 3. Procedure for IL recycling from IL/Organic/Acid mixture [Reproduced from Dal & Lancaster, 2005].

Recycle #	Yield [%]	Mass loss of [bmpy][N(Tf)$_2$] recovered [g]
1	76	0.22
2	70	0.30
3	71	0.38
4	76	0.62
5	74	0.88

Table 1. Yield of chlorobenzene and recyclability of IL [bmpy][N(Tf)$_2$]

Dal & Lancaster (2005) admit that there is one major problem with the recycling of the IL as described above; the reliance on dichloromethane does not make this process appear green at first glance. Dal & Lancaster (2005) added that the role of dichloromethane in their work was to enable the determination of both product and recovery of the IL. In an industrial setting, it would be quite possible to perform a steam distillation of the entire reaction mixture. The organics will form a separate phase from the aqueous acids distilled and could probably be removed by simple decantation. And because this ([bmpy][N(Tf)$_2$]) IL is immiscible in water, it should be possible to decant any water from the IL and reuse it immediately. Thus recycling the IL does not require the use of dichloromethane [Dal & Lancaster, 2005].

3. Extraction with supercritical CO_2

Extraction of solutes from ILs with VOC solvents obviate some of the advantages of using ILs, but supercritical CO_2 ($scCO_2$) extraction may offer a promise. The "green" solvent $scCO_2$ may be a good choice for the recovery of solutes and the regeneration of ILs. Blanchard et al. [1999] demonstrated that $scCO_2$ exhibits high solubility in ILs whereas ILs show generally no detectable solubility in CO_2: The solubility of $scCO_2$ in ILs was determined to be less than 5×10^{-7} mole fraction [Blanchard & Brennecke, 2001]. This unique property of $scCO_2$ means that solutes can be isolated from ILs without cross-contamination of the gas phase with IL and also provides a means for recycling the ILs [Blanchard et al., 1999]. Recovery of the organic solute from ILs was near quantitative and achieved by simple expansion of the gas phase downstream. This efficient exchange between the two phases, coupled with the lack of mutual solubility seems to be ideally suitable for product separation and recycling of ILs [Gordon & Muldoon, 2008].

Blanchard et al. (1999) and Blanchard & Brennecke (2001) extracted aromatic and aliphatic compounds into a CO_2 phase from IL [bmim][PF_6]. Examples include separation of naphthalene from the [bmim][PF_6] [Blanchard & Brennecke, 2001] and methanol [bmim][PF_6] using pressurized CO_2 [Scurto et al., 2002]. A two-step extraction system (water/RTIL/CO_2) for trivalent lanthanum and europium was reported by Mekki et al. (2006) where the metal ions were extracted from the aqueous phase into $scCO_2$ via a RTIL/fluorinated β-diketonate mixture with high extraction efficiencies.

Taking into account the specific properties of carbohydrates, only hydrophilic ILs can be used [Murugesan and Linhardt, 2005]. This means that only apolar solvents or $scCO_2$ may be used to extract reaction products. Unfortunately, 5-hydroxy methyl furfural (HMF) was found to have a larger affinity for the IL 1-H-3-methyl imidazolium chloride than for the organic solvent (such as diethyl ether) or $scCO_2$. For the dehydration of fructose, Moreau et al. (2006) showed that HMF could be completely extracted with diethyl ether in a continuous or stepwise manner. But, their preliminary results using $scCO_2$ as extraction solvent do not seem to have better expected extraction properties.

Scurto et al. (2003) studied CO_2 induced separation of IL and water. If a pressure of CO_2 is placed upon a mixture of the IL and water, a second liquid phase can appear. In Fig. 4, the densest liquid is rich in IL (L_1), the next phase is rich in water (L_2) and the third phase is vapor (V), which is mostly CO_2 with a small amount of dissolved water. Scurto et al.(2003) demonstrated that water/[C_4mim]BF_4 solutions can be induced to form three phases in the presence of $scCO_2$.

Although the original solution of [C_4mim]BF_4 was quite dilute, the application of $scCO_2$ induced the formation of an additional liquid phase rich in [C_4mim]BF_4, indicating that the separation of [C_4mim]BF_4 from aqueous solutions by the application of $scCO_2$ is possible. In Fig. 4, the applied CO_2 pressure at which the second liquid phase appears (at a given temperature and initial concentration of IL in water) is called the *Lower Critical End Point* (LCEP). However, Wu et al. (2009) considered this effect as a pressure phenomenon, as well as due to a slight decrease in the dielectric constant upon addition of CO_2. In the opinion of Wu et al. (2009), a salting-out effect due to the formation of carbonate in the solution may also played a key role, since CO_2 reacts with water to form carbonic acid and its dissociation products.

Recycling of ILs with high efficiency is of key importance on going from the laboratory-scale to large-scale industrial application of these solvents [Wu et al., 2009]. It is noteworthy that the efficiency of the recycling process for various ILs varies from quite poor [Seddon et al., 1999] to very good [Welton, et al., 2000; Handy & Zhang, 2001; Fukuyama et al., 2002]. The quality of the recovered IL should be assessed by some suitable technique [see for example, Wagner & Hilgers, 2008]. Even though ILs in some systems could be recovered efficiently, the yield of the product might decrease gradually with time (upon multiple usages) due to some impurities in the recovered ILs.

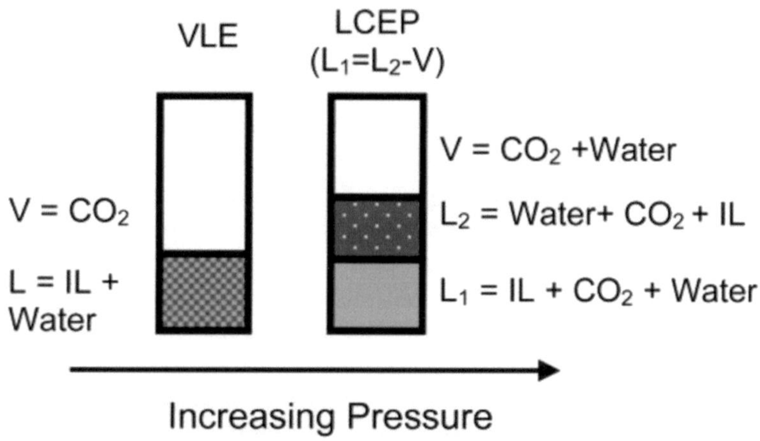

Fig. 4. Phase behaviour of IL/water/CO_2 mixture at increasing pressure at near-ambient temperatures (Adapted from Scurto et al., 2003)

Several procedures for the recycling of ILs have been reported in the literature. Depending on the IL used and the application where it is used, a variety of methods for recycling are possible. By picking the right purification steps, an individually optimized work-up procedure can be obtained [Wasserscheid & Welton, 2008]. However, if the customer does not feel comfortable with the task of recycling the IL there is a further option at hand: why not rent or lease the IL rather than buy it? The customers would in this case perform their application with the IL and send, the probably impure, IL back to the supplier, who has the expertise to recycle and clean it up [Wasserscheid & Welton, 2008]. This scenario could be interesting from an economic point of view for truly commercial applications on a large scale. *But will this be a practical solution?*

4. Ionic liquids as reaction media

Because of their negligible vapor pressure, thermal stability and easy recyclability, neutral ILs have been referred to as environmentally benign solvents. These ILs have been employed as excellent and recyclable medium for a wide array of reactions; e.g., Heck reaction [Park & Alper, 2003], Bischler–Napierlaski cyclisation [Judeh et al., 2002], Beckmann rearrangement [Ren et al., 2001], addition of thiols to unsaturated ketones [Yadav et al., 2003], l-proline catalysed aldol reaction [Loh et al., 2002], and Pechmann condensation of phenols and ethyl acetoacetate (EAA) catalyzed by $POCl_3$ in [bmim]PF_6 and [bmim]BF_4 ILs [Potdar et al., 2005]. Other examples where ILs have been used as a reaction media include, but not limited to, Diels-Alder reactions [Reinhardt, 2009; Doherty, 2004; Song, 2001], Friedel-Crafts alkylation and acylation reactions [Xiao, 2006; Xiao & Malhotra, 2005;

Ross & Xiao, 2002; Itoh, 2001; Stark et al., 1999], hydroformylation reactions [Magna et al., 2007; Sharma, 2009], Pd-mediated C-C bond formation [Park & Alper, 2003], alkene polymerisation [Hardacre et al., 2002], and biotransformations [Bornscheuer & Kazlauskas, 2006].

Bortolini et al. (2002) accomplished the recycling of [bmim][PF_6] by washing with an aqueous solution of $NaHSO_3$ and then with water. The uncoloured recovered IL was dried, analyzed by NMR and reused. Recycling of trihalide-based ILs was performed, after removal of the solvent used for product extraction, by addition of the consumed halogen, evaluated on the basis of the product yields.

The inability to successfully recycle morpholinium-based IL was reported by Kim & Shreeve (2004). Similar results were reported by Murray et al. (2003) after the fluorination of benzyl bromide using cesium fluoride (CsF) in [bmim][PF_6]. This suggests that recycling of ILs is limited for nucleophilic fluorination reactions using alkali metal fluorides. Although this system was not investigated in detail but it indicates that ILs cannot be recycled effectively when reaction is carried out under highly-basic conditions and higher temperatures.

Chrobok (2010) reported a new and efficient method for the synthesis of lactones involving the application of an oxygen/benzaldehyde system as the oxidant and ILs as solvents with the possibility of effective ILs recycling. After the Baeyere-Villiger (B-V) reaction, the recycling of the highly hydrophobic IL [bmim]NTf_2 was purified first with a saturated aqueous $NaHCO_3$ solution, and then with diethyl ether. The [bmim]NTf_2 was then concentrated and dried (60 °C, 8 h) under vacuum.

On the other hand, when hydrophilic ILs were used as reaction solvents, the product was first extracted with diethyl ether. Next, the organic extract was washed with a saturated aqueous solution $NaHCO_3$, dried and concentrated. Table 2 shows the results of reactions using three recycles of [bmim]NTf_2 in the oxidation of cyclohexanone. Alternatively, it is possible to distil the product from the crude reaction mixture [Chrobok, 2010].

Recycle #	Yield of ε-caprolactone [%], as determined by GC	IL [bmim]NTf_2 recovery [%]
0	90	91
1	88	92
2	88	90
3	87	89

Table 2. Recyclability of IL [bmim]NTf_2 in the oxidation of cyclohexanone in the presence of benzaldehyde with oxygen and ACHN additive at 90 °C over 2.5 h reaction

The possibility of recycling the highly hydrophobic ILs [bmim]BF_4 and [Hmim]OAc used in the Baeyere-Villiger (B-V) oxidation of ketones with Oxone was also checked [Chrobok, 2010]. After the filtration of the post reaction mixture and the extraction of the product with ethyl acetate (for [bmim]BF_4) or dibutyl ether (for [Hmim]OAc), the ILs were concentrated, dried under vacuum (60 °C, 5 h) and reused. Table 3 shows four cycles of the oxidation of 2-adamantanone in ILs that are recovered and reused for further B-V reactions. No significant loss of activity of these solvents was observed. Additionally, ILs were recycled three times in a model reaction without significant loss of activity [Chrobok, 2010].

Recycle #	Yield [%] for 3 h reaction time using [bmim]BF_4	IL Recovery [%]	Yield [%] for 5.5 h reaction time using [Hmim]OAc	IL Recovery [%]
0	95	92	97	91
1	94	90	97	91
2	93	91	97	90
3	93	89	97	89

Table 3. Recyclability of ILs in the oxidation of adamantanone in the presence of Oxone at 40°C

Gordon & McCluskey (1999) carried out the recycling of the IL used in the allylation reaction of benzaldehyde in the presence of tetraallylstannane catalyst following an extremely straight-forward protocol: The procedure employed was to dissolve the IL in ethyl acetate (10 ml), and wash with water (2x5 ml) and brine (5 ml). Addition of diethyl ether (20 ml) caused two layers to form, the lower being essentially pure IL. No decrease in the yield was observed in the runs carried out using the 'old' IL (see Table 4). Furthermore the products obtained were of the same purity as in the first run. In the case of the recycled [bmim][PF_6] salt, the purification removed all of the cloudy residue, while some cloudiness was observed in the recycled [bmim][BF_4] salt, but this did not seem to impair the performance of the IL [Gordon & McCluskey, 1999].

Recycle #	Isolated Yield [%] using [bmim][BF_4]	Isolated Yield [%] using [bmim][PF_6]
1	79	82
2	82	81
3	78	83

Table 4. Isolated yield using recycled ILs in the allylation reaction of benzaldehyde in the presence of tetraallylstannane catalyst

Potdar et al. (2005) employed *neutral* ILs such as [bmim]PF_6 or [bmim]BF_4 with catalytic amount of acid for synthesis of coumarin via Pechmann condensation of phenols and ethyl acetoacetate (EAA) under ambient conditions. They also carried out the reaction at high temperature in IL [bmim]PF_6, without the use of any acid catalyst and investigated the recycling of the ILs for reuse.

Potdar et al. (2005) also presented a general work up procedure for recycling of ILs: The reaction mixture was quenched with 10% aqueous NaOH solution and settled. The settled down IL was extracted with dichloromethane. Further, the dichloromethane layer was stirred vigorously with 10% aqueous NaOH in order to remove coumarin and acid traces. Further, dichloromethane layer was washed with water. The solvent was then evaporated and the resultant IL was subjected to consecutive extractions with diethyl ether to remove any organic impurities present. The resultant IL was then dried under vacuum at 60 °C and reused for subsequent reactions. When the IL [bmim]BF_4 was used, the reaction mixture was quenched by addition of water. The coumarin derivative separated as a solid was isolated by filtration. The IL was recovered by evaporation of water in a rotary evaporator. The above ILs have been reused for consecutive four cycles without loss in their efficiency [Potdar et al., 2005].

Potdar et al. (2005) found the yields of the above reactions obtained by employing chloroaluminate ILs or neutral ILs with catalytic amount of acid at room temperature compare well to those obtained at high temperature. They also found that the IL [bmim]PF_6 can be recycled efficiently and used consecutively for three runs without any considerable loss in its activity (see Table 5).

Recycle #	Yield [%] using [bmim]PF_6
1	95
2	92
3	94

Table 5. Recyclability of [bmim]PF_6 IL in Pechmann condensation performed at 100 °C

Xie & Shao (2009) studied the phosphorylation of corn starch was by reaction with sodium dihydrogenphosphate or disodium hydrogenphosphate using IL 1-butyl-3-methyl-imidazolium chloride (bmimCl) as reaction medium. The bmimCl was recovered by separating the product, and distilling the bmimCl under reduced pressure to remove the volatile compounds before being recycled and reused in the next phosphorylation reaction. The recycling test result is illustrated in Fig. 5 where it can be seen that a lower degree of solution (DS) of the starch was observed when the recovered IL was used for the subsequent cycle. However, the decline in the DS value was small even after five cycles. Therefore, the IL could be recovered and recycled without problem.\

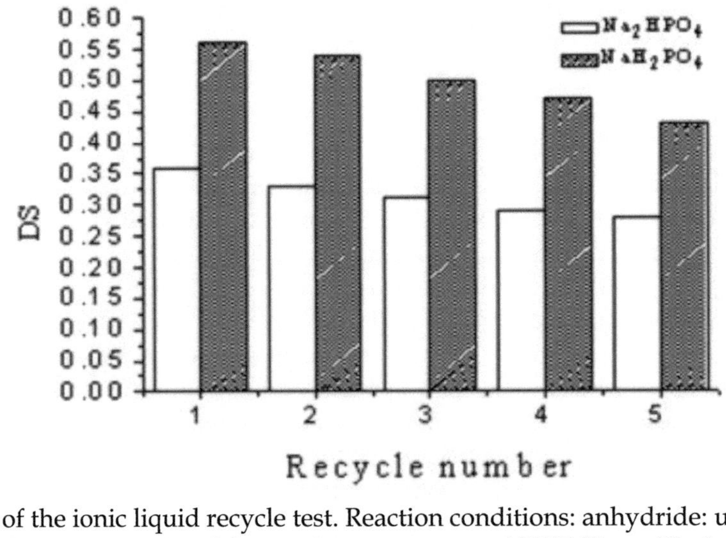

Fig. 5. Result of the ionic liquid recycle test. Reaction conditions: anhydride: urea: AGU molar ratio 3: 3: 1, reaction time 3 h, reaction temperature 1207C [From Xie & Shao, 2009]

The most studied examples are the 1,3-dialkylimidazolium-based ILs where the ionic solvent is usually recycled through several cycles of the reaction. Numerous examples describe the immobilization of a transition metal catalyst in the IL phase of a biphasic system [WO 01/77081, WO 05/019183, WO 01/15175, WO 2005/068404, Earle et al., 2006].

Faizee et al. (2007) presented the recovery of ILs used for the pretreatment of cellulose and lignocellulosic substrates using an ion exchange bed. This was carried out using Solid Phase Extraction (SPE) using Strong Cationic Exchange resins (SCX) or liquid-liquid extraction. As it is known, ionic liquids consist of an organic cation and an inorganic anion (chloride, in this case). The SCX is strongly acidic and exchanges the cation. The ion exchange bed then gets saturated with the cations and can be regenerated by passing hydrochloric acid-a process which allows for recovery of IL via the reverse ion exchange process.

Several groups have tested recycling of ILs as well as separation methods that would be suitable on an industrial scale, such as extraction with benign solvents like water, or $scCO_2$ or distillation [Itoh et al., 2003]. However the economics of commercially using ILs in industrial processes are still far behind and not clear. The currently published data on IL recovery do not allow for reliable extrapolation to industrial applications scale. However, it is clear that the ionic liquids which are proposed to be used as solvents need to be recycled and reused in order to have commercially attractive processes.

However, the scheme presented in Fig. 6 comprises a possible industrial application of the separation of ILs from other compounds. It can be used in breaking of azeotropes such as tetrahydrofuran (A) - water (B) mixture in the presence of an IL. Here the IL acts as an extractant (or entrainer) to wash down compound B and removed at the bottom of the column while compound A is released and distilled as a pure compound at the top of the column. In this case, the IL can be recycled into the process after removing residual B in a falling film evaporator [BASF, 2011; Maase, 2008].

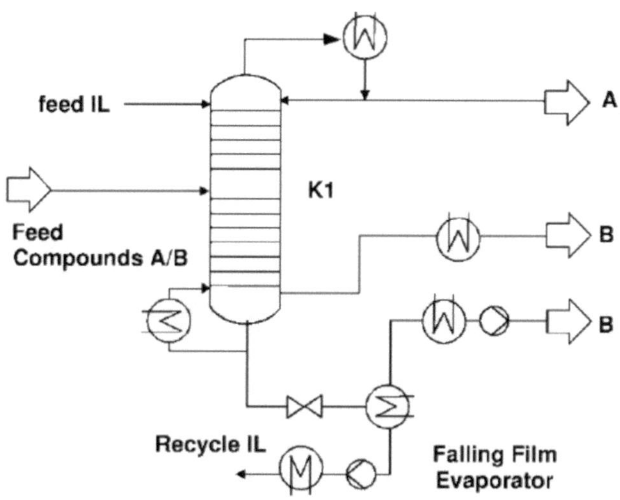

Fig. 6. Generic flow chart for the IL-based separation process [BASF, 2011; Maase, 2008]

Meindersma and Co-workers (2011) have their own experience with regenerating ILs, both on laboratory and pilot scales. They used a rotary evaporator (75°C, vacuum, 15 h) at the laboratory level and extraction column (6-m high, 0.06-m diameter, flux rate 8 $m^3/m^2.h$) in the pilot plant as a regeneration unit, which was, of course, not optimal. After the extraction experiments and after a phase separation between the IL and the organic extract phase, they bubbled hot nitrogen, 75°C, through the IL phase in the column. The complete regeneration of the IL was realized in (20-22) h. In a proper regeneration column, e.g. a falling film column, they expect that the process will be much shorter. They removed any remaining IL from the raffinate phase by a one step washing with water. Both in the laboratory as in our pilot plant, they have used regenerated ILs, without loss of any activity. The colour of the IL changed, of course, from the original colour (e.g. yellow) to dark brown or black, probably due to some impurities. In one occasion, the fresh pyridinium-based IL smelled of pyridine. Table 6 compares several physical properties of the ILs before and after several regenerations [Meindersma & Co-workers, 2011].

Characteristic	New [3-mebupy]$N(CN)_2$	Regenerated [3-mebupy]$N(CN)_2$
Toluene distribution coefficient (10 wt%, kg/kg)	0.39	0.42
Viscosity (cP)	20	21
Density (kg/m^3)	1041	1040
Interfacial tension (mN)	13.3	13.4
Colour	Light red	Dark brown

Table 6. New IL vs. regenerated IL used for the extraction of toluene from alkanes at 313 K [Meindersma & Co-workers, 2011]

A review by Olivier-Bourbigou and Magna (2002) indicates that ILs have been successfully recycled in many reactions. Examples of reactions using recyclable IL included palladium-catalyzed Heck reactions (Ramnial, 2005; Carmichael, 1999] oxidations of benzaldehyde and alkylbenzene [Seddon & Stark, 2002], Friedel-Crafts reactions (Adams et al., 1998; Stark et al., 1999; Ross & Xiao, 2002], copolymerization of styrene and carbon monoxide [Hardacre et al., 2002], and dimerization of propene [Chauvin et al., 1990].

Recyclability requires rates and yields of reactions to be maintained at a reasonable level after repeated reactions. In particular, reactions containing a transition metal catalyst immobilized into the IL of a biphasic reaction system have proved to be recyclable. Generally, recycling is based on the non-volatile nature of ILs and the solubility differences between ILs, organic compounds and water. Products can be extracted from ILs with a non-polar solvent or they can be separated by distillation. A water-immiscible IL can be washed with water to get a water-soluble product or side products out of the reaction mixture [Kärkkäinen, 2007].

The major benefit of using ILs in the dimerization reaction is simplified separation of the products. ILs are polar, thus non-polar products are not miscible into them. The biphasic procedure enables separation of the products by decantation. The cationic nickel catalysts used showed good activity in ILs and they do not seem to leach from the IL phase to the product phase. This makes recycling of the IL and the catalyst possible. However, the sensitivity to air and water limits the use of $AlCl_3$-based ILs [Kärkkäinen, 2007].

Recycling studies indicated that the catalytic properties of an IL could be maintained for several hours. Kärkkäinen (2007) carried out the batch reaction in [C_4mim]Cl/$InCl_3$, with $InCl_3$ 0.58 mol%, at 120 °C for 1 hour. In these experiments the product layer was separated on the surface of the IL and decanted before a new batch of 2-methylpropene was fed in. The conversions of 2-methylpropene stayed at the same level in first three reactions, but the fourth one gave a 40% lower conversion.

Kärkkäinen (2007) carried out the dimerization reaction with a continuous feed of 2-methylpropene (1.5 mmol/min) in [C_6mim]Cl/$InCl_3$, $x(InCl_3)$ = 0.55 for 6 hours at 160 °C.

The amounts of the main dimers as a function of reaction time are shown in Table 7. The colour of the IL changed during the reaction; at the beginning the solution was clear, but after 6 hours it changed *from* yellow and orange to brown.

		Reaction time [h]					
Products	%	1	2	3	4	5	6
2,4,4-trimethyl-1-pentene	8	63	66	64	61	56	53
2,4,4-trimethyl-2-pentene	9	18	17	14	16	14	14
2,3,4-trimethyl-1-pentene	33	5	5	6	9	8	9
2,3,4-trimethyl-2-pentene	34	8	5	5	4	4	4

Table 7. Amount of main dimers formed as a function of reaction time in the continuous reaction of 2-methylpropene in $[C_6mim]Cl/InCl_3$, $x(InCl_3) = 0.55$, IL at 160 °C

The used catalytic IL was washed several times with hexane to purify it for the next reaction. After drying in a high vacuum it was analyzed with NMR. The 1H NMR spectrum indicated that the IL did not contain any products. In addition, indium content was analyzed from the hexane extract by elemental analysis and it showed the absence of $InCl_3$ indicating that it stayed in the IL and did not leach out from it. The IL was used again in a dimerization reaction and it catalyzed successfully the reaction. Excess $InCl_3$ did not leach out from the IL and recycling of the IL was possible. This method should be suitable for dimerizing other light olefins as well.
Xiao (2006) investigated the reusability and efficiency of ILs with and without catalysts. The recycling process involved washing the used ILs with diethyl ether. Any organic residue left in the IL layer could be separated by the ether wash. Two layers formed (IL and organics). The resulted IL layer was then separated and dried at 65 °C under reduced pressure (see Fig. 7).

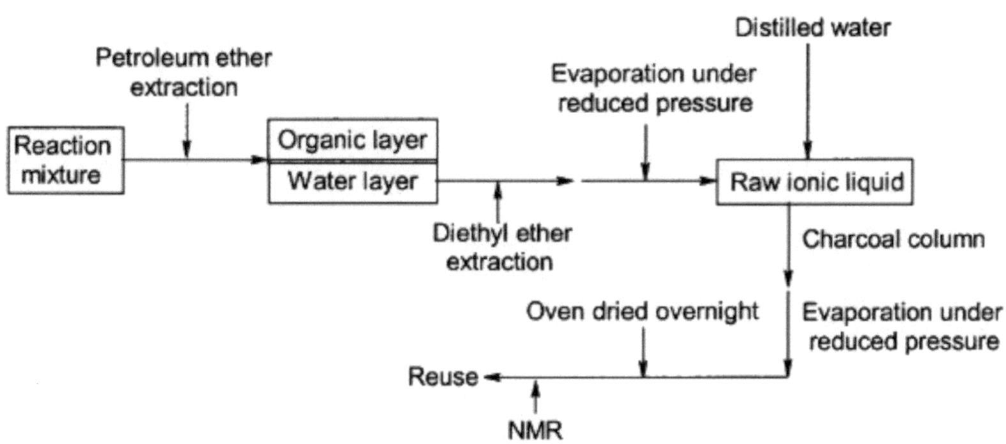

Fig. 7. Flow chart of recycling process of ILs [Xiao, 2006]

Successive runs were performed with the recovered IL, $[EtPy]^+[BF_4]^-$ or $[EtPy]^+[CF_3COO]^-$, for the Friedel-Crafts alkylation between benzene and 1-chlorobutane at 50 °C for four hours. Table 8 shows that both ILs could be recovered quantitatively and almost without loss of activity and selectivity. This was evident from the fact that the conversion of the Friedel-Crafts alkylations was not affected even after the third run with the recovered IL.

Recycle #	Conversion [%] using $[EtPy]^+[BF_4]^-$	IL recovery [wt%]	Conversion [%] using $[EtPy]^+[CF_3COO]^-$	IL recovery [wt%]
0	45 (75)	-	55 (80)	-
1	43 (73)	94	53 (75)	93
2	44 (74)	92	51 (78)	92
3	41 (72)	93	50 (76)	93

Table 8. Recyclability of IL in the Alkylation of Benzene and 1-Chlorobutane. [Xiao, 2006]. Numbers in parenthesis isolated yields

However, yields of the IL recovery from the used $FeCl_3$ - IL system (Fig. 8) were relatively low. Table 9 shows that, although the ILs in $FeC1_3$-IL system can be recovered efficiently, the conversion dropped dramatically even with fresh $FeC1_3$. This was attributed to possible presence of some impurities in the recovered ILs. However, the product selectivity remained nearly the same [Xiao, 2006].

Recycle #	Conversion [wt%] using $FeC1_3$ - $[EtPy]^+[BF_4]^-$	IL recovery [wt%]	Conversion [wt%] using $FeC1_3$ - $[EtPy]^+[CF_3COO]^-$	IL recovery [wt%]
0	88 (87)	-	94 (94)	-
1	75 (84)	88	82 (92)	90
2	67 (83)	85	76 (90)	86
3	65 (81)	89	75 (91)	87

Table 9. Recyclability of $FeC1_3$-ILs in the Alkylation of Benzene and 1-Chlorobutane [Xiao, 2006]. Numbers in parenthesis isolated yields

[Xiao, 2006] also conducted successive runs for acylation between benzene and acetic anhydride at 50°C for four hours. Table 10 shows that both ILs could be recovered quantitatively with negligible loss of activity. Moreover, the acylation was not affected even after the third run with the recovered IL.

A similar study was also carried out with recovered $FeC1_3$-IL systems for the acylation of benzene with acetic anhydride [Xiao, 2006]. The IL recovery scheme is as presented in Fig. 8. The results of the acylation conversion using fresh IL and reused IL are shown in Table 11.

In this case, even though ILs in $FeC1_3$-IL system could be recovered efficiently, the yield of acylation product (Acetophenone) decreased gradually. This may be due to some impurities in recovered ILs.

Xiao (2006) also investigated the reusability of IL for the Friedel-Crafts reaction of N, N-dimethylaniline with ethyl glyoxylate catalyzed by the complex of (R)-BINOL-Br-Ti$(O^iPr)_4$. The recycling process involved washing the used ILs with diethyl ether. Any organic residue left in the IL layer could be separated by the ether wash. The IL layer was decanted and evaporated at 65°C under reduced pressure, then purified according to the schematic procedure shown in Fig. 9. Successive runs were performed with the recovered IL $[EtPy]^+[BF_4]^-$ or $[EtPy]^+[CF3COO]^-$ at room temperature for 24 hours.

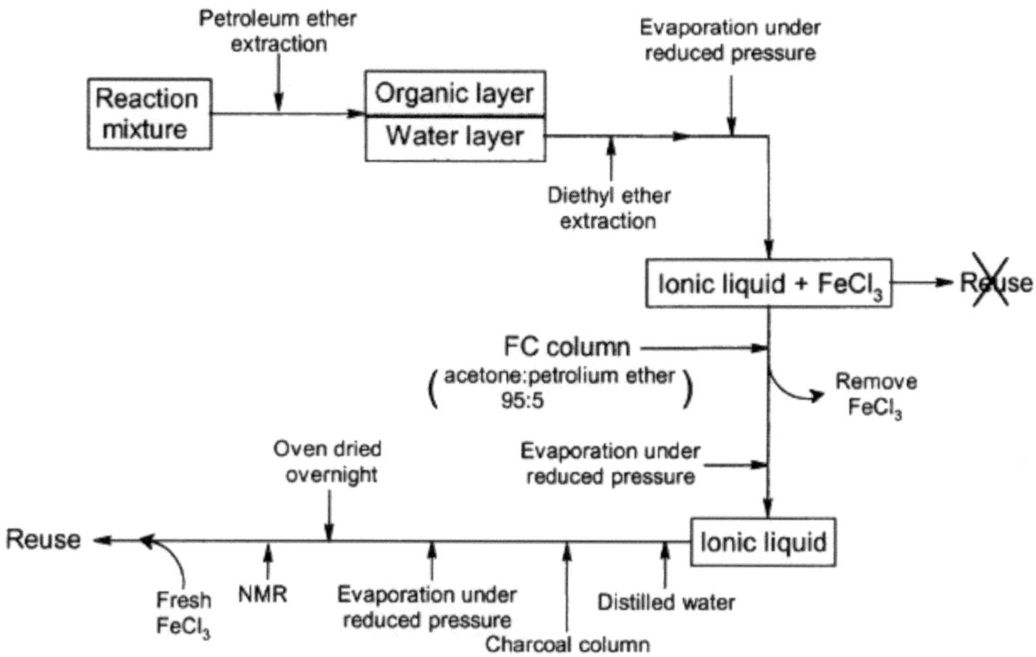

Fig. 8. Flow chart of recycling process of $FeCl_3$ - IL system [Xiao, 2006]

Recycle #	Conversion [wt%] using $[EtPy]^+[BF_4]^-$	IL recovery [wt%]	Conversion [wt%] using $[EtPy]^+[CF_3COO]^-$	IL recovery [wt%]
0	72 (97)	-	77 (99)	-
1	70 (97)	93	76 (98)	96
2	70 (95)	93	74 (97)	94
3	68 (96)	94	74 (98)	95

Table 10. Recyclability of Ionic Liquids in the Acylation of Benzene and Acetic anhydride [Xiao, 2006]. Numbers in parenthesis isolated yields

Recycle #	Conversion [wt%] using $FeCl_3$ - $[EtPy]^+[BF_4]^-$	IL recovery [wt%]	Conversion [wt%] using $FeCl_3$ - $[EtPy]^+[CF_3COO]^-$	IL recovery [wt%]
0	90 (94)	-	97 (97)	-
1	88 (91)	90	96 (94)	88
2	84 (89)	87	91 (91)	91
3	83 (88)	89	88 (92)	88

Table 11. Recyclability of $FeCl_3$-ILs in the Acylation of Benzene and Acetic Anhydride [Xiao, 2006]. Numbers in parenthesis isolated yields

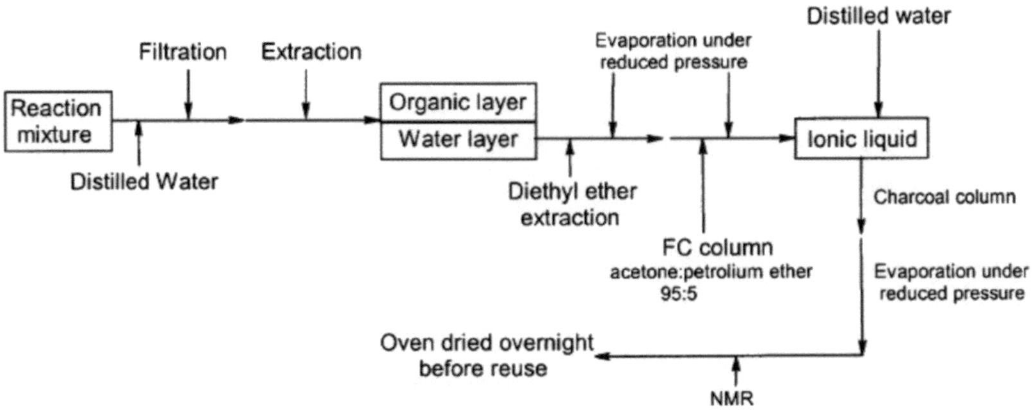

Fig. 9. Flow chart of recycling process of ILs used in the Friedel-Crafts reaction of N, N-dimethylaniline with ethyl glyoxylate catalyzed by the complex of (R)-BINOL-Br-Ti(O^{i}Pr)$_4$ [Xiao, 2006]

Table 12 shows that both ILs could be recovered efficiently and utilized almost without loss of activity and selectivity. This is evident from the fact that the yield of the Friedel-Crafts acylations are not much affected even after the 5th run with the recovered IL.

Recycle #	Yield [wt%] using $[EtPy]^+[BF_4]^-$	IL recovery [wt%]	Yield [%] using $[EtPy]^+[CF_3COO]^-$	IL recovery [wt%]
0	74	-	86	-
1	72	92	84	93
2	73	94	83	93
3	72	91	82	92
4	71	93	83	94
5	69	92	82	92

Table 12. Recyclability and Reuse of Ils used in the Friedel-Crafts reaction of N, N-dimethylaniline with ethyl glyoxylate catalyzed by the complex of (R)-BINOL-Br-Ti(O^{i}Pr)$_4$

Xiao (2006) also investigated the reusability of ILs for the asymmetric reduction of acetophenone catalyzed by the complex of (R)-BINOL-LAH. The recycling process involved washing the used ILs with diethyl ether. Any organic residue left in the IL layer could be removed by the ether wash. The IL layer was decanted and evaporated under reduced pressure at 65 °C, then purified according to the schematic procedure shown in Fig. 10. Successive runs were performed with the recovered IL $[EtPy]^+[BF_4]^-$ or $[EtPy]^+[CF_3COO]^-$ at room temperature for four hours.

The results show IL $[EtPy]^+[BF_4]^-$ could be recovered efficiently and almost without loss of activity and selectivity. This is evident from the fact that the yield was not affected much even after the 3rd run. In this reaction, however, $[EtPy]^+[CF_3COO]^-$ could not be recovered as efficiently as the previous cases. It may be because lithium aluminum hydride, as a powerful reductant, might react with carbonyl group of trifluoroacetate anion. The evidence is that a color change of $[EtPy]^+[CF_3COO]^-$ has been observed during experiments, which changed from colorless/light yellow to yellow, then to brown. See Table 13.

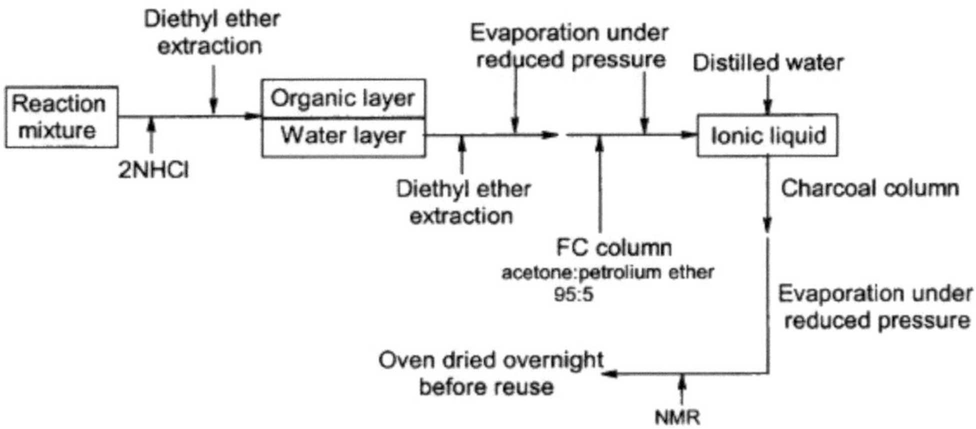

Fig. 10. Flow chart of recycling process of ILs used in the asymmetric reduction of acetophenone catalyzed by the complex of (R)-BINOL-LAH [Xiao, 2006]

Several aziridines in high yields with *cis* selectivity have been preparaed by using ILs [bmim][PF_6] and [bmim][BF_4] as reaction media. After each reaction the products were extracted with petroleum ether and ethyl acetate (5:1). It was found that the IL [bmim][PF_6] was easily recycled for five times with good conversions and high *cis* selectivities, although a gradual decrease in its activity was observed [Candeias et al. , 2009].

Recycle #	Yield [wt%] using IL $[EtPy]^+[BF_4]^-$	IL recovery [wt%]	Yield [wt%] using IL $[EtPy]^+[CF_3COO]^-$	IL recovery [wt%]
0	99	-	99	-
1	97	93	98	87
2	98	92	99	76
3	98	90	99	68

Table 13. Recyclability and Reuse of ILs used in the asymmetric reduction of acetophenone catalyzed by the complex of (R)-BINOL-LAH [Xiao, 2006]

In their reviw on the recovery of solutes extracted in ILs and regeneration of ILs, Han & Armstrong (2007) stated the following: Compounds with ionizable groups can be back-extracted into water at suitable pH values. Organic solvents that form two-phase systems with ILs can be used for the back extraction of neutral, thermally-unstable compounds [Blanchard et al., 1999; Blanchard & Brennecke, 2001]. Also the "green" solvent $scCO_2$ is a good choice for the recovery of solutes and the regeneration of ILs. $ScCO_2$ can dissolve in the ILs, while ILs cannot dissolve in $scCO_2$. This unique property means that solutes can be isolated from ILs without IL contamination. It also provides a means for recycling the ILs [Blanchard et al., 1999]. A methanol/[bmim][PF_6] mixture can be separated using pressurized CO_2 [Scurto et al., 2002]. Aromatic and aliphatic compounds were extracted into a CO_2 phase from [bmim][PF_6] [Blanchard et al., 1999; Blanchard & Brennecke, 2001].

Mekki et al. (2006) reported a two-step extraction system (water/RTIL/CO_2) for trivalent lanthanum and europium. The metal ions were extracted from the aqueous phase into $scCO_2$ via a RTIL/fluorinated β-diketonate mixture with high extraction efficiencies.

Docherty & Kulpa (2005) stated that the environmental benefits of ILs need to be carefully considered. Recent work demonstrated the toxicity of some ILs. More research is also needed on the long-term stability and recyclability of ILs. New ILs or novel applications of existing ILs can be developed that will further improve the capabilities of ILs for chemical separations. For example, the loss of ILs into aqueous media in metal ion extractions needs be suppressed.

New ILs and coating methods are being developed for high efficiency and high thermo-stability GC columns. Chiral GC stationary phases with high thermo-stability and broad enantiomeric selectivity are needed. One area that will continue to grow in importance is the use of ILs as absorbents in solid-phase extractions (SPE) and solid-phase micro-extractions (SPME). It is likely that ILs will fill the role of a polar absorbent for these techniques [Han & Armstrong, 2007].

Zhang et al. (2009) developed (3-aminopropyl) tributylphosphonium amino acid ILs for CO_2 capture, leading to 1 mol of CO_2 captured/1 mol of IL, and the ILs can be repeatedly recycled for CO_2 uptake. Nevertheless, the use of amine-functionalized ILs for CO_2 uptake shows several problems, with the main one being their high viscosities at ambient temperature, which are even larger upon complexation with CO_2 [Gutowski & Maginn, 2008]; this fact would hinder the CO_2 diffusion and uptake rate. Likewise, the synthesis of these ILs would require several synthetic steps [Bates et al., 2002], thus limiting their commercial viability [Karadas et al., 2010].

After the extraction of 5-hydroxy methyl furfural (HMF) with diethyl ether, the IL is recycled as such or after removal of the water formed during the course of the reaction (see Table 14). Such operation is rather important since three water molecules are produced during the course of the reaction. Under the operating conditions used, 5.55 mmol of fructose and 25.3 mmol of IL at 90 °C for 1 h, and although the amount of water formed after three cycles was twofold higher than that of the IL, the yield in HMF is not affected in a spectacular manner, but is more affected than after water removal where no significant loss in activity is observed within the experimental error [Moreau et al., 2006].

Recycle #	Yield of HMF [%] using as such IL	Yield of HMF [%] using after water removal IL
1	86	86
2	81	88
3	70	85
4	75	84
5	62	82

Table 14. Yield of 5-hydroxy methyl furfural (HMF) from fructose dehydration after recycling of IL 1-H-3-methyl imidazolium chloride

In order to have ILs that represent an ecological and further economic alternative to conventional solvent systems, their process advantages like solvent recovery are of particular importance. Ilgen (2009) investigated a particular process advantage of the IL $[C_6mim][BF_4]$ which was recycled and reused for 3 times. $[C_6mim][BF_4]$ was recycled via addition of cyclohexane or diethyl ether after reaction completion. The extract phase was afterwards separated. The extraction process was examined by gas chromatography, until no products or wastes were detected in the IL. Before each run, the IL was dried *in vacuo.* Mass losses and performance properties are listed in Table 15. After three recycling steps, a mass loss of 5% was determined, while the solvent performance seems to remain unchanged [Ilgen, 2009]. Very similar results were reported for the use of the IL [C_6act like][BF_4] [Reinhardt, 2009]. See Table 16.

Recycle #	Mass loss [g]	Mass deviation [%]	Conversion of methyl acrylate [%] as determined by GC
1	-	-	92
2	0.52	3.1	96
3	0.72	4.3	97
4	0.84	5.0	97

Table 15. Mass loss and reaction conversion with recycled IL [C_6mim][BF_4] used in Diels-Alder reaction of methyl acrylate and cyclopentadiene [Ilgen, 2009]

Kralisch et al. (2007) carried out the work-up by distillation, as well as the recycling of solvents, using a rotary evaporator fitted with a water-bath, by placing 500 g of the respective solvent in a 1000 mL round bottom flask. For ethanol, n-heptane, xylene and cyclohexane, a temperature setting of T = 65 °C was chosen. Water and diethyl ether were distilled at T = 80 °C and T = 40 °C, respectively. The pressure decay was adjusted to the boiling point of the solvent.

Recycle #	Mass loss [g]	Mass Deviation [%]	Conversion methyl acrylate [%] as determined by GC
1	-	-	92.4
2	0.52	3	95.8
3	0.72	4	97.3
4	0.84	5	97.8

Table 16. Mass loss and conversion with recycled IL [C_6act like][BF_4] used in Diels-Alder reaction of methyl acrylate and cyclopentadiene [Reinhardt, 2009]

After the acetylation of cellulose, effective recycling of the solvent ILs is absolutely required. For example, at the end of the acetylation cornhusk cellulose in the presence of the IL AmimCl, the cornhusk cellulose acetates (CCA) product was precipitated with a large excess of water. The polymer was filtered off, and the residual IL in the filtrate was recovered by a simple evaporation, giving clean AmimCl. The purity of the recovered AmimCl was above 99%, confirmed by 1H NMR spectroscopy [Cao et al., 2007].

However, for the future industrial applications on a large scale, other methods suggested for removal of water may prove to be more practical [Wong et al., 2006; Swatloski et al., 2003]. These alternative methods include nanofiltration, reverse osmosis, pervaporation, and salting out of the IL. Apparently, the advantage of easy recycling of ILs will promote their industrial application in this field [Cao et al. (2007]. Based on the fact that the IL AmimCl can be effectively recycled after each acetylation, Swatloski et al. (2003) provides a technically feasible and environmentally acceptable method to prepare acetone-soluble cellulose diacetates in one step using relatively cheap cornhusk as cellulose resource.

The separation, post-reaction, of ILs and catalysts from reaction products is a challenge in the application of ILs to organometallic catalysis [Wong et al., 2006]. Suzuki reaction is a key step in the synthesis of active pharmaceutical ingredients. Separation of Suzuki reaction' products from IL and homogeneous transition metal catalyst (TMC) via solvent extraction is the conventional choice [McNulty et al., 2002; Mathews et al., 2000; McLachlan et al., 2003] for obtaining IL (and catalyst in some cases) in a recyclable form (see Fig. 11.a). However, while solvent extraction is suitable for separation of apolar products, its use in separation of polar products is more problematic. For the latter case, a moderately polar extracting organic solvent is required, but since ILs tend to have significant partition coefficients into polar solvents, or are completely miscible with them, this can result in loss of IL to the product stream [Wong et al., 2006].

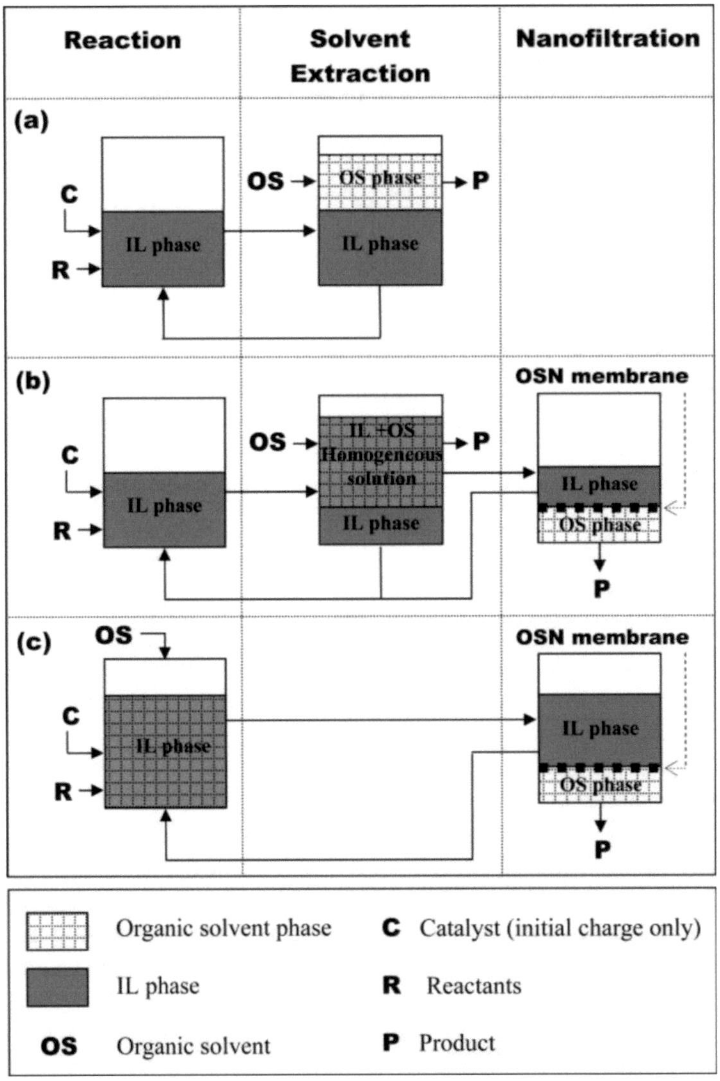

Fig. 11. Ionic liquid-transition metal catalyst recycle schemes for coupling reactions: (a) conventional product isolation via solvent extraction, (b) Organic Solvent Nanofiltration (OSN) used with a biphasic IL/organic system, and (c) Organic Solvent Nanofiltration (OSN) used with a single phase IL–organic solvent system [Wong et al., 2006].

Wong et al. (2006) stated that Organic Solvent Nanofiltration (OSN) can provide a separation solution which reduces these problematic losses. In this context, they envisaged two schemes for the use of OSN. In systems in which the IL and the moderately polar extracting solvent form a biphasic mixture, a significant amount of IL can partition into the organic solvent phase. The first scheme involves (see Fig. 11.b) separating the phases, and then applying OSN to the organic solvent phase to separate the product from the IL. The IL can be retained by the membrane, and then recycled to the reaction system. The second scheme (see Fig. 11.c) dispenses with the need for an intermediate solvent extraction. The reaction is carried out in a homogeneous solution of IL and organic solvent. The post reaction mixture is subjected to OSN, resulting in a permeate stream containing solvent and product, and a retentate stream containing solvent, TMC and IL.

Pure ILs can lower reaction rates due to unfavourable heat and mass transfer characteristics arising from their high viscosities. Using 50:50 wt% mixtures of organic solvent and miscible ILs avoided the problems associated with these high viscosities, while maintaining the beneficial effects of IL on the reaction [Wong et al., 2006]. Crucially, this means that IL can be recycled for use in subsequent reactions as a single phase mixture with organic solvent, as opposed to in the form of pure IL. In these cases, the scheme shown in Fig. 11.c is directly applicable.

Rasalkar et al. (2005) extracted the system of IL and l-proline with diethyl ether to remove all organic impurities and recharged it with a new batch of substrates. The IL acted as a reservoir for the catalyst. [MOEMIM]OMs containing l-proline was used for three runs consecutively. Though the yield of the product was comparable to that obtained in the first run, a decrease was observed in the *enantiomeric excess* of the *syn* isomer [%ee] in subsequent runs as shown in Table 17.

Recycle #	Isolated yield [%]	ee [%]
1	75	73
2	74	47
3	74	26

Table 17. Recyclability of IL and l-proline

In recent years, the use of ILs as a recyclable reaction medium, to replace volatile organic solvents has received considerable attention [Welton, 1999; Wasserscheid & Keim, 2000; Sheldon, 2001; Dupont et al., 2002]. Khan et al. (2003) recovered the ILs [bmim][PF_6] and [bmim][BF_4] used in chemoselective reduction of aromatic nitro and azo compounds using zinc and ammonium salts. After each cycle, the residue containing IL was dissolved in minimum amount of dichloromethane (CH_2Cl_2), or ethyl acetate (EtOAc), filtered through a filter paper, dried over Na_2SO_4 and the solvent was removed on a rotary evaporator. The IL was further dried at 70°C and 0.1 mmHg for 4 h. This reaction was repeated for 3 more cycles using the IL recovered from the previous run under identical conditions as for cycle 1. It was found that no decrease in yield or purity of the products when the ionic liquids were both reused. The results are listed in Table 18. The reduction of nitroarenes was performed with Zn (4 equiv.)/NH_4Cl (3 equiv.) under an inert atmosphere (in 10:1 [bmim][PF_6]/water for Method A) and (in 10:1 [bmim][BF_4]/water for Method B). The IL [bmim][BF_4] was also recycled effectively and reused for the reduction of azobenzene as shown in Table 19.

There are some disadvantages associated with the use of chloroaluminate ILs; they are moisture sensitive and cannot be recycled after the reaction. Potdar et al. (2005) adopted two

Recycle #	Yield [%], using [bmim][PF_6] (Method A)	Yield [%], using [bmim][BF_4] (Method B)
1	93	87
2	95	93
3	92	86
4	87	--

Table 18. Recyclability of ILs for the reduction of nitrobenzene to aniline

Recycle #	Yield [%] using [bmim][BF_4]
1	98
2	94
3	100
4	96

Table 19. Recyclability of the IL [bmim][BF_4] for the reduction of azobenzene

different approaches to the Pechmann condensation employing neutral ILs to establish cleaner synthetic methodologies. The IL [bmim]PF_6 that settled down, after quenching the reaction mixture with 10% aqueous NaOH solution, was extracted with dichloromethane. The dichloromethane layer was stirred vigorously with 10% aqueous NaOH in order to remove coumarin and acid traces. Further, dichloromethane layer was washed with water. The solvent was then evaporated and the resultant IL was subjected to consecutive extractions with diethyl ether to remove organic impurities present. The resultant IL was then dried under vacuum at 60 °C and reused for subsequent reactions. When [bmim]BF_4 IL was used for the reaction, the reactions were quenched by addition of water. The coumarin derivative separated as a solid was isolated by filtration. The IL was recovered by evaporation of water on a rotary evaporator. The IL was further purified *as per the procedure used* for the purification of IL [bmim]PF_6. Potdar et al. (2005) found that IL [bmim]PF_6 can be recycled efficiently and it was used consecutively for 3 runs without any considerable loss in its activity as shown in Table 20.

Recycle #	Yield [%] using [bmim]PF_6
1	95
2	92
3	94

Table 20. Recyclability of IL in Pechmann condensation performed at 100 °C

Because of their negligible vapour pressure, thermal stability and easy recyclability, neutral ILs have been referred to as environmentally benign solvents. These ILs have been employed as excellent and recyclable medium for a wide array of reactions; e.g., Heck reaction [Park & Alper, 2003], Bischler-Napierlaski cyclisation [Judeh et al., 2002], Beckmann rearrangement [Ren et al., 2001] addition of thiols to unsaturated ketones [Yadav et al., 2003] and l-proline catalysed aldol reaction [Loh et al., 2002].

5. Hydrophobic ionic liquid recycling

Among the published literature, it is generally recognized that ILs are easily recyclable. Undoubtedly, this is true for some specific biphasic systems that contain ILs, in particular for hydrophobic ILs, such as $[PF_6]^-$ and $[(CF_3SO_2)_2N]^-$ ILs [Wu et al. 2009]. In this case, liquid–liquid extraction is an efficient method for separation and has been used for recycling both the catalyst and the IL solvent in the course of palladium coupling reactions [Handy & Zhang, 2001; Mathews et al., 2000; Fukuyama et al., 2002]. The recyclability of these ILs is largely a result of their lack of solubility in some key organic solvents, such as diethyl ether. This lack of solubility allows products and residual organics to be extracted by using an organic solvent, while byproducts present in water-immiscible ILs can be washed out by using water with minimal loss of ILs [Wu et al. 2009].

In order to provide significant advantages over conventional technologies for extraction processes, Birdwell et al. (2006) developed a centrifugal solvent-extraction contactor recycle method successfully used for separation of dispersions containing immiscible organic IL/hydrocarbon/and aqueous systems. The systems tested are NaCl(aq)/$[C_4mim]NTf_2$, $H_2O/[C_4mim]BETI$ and cyclohexane/$[C_4mim]BETI$ (which was the most efficient, but still with slight loss of ILs).

6. Hydrophilic ionic-liquid recycling

Generally, *hydrophobic ILs* can be extracted with water to separate water-soluble solutes from the IL into the aqueous phase; however, this method is not suitable for *hydrophilic ILs*. Recovery of the hydrophilic ILs is more difficult in comparison to hydrophobic ILs, and the study in this field is in its infancy. To avoid cross-contamination of IL with water or vice versa, novel ways must be exploited. $ScCO_2$ was found to have the potential for the regeneration of hydrophilic ILs [Wu et al. 2009].

7. Membrane technology for ionic-liquid recovery

Membrane techniques have been applied to perform fine separation of undesirable constituents and these show promising results for the selective removal of volatile solutes from ILs [WO 2003/039719]. Haerens et al. (2010) investigated the use of pressure-driven membrane processes, nano-filtration, reverse osmosis and pervaporation, as a possibility to recycle ILs from water. They used Ethaline200 {a deep eutectic formed between choline chloride (a quaternary amine salt) and ethylene glycol} to perform these tests and compared their results with those found in the literature.

For concentrating ILs, the osmotic pressure was found to be the limiting factor when using nano-filtration or reverse osmosis. [Haerens et al., 2010] investigated pervaporation as an alternative and found it to have limited usability for this application as the water content was too high. For low water contents, pervaporation can be applied although the flux is very low due to the presence of the IL, which decreases the activity of the water and thus the flux through the membrane. The necessary membrane area would be very high and makes pervaporation rather impractical [Haerens et al., 2010].

The key among all performance advantages is that for the removal of salt and other materials that are in the ionic size range, only distillation processes can provide a similar level of fluid purity. Hence, this technology is vitally important in the case of recycling of ILs from nonvolatile compound/IL systems that are not suitable for distillation. Therefore, it is necessary to explore novel membrane treatment process for IL recovery. As it is known, ILs exhibit distinctive fluid properties with high viscosity typically 2–3 orders of magnitude higher than water, rendering their separation, recovery and reuse from some reaction mixtures a difficulty [Wu et al. 2009].

ILs have been widely employed in extractions, as GC stationary phases, and as supported liquid membranes. In some cases, they can be considered "green" solvents, but they have many other benefits including unusual selectivities, high extraction efficiencies, dual-nature GC properties, durability, and resistance to thermal degradations [Han & Armstrong, 2007].

Since the phase behavior of ILs with gases will affect the attractiveness of using ILs in reactions and separations, understanding this phase behavior is of utmost importance. Anthony et al. (2002) presented the solubility of several permanent and reactive gases in various ILs and the separation of CO_2 from methane and nitrogen using ILs in a supported-liquid membrane system. They also showed that ILs can be recovered from organic or aqueous solutions using CO_2 as an anti-solvent. Solutions of ILs with organic and water can be induced to form three phases in the presence of CO_2, leading to the recovery of the ILs.

From their discussion, Wu et al. (2009) found that hydrophobic and hydrophilic ionic-liquid recycling techniques have their advantages and disadvantages. Their common drawback is that they are not amenable to industrial implementation. Membrane separation technology may be another alternative, since it is a mature technology and commercially available. Membrane separation has been widely applied in the field of water and wastewater treatment, food and beverage processing, and pharmaceutical and medical processes due to a number of performance and cost advantages over competitive technologies. Key among all performance advantages is that for the removal of salt and other materials that are in the ionic size range, only distillation processes can provide a similar level of fluid purity. Hence, this technology is vitally important in case of recycling ILs from non-volatile compound/IL systems that are not suitable for distillation. Therefore, it is necessary to explore novel membrane treatment process for IL recovery. As it is known, ILs exhibit distinctive fluid properties with high viscosity typically 2–3 orders of magnitude higher than water, rendering their separation, recovery and reuse from some reaction mixtures a difficulty [Wu et al. (2009)].

The dissolution of the hydrophilic ILs in the aqueous phase could represent a wastewater treatment challenge. Based on the inherent amphiphilic character of ILs, many measurements have been carried out to investigate the aggregation behaviour of ILs in water, which is very necessary for developing IL-regeneration methods, typically for membrane treatment technique [Wu et al. 2009].

Based on the different features of IL/H_2O mixtures with different concentrations, one can choose the most appropriate way for different IL recovery stages, leading to a combination of methods. For instance, in water- and ion-rich regions, in which the hydrated ions have a small radius but large ionization degree, the first choice may be electrodialysis technology, because this technology has been well developed, and it is energy-saving when the solution concentration is low. This also makes use of the property of ILs, that is, they composed entirely of ions. After the solution is concentrated in the salt-region, in which the target compound has a larger radius, then membrane separation can be employed; one can even use vacuum distillation at this stage. There are several different possible combinations and how to put every unit organically together to form a continuous process is open for debate. Certainly, the choice of any combination must be made on the basis of economical comparison and analysis [Wu et al. 2009] and more work is needed.

Gordon (2001) attempted to recycle the catalyst/IL system, but after the third cycle a decrease in yield and increase in reaction time was observed, suggesting that catalyst decomposition or leaching was occurring. Xiao & Malhotra (2005) recycled catalyst/IL systems to investigate the reusability and efficiency of ILs, with or without catalyst. Two ILs were examined, namely, $[EtPy]^{+}[BF_4]^{-}$ and $[EtPy]^{+}[CF_3COO]^{-}$. The recycling process involved washing the used ILs with diethyl ether to remove any leftover organic residues. Two layers formed (IL and organic). The resulted ILs were separated and the IL dried under reduced pressure at 65°C. Successive runs were used for acylation of benzene with acetic anhydride at 50 °C for 4 h. It is clear from the results are shown in Table 21 that both ILs could be recovered quantitatively with negligible loss of activity. Moreover, the acylation was not affected even after third run with the recovered IL.

Recycle #	Isolated yield of Acetophenone [%] using $[EtPy]^{+}[BF_4]^{-}$	IL Recovery [wt%]	Isolated yield of Acetophenone [%] using $[EtPy]^{+}[CF_3COO]^{-}$	IL Recovery [wt%]
0	70	–	76	–
1	68	93	75	96
2	67	93	72	94
3	65	94	73	95

Table 21. Recycling of ILs in the acylation of benzene with acetic anhydride [Xiao & Malhotra, 2005]

Xiao & Malhotra (2005) also carried out a similar study with the recovered $FeCl_3$-IL system, for the acylation of benzene with acetic anhydride. The results in Table 22 show that even though $FeCl_3$-IL system could be recovered efficiently, the conversion to acylation product dropped dramatically and the conversions obtained after the third trial were nearly the same as seen with pure IL alone (Table 21). This could be due to a loss in the catalytic activity of the $FeCl_3$-IL system [Xiao & Malhotra, 2005].

Recycle #	Isolated yield of Acetophenone [%] using $FeCl_3$ - $[EtPy]^{+}[BF_4]^{-}$	IL Recovery [wt%]	Isolated yield of Acetophenone [%] using $FeCl_3$ - $[EtPy]^{+}[CF_3COO]^{-}$	IL Recovery [wt%]
0	85	–	94	–
1	80	90	90	88
2	75	87	83	91
3	73	89	81	88

Table 22. Recycling of ILs-$FeCl_3$ in the acylation of benzene and acetic anhydride

Based on their good thermal stability, no vapor tension and no azeotrope formed, ILs may be separated from the products. Xie & Shi (2010) attributed the suitability of ILs for wood liquefaction to the functional groups of the ILs. After wood liquefaction in [BAIM][Cl/$AlCl_3$] or [BAIM][Cl] ILs, the product solution was extracted three times with ethyl ether at room temperature, and then dissolved in 1:1.6–1.8 volume ratio of acetonitrile: ethyl acetate and frozen at -20 °C for 24 h. Coarse ILs were dried at 90 °C under vacuum for 8 h. The results of reuse of the recycled ILs in wood liquefaction are shown in Table 23. It was found that both [BAIM][Cl] and [BAIM][Cl/$AlCl_3$] can be recycled (no less than five times) to maintain the original level of wood liquefaction rate with hardly any change in the activity of liquefaction [Xie & Shi, 2010].

Recycle #	Wood residue [%] using [BAIM][Cl] at 90 °C	Wood residue [%] using [BAIM][Cl/$AlCl_3$] at 70 °C
1	97.3	14.4
2	97.4	14.7
3	97.7	14.8
4	97.9	14.7
5	98.2	14.9

Table 23. Results of wood liquefaction using recycled ILs at a liquid/wood ratio = 6.5 and after 30 min

Li et al. (2010) reported on the dissolution of wood species in the ionic liquid [AMIM]Cl. The filtrate (after separating the regenerated wood) was condensed by rotary evaporation, dried under vacuum at 40 °C overnight to recycle the IL. It was found that as the reuse cycles of the IL increased, the wood regeneration yield increased, while certain wood components enriched within the recycled IL and the efficiency of cellulase enzymatic hydrolysis on the regenerated wood decreased.

8. Conclusions

Several procedures for the recycling of ILs have been reported in the literature. Depending on the IL used and the application performed on it, a variety of recycling methods were possible. By picking the right purification steps, an individually optimized work-up procedure can be obtained [Wasserscheid & Welton, 2008]. In this survey several procedures have been reported on how to recover ionic liquids from their solution. Among these methods are distillation/stripping at some suitable temperature (< 300 °C) and under vacuum, liquid-liquid extraction (using objectionable VOCs!), supercritical fluids (using CO_2 at room temperature and several MPa), and membrane separation (to separate nano size particles from ILs).

It was clear that ionic liquids will not provide advantages in all systems, but improvements in reactivity or selectivity are observed in many cases when the appropriate combination of cation and anion are selected [Gordon, 2001]. Despite this, the two factors that will ultimately decide whether reaction systems, when ILs are used, are viable on a larger scale are likely to be the ability to reuse the catalyst without a decrease in its activity, and whether the products can be separated efficiently without contamination from the ionic liquid or the catalyst. The combination of $scCO_2$ and ionic liquids seems to be a very promising approach to attain this goal. The simultaneous efficient nano-filtration recycling of ILs and homogeneous catalysts extends the possibilities of the practical application of these media in organic synthesis [Volkov et al. 2008]. Destruction of ILs and removal of extracted organic contaminants by photolytic degradation is possibly another method of recycling ILs [Yang & Dionysiou, 2004; Khodadoust et al., 2006].

Due to the current market size and the relatively high cost of ILs, the industrial production of ionic liquids is still small or limited to lab-scale (and sometimes pilot scale) applications and hence no industrial technology is yet available for ionic liquids recycling with the objective of reuse. Most of the published works on ionic liquids recycling are made at the bench scale where the investigators are trying to extract a solute from a mixture using some suitable ionic liquid then recover the ionic liquid for further reuse. The number of trial cycles in most cases was very few (3 to 5). Some other researchers were trying to recover a catalyst from a reaction mixture where some ionic liquid was used. In this case, both the catalyst and the ionic liquid have to be recovered and recycled to the process for further cycles of the reaction. Again, the number of cycles was, in most cases, between 3 and 5.

Someone might say: If the customer does not feel comfortable with the task of recycling the IL, then why not rent or lease the IL rather than buy it? The customers, in this case, perform their application with the IL and send, the probably impure, IL back to the supplier, who has the expertise to recycle and clean it up [Wasserscheid & Welton, 2008]. This scenario could be interesting from an economic point of view for truly commercial applications on a large scale. But transferring of such possibly hazardous and contaminated materials from one place, or country, to another is not safe. Thus I think the recycling methods must be available and practicable for both parties; the customer and the supplier.

In summary, many authors and researchers agree that we are only at the very beginning of understanding the recyclability of ILs based on the available literature in the various application fields. Understanding of ILs volatility, purity, stability, biodegradability and toxicity is necessary for their recovery, since this determines whether an IL can be sustainably developed. In other words, there is a long way to go before large-scale implementation of ILs. Hopefully, this review could provide some clues to support a great deal of future research on ILs recycling and reuse.

9. Acknowledgement

The author likes to thank the many researchers who responded to his request about their experience with ionic liquids recycling and reuse. Very deep thanks should go to Dr. G. W. Meindersma (Eindhoven University of Technology, The Netherland) for the invaluable information he supplied.

10. References

Adams, C.J., Earle, M.J., Roberts, G. & Seddon, K.R. (1998). Friedel-Crafts Reactions in Room Temperature Ionic Liquids. *Chem. Commin.*, pp. 2097-2098

Anonymous (2004). Accelerating Ionic Liquid Commercialization– Research Needs to Advance New Technology, Prepared by BCS Incorporated, Approved and Issued by the Chemical Industry Vision2020 Technology Partnership. June 2004, Available from http://www.chemicalvision2020.org/ionic_liquids.html

Anthony, J.L., Scurto, A.M., Crosthwaite, J.M., Aki, S.N.V.K., Maginn, E.J. & Brennecke, J.F. (2002). Processes Using Ionic Liquids and Permanent Gases, In: Ionic Liquids as Green Solvents: Progress and Prospects, *224th ACS National Meeting*, Boston, MA, August 18-22, 2002

Attri, P., Reddy, P.M. & Venkatesu, P. (2010). Density and Ultrasonic Sound Speed Measurements for N,N-Dimethylformamide with Ionic Liquids, *IJCA*, 49A (5-6), pp. 736-742

Attri, P., Reddy, P.M., Venkatesu, P., Kumar, A. & Hofman, T. (2010). Measurements and Molecular Interactions for N,N-Dimethylformamide with Ionic Liquid Mixed Solvents, *J. Phys. Chem. B*, 114, pp. 6126–6133

BASF (2011), Frequently Asked Questions: How can Ionic Liquids be recycled? Available from http://www.basionics.com/en/ionic-liquids/faq.htm#2a

Bates, E.D., Mayton, R.D., Ntai, I. & Davis, J. H. (2002). CO_2 capture by a task-specific Ionic Liquid. *J. Am. Chem. Soc.,* 124, pp. 926–927

Birdwell, J.F., McFarlane, J., Hunt, R.D., Luo, H., DePaoli, D.W., Schuh, D.L. & Dai, S. (2006). Separation of Ionic Liquid Dispersions in Centrifugal Solvent Extraction Contactors, *Sep. Sci. Technol.*, 41(10), pp. 2205–2223.

Blanchard, L.A., Hancu, D., Beckman, E.J. & Brennecke, J.F. (1999). Green Processing Using Ionic Liquids and CO_2. *Nature,* 399, pp. 28–29

Blanchard, L.A. & Brennecke, J.F. (2001). Recovery of Organic Products from Ionic Liquids Using Supercritical Carbon Dioxide. *Ind. Eng. Chem. Res.,* 40, pp. 287–292

Bortolini, O., Bottai, M., Chiappe, C., Conte, V. & Pieraccini, D. (2002). Trihalide-based Ionic Liquids. Reagent-Solvents for Stereoselective Iodination of Alkenes and Alkynes, *Green Chemistry,* 4, pp. 621–627

Bornscheuer, U.T. & Kazlauskas, R.J. (Eds.) (2006). *Hydrolases in Organic Synthesis: Regio- and Stereoselective Biotransformations,* 355 pages, Wiley-VCH Verlag GmbH & Co. KGaA, Weinheim, ISBN-10: 3-527-31029-0

Bowers, J., Butts, C.P., Martin, P.J., Vergara-Gutirrrez, M.C. & Heenan, R.K. (2004). Aggregation Behaviour of Aqueous Solutions of Ionic Liquids. *Langmuir,* 20, pp. 2191–2198

Candeias, N.R., Branco, L.C., Gois, P.M.P., Afonso, C.A.M. & Trindade, A.F. (2009). More Sustainable Approaches for the Synthesis of N-Based Heterocycles, *Chem. Rev.*, 109, pp. 2703–2802

Cao, Y., Wu, J., Meng, T., Zhang, J., He, J., Li, H. & Zhang, Y. (2007). Acetone-Soluble Cellulose Acetates Prepared by One-Step Homogeneous Acetylation of Cornhusk Cellulose in an Ionic Liquid 1-Allyl-3-Methylimidazolium Chloride (AmimCl), *Carbohydrate Polymers,* 69, pp. 665–672

Carmichael. A.J., Earle. M.J., Holbrey, J.D., McCormac, P.B. & Seddon, K.R. (1999) The Heck Reaction in Ionic Liquids: A Multiphasic Catalyst System. *Org Lett.,* 7, pp. 997-1000

Chauvin, Y., Gilbert, B. & Guilbard, I. (1990). Catalytic Dimerization of Alkenes by Nickel Complexes in Organoaluminate Molten Salts. *J Chem. Soc. Chem. Commun.*: 1715-1716

Chrobok, A. (2010). Baeyer–Villiger Oxidation of Ketones in Ionic Liquids Using Molecular Oxygen in the Presence of Benzaldehyde, *Tetrahedron,* 66, pp. 2940–2943Chrobok, A. (2010). The Baeyere-Villiger Oxidation of Ketones with Oxone in the Presence of Ionic Liquids as Solvents, *Tetrahedron,* 66, pp. 6212–6216

Dal, E. & Lancaster, N.L. (2005). Acetyl Nitrate Nitrations in [bmpy][$N(Tf)_2$] and [bmpy][OTf], and the Recycling of Ionic Liquids, *Org. Biomol. Chem., 3,* pp. 682–686

Docherty, K.M. & Kulpa, C.F., Jr. (2005).Toxicity and Antimicrobial Activity of Imidazolium and Pyridinium Ionic Liquids. *Green Chem.,* 7, pp. 185–189

Doherty, S., Goodrich, P., Hardacre, C., Luo, H. K., Rooney, D.W., Seddon, K.R. & Styring, P. (2004). Marked Enantioselectivity Enhancements for Diels–Alder reactions in Ionic Liquids Catalysed by Platinum Diphosphine Complexes, *Green Chem.*, 6, pp. 63-67

Dorbritz, S., Ruth, W. & Kragl, U. (2005). Investigation on Aggregate Formation of Ionic Liquids. *Adv. Synth. Catal.,* 347, pp. 1273–1279

Dupont, J., de Souza, R.F. & Suarez, P.A.Z. (2002). Ionic Liquid (Molten Salt) Phase Organometallic Catalysis. *Chem. Rev.,* 102, pp. 3667-3691

Dupont, J. (2004). On the Solid, Liquid and Solution Structural Organization of Imidazolium Ionic Liquids. Braz, J. *Chem. Soc.*, 15(3), pp. 341-350

Earle, M.J., Esperanca, J.M.S.S., Gilea, M.A., Canongia Lopes, J.N., Rebelo, L.P.N., Magee, J.W., Seddon, K.R. & Widegren, J.A. (2006). The Distillation and Volatility of Ionic Liquids, *Natur, 439*, pp. 831-834

Faizee, N., Rao, K., Dadi, A.P., Varanasi, S. & Schall, C.A. (2007). Ion Exchange Process For The Recovery Of Ionic Liquids, *The 2007 Annual Meeting*, Salt Lake City, UT, November 7, 2007

Fioroni, G., Fringuelli, F., Pizzo, F. & Vaccaro, L. (2003). Epoxidation of α,β-Unsaturated Ketones in Water. An Environmentally Benign Protocol, *Green Chemistry*, 5, pp. 425–428

Fukuyama, T., Shinmen, M., Nishitani, S., Sato, M. & Ryu, I. (2002). A Copper-Free Sonogashira Coupling Reaction in Ionic Liquids and Its Application to a Microflow System for Efficient Catayst Recycling, *Org. lett.*, 4, pp. 1691–1694

Galán Sánchez, L.M. (2008). Functionalized Ionic Liquids - Absorption Solvents for Carbon Dioxide and Olefin Separation, PhD Thesis, University of Twente, The Netherlands

Gonzales, E.J., Alonso, L. and Dominguez, A. (2006). Physical Properties of Binary Mixtures of the Ionic Liquid 1-Methyl-3-octylimidazolium Chloride with Methanol, Ethanol, and 1-Propanol at T = (298.15, 313.15, and 328.15) K and at P = 0.1 MPa, *J. Chem. Eng. Data,* 51, pp. 1446-1452

Goodchild, I., Collier, L.L., Millar, S.J., Prokes, I., Lord, J.C.D., Butts, C.P., Bowers, J. & Webster, J.R.P. (2007). Structural Studies of the Phase, Aggregation and Surface Behaviour of 1-Alkyl-3-Methylimidazolium Halide plus Water Mixtures, J. *Colloid Interface Surf.*, 307(2), pp. 455–468

Gordon, C.M. & McCluskey, A. (1999). Ionic Liquids: A Convenient Solvent for Environmentally Friendly Allylation Reactions with Tetraallylstannane, *Chem. Commun.*, pp. 1431–1432

Gordon, C.M. (2001). New Developments in Catalysis Using Ionic Liquids, *Appl. Catal., A General,* 222, pp. 101–117

Gordon, C.M. & Muldoon, M.J. (2008). Synthesis of Ionic Liquids, In: Ionic Liquids in Synthesis, P. Wasserscheid & T. Welton (Eds.), Vol. 1, Ch. 2, p. 43, Wiley-VCH Verlag GmbH & Co. KGaA, Weinheim, ISBN-13 978-1-4020-4087-0 (e-book)

Gutowski, K. E. & Maginn, E.J. (2008). Amine-Functionalized Task-Specific Ionic Liquids: A Mechanistic Explanation for the Dramatic Increase in Viscosity upon Complexation with CO_2 from Molecular Simulation, *J. Am. Chem. Soc.,* 130, pp. 14690–14704

Haerens, K., Van Deuren, S., Matthijs, E. & Van der Bruggen, B. (2010). Challenges for Recycling Ionic Liquids by Using Pressure-Driven Membrane Processes, *Green Chem.*, 12, pp. 2182-2188

Han, X. & Armstrong, D.W. (2007). Ionic Liquids in Separations, *Acc. Chem. Res.*, 40 (11), pp. 1079–1086

Handy, S.T. & Zhang, X.L. (2001). Organic Synthesis in Ionic Liquids: The Stille Coupling, *Org. Lett.*, 3, pp. 233–236

Hardacre, C., Holbrey, J.D., Katdare, S.P. & Seddon, K.R. (2002). Alternating Copolymerisation of Styrene and Carbon Monoxide in Ionic Liquids, *Green Chem.*, 4, pp. 143-146

Hassan, M.A., Batterjee, S. & Taib, L.A. (2006). Novel Synthesis of 1*H*-Inden-1-Ones and Thienylpropenones in Aqueous, Medium, *J. Chinese Chem. Soc.*, 53, pp. 939–944

Heintz, A. (2005). Recent Developments in Thermodynamics and Thermophysics of Non-Aqueous Mixtures Containing Ionic Liquids - a Review, *J. Chem. Thermodyn.*, 37, pp. 525–535

Holbrey, J.D. & Seddon, K.R. (1999). Ionic Liquids, *Clean Prod. Proc.*, 1, pp. 223–236

Holbrey, J.D. (2004). Industrial Applications of Ionic Liquids, *Chimica Oggi.*, 22 (6), pp. 35–37

Ilgen, F. (2009). Low Melting Carbohydrate Mixtures as Solvents for Chemical Reactions and the Conversion of Carbohydrates, PhD Thesis, Institute of Organic Chemistry, University of Regensburg, Germany

Itoh, T., Ouchi, N., Hayase, S. & Nishimura, Y. (2003). Lipase-Catalyzed Enantioselective Acylation in a Halogen Free Ionic Liquid Solvent System, *Chem. Lett.*, 32, pp. 654–655

Judeh, Z.M.A., Ching, C.B., Bu, J. & McCluskey, A. (2002). The First Bischler–Napieralski Cyclization in a Room Temperature Ionic Liquid, *Tetrahedron Lett.*, 43, pp. 5089-5091

Kanel, J.S. & Associates, LLC (2003). Overview: Industrial Application of Ionic Liquids for Liquid Extraction, *Chemical Industry Vision2020 Technology Partnership Workshop*, New York, NY, Sept 11, 2003

Karadas, F., Atilhan, M. & Aparicio, S. (2010). Review on the Use of Ionic Liquids as Alternative Fluids for CO_2 Capture and Natural Gas Sweetening, *Energy & Fuels, 24* (11), pp. 5817–5828

Kärkkäinen, J. (2007). Preparation and Characterization of Some Ionic Liquids and Their Use in the Dimerization Reaction of 2-Methylpropene, PhD Dissertation, Acta Univ. Oul. A 480, Oulu, Finland

Khan, F.A., Dash, J., Sudheer, C. & Gupta, R.K. (2003). Chemoselective Reduction of Aromatic Nitro- and Azo-Compounds in Ionic Liquids Using Zinc and Ammonium Salts, *Tetrahedron Lett.*, 44, pp. 7783–7787

Khodadoust, A.P., Yachandrasekaran, S. & Dionysiou, D.D. (2006). Preliminary Assessment of Imidazolium-Based Room-Temperature Ionic Liquids for Extraction of Organic Contaminants from Soils, *Environ. Sci. Technol.*, 40, pp. 2339–2345

Kim, J. & Shreeve, J.M. (2004). The First Cu(I)-Mediated Nucleophilic Trifluoromethylation Reactions Using (Trifluoromethyl) Trimethylsilane in Ionic Liquids, *Org. Biomol. Chem.*, 2, pp. 2728–2734

Kralisch, D., Reinhardt, D. & Kreisel, G. (2007). Implementing Objectives of Sustainability into Ionic Liquids Research & Development, *Green Chem.*, 9, pp. 1308-1318

Laali, K.K. & Gettwert, V.J.J. (2001). Fluorodediazoniation in Ionic Liquid Solvents: New Life for the Balz-Schiemann Reaction, *J. Fluorine Chem.*, 107(1), pp. 31-34

Lee, S.-G., Park, J.H., Kang, J. & Lee, J.K. (2001). Lanthanide Triflate-Catalyzed Three Component Synthesis of α-Amino Phosphonates in Ionic Liquids. Reactivity and Reusability Study, *Chem. Commun.*, pp. 1698-1699

Li, B., Asikkala, J., Filpponen, I. & Argyropoulos, D.S. (2010). Factors Affecting Wood Dissolution and Regeneration of Ionic Liquids, *Ind. Eng. Chem. Res.*, 49 (5), pp. 2477-2484

Liu, W.W., Zhang, Y.M., Wang, H.P. & Yu, M.F. (2008). The Physical Properties of Aqueous Solution of Room-Temperature Ionic Liquids Based on Imidazolium: Database and Evaluation, *J. Mol. Liq.*, 140, pp. 68–72

Loh, T.-P., Feng, Li.-C., Yang, H.-Y. & Yang, J.-Y. (2002). L-Proline in an Ionic Liquid as an Efficient and Reusable Catalyst for Direct Asymmetric Aldol Reactions, *Tetrahedron Lett.*, 43, pp. 8741-8743

Maase, M. & Emanuel, C.J. (2006). Not Always as Black as Coca Cola - Ionic Liquids are Bright Solutions to Many Kinds of Industrial Problems, 231st ACS National Meeting, Atlanta, GA, March 26-30, 2006

Maase, M. (2008). Industrial Applications of Ionic Liquids. In: Ionic Liquids in Synthesis, P. Wasserscheid & T. Welton (Eds.), Vol. 1, p. 669-670, Wiley-VCH Verlag GmbH & Co. KGaA, Weinheim, ISBN-13 978-1-4020-4087-0 (e-book)

Maase, M. (2008). Industrial Applications of Ionic Liquids. In: Ionic Liquids in Synthesis, P. Wasserscheid & T. Welton (Eds.), Vol. 1, p. 685-686, Wiley-VCH Verlag GmbH & Co. KGaA, Weinheim, ISBN-13 978-1-4020-4087-0 (e-book)

Magna, L., Harry, S., Proriol, D., Saussine, L. & Olivier-Bourbigou, H. (2007). Hydroformylation of 1-Hexene with a Cobalt Catalyst in Ionic Liquids: a New Efficient Approach for Generation and Recycling of the Catalyst, *Oil & Gas Science and Technology - Rev. IFP*, 62(6), pp. 775-780

Marsh, K.N., Boxall, J.A. & Lichtenthaler, R. (2004). Room Temperature Ionic Liquids and Their Mixtures, a Review, *Fluid Phase Equilib.*, pp. 219, 93–98

Mathews, C.J., Smith, P.J. & Welton, T. (2000). Palladium Catalysed Suzuki Cross-Coupling Reactions in Ambient Temperature Ionic Liquids, *Chem. Commun.*, 14, pp. 1249-1250

McNulty, J., Capretta, A., Wilson, J., Dyck, J., Adjabeng, G. & Robertson, A. (2002). Suzuki Cross-Coupling Reactions of Aryl Halides in Phosphonium Salt Ionic Liquid under Mild Conditions, *Chem. Commun.*, 17, pp. 1986-1987

McLachlan, F., Mathews, C.J., Smith, P.J. & Welton, T. (2003). Palladium-Catalyzed Suzuki Cross-Coupling Reactions in Ambient Temperature Ionic Liquids: Evidence for the Importance of Palladium Imidazolylidene Complexes, *Organometallics*, 22, pp. 5350-5357

Meindersma, G.W., Podt, J.G., Gutiérrez Meseguer, M. & de Haan, A.B. (2005). Ionic Liquids as Alternative to Organic Solvents in Liquid-Liquid Extraction of Aromatics, *ACS Symposium Series* 902, In: Ionic Liquids IIIB: Fundamentals, Progress, Challenges and Opportunities, Chap. 5, p. 57–71, R.D. Rogers & K.R. Seddon (Eds.), ACS, Washington, D.C. ISBN-13: 9780841238947

Meindersma, G.W. (2011). Eindhoven University of Technology, Netherlands (Private communication)

Mekki, S., Wai, C. M., Billard, I., Moutiers, G., Burt, J., Yoon, B., Wang, J. S., Gaillard, C., Ouadi, A. & Hesemann, P. (2006). Extraction of Lanthanides from Aqueous Solution by Using Room-Temperature Ionic Liquid and Supercritical Carbon Dioxide in Conjunction, *Chem. – Eur. J.*, 12, pp. 1760–1766

Moreau, C., Finiels A. & Vanoye, L. (2006). Dehydration of Fructose and Sucrose into 5-Hydroxymethylfurfural in the Presence of 1-H-3-Methyl Imidazolium Chloride Acting both as Solvent and Catalyst, *J. Mol. Catal. A: Chemical*, 253, pp. 165-169

Murugesan, S. & Linhardt, R.J. (2005). Ionic Liquids in Carbohydrate Chemistry - Current Trends and Future Directions, *Current Organic Synthesis*, 2, pp. 437-451

Murray, C.B., Sandford G. & Korn, S.R. (2003). Ionic Liquids as Media for Nucleophilic Fluorination, *J. Fluorine Chem.*, 123, pp. 81–84

Olivier-Bourbigou, H. (1999). Recent Developments in the Use of Non-Aqueous Ionic Liquids for Two-Phase Catalysis, *J. Mol. Catal. A: Chem.*, 146, pp. 285–289

Olivier-Bourbigou, H. & Magna, L.J. (2002). Ionic Liquids: Perspectives for Organic and Catalytic Reactions, *Mol. Catal. A: Chem.*, 182-183, pp. 419-437

Park, S.B. & Alper, H. (2003). Highly Efficient, Recyclable Pd(II) Catalysts with Bisimidazole Ligands for the Heck Reaction in Ionic Liquids, *Org. Lett.*, 5, pp. 3209-3212

Potdar M.K., Rasalkar M.S., Mohile S.S., Salunkhe M.M. (2005). Convenient and Efficient Protocols for Coumarin Synthesis via Pechmann Condensation in Neutral Ionic Liquids, *J Mol. Catal. A: Chemical*, 235, pp. 249–252

Rakita, P.E. (2003). Challenges to the commercial production of Ionic Liquids, *ACS Symposium Series 856*, In: Ionic Liquids as Green Solvents - Progress and Prospects, Chap. 3, p. 32-40, R.D. Rogers & K.R. Seddon (Eds.), ACS, Washington, D.C. ISBN-13: 9780841238565

Ramnial, T., Ino, D.D. & Clyburne, J.A.C. (2005) Phosphonium Ionic Liquids as Reaction Media for Strong Bases, *Chem. Commun.*, pp. 325-327

Rasalkar, M.S., Potdar, M.K., Mohile, S.S. & Salunkhe, M.M. (2005). An Ionic Liquid Influenced L-Proline Catalysed Asymmetric Michael Addition of Ketones to Nitrostyrene, *J. Mol. Catal. A: Chemical*, 235, pp. 267–270

Rayner, C. (N.D.) Green Solvents, Fundamentals and Industrial Applications, University of Leeds, Available from: http://www.chem.leeds.ac.uk/People/Rayner.html

Reetz, M.T., Wiesenhofer, W., Francio, G. & Leitner, W. (2002). Biocatalysis in Ionic Liquids: Batchwise and Continuous Flow Processes Using Supercritical Carbon Dioxide as the Mobile Phase, *Chem. Commun.*, pp. 992-993

Reinhardt, D. (2009). Implementation of Sustainability Criteria in the Chemical Research, Development and Theory by the Example of Ionic Liquids (in German Language), PhD Thesis, Chemical-Geoscientific Faculty, Friedrich-Schiller-University Jena, Germany

Ren, R.X., Zueva, D.L. & Ou, W. (2001). Formation of ε-Caprolactam via Catalytic Beckmann Rearrangement Using P_2O_5 in Ionic Liquids, *Tetrahedron Lett.*, 42, pp. 8441- 8443

Ross, J. & Xiao, J. (2002) Friedel-Crafts Acylation Reactions Using Metal Triflates in Ionic Liquids, *Green Chem.*, 4, pp. 129-133

Scammells, P.J., Scott, J.L. & Singer, R.D. (2005). Ionic Liquids: The Neglected Issues, *Australian Journal of Chemistry*, 58(3), pp. 155–169

Scurto, A.M., Aki, S.V.K. & Brennecke, J.F. (2002). CO_2 as a Separation Switch for Ionic Liquid/Organic Mixtures, *J. Am. Chem. Soc.*, 124, pp. 10276–10277

Scurto, A.M., Aki, S.N.V.K. & Brennecke, J.F. (2003). Carbon Dioxide Induced Separation of Ionic Liquids and Water, *Chem. Commun.*, pp. 572–573

Seddon, K.R., Stark, A. & Torres, M.-J. (2000). Influence of Chloride, Water, and Organic Solvents on the Physical Properties of Ionic Liquids, *Pure Appl. Chem., 72*, pp. 2275–2287

Seddon, K.R. & Stark, A. (2002) Selective Catalytic Oxidation of Benzyl Alcohol and Alkylbenzenes in Ionic Liquid, *Green Chem.*, 4, pp. 119-123

Seddon, K.R., Stark, A. & Torres, M.J. (2002). Viscosity and Density of 1-Alkyl-3-Methylimidazolium Ionic Liquids, *ACS Symposium Series 819*, In: Clean Solvents:

Alternative Media for Chemical Reactions and Processing, M.A. Abraham & L. Moens (Eds.), Chap. 4, 34–49, ACS, Washington, D.C. ISBN-13: 9780841237797

Sen, D.J. (2006). Reflection of Green Chemistry, *Pharma Times,* 38(7), pp. 35-39

Sharma, A. (2009). Catalytic Reaction Engineering Using Ionic Liquids: Hydroformylation of 1-Octene, PhD Thesis, University of Toulouse, France.

Sheldon, R.A. (2001). Catalytic Reactions in Ionic Liquids, *Chem. Commun.*, p. 2399–2407

Singh, T. & Kumar, A. (2007). Aggregation Behavior of Ionic Liquids in Aqueous Solutions: Effect of Alkyl Chain Length, Cations, and Anions, *J. Phys. Chem. B,* 111, pp. 7843–7851

Song, C.E., Shim, W.H., Roh, E.J., Lee, S.C. & Choi, J.H. (2001). Ionic Liquids as Powerful Media in Scandium Triflate Catalysed Diels–Alder Reactions: Significant Rate Acceleration, Selectivity Improvement and Easy Recycling of Catalyst, *Chem. Commun.*, pp. 1122-1123

Stark, A, MacLean, B.L. & Singer, R.D. (1999). 1-Ethyl-3-Methylimidazolium Halogenoaluminate Ionic Liquids as Solvents for Friedel-Crafts Acylation Reactions of Ferrocene, *J. Chem. Soc. Dalton Trans.*, pp. 63-66

Stark A. & Ondruschka, B. (2011). Friedrich-Schiller-University, Institute for Technical Chemistry & Environmental Chemistry, Jena, Germany. Available from http://www.sigmaaldrich.com/sigma-aldrich/technical-documents/articles/chemfiles/ionic-liquids-for0.html

Swatloski, R.P., Spear, S.K., Holbrey, J.D. & Rogers, R.D. (2002). Dissolution of cellulose with Ionic Liquids, *J. Am. Chem. Soc.*, 124, pp. 4974–4975

Swatloski, R.P., Rogers, R.D. & Holbrey, J.D. (2003). Dissolution and Processing of Cellulose Using Ionic Liquids, US Patent 2003/0157351

Tubbs, J.M. & Hoffmann, M. (2004). Ion-Pair Formation of the Ionic Liquid 1-Ethyl-3-Methylimidazolium bis(trifyl)imide in Low Dielectric Media, *J. Solution Chem.*, 33, pp. 381–394

Verma, A.K., Attri, P., Chopra, V., Tiwari, R.K. & Chandra, R. (2008). Triethylammonium Acetate (TEAA): A Recyclable Inexpensive Ionic Liquid Promotes the Chemoselective Aza- and Thia-Michael Reactions, *Monatsh. Chem.,* 139(51), pp. 1041–1047

Volkov, A.V., Korneeva, G.A. & Tereshchenko, G.F. (2008). Organic Solvent Nano-Filtration: Prospects and Application, *Russian Chem. Rev.*, 77 (11), pp. 983–993

Wagner, M. & Hilgers, C. (2008). Quality Aspects and Other Questions Related to Commercial Ionic Liquid Production. In: Ionic Liquids in Synthesis, P. Wasserscheid & T. Welton (Eds.), Vol. 1, p. 27-28, Wiley -VCH Verlag GmbH & Co. KGaA, Weinheim, ISBN-13 978-1-4020-4087-0 (e-book)

Wasserscheid, W. & Keim, W. (2000). Ionic Liquids - New Solutions for Transition Metal Catalysis, *Angew. Chem. Int. Ed.,* 39 (21), pp. 3772- 3789

Wasserscheid, W. & Welton, T. (2003). Ionic Liquids in Synthesis, Wiley-VCH Verlag: Weinheim, Germany, ISBN-13 978-1-4020-4087-0 (e-book)

Wasserscheid, P. & Haumann, M. (2006). Catalyst Recycling Using Ionic Liquids, Chap. 7, pp. 183-213, In: Catalyst Separation, Recovery and Recycling, D.J.Cole-Hamilton & R.P. Tooze (Eds.), Springer, ISBN 1402040865

Weatherley, L.R. (N.D.) Sustainable Processing via Process Intensification. Available from www.nysp2i.rit.edu/PublicDocs/Training/GreenEng/LaurenceWeatherley.pdf

Welton, T. (1999). Room-Temperature Ionic Liquids: Solvents for Synthesis and Catalysis, *Chem. Rev.*, 99, pp. 2071-2083

Welton, T. (2004). Review Ionic Liquids in Catalysis, *Coord. Chem.Rev.*, 248, pp. 2459–2477

Werner, S., Haumann, M. & Wasserscheid, P. (2010). Ionic Liquids in Chemical Engineering, *Annu. Rev. Chem. Biomol. Eng.*, 1, pp. 203–230

Wishart, J.F. & Castner, E.W. (2007). The Physical Chemistry of Ionic Liquids, *J. Phys. Chem. B*, 111 (18), pp. 4639–4649

Wong, H., Pink, C.J., Ferreira, F.C. & Livingston, A.G. (2006). Recovery and Reuse of Ionic Liquids and Palladium Catalyst for Suzuki Reactions Using Organic Solvent Nanofiltration, *Green Chem.*, 8, pp. 373–379

WO 01/77081 (2001), to Quill, Preparation of Imidazole Carbenes and the Use Thereof for the Synthesis of Ionic Liquids, {Inventors: Earle, M.J. & Seddon, K.R.}

WO 01/15175 (2001), to British Nuclear Fuels plc, Process for Recycling Ionic Liquids, {Inventors: Jeapes, A.J., Thied, R.C., Seddon, K.R., Pitner, W.R. & Rooney, D.W.}

WO 2003/039719, to Solvent Innovation Gmbh. Method for Separating Substances from Solutions Containing Ionic Liquids by Means of a Membrane, {Inventors: Wasserscheid, P., Kragl, U. & Kröckel, J.}

WO 05/019183 (2005), to BASF, {Inventors: Maase, M. & Massonne, K.}.

WO 2005/068404, to BASF AG, Distillation of Ionic Liquids, {Inventor: Maase, M.}

Wu, B., Liu, W.W., Zhang, Y.M. & Wang, H.P. (2009). Do We Understand the Recyclability of Ionic Liquids? *Chem. Eur. J.*, 15, pp. 1804–1810.

Xiao, Y. & Malhotra, S.V. (2005). Friedel-Crafts Acylation Reactions in Pyridinium-Based Ionic Liquids, *J. Organometallic Chemistry*, 690, pp. 3609–3613

Xiao, Y. (2006). Study of Organic Reactions in Pyridinium-Based Ionic Liquids, PhD Dissertation, Department of Chemistry & Environmental Science, New Jersey Institute of Technology, N.J., USA.

Xie, W. & Shao, L. (2009). Phosphorylation of Corn Starch in an Ionic Liquid, *Starch/Stärke*, 61, pp. 702–708

Xie, H. & Shi, T. (2010). Liquefaction of Wood (Metasequoia Glyptostroboides) in Allyl Alkyl Imidazolium Ionic Liquids, *Wood Sci. Technol.*, 44, pp. 119–128

Yang, Q. & Dionysiou, D.D. (2004). Photolytic Degradation of Chlorinated Phenols in Room-Temperature Ionic Liquids, *J. Photochem. Photobiol.*, 165, pp. 229–240

Yadav, J.S., Reddy, B.V.S. & Baishya, G. (2003). Green Protocol for Conjugate Addition of Thiols to α,β-Unsaturated Ketones Using a [Bmim]PF6/H2O System, *J. Org. Chem.*, 68, pp. 7098-7100

Zhang, Y., Zhang, S., Lu, X., Zhou, Q., Fan, W. & Zhang, X. (2009). Dual Amino-Functionalised Phosphonium Ionic Liquids for CO_2 Capture, *Chem.: Eur. J.*, 15, pp. 3003–3011

12

Ionic Liquids in Green Carbonate Synthesis

Jianmin Sun[1], Ruixia Liu[2], Shin-ichiro Fujita[2] and Masahiko Arai[2]
[1]*State Key Laboratory of Urban Water Resource and Environment*
The Academy of Fundamental and Interdisciplinary Science
Harbin Institute of Technology, Harbin
[2]*Division of Chemical Process Engineering*
Faculty of Engineering, Hokkaido University
Sapporo
[1]*China*
[2]*Japan*

1. Introduction

Ionic liquids (ILs), made of relatively large organic cations and inorganic anions, could contribute as solvents and catalysts to green organic synthetic reactions [1-5]. The reaction systems using no organic solvents, dense phase carbon dioxide (supercritical and carbon dioxide-dissolved expanded liquid phases), and/or water are obviously advantageous from environmental viewpoints [1]. ILs would be one of attractive components for environment-friendly reaction and separation processes because of their physicochemical properties. But we should also note the possibility of negative effects caused by their toxicity on environment and products when we use ILs for practical applications.

There are a number of ILs using different cations and anions, and a few typical examples are shown in Fig. 1. It is evidenced that ILs act as good catalysts for several organic synthetic reactions, as noted in recent review articles [1–5], including Friedel-Crafts reactions, Diels-Alder reactions, hydrogenation, polymerization, and other reactions. Haumann and Riisager discuss various aspects (structure of substrates, type of ILs, form of ILs catalysts) in hydroformylation reactions using ILs [4]. Recently Baiker et al. have published a review on multiphase catalytic systems including ILs and carbon dioxide [5]. They point out the significance of the combination of ILs and carbon dioxide for chemical and separation processes. This review deals with physicochemical features of those multiphase reaction systems and catalytic reactions therein including those using carbon dioxide as a reactant.

After considering those previous reviews and other chapters of this book, we will concentrate our attention on the application of ILs to the synthesis of carbonate and related compounds using carbon dioxide. Those compounds are of practical importance as solvents, reagents, fuel additives, and intermediates in the production of pharmaceuticals and fine chemicals [6,7]. The authors previously reviewed the potential application of ILs for the synthesis of cyclic carbonates from carbon dioxide [8]. The present chapter will deal with the synthesis of cyclic carbonates, dialkyl carbonates, other related compounds, and polycarbonates using ILs as solvents and catalysts.

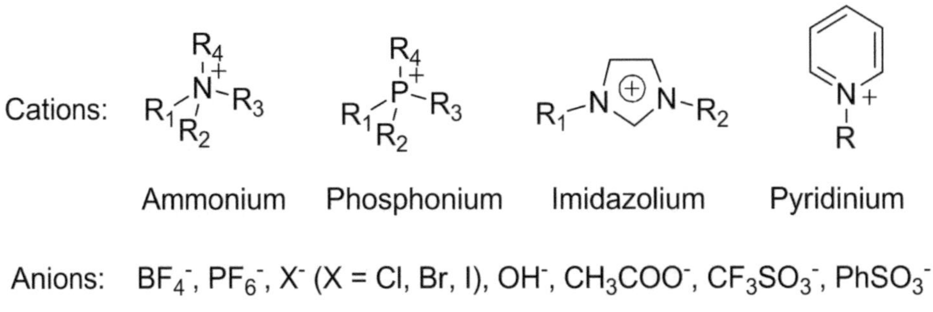

Fig. 1. Representative anions and cations of ILs.

2. Synthesis of cyclic carbonate in/by ILs

2.1 Synthesis of cyclic carbonate via CO_2 cycloaddition to epoxide

2.1.1 CO_2 cycloaddition catalyzed by ILs without additives

The synthesis of cyclic carbonates from epoxides and CO_2 (Scheme 1) is one of the major ways to transform CO_2 to valuable chemicals. The first successful synthesis of cyclic carbonate using room temperature ILs of imidazolium and pyridinium salts was reported by Peng and Deng in 2001 [9]. It has been shown that cycloaddition of CO_2 to propylene oxide (PO) producing propylene carbonate (PC) is effectively catalyzed by [BMIm]BF_4. PO was quantitatively converted to PC with 2.5 MPa of CO_2 at 110°C for 6 h in the presence of 2.5 mol% [BMIm]BF_4. After the reaction, PC was distilled from the reaction mixture and the catalyst was recycled up to four times with only a minor loss in activity. The ionic liquid catalyst is recyclable for the cycloaddition of CO_2 to PO. They also showed that the type of either cation or anion affect the activity of the ILs. The activity decreased in the orders of $[BMIm]^+ > [BPy]^+$ and of $BF_4^- > Cl^- > PF_6^-$.

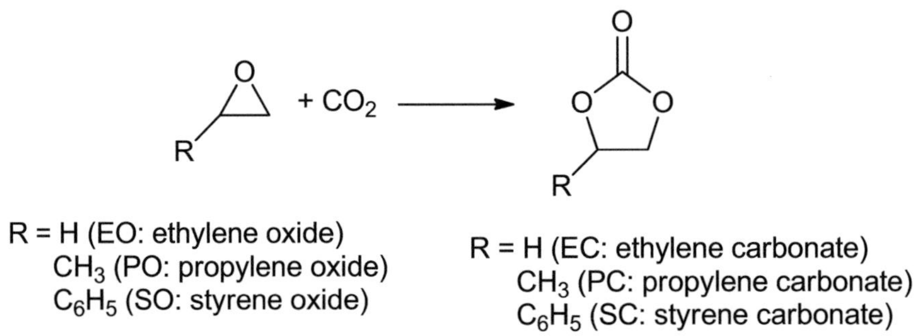

Scheme 1. Cycloaddition of CO_2 to epoxide. Abbreviations for some representative substrates and products are also given.

Kawanami and coworkers reported the PC synthesis from PO using various imidazolium salts under supercritical carbon dioxide ($scCO_2$) that is highly miscible with ILs [10]. The experimental results showed that both the type of anion and the length of alkyl chain in the cation have decisive effects on the conversion and selectivity. The performance of different ILs of 1-ethyl-3-methylimidazolium cation ($[EMIm]^+$) with NO_3^-, $CF_3SO_3^-$, BF_4^- and PF_6^- was investigated, and BF_4^- was again found to be the most effective. In addition, the carbonate yield increased markedly with an increase of the alkyl chain length (from C_2 to C_8). In the presence of 1-octyl-3-methylimidazolium ([OMIm]BF_4), 98% yield and 100% selectivity for the production of PC from CO_2 and PO were achieved at 100°C and 14 MPa of CO_2 in a short reaction time of 5 min. On the other hand, the yield and the selectivity obtained with [EMIm]BF_4 were 61% and 87%, respectively, even for a longer reaction time of 2 h. This enhancement was attributed to higher solubility of both the epoxide and CO_2 in the IL having longer alkyl chain under the same pressure and temperature. Interestingly, because of the difference in the solubility, influence of CO_2 pressure on the yield also depends on the alkyl chain length. In the case of [OMIm]BF_4, when the CO_2 pressure was reduced from 14 MPa to a sub-critical pressure of 6 MPa, a remarkable decrease in the yield was observed. In contrast to this, [EMIm]BF_4 gave a higher yield at 6 MPa than 14 MPa. It was also shown that the [OMIm]BF_4–scCO_2 reaction media gave satisfactory yields for the synthesis of other carbonates.

As described above, imidazolium-based ILs containing BF_4^- such as [BMIm]BF_4 and [EMIm]BF_4 were more active than the ones containing another type anion of PF_6^-, and [BMIm]BF_4 was more active than butylpyridinium IL [BPy]BF_4. Recently, Seki et al. studied the molecular interactions of pressurized CO_2 with imidazolium and pyridinium salts ([BMIm]BF_4, [BMIm]PF_6, [BMIm]Tf_2N, [BPy]BF_4) by in situ ATR infrared spectroscopy [11]. They suggested that Lewis acid-base interaction between CO_2 and BF_4^- leads to the formation of a new anion species, $[BF_4–CO_2]^-$, which is more basic than the original anion of BF_4^-. Similar new anion species were also produced from PF_6^- and Tf_2N^-; however, they were less basic than $[BF_4–CO_2]^-$ because of the difference in the interaction strength of CO_2 with the anion. It was also shown that the basicity of $[BF_4–CO_2]^-$ was stronger for [BMIm]BF_4 than [BPy]BF_4, revealing that the interaction between CO_2 and BF_4^- also depends on the cation species of ILs. Thus, the basicity of the new anion species produced from CO_2 and the original anion of IL may explain the effectiveness orders of $BF_4^- > PF_6^-$ and of $[BMIm]^+ > [BPy]^+$ for the cyclic carbonate synthesis observed in the previous studies.

Quaternary ammonium salts can also catalyze CO_2 fixation into cyclic carbonate. Caló et al. reported that this reaction was effectively promoted by molten tetrabutylammonium bromide ([Bu_4N]Br) and/or tetrabutylammonium iodide ([Bu_4N]I) under atmospheric pressure of CO_2 at 120 or 60°C [12]. Using these salts as solvents and catalysts, a less active substrate (styrene oxide) and a polymerisation-sensitive oxirane (glycidyl methacrylate oxirane) were converted to the corresponding cyclic carbonates in satisfactory yields. [Bu_4N]I showed higher activity than [Bu_4N]Br because of the difference in the nucleophilicity of the halide ions. The reaction rate depended greatly on the structure of the cation as well as the nucleophilicity of anion. They also showed significant difference in the activity among [BMIm], N-methyl-pyridinium ([MPy]) and [Bu_4N] iodides. The first two salts indicated inefficient catalytic performance for the cyclic carbonate synthesis. They suggested that the effective activity of [Bu_4N]I comes from the bulkiness of the tetrahedral ammonium ion, which forces the halide ion away from the cation easier, resulting in less electrostatic interaction between anion and cation and consequently in more nucleophilicity of the anion. It was also shown that the ammonium salt was easily recyclable by vacuum distillation or extraction with ethyl acetate, in which the salt is insoluble.

A plausible reaction mechanism for CO_2 cycloaddition to epoxide catalyzed by IL is illustrated in Scheme 2. The reaction is initiated by the ring opening of epoxide that is made by a nucleophilic attack of the anion of IL to the less hindered carbon atom of the epoxide ring; then, an oxy anion species **2** is formed. The carbon atom of CO_2 interacts with the oxy anion of complex **2**, producing an alkylcarbonate anion **3**. This species is further converted to the cyclic carbonate through the intermolecular cyclic elimination. Thus, the nucleophilic nature of the IL anion is significant for the reaction. Seki et al. who suggested the formation of the anion species of $[Y\text{–}CO_2]^-$ proposed a different mechanism [11], in which this new anion reacts with the epoxide, affording another intermediate (Scheme 3).

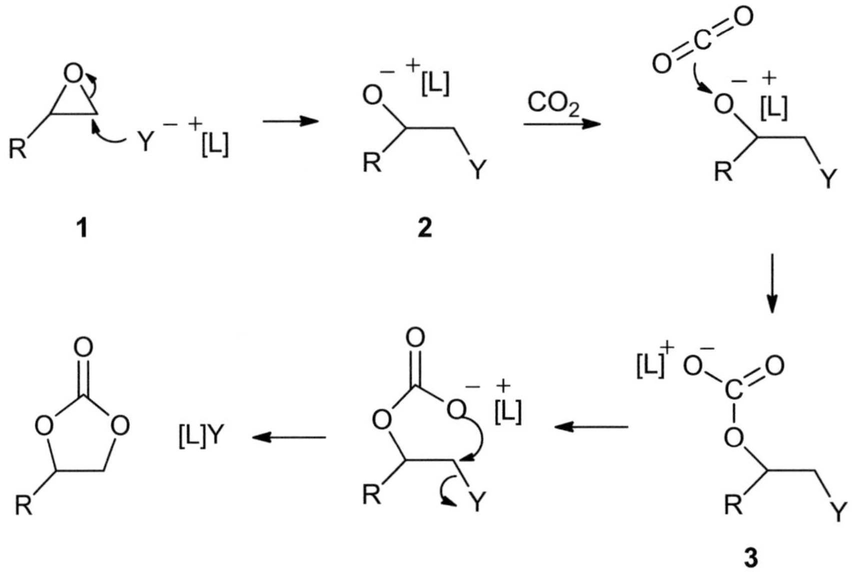

Scheme 2. A plausible mechanism for CO_2 cycloaddition to epoxide catalyzed by IL.

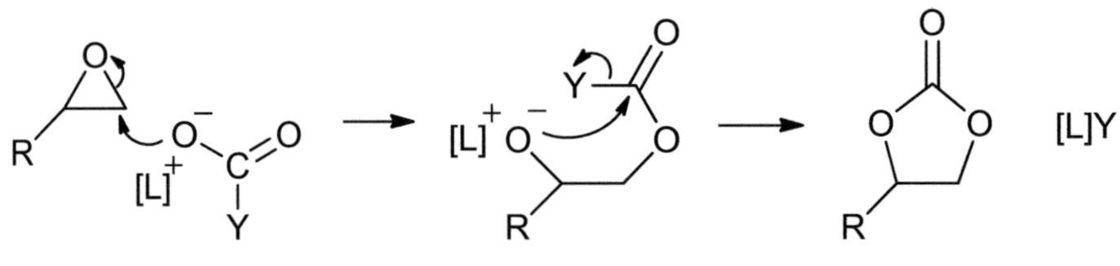

Scheme 3. A proposed mechanism involving an anion species of $[Y\text{–}CO_2]^-$ for CO_2 cycloaddition to epoxide [11].

The reaction of CO_2 and propargyl alcohol also produces a cyclic carbonate having a methylene group (Scheme 4). Gu et al. used the catalyst system of CuCl/[BMIm][PhSO$_3$] for this reaction [13]. 2-Methyl-3-butyn-2-ol (R_1, R_2 = CH_3) was almost qualitatively converted to the corresponding cyclic carbonate at 100°C under 1 MPa of CO_2 for 8 h. Similar results were obtained with CuBr and CuI. $CuCl_2$ also gave a good yield of 84%. The activities of iron and cobalt salts were poor, and noble metal salts gave only polymeric products. Solvent screening showed that [BPy][[PhSO$_3$] gave a slightly lower yield of 80%, and moderate yields were obtained with [BPy]BF$_4$ and [BMIm]NO$_3$. Moderate product yields were obtained even in organic solvents. However, [BMIm]PF$_6$ gave only polymeric products. Thus, the IL seems to act as a solvent, but there would be some interactions between the IL and CuCl and/or the alcohol. Interestingly, the reaction system was specific for tertiary alcohols; no desired products were detected from secondary and primary alcohols.

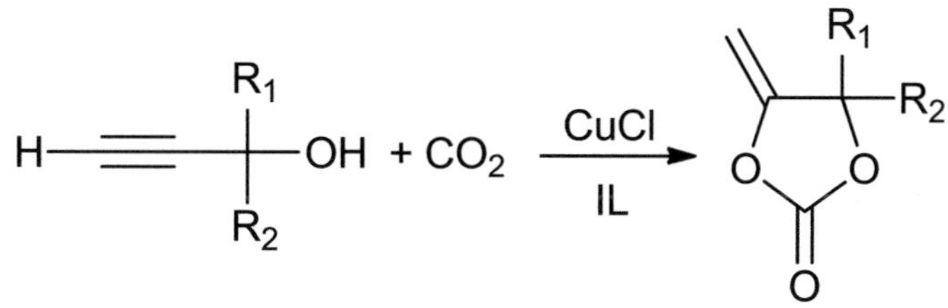

Scheme 4. Reaction of CO_2 with propagyl alcohol.

2.1.2 CO_2 cycloaddition catalyzed by ILs with Lewis acids

The presence of Lewis acidic compounds as co-catalysts greatly enhances the activity of ILs for the cyclic carbonate synthesis. In 1986, surprisingly, Kisch et al. already pointed out the significance of Lewis acid-base bi-functional system for the cyclic carbonate synthesis [14]. They showed that mixtures of $ZnCl_2$ and various ammonium and phosphonium halides can catalyze the PC synthesis even at room temperature with CO_2 at atmospheric pressure, although very long periods of reaction time were required to get high yields. This result was significant; however, it unfortunately did not attract attention of other researchers for a long time.

Lately, Kim et al. showed that the catalytic activities of ILs such as [BMIm]Cl and [BMIm]Br for the reactions of CO_2 with ethylene oxide (EO) and propylene oxide (PO) can surprisingly be improved by the combination of them with zinc bromide ($ZnBr_2$), although $ZnBr_2$ has no activity for the reactions [15]. For example, a TOF value of 37 h^{-1} for PC production was obtained with [BMIm]Br alone at 100°C under CO_2 of 3.5 MPa; this value was increased to 1679 h^{-1} by the co-presence of $ZnBr_2$. They concluded that the high activities are attributed to the in situ formation of $[BMIm]_2ZnBr_2Cl_2$ and $[BMIm]_2ZnBr_4$, because these tetrahalides separately prepared showed similar activities to the corresponding catalyst systems of [BMIm]Cl/$ZnBr_2$ and of [BMIm]Br/$ZnBr_2$. The catalytic activity of imidazolium zinc tetrahalide was greatly influenced by the nature of halide groups bonded to the zinc center. The activity was found in an order of $[ZnBr_4]^{2-}$ > $[ZnBr_2Cl_2]^{2-}$ >> $[ZnCl_4]^{2-}$, suggesting the importance of the nucleophilicity of halide ligands. They propose that a halide ion is dissociated from the zinc tetrahalide and subsequently attacks the carbon atom of epoxide. On the contrary, the alkyl group attached on the imidazolium cation was found to have a negligible effect on the activity. It is also shown that the imidazolium zinc tetrahalide is stable and reusable. They also determined the structure of the tetrahalide catalyst of (1,3-dimethylimidazolium)$_2ZnCl_2Br_2$ by XRD [16]. A similar catalyst system of [BMIm]Br/$ZnCl_2$ was reported by Li et al. [17]. NMR measurements of the catalyst in D_2O suggested the formation of a $ZnCl_2$–[BMIm]Br complex through the coordination of $ZnCl_2$ with the hydrogen atom on C(2) of imidazolium ring. This catalyst system gave excellent yields for the cyclic carbonate synthesis with several types of epoxides at 100°C for 1 h under CO_2 of 1.5 MPa.

Shortly thereafter, our group used a series of metal halides together with [BMIm]Cl for the synthesis of styrene carbonate (SC) from styrene oxide (SO), which is less reactive as compared with PO and EO [18]. It was found that the catalyst system comprised of $ZnBr_2$ and [BMIm]Cl can afford 93% yield with 100% selectivity to styrene carbonate at a low

reaction temperature of 80°C for 1 h. No activity was observed with $ZnBr_2$ alone and [BMIm]Cl itself gave a very low SC yield (6%). However, the combination of $ZnBr_2$ and [BMIm]Cl can exhibit an appreciably high activity for the carbonate synthesis. The type of metal cations had strong effect on the carbonate yield, the order of activity being $Zn^{2+} > Fe^{3+} > Fe^{2+} > Mg^{2+} > Li^{+} > Na^{+}$, which is the same order of Lewis acidity of the metal cations. When ZnI_2, $ZnCl_2$ or ZnO was used instead of $ZnBr_2$, the styrene carbonate yield was lowered. The influence of ILs showed that both the type of anion and the length of alkyl chain on the imidazolium cation had strong effects on the conversion and selectivity. Due to non-nucleophilic nature of BF_4^- and PF_6^- ions, when the [BMIm]BF_4 and [BMIm]PF_6 were used instead of [BMIm]Cl, low epoxide conversions were obtained. With these catalysts, selectivity for the styrene carbonate was also low because of the formation of oligomer of styrene oxide and/or styrene carbonate. When the IL having a longer alkyl chain, [OMIm]Cl, was used instead of [BMIm]Cl, the conversion was increased. Such an enhancement of the activity by lengthening the alkyl chain was also observed in the system of the imidazolium BF_4 under $scCO_2$, as described above. Interestingly, besides the influence of the types of metal halides and the ILs, the [BMIm]Cl/$ZnBr_2$ ratio also affected the carbonate yield, a ratio of 2 being optimum. The optimum CO_2 pressure was 4 MPa and elevated pressures had no positive effect on this SC synthesis. After the reaction, the catalyst phase of [BMIm]Cl and $ZnBr_2$ can be separated by an extraction with ethyl acetate and reused for another run without significant loss in activity.
Similar to the cases of the imidazolium salts, Lewis acid enhances the activities of ammonium salts. Our group also examined a series of metal halides together with $[Bu_4N]Y$ (Y = Cl, Br, I) for the SC synthesis [19]. The co-presence of zinc halides again appreciably enhanced the activity of $[Bu_4N]I$ by a factor of more than 25 times. The SC yield was in orders of $ZnBr_2 > ZnI_2 > ZnCl_2$ and of $[Bu_4N]I > [Bu_4N]Br > [Bu_4N]Cl$. The yield showed a maximum at a CO_2 pressure of 8 MPa. The most optimum ratio of $[Bu_4N]I/ZnBr_2$ was again 2. The activity of this catalyst system was very high; an almost quantitative yield was achieved even in a short reaction time of 30 min and at a mild reaction temperature of 80°C. It should be noted that, compared to the above mentioned catalyst system of [BMIm]Cl/$ZnBr_2$, the reaction time required to get the qualitative yield was half, although the types of halides were different.
As mentioned above, several bi-functional catalyst systems consisting of onium salts and zinc halides had been reported; however, direct comparison of their activities was difficult, because those catalysts were used for the reactions under different reaction conditions. Recently, our group has found that various zinc halide-based ILs whose general formula is $[L]_nZnX_2Y_n$ where L is a onium cation and X and Y are halides can be prepared by a simple heat treatment of mixtures of zinc halides and corresponding onium halides [20]. Under these circumstances, we prepared zinc halide-based ILs containing $[Bu_4N]^+$, $[BPy]^+$, $[BMIm]^+$, and $[Chol]^+$, and used them for the SC synthesis [21]. The effectiveness of the onium cation as the active catalyst component was in the order of $[Bu_4N]^+ > [BPy]^+ > [BMIm]^+ >> [Chol]^+$ (choline). Interestingly, the halide originating from the Zn halide (X) was more influential on the activity than that originating from the IL used (Y). Influence of the ratio of IL cation to Zn was different by the type of the IL cation. For $[Bu_4N]^+$, the activity increased linearly with the ratio up to 2; however, further increase in the ratio did not lead to the enhancement of the activity. For $[BMIm]^+$, the activity increased linearly with the ratio at least up to 4. The explanation for these observations remained speculative. With the best catalyst, an almost quantitative yield was obtained even in a short reaction time of 30 min

and at 80°C under CO_2 of 5 MPa. On the other hand, the use of analogous physical mixtures of $ZnBr_2$ and the onium bromides for the reaction resulted in a similar order of $[Bu_4N]^+$ ~ $[BPy]^+$ > $[BMIm]^+$, but the SC yields obtained with the mixtures were lower than those with the corresponding IL of $[L]_nZnX_2Y_n$.

The number of studies using mixtures of zinc halides and tetraalkylphosphonium halides as the catalysts for the CO_2 cycloaddition is limited [6,14,22,23]. Under the same conditions, the phosphonium catalysts showed lower activity compared with ammonium halide catalysts [14], but it was claimed that the former catalysts were more thermally stable than the latter ones [22].

As well as zinc halides, $CaCl_2$ [24] and $CoCl_2$ [25] can also promote the CO_2 cycloaddition to epoxide catalyzed by tetraalkylammonium or phosphonium halides. The optimum ratio of $CaCl_2$/IL was again 2 [24]. Unfortunately, these catalyst systems required relatively higher temperature to achieve high cyclic carbonate yields. Interestingly, $CoCl_2$ gave higher cyclic carbonate yields than $CoBr_2$ [25], being different from the results obtained with zinc halides.

Yokoyama and co-workers applied catalyst systems consisting of zinc salts and ILs for the PC synthesis under microwave radiation as heating source [26]. They employed zinc halides, zinc triflate ($Zn(OTf)_2$), and zinc phenosulfonate octahydrate (ZnPO), and five kinds of ILs. The PC yields under microwave were in the orders of $[Bu_4N]Br$ > [HMIm]Br ~ [BMIm]Br ~ [EMIm]Br > $[Bu_4N]I$ and of ZnPO > ZnI_2 > $ZnBr_2$ > $Zn(OTf)_2$. Those orders obtained under oil bath heating were slightly different, being $[Bu_4N]Br$ > [BMIm]Br > [HMIm]Br > $[Bu_4N]I$ > [EMIm]Br and ZnPO > ZnI_2 > $Zn(OTf)_2$ > $ZnBr_2$. It is interesting that rather unusual zinc salt of ZnPO was more effective than zinc halide conventionally used for the reaction. At optimum temperature of 120°C and CO_2 pressure of 5 MPa, a surprisingly high TOF of 20,371 h^{-1} was achieved with ZnPO/$[Bu_4N]Br$. Analysis of the reaction kinetics led to the conclusion that activation energy was lowered by the use of microwave irradiation.

Sun et al. reported an interesting catalyst system of IL/water [27]. The presence of water significantly enhanced the activities of ILs for the PC synthesis. At the optimal molar ratio of water to IL around 0.3, the PC formation rate could be increased 7 times. Almost 100 % yield was obtained at 120°C and CO_2 pressure of 2 MPa for 30 min. Further increase of this water/IL ratio caused the decrease in the PC selectivity because of the formation of 1,2-propylene glycol, which is produced by the side reaction between water and PO. This catalyst system was applicable for other epoxides. Such promotional effects were also observed with organic compounds containing OH group(s) such as phenol, acetic acid, ethanol, propylene glycol, and 2-propanol.

Synergistic effect also occurs by the combination of metal complex and onium salt. Lu et al. employed coupled catalyst systems of a tetradentate Schiff-base aluminium complex (salenAlCl) and tetraalkylammonium halides for the reactions of CO_2 and epoxides [28–30]. It was reported that various cyclic carbonates are produced from corresponding epoxides even at room temperature with good yields, although the reaction time required is slightly long [30]. Various salen metal complexes were also used for the EC synthesis. The EC yield was in the order of AlCl ~ CrCl > Co >> Ni > Mg ~ Cu ~ Zn [28]. They proposed the reaction mechanism in which ring-opening insertion of epoxide into Al–Cl bond and subsequent insertion of CO_2 into Al–O bond of the resulting alkoxy complex are involved. The ammonium halide is supposed to contribute to the activation of epoxide [29]. Very recently, North and Pasquale analyzed the kinetics of the PC synthesis catalyzed by a similar system of $[(salen)Al]_2O$/$[Bu_4N]Br$ [31]. The reaction was second-order on the

[Bu_4N]Br concentration, and induction periods were observed when the bromide concentration was very low. Furthermore, the formation of trace amounts of Bu_3N was detected after reaction runs. Based on these results, they proposed the involvement of a [Bu_3N-CO_2] complex in the catalytic cycle of the reaction, in which Bu_3N is produced from [Bu_4N]Br and interacts with CO_2, resulting in the formation of [Bu_3N-CO_2] that reacts with the ring opening product of hydroxyl anion species.
As described in this subsection, metal halides, metal complexes and water can improve the activities of ILs for the synthesis of cyclic carbonates from epoxides and CO_2. A generally accepted reaction mechanism for those bi-functional catalyst systems can be drawn as Scheme 5. Because these additives have Lewis acidic nature, they would interact with the oxygen atom of the epoxide. On the other hand, the basic anion of IL would attack the less hindered carbon atom of the epoxide ring. Such cooperative activation of the epoxide should make the ring opening easier, being the reason of the promotional effects of the additives. In some cases, the additives may have a role in the activation of CO_2.

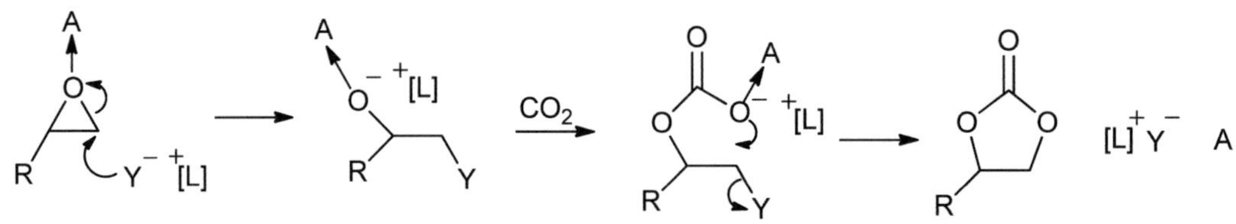

Scheme 5. A generally accepted reaction mechanism for CO_2 cycloaddition to epoxide catalyzed by IL/Lewis acid bi-functional systems. A = metal halides, metal-salen complexes, water. [L] = dialkyl-imidazolium, tetraalkyl-ammonium, tetraalkyl- phosphonium, alkyl-pyridinium. Y = Cl, Br, I.

For the catalyst systems using metal halides, the optimum ratio of metal/IL was sometimes observed to be 2 [18,19,24]. This may be explained by simultaneous activation of two epoxide molecules by one metal halide molecule and two IL molecules (Scheme 6). Zinc tetrahalide ILs also showed notable activity for the reaction. Similar cooperative actions of the Zn center of $[ZnX_2Y_2]^{2-}$ (Lewis acid) and a halide anion (Lewis base) liberated from the dianion were proposed for the epoxide ring opening [15,16,21].

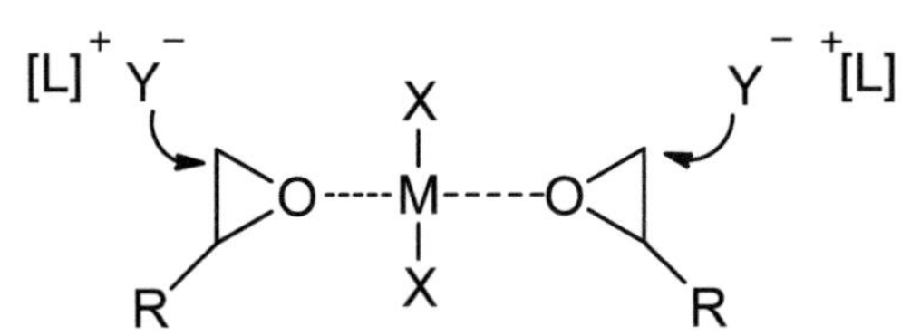

Scheme 6. Simultaneous activation of two epoxide molecules by one metal halide molecule and two IL molecules.

According to Scheme 5, if IL has hydroxyl-alkyl moiety, it can have high activity for CO_2 cycloaddition without additives, because of Lewis acidic nature of the OH group. Sun et al. [32] synthesized such task-specific ILs as hydroxyl-functionalized imidazolium and ammonium salts (Scheme 7). Among them, 1-(2-hydroxyethyl)-3-methylimidazolium bromide ([HEMIm]Br) had the highest activity. It gave almost quantitative yield for the PC synthesis at 120°C and CO_2 pressure of 2 MPa for 1 h, while non-modified IL of [EMIm]Br gave a lower PC yield of 83%. Similar results were obtained with (2-hydroxyethyl)-tributylammonium bromide ([HETB]Br) and the analogous unmodified ammonium bromide of [Bu_4N]Br. Thus, introducing a functional group of OH shows promotional effects for the activities of the ILs. Based on these results, they proposed the reaction mechanism in which the epoxide is activated by cooperative actions of the OH group and the halide anion (Scheme 8). [HETB]Br was precipitated by cooling after the reaction, separated by simple filtration, and could be reused four times with minor loss in the activity.

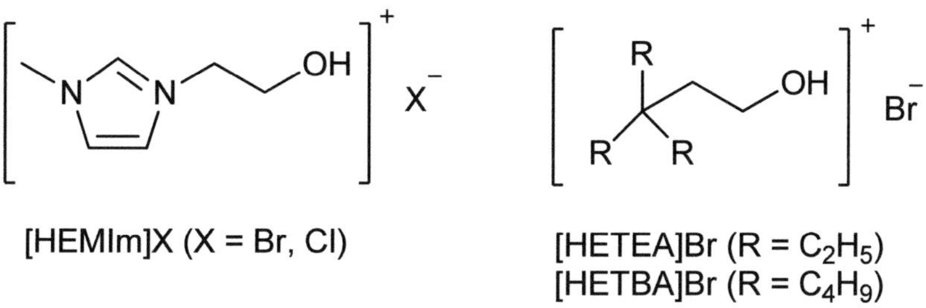

Scheme 7. Hydroxyl-functionalized ILs.

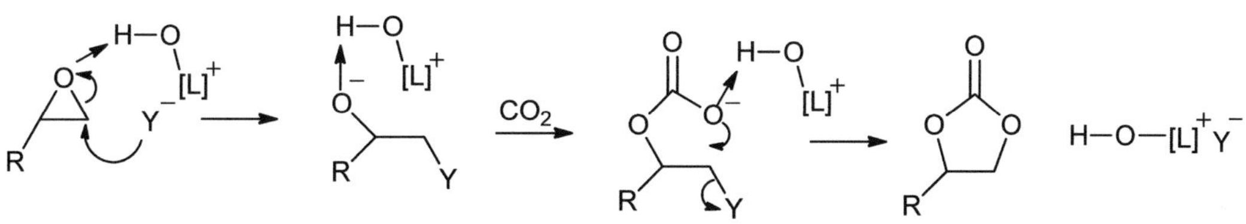

Scheme 8. CO_2 cycloaddition to epoxide with hydroxyl-functionalized IL. (H-O-[L] = [HEMIm], [HETB]. Y = Br, Cl.)

At almost the same time, Zhou et al. reported the PC synthesis using betaine-based ILs of [HBet]Y ([HBet] = 1-carboxy-N,N,N,-trimethylammonium, Y = Cl, Br, I), which have a carboxylic group (Scheme 9) [33]. Carboxyl acid is supposed to be more acidic than hydroxyl group; however, the difference in the activity between [HBet]Cl (having carboxylic acid group) and [Chol]Cl (having hydroxyl group) was small. Furthermore, [HBet]Y catalysts required higher temperature and CO_2 pressure and longer reaction time to obtain good PC yields, compared with the above mentioned hydroxyl-functionalized ILs of [HEMIm]Y and [HETB]Y. The reason of this may be lower nucleophilicity of halide anion of [HBet]Y than [HETB]Y.

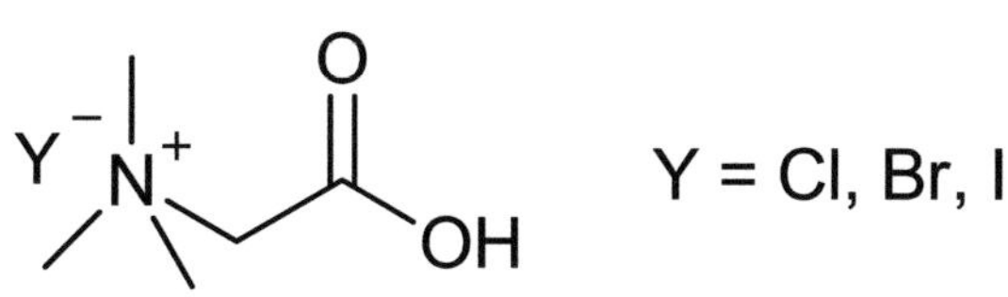

Scheme 9. Betaine-based ILs.

2.1.3 CO_2 cycloaddition catalyzed by immobilzed ILs

Heterogeneous catalysts are easily separated from reaction mixtures by simple filtration. This advantage can be added to IL catalysts by immobilization of them on solid materials. One easy way for the immobilization is the impregnation method. Wang et al. employed this method for the preparation of silica gel-supported ammonium and imidazoilium IL catalysts [34,35]. Similarly, Zhu et al. prepared a molecular sieve-supported [Chol]Cl-urea IL catalyst [36]. In these preparations, the support materials were just dispersed in an acetone or methanol solution containing the IL, and the solvent was removed by evaporation. The activities of these catalysts for PC synthesis were almost the same with those of corresponding unsupported IL catalysts. These supported IL catalysts were easily separated from the reaction mixtures by simple filtration, and could be recycled several times without significant loss in the activity for the PC synthesis under solventless conditions, although, over these catalysts, the active component of IL was fixed on the surface of the support by weak physical adsorption.

ILs can also be covalently immobilized on inorganic materials of silica, mesoporous silicates, aluminosilicate, and alumina. For that three ways are known (Scheme 10). The first method is the co-condensation of a trialkoxysilylalkyl onium salt and triethoxysilane (Scheme 10-A) [37–39]. The second one is the immobilization by the reaction between surface hydroxyl groups of the solid and the alkoxysilyl group (Scheme 10-B) [40,41]. For the last one (Scheme 10-C), the inorganic materials modified with alkyl halide (usually propylchloride) are used [42–46]; on them, corresponding amine, phosphine, or pyridine can be immobilized and quaternized. For the preparation of imidazolium-based immobilized IL, imidazole is fixed on the modified support in the presence of alkali compounds, followed by quaternization with alkyl halides.

The advantage of immobilizing ILs on solid materials is not only the easier separation of the catalyst, but also the usability of it for the continuous operation using a fixed-bed flow reactor. Takahashi et al. first reported the use of silica-immobilized phosphonium-based IL for a flow reactor [42]. They carried out the PC synthesis using 10 MPa of CO_2 for more than 1000 h. Unfortunately, the reaction temperature was required to be increased from the initial one of 90°C to 160°C for keeping the yield above 80%, suggesting that some leaching of IL from the support might occur; however, during the reaction run, the selectivity to PC was kept over 99.9%. In the same paper [42], they reported interesting synergistic effects of the immobilization of IL on silica. They also carried out the reaction runs in a batch reactor, and showed that the TOF of the immobilized IL was more than 300 times larger compared to that of corresponding phosphonium IL that was used homogeneously. However, the activity was not changed by the immobilization on polystyrene. The surprising enhancement of the activity by supporting on silica was ascribed to the co-operational activation of the epoxide by the acidic surface silanol groups and the halide anion of the immobilized phosphonium IL (Scheme 11). The enhancement of the activity by the immobilization was also observed with other ILs [44,46]. Thus, the activity of an inorganic solid-immobilized IL catalyst would be determined not only by its IL content, but also by the amount of surface hydroxyl group. This was pointed out by Udayakumar et al. [39] who prepared several silica-immobilized imidazolium IL catalysts for the reaction of CO_2 and allyl glycidyl ether. A higher TOF value was obtained with the catalyst containing a less amount of IL. This was ascribed to the presence of a larger amount of surface free silanol group on the catalyst.

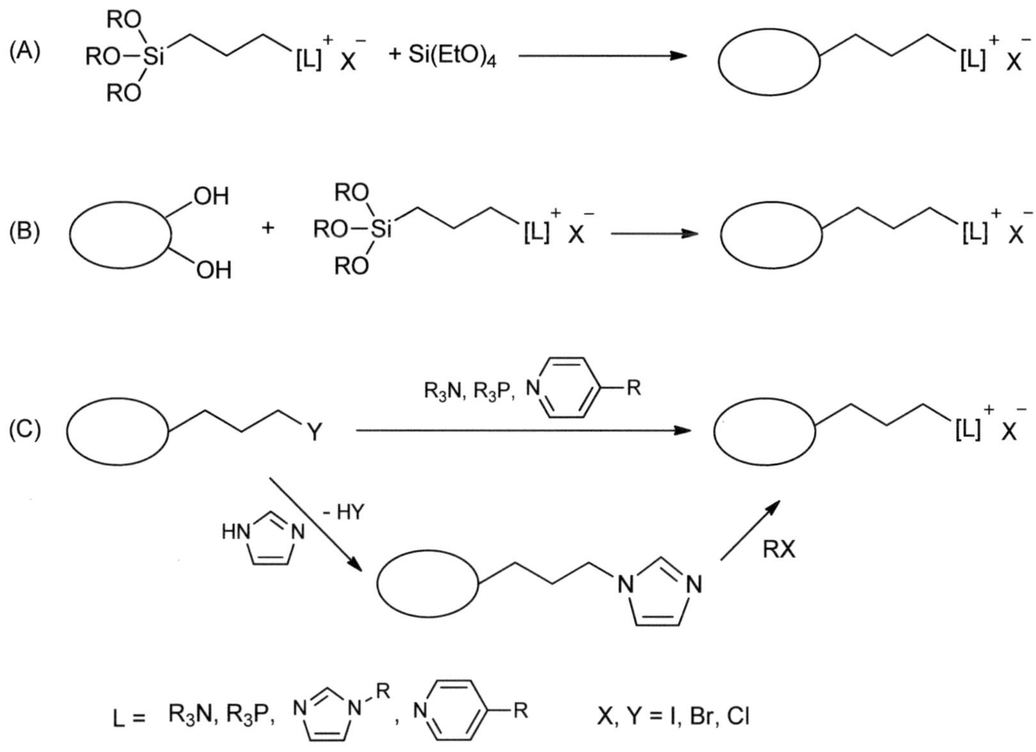

Scheme 10. Methods for the immobilization of ILs on an inorganic solid.

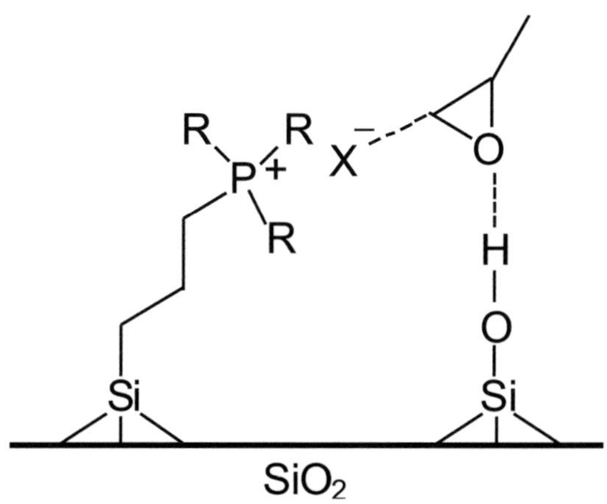

Scheme 11. Cooperative activation of an epoxide molecule on silica-immobilized IL.

Furthermore, the acidity of the OH group on the solid surface was also suggested to affect the activity of immobilized IL catalysts. Sakai et al. prepared immobilized phosphonium IL catalysts for the CO_2 addition to epoxyhexane using silica, aluminosilicate, and basic alumina as the support materials [40]. The conversion was in the order of silica > aluminosilicate > basic alumina. Because this order is the same with that of the acidity of the

OH group on the support surface, they ascribed the differences in the activity among the catalysts to those in the acidity of the OH group of the supports used.
Udayalumar et al. reported a series of papers in which amorphous silica [38,39] and mesoporous silicate of MCM-41 [43,45] were used as the supports for immobilizing imidazolium-based ILs. They investigated the effects of pore structures, IL content, type of counter anion, and the length of the alkyl group attached to the IL on the catalyst activity for the CO_2 cycloaddition to allyl glycidyl ether producing allyl gricidyl carbonate (AGC) (Scheme 12). Based on the reaction results obtained, they suggested that free space around the immobilized IL molecule is significant for the reaction, because the substrate and the intermediates are rather bulky [45].

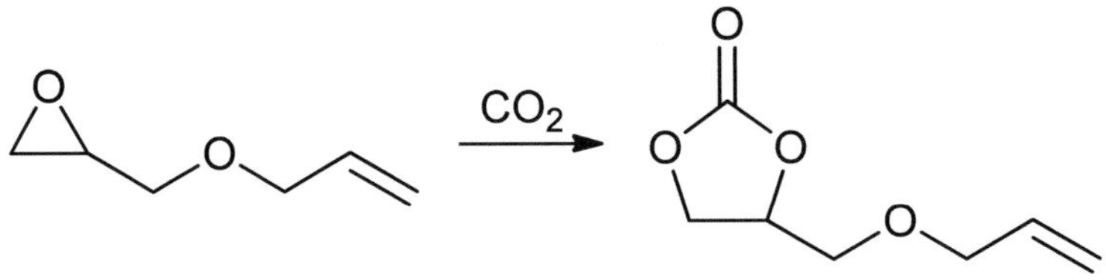

Scheme 12. Reaction of CO_2 and allyl glycidyl ether.

The same group also examined the influence of the co-presence of $ZnBr_2$ [38], which significantly improved the catalytic activity of un-immobilized imidazolium halides (see section 2.1.2.) [15,17,18]. $ZnBr_2$ also improved the activity of the immobilized IL catalyst, the AGC yield being increased from 52% to 78%. The promotional effect of $ZnBr_2$ for the immobilized IL catalyst was not drastic compared to those observed for the un-immobilized imidazolium catalysts whose activities were enhanced tens times by the co-presence of $ZnBr_2$.
As mentioned in section 2.1.2., the hydroxyl-functionalized IL has enhanced activity because of the cooperative action of the hydroxyl group (Scheme 7). Dai et al. [46] immobilized 3-(2-hydroxyethyl)-1-propylimidazolium bromide ([HEPIm]Br) on silica and mesoporous materials of SBA-15 and Al-SBA-15 (Scheme 13). The activity for the PC synthesis was in the order of SBA-15 > Al-SBA-15 > SiO_2. One possible reason for this difference was proposed to be the difference in the surface area. The PC yield obtained with unsupported [HEPIm]Br was almost the half of that with the IL supported on SBA-15 ([HEPIm]Br-SBA-15). Thus, the synergistic effect resulting from the presence of silanol is again observed, even though [HEPIm]Br itself has the OH group. Recycling of [HEPIm]Br-SBA-15 was tested seven times. Although the PC yield declined in the first three recycling, the catalyst activity remained unchanged after the forth run.

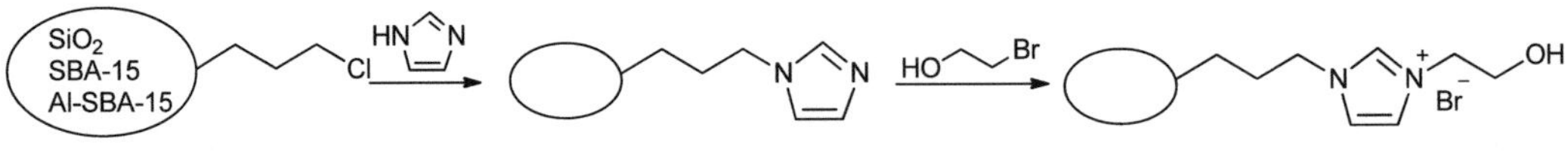

Scheme 13. Preparation of immobilized [HEPIm]Br.

ILs can also be immobilized on polystyrene (PS) including soluble and insoluble (cross-linked) ones. One method for the immobilization is the radical copolymerization of 1-vinyl-

3-alkylimidazolium halide with styrene (ST) and/or the cross-linker of divinylbenzene (DVB) using 2,2′-azobis(isobutyronitrile) (AIBN) (Scheme 14-A) [47,48]. Polymerization of vinylpyridine was also reported for the preparation of immobilized pyridinium-based IL (not shown) [49]. Another method to immobilize ILs on PS is the use of chloromethylated PS on which ILs can be immobilized by the reaction between the chloromethyl group and the IL precursor compound (Scheme 14-B) [50,51]. In this method, the quaternization with alkyl halide is also required for the preparation of immobilized imidazolium-based IL. The chloromethylated PS could be prepared by copolymerization of vinylbenzyl chloride (VBC) with ST and/or DVB.

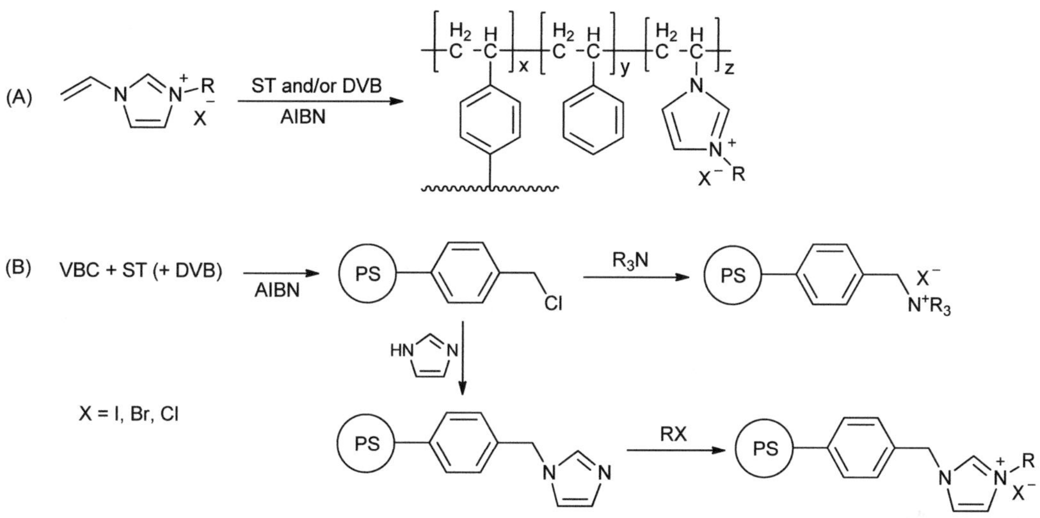

Scheme 14. Methods for the immobilization of ILs on polystyrene.

Xie et al. prepared a cross-linked PS supported imidazolium chloride catalyst by copolymerization of 1-vinyl-3-butylimidazolium chloride ([VBIm]Cl) and DVB [47]. This catalyst afforded 98% PC yield at 110°C and CO_2 pressure of 6 MPa for 7 h. The monomer of [VBIm]Cl gave the PC yield of 75%, and the polymer of [VBIm]Cl, which was prepared without the cross-linker DVB, gave 46% yield. The lower yield with [VBIm]Cl was ascribed to the immiscibility of the monomer with the substrate, because the copolymer of [VBIm]Cl and DVB was microparticles, and hence well dispersed in the substrate, resulting in the faster mass transfer. On the other hand, the lower yield of the [VBIm]Cl polymer was explained by insufficient use of the active IL site inside the polymer. Penetration of the substrates inside the polymer might be slow. Thus, using DVB for the mobilization is better to get more active catalyst.

In a similar way, Qiao et al. prepared soluble PS supported [VBIm] salt catalysts by copolymerization of ST and [VBIm] salts [48]. These PS supported IL catalysts were further reacted with $ZnBr_2$ ($ZnBr_2$/PS-[VBIm]X, X = Br, Cl, BF_4), to realize the high activity of zinc tetrahalide-base IL (see section 2.1.2.). $ZnBr_2$/PS-[VBIm]Br showed a high TOF of 3808 h^{-1} for the reaction of CO_2 and less active epoxide of SO at 110°C and CO_2 pressure of 6 MPa. This value was much faster than those with $ZnBr_2$/PS-[VBIm]Cl and $ZnBr_2$/PS-[VBIm]BF_4. Furthermore, the activity of $ZnBr_2$/PS-[VBIm]Br was almost the same with that of un-immobilized $ZnBr_2$/[VBIm]Br, because the polymer catalyst was completely dissolved in the

substrate of SO and in the product of SC. Although the miscible nature of the catalyst is disadvantage from the viewpoint of its separation, $ZnBr_2$/PS-[VBIm]Br was precipitated by addition of ethanol to the reaction mixture after completing the reaction and separated by centrifugation. The catalyst could be recycled without much loss in the activity for four times.
Kim et al. also reported the modification of polyvinylpyridine with $ZnBr_2$ [49]. Unfortunately, this polymer catalyst was less effective than homogeneous analogue of (pyridine)$_2$$ZnBr_2$.
Park et al. investigated the influence of the structure of the polymer on the activity of PS-immobilized IL [50]. They immobilized ammonium-based IL on three types of chloromethylated PS (PS1, PS2, and PS3), and used the immobilized IL catalysts for the reaction of CO_2 and glycidyl methacrylate. PS1 is a soluble copolymer prepared from ST and VBC. Both PS2 and PS3 are the cross-linked (insoluble) copolymers prepared from ST, DVB, and VBC, but the latter had macropores which were generated by addition of isooctyl alcohol to the mixture of ST, DVB and VBC before the copolymerization and the consecutive extraction of the alcohol by methanol after the polymerization. The product yield depended on the structure of the support polymer. It was in the order of PS1 > PS3 > PS2, which could be expected by considering the accessibility of the substrates to the IL active site.
Sun et al. reported the use of polymer supported hydroxyl-functionalized IL catalyst for the PC synthesis [51]. They prepared hydroxyethyl imidazolium-based IL immobilized on PS (PS-[HEIm]Br), and the corresponding un-functionalized immobilized IL (PS-[EIm]Br) for comparison (Scheme 15). An excellent PC yield of 98% was obtained with PS-[HEIm]Br at 120°C and CO_2 pressure of 2.5 MPa for 4 h, while that with PS-[EIm]Br was a lower value of 64%. Thus, introducing the OH group is effective for the CO_2 cycloaddition reaction even for the polymer immobilized IL system.

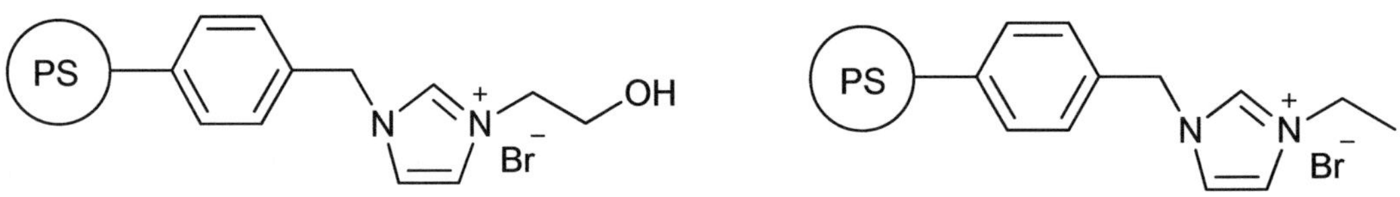

Scheme 15. Hydroxyl-functionalized and un-functionalized ILs immobilized on PS.

Zhao et al. first reported that a biopolymer of chitosan can also be employed for the immobilization of IL [52]. They immobilized ammonium salts on chitosan (CS-[R_3N]X) via the reaction between the amino group attached to the six-membered ring of chitosan and corresponding chloroisopropyl-trialkyl ammonium salts (Scheme 16). Interestingly, CS-[Me_3N]Cl revealed high activity for the PC synthesis, although the corresponding homogenous ammonium salt of [Me_4N]Cl produced only a trace amount of PC. This promotional effect by the immobilization was ascribed to the presence of the OH group of the ring; however, such effect was absent, when triethyl ammonium chloride was used instead of trimethyl one. Thus, the reason of the synergetic effect was still unknown.

2.2 Electrochemical synthesis of cyclic carbonates

Electrochemical synthesis of cyclic carbonates from CO_2 with epoxides, anilines, alcohols, and glycols was reported [53–60]. For example, Duñach et al. [54] conducted the reaction of epoxides with CO_2 to the corresponding cyclic carbonates by an electrochemical method in the presence of nickel(II) complexes. In these methods, harmful organic solvents, supporting

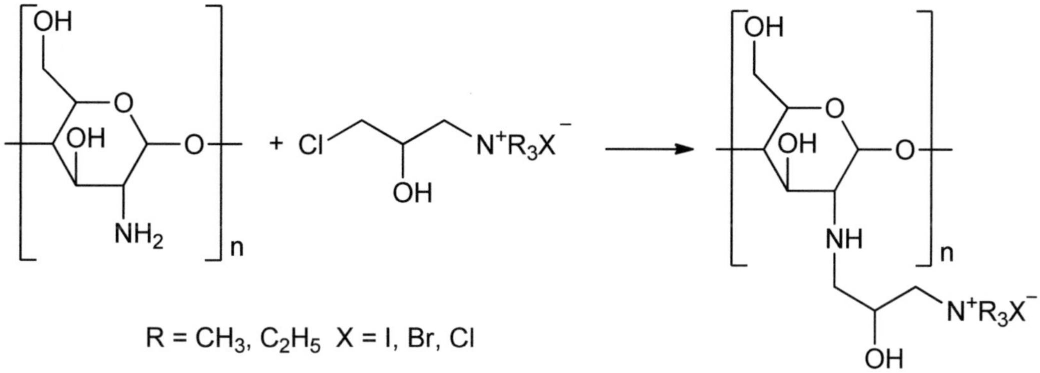

Scheme 16. Preparation of chitosan-immobilized ammonium ILs.

electrolytes, and/or catalysts were necessarily used, making the reaction systems complicated. Ionic liquids, because of their high ionic conductivities, wide electrochemical windows, and miscibility with many gaseous substrates, should be an effective medium for electrocatalytic organic synthesis from CO_2. Deng et al. [61,62] reported the first use of pure ILs as reaction media in the electrochemical activation of CO_2 for the synthesis of cyclic carbonates from several epoxides under mild conditions (Scheme 17). They tested different kinds of ILs including cations of [BMIm]+, [EMIm]+, [BPy]+ and anions of BF_4^-, PF_6^-. [BMIm]BF_4 was the best catalyst, giving conversions of 92% and 54% with 100% selectivity for PO and epichlorohydrin, respectively, at room temperature. For SO, a high conversion of 78% was obtained but the selectivity was 69% with a byproduct of benzylacetaldehyde. [EMIm]BF_4 and [BMim]PF_6 showed lower total conversions but the selectivity values remained almost unchanged. These results indicate that the activity of ILs depends on both cation and anion and the rate of conversion and the product selectivity also change with the structure of epoxide. It is noted that harmful organic solvent, additional supporting electrolyte, and catalyst were not used in these electrochemical reaction systems.

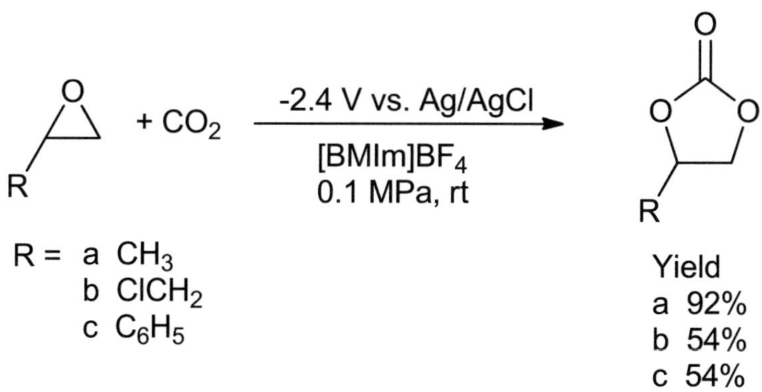

Scheme 17. Electrochemical cycloaddition of CO_2 to epoxides forming cyclic carbonates in the presence of IL.

Jiang et al. used a 0.05 M solution of [Bu_4N]Br in dimethylformamide (DMF) as electrolyte for the synthesis of cyclic carbonates at ambient temperature and at 4.0 MPa CO_2 [63]. The effects of such key factors as electrode materials, supporting electrolyte, CO_2 pressure, and solvent on the electrochemical synthesis of cyclic carbonates were investigated. The best results with PO were obtained with an aluminium anode and a platinum cathode (98% yield) or a zinc anode with a platinum cathode (95%). Lower, but still reasonable, yields were obtained when the aluminium anode was used with either copper, nickel or brass cathode (67–77%). Other terminal epoxides also showed moderate to high yields of 65–98% to the corresponding cyclic carbonates under the same conditions. In those reactions $[Bu_4N]Br$ acts as supporting electrolyte and catalyst. Similar work on electrocatalytic synthesis of PC was reported using ILs/DMF mixed solvents with Mg as a sacrificial anode [64]. It was also confirmed that the ILs including Br^- showing higher nucleophilicity were the best supporting electrolytes.

2.3 Synthesis of cyclic carbonates by oxidative carboxylation

Currently cyclic carbonates are extensively synthesized by CO_2 cycloaddition to epoxides, which has been commercialized. Although the above synthetic approach is quite atom-efficient, such a reaction usually requires the initial synthesis of epoxides; an additional step sometimes involves the use of expensive or toxic reagents and requires chemical separations [65]. Therefore, the direct synthesis of cyclic carbonates from olefins instead of epoxides, a so-called one-pot "oxidative carboxylation" of olefins, would be appealing. The oxidative carboxylation synthesis from olefins can be roughly viewed as the coupling of two sequential processes of epoxidation of olefins and CO_2 cycloaddition to epoxides formed (Scheme 18). The reaction uses easily available and low-priced chemicals of olefins as substrates and, moreover, preliminary synthesis and separation of epoxides would be avoided. So, the oxidative carboxylation would be a simpler and cheaper carbonate synthesis process with industrial potential from environmental and economic points of view. Although the three-component couplings have been known at least since 1962 [66], up to date, only a few works have been made on these reactions in contrast to extensive studies on the addition reactions of CO_2 to epoxides in ILs as catalyst/or solvent.

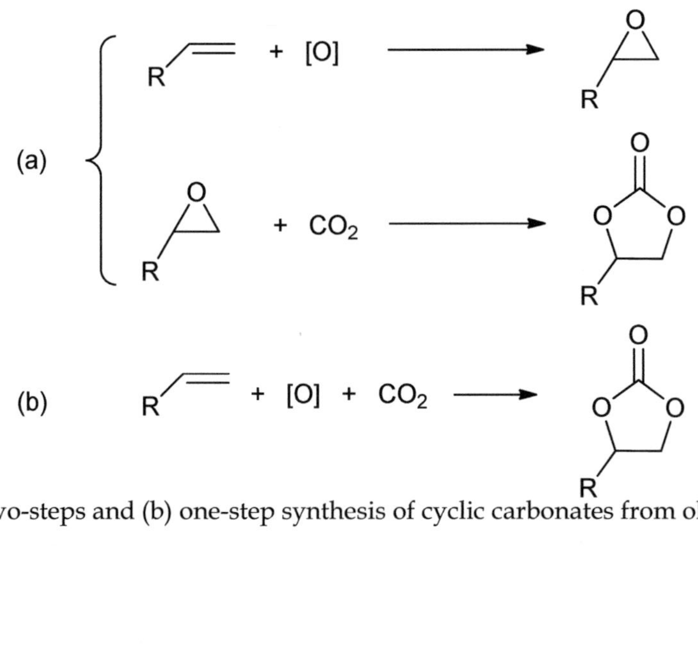

Scheme 18. (a) Two-steps and (b) one-step synthesis of cyclic carbonates from olefins. [O] = oxidant.

Our group reported the first example of direct oxidative carboxylation of styrene (ST) to SC in the presence of ionic liquid catalyst and an aqueous solution of *tert*-butyl hydroperoxide (TBHP) as an oxidant [67–69]. In this reaction system, ILs acted both as catalyst and solvent and no additional solvent was used. It was found that the structure of ILs and the nucleophilicity of anions strongly affected the yield of SC. $[BMIm]BF_4$, $[BMIm]PF_6$, and $[EMIm]BF_4$ were inactive, whereas $[Bu_4N]X$ (X = Cl, Br, I) was effective to the synthesis of SC, which could be ascribed to the differences in the structure and solubility of ILs [67]. With quaternary ammonium salt catalysts, the carbonate yield was improved with the increase in alkyl chain lengths [69]. For example, the yield of SC was only 10% with a large amount of styrene oxide intermediates in the presence of $[Me_4N]Br$ catalyst. However, the SC yield was improved to 38% with $[Bu_4N]Br$ having a longer alkyl chain. Moreover, the effectiveness of counter anion appeared in the order of $[Bu_4N]Br > [Bu_4N]I > [Bu_4N]Cl$. Although the nucleophilicity of iodide anion was stronger than that of bromide anion, the stronger reductive property of iodine anion led to its decrease in activity in the presence of the oxidant. Furthermore, CO_2 pressure played a crucial role in obtaining the high yield of SC. The change of the SC yield with the pressure was complicated; it was maximized at 1 MPa, 8 MPa and 15 MPa, which was discussed on the basis of the phase behavior of the multiphase reaction system [68]. Under mild reaction conditions (1 MPa CO_2, 80°C, 6 h), SC was obtained in a yield of 38%. Scheme 19 illustrates the proposed reaction mechanism, in which hypobromite (OBr^-) and bromide ion (Br^-) catalyze the epoxidation reaction and CO_2 cycloaddition reaction, respectively. The formation of SO intermediate was observed in this catalytic system.

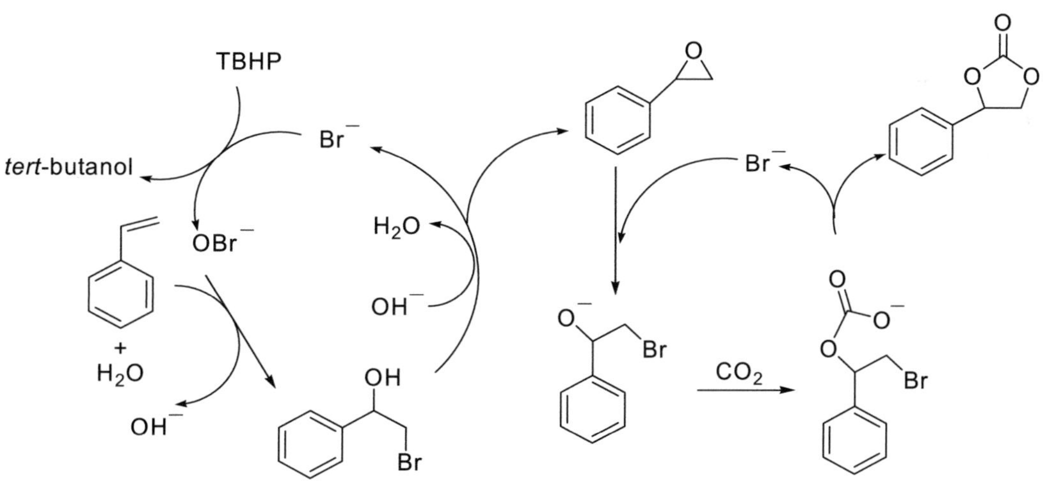

Scheme 19. A proposed reaction mechanism for one-pot synthesis of SC from styrene in the presence of Bu_4NBr.

Later on, Eghbali and Li investigated the direct conversion of alkenes and CO_2 into cyclic carbonates with hydrogen peroxide catalyzed by $[Bu_4N]Br$ in water [70]. They used a certain amount of such an organic base as 1,8-diazabicyclo[5.4.0] undecenc-7-ene (DBU) as a "CO_2 activator", which was considered to deprotonate the weakly acidic alcohol and neutralized hydrobromic acid formed. For ST, a good SC yield of 70% was obtained at a total olefin conversion of 89% in the presence of [Bu_4N]Br and DBU at about 50°C and around 20 MPa CO_2 for 15 h. A water-soluble substrate, 4-$SO_3C_6H_5$-$CHCH_2$, gave a better yield of the desired carbonate product of 89% at a total conversion of 98% under the same conditions. Because catalytic amounts of *N*-bromosuccinimde and NaBr (both are bromide ion sources) also gave moderate SC yields, the presence of bromide ions were suggested to be important for the reaction. It was speculated that the bromide ion could be readily oxidized to bromine or hypobromous acid, a reagent known to react with olefins in water to form the bromohydrin [71]. Indeed, styrene bromohydrin was formed with 8% yield in the case of SC synthesis under the conditions described above. After the formation of the bromohydrin intermediate and its subsequent reaction with CO_2, the bromide ions were regenerated, leaving water as the only byproduct. Bromohydrin would be assumed to be a crucial intermediate from an olefin, thereby leading to the formation of cyclic carbonate. After the in situ generation of the "bromohydrin" intermediate, the product was formed by attacking by base and the reaction with CO_2 (Scheme 20), which could avoid the preparation of epoxide. This method was also applicable for other alkenes; for example, 1-hexene and 1-octene were also effectively converted to the corresponding carbonates in 47% and 27% yields at temperatures of 45–55°C in 20 h. The ratio of reagents, particularly Br^-/H_2O_2, was found to be important for the reaction because bromide ion can catalyze the unwanted and non-productive decomposition of hydrogen peroxide to generate O_2 and water. An aqueous solution of sodium persulfate was also used as the oxidant but less effective compared to hydrogen peroxide.

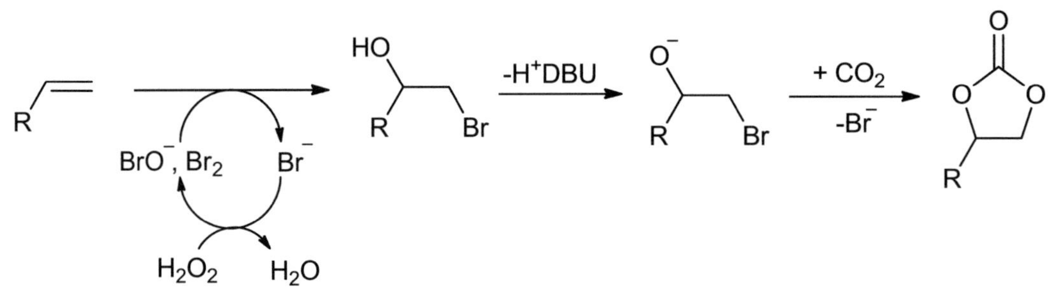

Scheme 20. Bromine-catalyzed direct conversion of alkenes to cyclic carbonates with H_2O_2.

Following the "bromohydrin" intermediate principle for the synthesis of cyclic carbonates via oxidative carboxylation [70], He and coworkers [72] developed a binary catalyst system composed of sodium phosphotungstate and [Bu_4N]Br for facile synthesis of cyclic carbonates in a single operation from alkenes and CO_2 with an inorganic base (Scheme 21). The direct synthesis of cyclic carbonate was conducted through in situ generation of "bromohydrin" intermediate with Bu_4NBr and H_2O_2 catalyzed by the phosphotungstate, followed by its subsequent reaction with CO_2 in the presence of the base acting as a "deprotonation reagent". Indeed, the key intermediate 2-bromo-1-phenylethanol was isolated in 16% yield in the absence of any base. The presence of a base as "deprotonation reagent" markedly improved the formation of SC; a SC yield of 57% was achieved with phenacyl benzoate in 14% yield under mild reaction conditions (50°C, 2.4 MPa, in aqueous medium). Interestingly, by subtly tuning the quantities of CO_2 and H_2O_2, the selective formation of phenacyl benzoate and cyclic carbonate could be controlled. Phenacyl benzoate, which is also an important intermediate and reagent in organic synthesis, could be obtained in good yields directly from ST in the absence of CO_2. This methodology was also applicable to several styrene derivatives producing the corresponding cyclic carbonates.

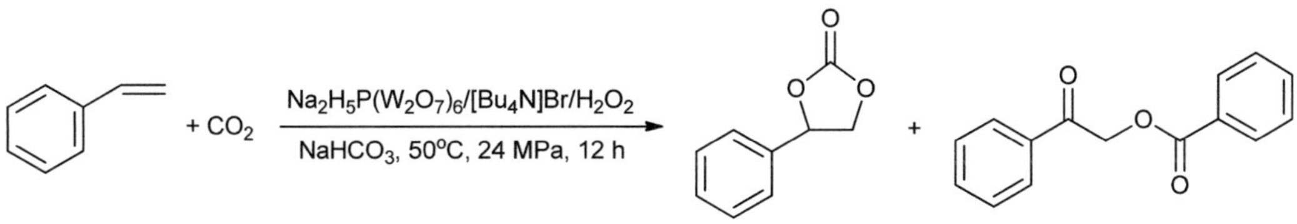

Scheme 21. A direct synthesis of SC from ST and CO_2 through "oxybromine" in situ generated.

The oxidative carboxylation of olefins to cyclic carbonates can proceed through the first step of epoxidation of olefins and the subsequent cycloaddtion of CO_2 to epoxides formed (Scheme 18). Thus it is supposed that a system of combining catalysts effective for the first step and for the second one would be effective for the direct synthesis of cyclic carbonates via the oxidative carboxylation of olefins. Indeed the direct preparation of carbonates was successfully achieved with a few catalyst systems including ILs coupled with oxidation catalysts. One patent [66] reported that the cyclic carbonate was formed from an olefin, CO_2, and oxygen in the presence of dual catalysts. The catalyst system includes a heavy metal compound and a quaternary ammonium hydroxide or halide. However, the heavy metal compounds would easily induce the corrosion of equipments and result in the undesired reduction of activity and selectivity.

Our group reported that $ZnBr_2$/[Bu_4N]X catalysts showed excellent activity and selectivity for CO_2 coupling with SO, as mentioned in section 2-1 [18,19]. By combining $ZnBr_2$/[Bu_4N]X with such an epoxidation catalyst as Au/SiO_2, the resultant three-component catalysts were efficiently applied for the direct oxidative carboxylation of ST to SC [73]. With Au/SiO_2-$ZnBr_2$/Bu_4NBr catalyst, the highest carbonate yield of 45% was obtained at 80°C and at 1 MPa CO_2 for 4 h with cumene hydroperoxide (CHP) as an oxidant. The influence of oxidants used was obvious and anhydrous TBHP and CHP were more active than aqueous TBHP, indicating that the water contained in the oxidant was detrimental to the carbonate synthesis. The lower activity with aqueous TBHP was possibly due to hydrolysis of epoxide to phenyl glycol, which would decrease the quantity of the starting material for the second step of carbonate synthesis. One of the authors further studied the influence of the support on the catalytic performance of Au nanoparticles [74]. With Au/$Fe(OH)_3$ and the $ZnBr_2$/Bu_4NBr catalyst, a SC yield of 53% was obtained at 80°C and at 4 MPa CO_2 in 10 h. The reaction pathways were postulated as illustrated in Scheme 22 in which the supported Au was active for the epoxidation step, while $ZnBr_2$/Bu_4NBr cooperatively catalyzed the subsequent step of CO_2 cycloaddition to epoxide formed. Furthermore, the epoxidation reaction was the rate-determining step in the one-pot oxidative carboxylation synthesis of cyclic carbonate, and the efficiency of the one-pot carbonate synthesis would be determined by the catalytic performance of the epoxidation catalyst. Hence, if we could explore more active and selective catalysts for the epoxidation reaction, more effective catalyst systems would be realized for the direct one-pot carbonate synthesis via the oxidative carboxylation from olefins.

Yokoyama et al. [75] further examined the combination of an epoxidation catalyst system of MTO/UHP/[BMIm]BF_4 (MTO, methyltrioxorhenium; UHP, urea hydrogen peroxide) [76] and a CO_2 cycloaddition catalyst of $[EMIm]_2ZnBr_4$ [15] for the one-pot synthesis of SC from ST. The epoxidation catalyst system of MTO/UHP/[BMIm]BF_4 offered a high SO yield of 93% at a ST conversion of 95% at 30°C in 2 h with a UHP/ST ratio of 1.1 in the absence of CO_2. This PO yield was decreased to 71% by the presence of 3 MPa CO_2, revealing the negative effect of CO_2 on the epoxidation reaction. It was also found that, in the presence of peroxide, $Zn[EMIm]_2Br_4$ was instable and hence not effective for the CO_2 cycloaddition, although it was reported to be very active for this reaction in the absence of peroxide. Based on these results, they employed multistep protocol for the one-pot synthesis of SC. In this method, after the epoxidation reaction was completed with MTO/UHP/[BMIm]BF_4 at 30°C in 2 h, CO_2 and $Zn[EMIm]_2Br_4$ were charged into the same reactor, and then the CO_2 cycloaddition to SO produced was carried out for 2 h at an elevated temperature (80 or 110°C). This change in the operation improved the SC yield to 83% (at 110°C and 3 MPa CO_2 for the second step), being much higher than previously reported results. Other improvements such as better utilization of oxidant, easy separation of the product, and reusability of reaction medium could also be achieved through this method.

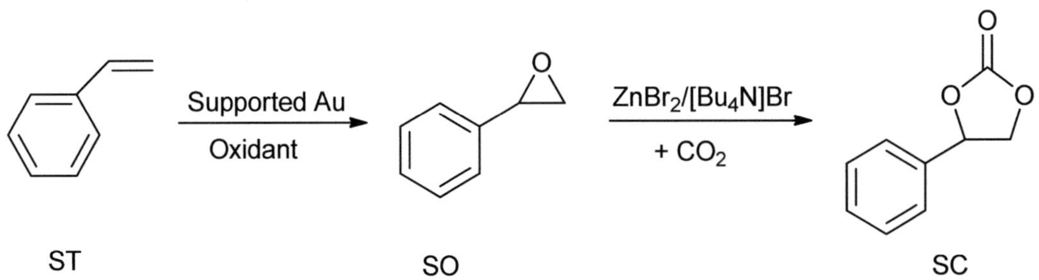

Scheme 22. Reaction pathways for the direct synthesis of SC from ST.

From the viewpoint of the cost, the use of molecular oxygen for the oxidative carboxylation is more preferable than hydrogen peroxide and organic peroxides. Aresta et al. [77,78] first reported the use of molecular oxygen for the one-pot synthesis of SC from ST with Rh complexes, which, however, suffered from short lifetime and low SC yields. Recently, Bai and Jiang [79] developed a catalyst system of $Ru(TPP)(O)_2$ (dioxo(tetraphenylporphyrinato)ruthenium) and $[Bu_4N]I$ for the carbonate formation from olefin, O_2 and CO_2, following the result that ruthenium porphyrin complexes can catalyze the epoxidation. SC was obtained in 76% yield with 100% selectivity using 4 mol% of $Ru(TPP)(O)_2/[Bu_4N]I$ at 30°C and low pressure (0.5 MPa O_2 and 1.1 MPa CO_2) for 48 h, which was the first example of 100% selectivity to SC with homogeneous catalyst. The molar ratio of $Ru(TPP)(O)_2/[Bu_4N]I$ was an important factor determining the catalyst activity. An SC yield of 76% was obtained at the ratio of 2. Larger or smaller ratio decreased the SC yield. The reactions of other olefins were also tested using $[Bu_4N]I$ or PTAT (phenyltrimethylammonium tribromide). Interestingly, $[Bu_4N]I$ gave a good yield of cyclic carbonate from chloropropylene, but poor yields from hexane, octane, cyclohexene, and cyclooctene. In contrast, PTAT gave good yields from the latter four olefins, but poor yield from the first olefin. Scheme 23 depicts the possible mechanism proposed by those authors in which the participation of two-type oxoruthenium complexes and the activation of CO_2 by the anion of IL are involved.

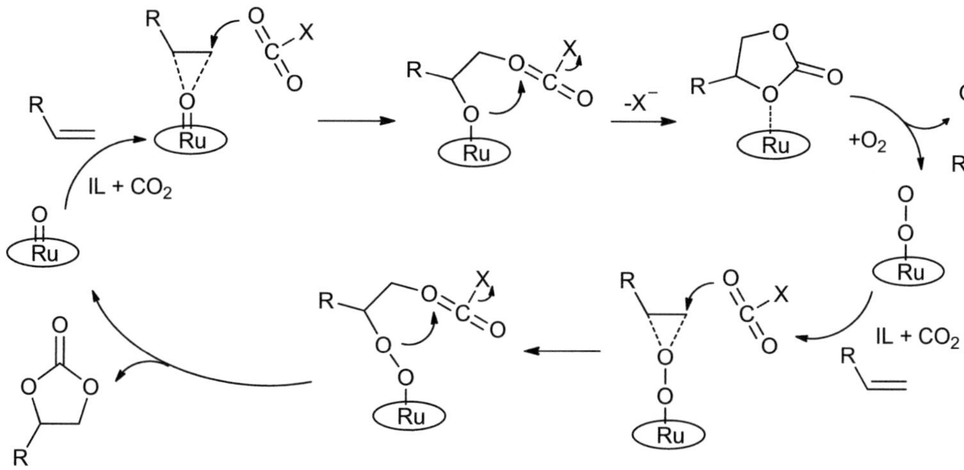

Scheme 23. Proposed mechanism of aerobic oxidative carboxylation of olefins catalyzed by Ru(TPP)(O)$_2$/IL. X = I^-, Br^-

3. Synthesis of dimethyl carbonate (DMC) in/by IL

Dimethyl carbonate (DMC) has been drawing much attention as a safe, non-corrosive, and environmentally friendly building block for the production of polycarbonate as well as useful methylating and methoxycarbonylating agents replacing toxic phosgene, dimethyl sulfate, and methyl iodide [80,81]. And DMC is also of interest as an additive to fuel oil replacing methyl *tert*-butyl ether to meet the oxygenate specifications, owing to a high octane number, reducing particulate emission from diesel engines [81−83]. In addition, it has been used as an electrolyte in lithium batteries due to its high dielectric constant [81,84]. DMC is industrially synthesized by the phosgenation of methanol, by the oxidative carbonylation of methanol using CO and O_2 with copper (II) and/or palladium (II) as catalyst, and by the reaction of CO and methyl nitrite [80,85,86]. These routes suffer from a low production rate, need for corrosion resistant reactors, and toxicity of phosgene and CO, and potential explosive hazard in the case of oxidative carbonylation. As the requirement of green chemistry, the utilization of CO_2 as a readily available, inexpensive, nonflammable, and environmentally acceptable starting material for DMC synthesis has been attempted. For this, two reactions are known. One is (A) direct synthesis of DMC from CO_2 and methanol, and the other one is (B) the transesterification of cyclic carbonate with methanol (Scheme 24). The latter can be called as an indirect synthesis of DMC from CO_2, because the cyclic carbonate can be produced from CO_2 and epoxide, as described in the earlier section. Some examples of these reactions using ILs will be given in this section.

3.1 Direct synthesis of DMC from methanol and CO_2

For the synthesis of DMC from methanol and CO_2, several metal complexes, metal oxides, and metal halides are known as effective catalysts. These catalysts generally require reaction temperatures around 150°C or higher [81]. Alkali salts can also catalyze the reaction in the presence of CH_3I as a reaction promoter [87,88]. With these catalysts, the required reaction temperatures were rather lower.

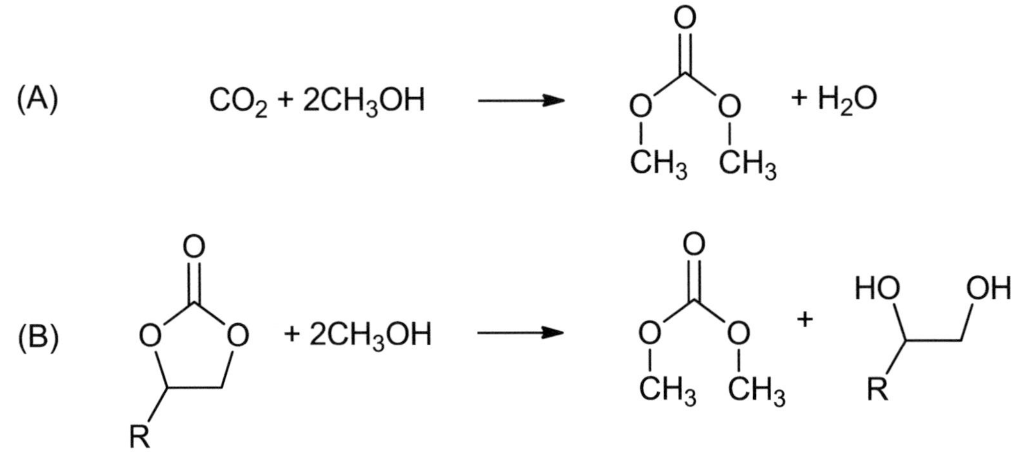

Scheme 24. Synthesis of DMC (A) from CO_2 and methanol, and (B) by the transesterification of cyclic carbonate with methanol.

To our knowledge, there is only one paper reporting the use of IL for the direct synthesis of DMC, which was published by Wang et al. [89]. They studied the effects of the presence of [EMIm]Br on the synthesis of DMC using the catalyst of KOH and the promoter of CH_3I (Scheme 25). Without [EMIm]Br, the DMC yield was 8.5% in the presence of 55 mmol KOH and 48 mmol CH_3I (methanol 850 mmol) at CO_2 pressure of 2 MPa. The yield was gradually increased to 11% by adding [EMIm]Br into the reaction system up to 21 mmol. With further increase of the IL amount, however, the DMC yield remained unchanged. They proposed that the promotional effect of the IL may be ascribed to the strong polarity and electrostatic field of the IL, which may stabilize the charged intermediate. Effects of the amounts of KOH and CH_3I and of CO_2 pressure were also investigated. Interestingly, as the CO_2 pressure was raised, the DMC formation showed two maxima near 2.0 MPa and 7.3 MPa. This trend is similar to that reported in our previous work [88], in which two maxima of the DMC yield were observed near 4.5 and 8.0 MPa in the presence of K_2CO_3 as the catalyst. Thus, the high CO_2 pressures were not required for the DMC formation and supercritical conditions were detrimental for the reaction.

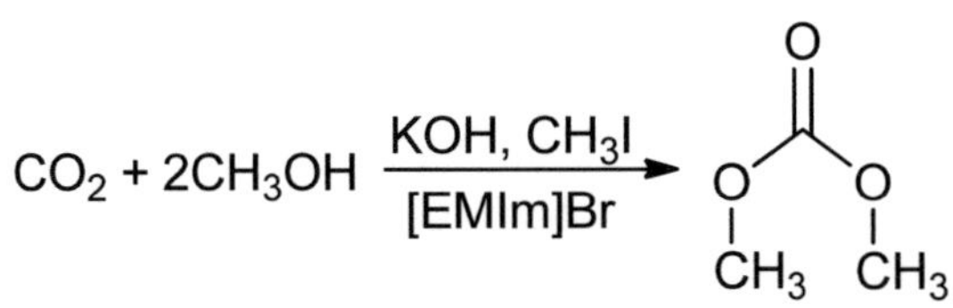

Scheme 25. Direct synthesis of DMC from CO_2 and methanol catalyzed by KOH.

3.2 DMC synthesis *via* transesterification of cyclic carbonate with methanol

The above mentioned direct synthesis of DMC requires organic or inorganic dehydration reagents to get high DMC yields because of the reaction equilibrium. On the other hand, the transesterification of cyclic carbonate with methanol relatively easily gives high DMC yields. ILs can also be used as the catalysts for this reaction. Scheme 26 illustrates the generally accepted reaction mechanism for the base catalyzed transesterification of methanol with cyclic carbonate. A methoxy anion produced from methanol reacts with the cyclic carbonate, producing an intermediate species of methyl-hydroxyalkyl carbonate. This species further reacts with another methoxy anion molecule, producing DMC and glycol.

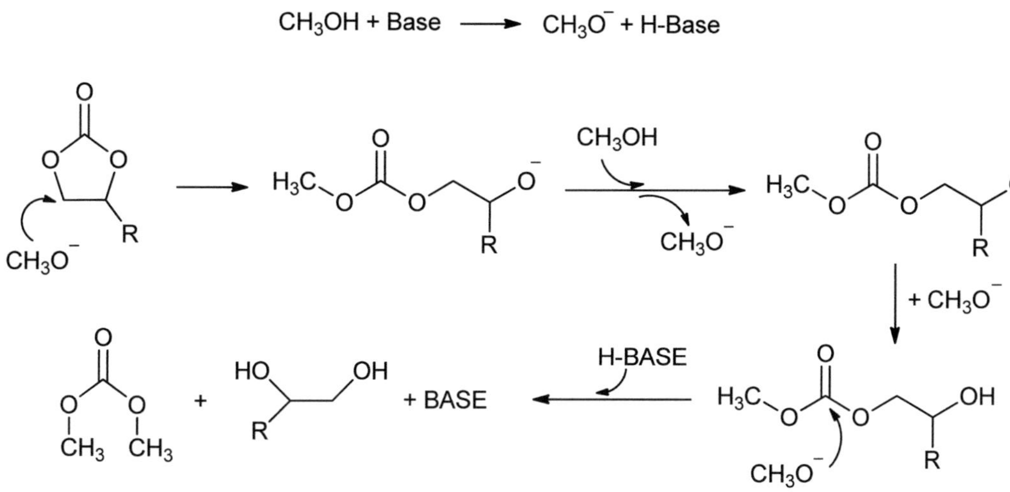

Scheme 26. Generally accepted reaction mechanism for the base catalyzed transesterification of methanol with cyclic carbonate.

Park's group did thorough investigation of the synthesis of DMC through transesterification using homogenous and heterogeneous IL catalysts [90−96]. At first, they used several tetraalkyl ammonium salts of $[(C_nH_{2n+1})_4N]X$ (n = 3, 4, 6, 8, 12; X = Cl^-, Br^-, I^-) for the reaction of methanol and PC (R = CH_3 in Scheme 24-B) in the presence of pressurized CO_2 [90]. The presence of CO_2 was required to prevent the undesired decomposition of PC to PO and CO_2. The PC conversion increases with the increase of alkyl chain length of IL, resulting from that bulkier ammonium ion makes the halide ion away from the cation easier, and, hence, more nucleophilic. The type of the halide anion also affected the activity of IL. The PC conversion was in the order of $Cl^- > Br^- > I^-$. This order is different from that observed for the cycloaddition of CO_2 to epoxide (see the section 2). The selectivity to DMC was slightly changed by the IL catalysts. Those values were around 75 %, although the authors did not mention what was the by-product, and the highest one was obtained with $[Bu_4N]Cl$. Consequently, this IL gave the highest DMC yield of 48% at 140°C for 6 h.
They also examined the use of a series of imidazolium-based ILs consisting of cations of $[EMIm]^+$, $[BMIm]^+$, $[HMIm]^+$ and $[OMIm]^+$, and of anions of Cl^-, BF_4^- and PF_6^- [91,92]. The DMC yield from the reaction of methanol and EC (R = H in Scheme 24-B) was in the orders of $[EMIm]^+ > [BMIm]^+ > [HMIm]^+ > [OMIm]^+$, and of $Cl^- > BF_4^- > PF_6^-$ [91]. The most active IL of [EMIm]Cl gave the largest DMC yield of 76% at 140°C for 6 h. The difference in the DMC yield by the alkyl chain length was ascribed to that in the solubility of ILs. IL having shorter alkyl chain is more soluble to methanol. The same orders of the DMC yield were observed for the reaction of methanol and PC [92]. However, the selectivity to DMC was significantly different by the type of the cyclic carbonate used. The selectivity values for DMC from EC were higher than 90%, while those from PC were in a range between 21 and 67%. Furthermore, the latter values were lower than those obtained with the above mentioned ammonium ILs. The conversions of PC obtained with the imidazolium ILs were

also lower than those with the ammonium ILs. Thus, the imidazolium ILs were less effective than the ammonium ILs; however, the authors claimed that the former ILs are more thermally stable than the latter ones at high reaction temperatures. Kinetic studies indicated that the reaction rate was pseudo first order with respect to the concentration of EC or PC.

They further investigated the reaction of EC and methanol under microwave heating using the catalysts of imidazolium-based and ammonium-based ILs [93]. Reaction experiments carried out for various periods of reaction time revealed that a high DMC yield with high selectivity was achieved in a short reaction time of 15 min. After that, the increase in the DMC yield was marginal, while the EC conversion increased gradually because of its decomposition. The optimum microwave power was in a range between 200 and 300W. Under microwave irradiation at 500 W, the DMC yield was significantly decreased. The EC conversion was in the order of $[HMIm]^+$ > $[BMIm]^+$ > $[OMIm]^+$ > $[EMIm]^+$, being different from the order observed with the conventional heating. On the other hand, when the ammonium ILs were used for the reaction, the EC conversion increased with increasing the alkyl chain length of the ammonium IL up to C_6, being consistent with the result obtained under the conventional heating. Thus, the influence of the catalyst structure on the reaction is different by the heating mode only for the imidazolium ILs. In most cases, the temperature of the reaction mixture reached around 120°C within 5 min under the microwave irradiation at 100 W. Under this irradiation condition, [BMIm]Br gave a DMC yield of 57%. Under the conventional heating, however, the same IL catalyst required a longer reaction time than 6 h to get the same yield. Thus, the activity of the IL is much better under the microwave irradiation. The reason of this was proposed to be rapid polarization of ionic species (the catalyst, the substrate, and the intermediates).

The same group also carried out the reaction of EC and methanol using imidazolium-based ILs and ammonium-based ILs covalently immobilized on amorphous silica [94,95] and on MCM-41, respectively (see section 2.1.3.) [96]. To get similar conversion levels, the immobilized IL needed slightly higher reaction temperatures than corresponding homogeneous ILs. The immobilized IL catalysts could be reused up to three consecutive runs without any considerable loss of their initial activity or with only a slight loss in the activity and the selectivity to DMC.

Another example of the DMC synthesis by the transesterification reaction of EC with methanol was recently reported by He and co-workers [97]. They developed DABCO-based (1,4-diazabicyclo[2.2.2]octane) Lewis basic ILs (Scheme 27). The prepared DABCO-based ILs were active under mild conditions. They could catalyze the reaction even at 80°C, which was much lower than the temperatures required for the above mentioned ammonium- and imidazolium-based ILs. The catalytic efficiency increased in the order of Cl^- < Br^- < OH^- (n=3). $[C_4DABCO]OH$ gave a DMC yield of 81% with 90% selectivity at 80°C for 6 h. This IL catalyst could be recovered after separation of DMC and ethylene glycol (EG) from the reaction mixture by distillation and reused for four times without loss in either DMC yield or selectivity. Furthermore, in comparison with the common basic catalysts, $[C_4DABCO]OH$ showed higher activity than $N^+(CH_3)_3HO^-$-functionalized PS, [BMIm]Br and $[Me_4N]Br$. Although DABCO itself and NaOH showed higher catalytic activity for the reaction, but these catalysts were hardly recyclable. The authors suggested that it could effectively activate methanol through tertiary nitrogen in the cation part of the IL in combination with hydroxide anion, producing methoxy anion.

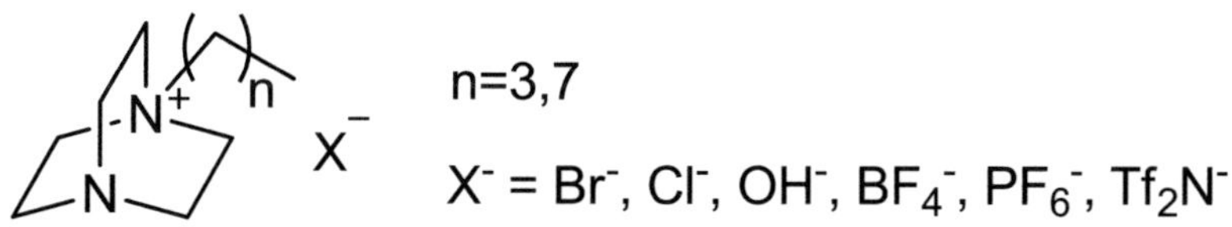

Scheme 27. DABCO-based IL of [C_{n+1}DABCO].

By combining the CO_2 cycloaddition to epoxide and the transesterification of cyclic carbonate with methanol, one-pot synthesis of DMC from epoxide, CO_2 and methanol is possible (Scheme 28). He and co-workers conducted the one-pot synthesis of DMC from SO (R = C_6H_5 in Scheme 28), CO_2 and methanol using a mixed catalyst system of [Bu_4N]Br/Bu_3N [98]. This catalyst system gave an SO conversion of 98% and a DMC yield of 84% at 150°C and 15 MPa CO_2 for 8 h. It should be noted that the selectivity for the methanolysis of epoxide (Scheme 29), which is sometimes problematic for the one-pot synthesis [99,100], was very low with this catalyst. The time course of the reaction clearly showed that DMC was formed through the formation of SC. The optimum CO_2 pressure was 15 MPa for this catalyst system. At lower pressures, the SO conversions were lower probably because of slower rate of the CO_2 cycloaddition step. At higher pressures, the conversion of SO was high, but the selectivity to DMC was lower because of the suppression of the transesterification of SC formed. The activities of either [Bu_4N]Br or Bu_3N were evaluated for the CO_2 cycloaddition, the transesterification reaction, and the one-pot reaction. [Bu_4N]Br was very active for the CO_2 cycloaddition, but inactive for the transesterification. On the other hand, Bu_3N had low activity for the former reaction, while significant activity for the latter one. Based on these results, the authors proposed a reaction mechanism, in which the CO_2 cycloaddition and the transesterification are catalyzed by [Bu_4N]Br and Bu_3N, respectively. They also suggested that the interaction of CO_2 with the amine caused the suppression of the transesterification at higher pressure than 15 MPa. Probably, CO_2 converted the amine to inactive carbamate species, although it was not stated.

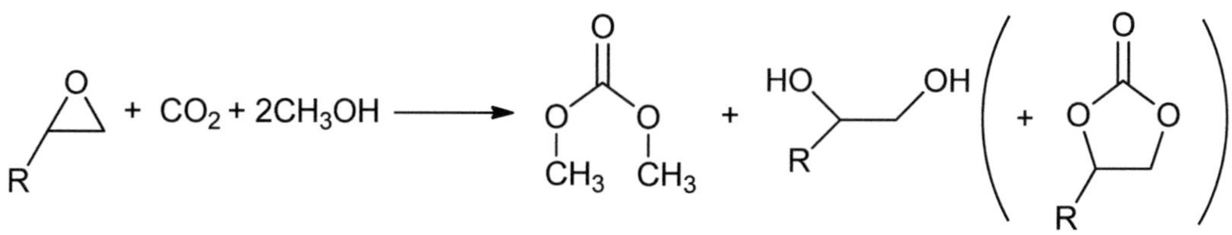

Scheme 28. One-pot synthesis of DMC from epoxide, CO_2 and methanol.

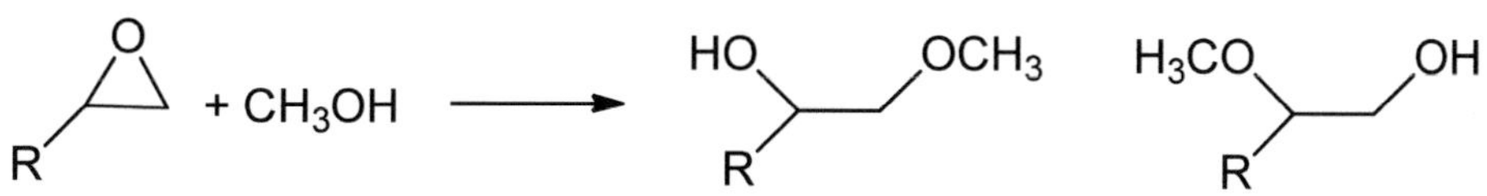

Scheme 29. Methanolysis of epoxide.

At almost the same time, Chen et al. reported a similar catalyst system of [BMIm]BF_4/CH_3ONa for the reaction of PO (R = CH_3 in Scheme 6), CO_2 and methanol [101].

This catalyst system gave PO conversion and DMC yield of 96% and 53%, respectively, under the conditions of 4 MPa, 150°C, and 5 h. Unfortunately, the selectivity for the methanolysis products was 19% with this reaction system, probably because of the nature of both the catalyst and the epoxide. When only CH_3ONa was used, the PO conversion, the DMC yield, and the selectivity for the methanolysis were changed to 96%, 30%, and 29%, respectively. Thus, there is some synergistic effect between [BMIm]BF_4/CH_3ONa. The ratio of [BMIm]BF_4/CH_3ONa was also important. With the ratio of 1:1 by weight, the DMC yield was 56%. Increasing only the amount of [BMIm]BF_4 three times resulted in the improvement of the DMC yield to 68%. However, further increase in this amount caused decrease of the DMC yield to below 20%. The amount of [BMIm]BF_4 had little effect on the conversion of PO. It was also observed that the DMC yield increased with the increase in CO_2 pressure, reached a maximum at 4 MPa, and then decreased when the CO_2 pressure was further increased. Too high CO_2 pressure again suppressed the transesterification reaction.

3.3 Electrochemical synthesis of DMC

Section 2.2. mentioned electrochemical synthesis of cyclic carbonate from CO_2 and epoxide. Electrochemical synthesis of DMC from CO_2 and methanol is also possible. Lu et al. used an electrochemical procedure for the synthesis of various dialkyl carbonates from CO_2 and alcohols in CO_2-saturated [BMIm]BF_4 solution (Scheme 30) [102]. Cyclic voltammetry of CO_2 in [BMIm]BF_4 with a Ti, Cu, Ni or Ag cathode and a Mg anode showed that [BMIm]BF_4 could act as a medium for CO_2 reduction, and the CO_2 reduction was the easiest on the Cu cathode. It was also found that the reduction of CO_2 in the IL was easier than that in organic solvents. The easiness of the CO_2 reduction was suggested to result from stabilization of CO_2 anion radical by the ion-paring with [BMIm] cation. For the synthesis of dialkyl carbonate, the electrolyses in the IL using the Cu cathode and the Mg anode was first performed under an atmospheric CO_2 stream. At the end of the electrolyses, alcohol was added to the IL and the solution was kept for 1 h under stirring. Then, alkyl iodide was further added to the solution and the mixture was stirred for 5 h. By this protocol, the primary and secondary alcohols were converted to dialkyl carbonates in good yields, whereas tertiary alcohol and phenol were unreactive. Effects of temperature, alcohol concentration, and the charge passed on the DMC yield were also investigated. At the optimum conditions, DMC was obtained with 75% yield. The IL could easily be separated from DMC and alcohol by simple distillation after the reaction and used directly for the next run. The reuse of the IL was tested for 5 times. After the second reuse, the DMC yield was decreased from 75% to about 50%, and then maintained at this level. After that, they further investigated the use of mixtures of [BMIm]BF_4/MeCN as the solvent for the DMC synthesis [103]. It was shown that the [BMIm]BF_4/MeCN mixture was better than the pure IL as the solvent, and the highest DMC yield was obtained at the volumetric ratio of 7/3.

Liu et al. employed the same system of [BMIm]BF_4−CH_3I for the electrochemical DMC synthesis but using a silver-coated nanoporous copper (NPC-Ag) cathode [104]. NPC-Ag showed much higher yields of DMC compared with silver and NPC cathode materials, and the highest yield of DMC was reached up to 80% under the optimal reaction conditions.

Yuan et al. conducted the electrochemical synthesis of DMC with two Pt electrodes in dialkyl imidazolium ILs−basic compounds−methanol solutions [105]. It is important to note that their system required neither alkyl iodide nor organic solvents, resulting in the greener reaction system. In it, IL served as electrolytes, and the base acted as a co-catalyst. The solution was saturated with CO_2 bubbling at ambient pressure, and the reaction was carried out for 48 h under a 5.5 V potential. The yield of DMC increased from 0.2% to 3.9% as following the order of $[EMIm]BF_4$ < $[BMIm]BF_4$ < [BMIm]OH < [BMIm]Cl < [BMIm]Cl < [EMIm]Br < [BMIm]Br. These findings indicated that the effect of the anions on the reaction was very significant, while the dialkylimidazolium cations had a negligible impact on the electrolysis.

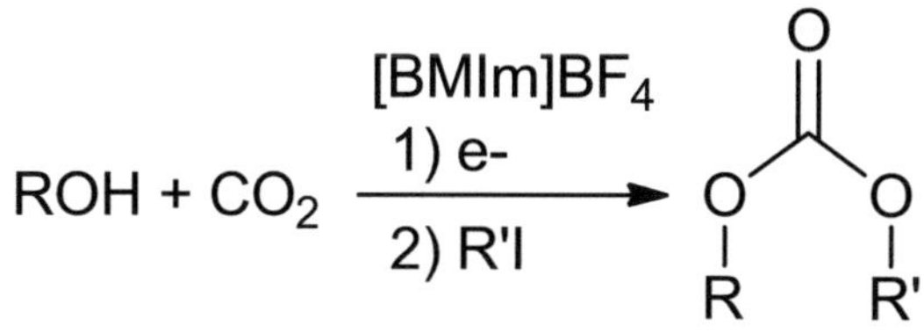

R= (a) CH_3, (b) C_2H_5, (c) *n*-C_4H_9, (d) *sec*-C_4H_9,
(e) *tert*-C_4H_9, (f) $PhCH_2$, (g) Ph $(CH_2)_2$, (h) Ph
R'= (a) CH_3, (b-h) C_2H_5

Scheme 30. Electrochemical synthesis of dialkyl carbonates from CO_2 and alcohols in CO_2-saturated $[BMIm]BF_4$ in the presence of alkylating agent.

The type of the basic compounds was also significant for the reaction. The DMC yield was in the order of CH_3OK > K_2CO_3 > KOH > NaOH >> CH_3ONa. CH_3ONa was almost completely inactive. Thus the type of alkali cation was very important for the reaction. They suggested that the small cation Na^+ is hardly adsorbed at the electrode surface because it is strongly solvated with the solvent. Conversely, the less solvated and bulky cation K^+ is preferentially adsorbed on the cathode. The specifically adsorbed K^+ at the electrode surface can directly interact with nucleophilic sites of CO_2 and methanol. Reaction runs for the optimization of the reaction conditions were performed. Among the results obtained, an interesting one was that the DMC yield showed a maximum value of 3.9% at the reaction time of 50 h. Further increasing the reaction time, the yield gradually decreased to 0.2% because of further transformation of DMC to trimethoxyl methane and tetramethoxyl orthocarbonate. Unlike CH_3OK, the addition of CH_3I to the system has basically no impact on the reaction. It is indicated that CH_3I does not participate as a methylating agent in the electrochemical reaction. They proposed a reaction mechanism as Scheme 31. CO_2 is adsorbed on the cathode via interaction with previously adsorbed K^+ on the cathode and, then, it is reduced into CO_2^-(ad) when it accepts an electron from the electrode surface. This species interacts with CH_3OH(ad), which is adsorbed on the electrode, generating a CH_3OCO^+ cation and K·OH(ad). The cation successively reacts with methoxide CH_3O^- in the bulk to give DMC. The K·OH(ad) formed in the course of the elemental step further reacts with bulk CH_3OH, producing CH_3OK. Thus, the catalyst is recyclable.

4. Synthesis of nitrogen containing carbonyl compounds from CO_2 using ILs

CO_2 can be one of the substrates for the synthesis of dialkyl urea and cyclic urethane compounds that have found extensive applications as dyes, antioxidants, and intermediates for the preparation of pharmaceuticals and agricultural chemicals. In this section, some examples where ILs could successfully be used for these reactions will be described.

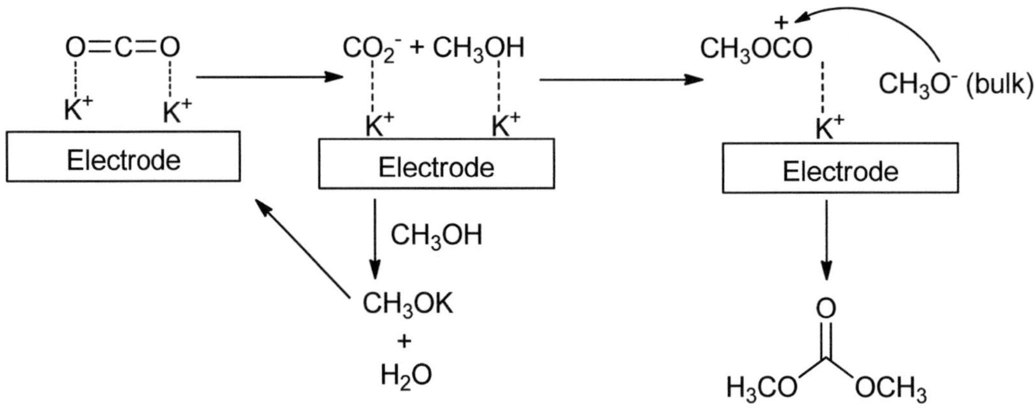

Scheme 31. A mechanism for electrosynthesis of DMC from CO_2 and methanol with Pt electrodes in the IL–CH_3OK–methanol system.

Deng and co-workers [106] reported the synthesis of symmetric urea derivatives from CO_2 with aliphatic and aromatic amines using a catalyst system consisting of [BMIm]Cl and CsOH (Scheme 32). This catalyst system gave excellent yields of dialkyl ureas at 170°C under 6 MPa CO_2 for 4–6 h without dehydration reagents. Although longer time was needed, good yields were also obtained with less reactive aniline compounds which were not directly carbonylated with CO_2 so far. For this catalyst system, the presence of the IL was indispensable, because no urea was obtained only with CsOH. KOH was less effective for the reaction than CsOH. After the reaction, the product was precipitated by adding water to the reaction mixture and separated by filtration. The IL and the base in the filtrate were recovered by removing water by distillation and could be reused three times with small loss in the activity.

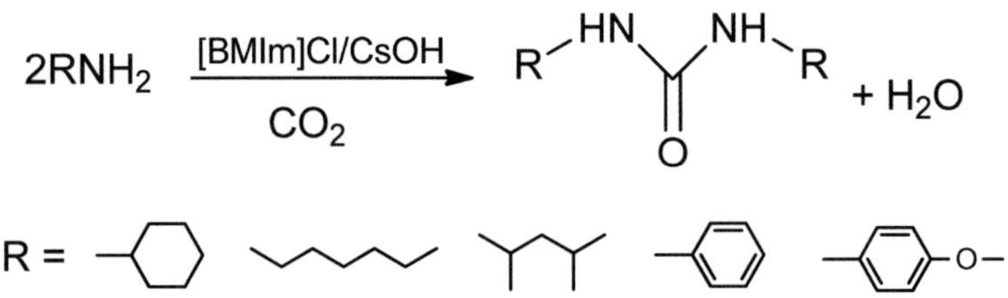

Scheme 32. Synthesis of dialkyl urea from amine and CO_2.

Lately, Jiang et al. [107] used [BMIm]OH for the dialkyl ureas under conditions similar to those employed by Deng et al. This catalyst also gave good yields of dialkyl ureas, although slightly longer reaction time was required. Unfortunately, no urea was obtained from aniline, which could afford *N,N′*-diphenyl urea with CO_2 by [BMIm]OH/CsOH. No ureas were produced from dibutyl amine and from tributyl amine. Scheme 33 illustrates a reaction mechanism proposed by the authors.

The reaction of CO_2 with aziridine producing 2-oxazolidinone is analogous one to the cyclic carbonate synthesis from CO_2 and epoxide. If an alkyl or aryl group is attached to the aziridine ring, two oxazolidinone products could come from the ring opening across N−CH_2

and across N−CH (Scheme 34). Nomura et al. reported the reactions of CO_2 with 1-phenyl-2-methyl (R_1 = C_6H_5, R_2 = CH_3) and 1-phenyl-2-ethyl (R_1 = C_6H_5, R_2 = C_2H_5) azilidines using $[Bu_4N]Br$ in an organic solvent of acetonitrile at 60°C under 5 MPa CO_2 [108]. The obtained products were nearly equimolar mixtures of 3,4-disubstituted and 3,5-disubsittuted oxiazolidinones.

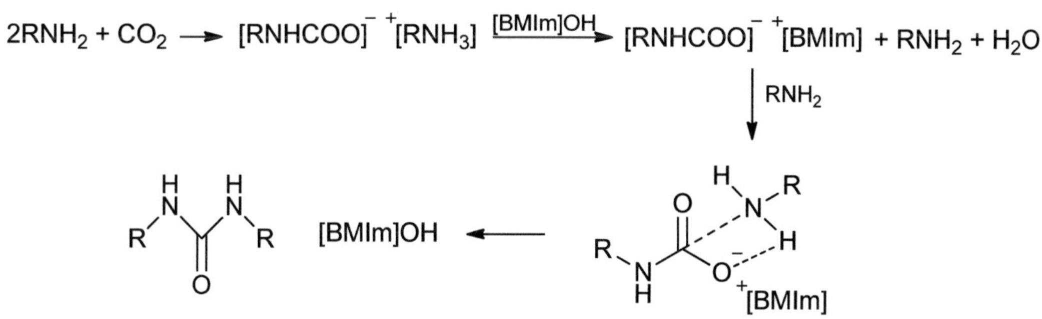

Scheme 33. A proposed mechanism for the synthesis of dialkyl urea from amine and CO_2 catalyzed by [BMIm]OH [107].

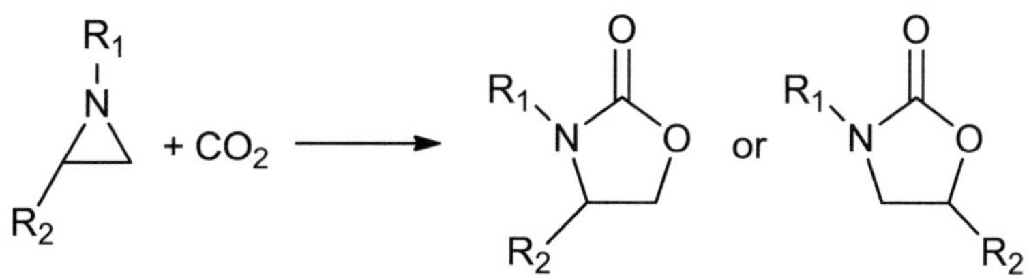

Scheme 34. Synthesis of 2-oxazolidinone from CO_2 and aziridine.

Kawanami et al. employed I_2/IL systems for the reaction of CO_2 and 2-methyl aziridine (R_1 = H, R_2 = CH_3) [109]. Surprisingly, the reaction system of $[Oc_4N][TFSI]$ (tetraoctylammonium bis(trifluoromethylsulfonyl)imide) regioselectively and almost quantitatively gave 4-methyl-oxiazolidione at 60°C under 5 MPa CO_2 for a very short reaction time of 5 min, although the reason for the high selectivity to the 4-substituted product was unknown. [EMIm][TFSI] was less effective for the reaction than $[Oc_4N][TFSI]$. The optimum pressure of CO_2 was around 10 MPa. This was discussed by the solubility of CO_2 in IL and that of the aziridine to $scCO_2$. At lower pressures, the lower concentration of CO_2 in the IL phase limited the reaction, while the lower yield at higher pressures was ascribed to the lower concentration of the substrate in the IL because of its dissolution into $scCO_2$ phase. When ethanol was used as the solvent instead of $[Oc_4N][TFSI]$, the selectivity to 4-methyl-2-oxazolidione was lower (70%) because of the formation of the regioisomer of 5-methyl-2-oxazolidinone, and a long period of reaction time was required to get a reasonable conversion. The TOF value obtained in $[Oc_4N][TFSI]$ was 116 times larger than that in ethanol.

Another way to produce cyclic urethane is the reaction of CO_2 and aminoethanol. Aminoethanols are much cheaper than aziridines, so this reaction is preferable for large scale synthesis of cyclic urethane. Our group examined the use of [BMIm]-based ILs,

[Bu_4N]Br and [Bu_4P]Br for this reaction at 150°C under 10 MPa CO_2 [110]. The reaction yielded not only 2-oxazolidinone but also 1-(2-hydroxyethyl)-2-imidazolidinone and oligomeric products (Scheme 35). Among the ILs used, [Bu_4N]Br gave the highest conversion of aminoethanol, and [BMIm]BF_4 produced almost only the substituted imidazolidinone with a high yield. On the basis of the work by Deng et al. [106] described above, effects of adding alkali metal carbonates or hydroxides on the reaction with [Bu_4N]Br were further investigated. Among the alkali salts tested, K_2CO_3 inhibited the formation of the oligomers. Other alkali additives did not show beneficial influence. The reaction runs for different periods of reaction time revealed that the oxiazolidione formed is consecutively converted to the substituted imidazolidinone by the reaction with aminoethanol. On combining this result with those of reaction runs using several types of amino alcohols, the reaction mechanisms for the oxazoilidionone and the substituted imidazolidione were proposed. In these mechanisms, IL was considered to be Lewis acid on the basis of reported IR and NMR studies, which showed the Lewis acidic nature of the hydrogen atom on the C(2) of the imidazolium ring [111, 112].

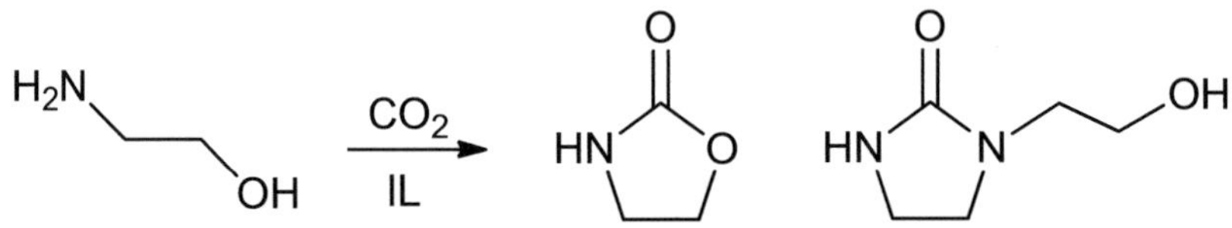

Scheme 35. Reaction of CO_2 with aminoethanol in the presence of ILs.

Deng's group who carried out the reaction of CO_2 and propargylic alcohol (see section 2.1.1.) also reported a similar reaction of the synthesis of methylene oxazolidionone from amine, propargylic alcohol, and CO_2 using the reaction system of CuCl/[BMIm]BF_4 (Scheme 36) [113]. Various methylene oxazolidinoes could be obtained from the corresponding amines and proprgylic alcohols with excellent yields at 100°C under 2.5 MPa of CO_2 for 10−15 h. Only a small amount of the product was obtained in the absence of CuCl, and good product yields were obtained even in organic solvents. Thus, the IL again seems to simply act as a solvent. Catalyst and IL screening revealed that the yield was in the orders of CuCl > CuI ~ CuBr > $CuCl_2$ and of [BMIm]BF_4 > [BMIm]PF_6 > [BPy]BF_4. It should be noted that [BMIm]PF_6 gave only polymeric products for the reaction of CO_2 and propargylic alcohol, i.e. in the absence of amine. The reaction system was again specific for tertiary alcohols; no desired products were detected from secondary and primary alcohols. The same results were obtained for the reaction of CO_2 and propargylic alcohol.

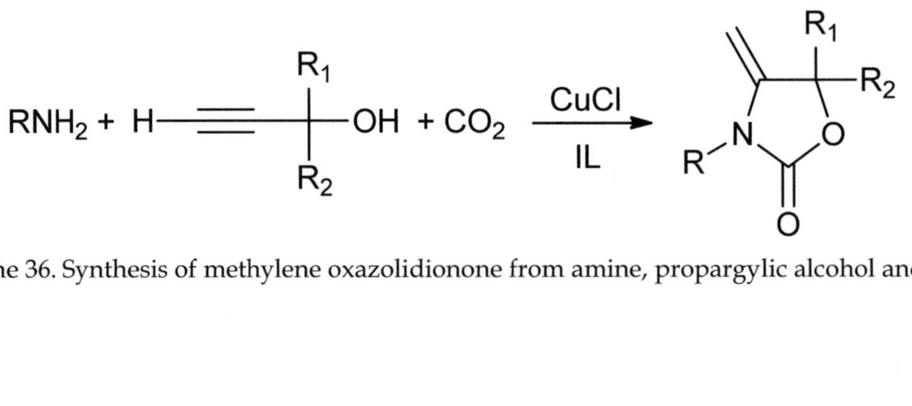

Scheme 36. Synthesis of methylene oxazolidionone from amine, propargylic alcohol and CO_2.

Recently, Bhanage's group reported another interesting example of using IL for a reaction converting CO_2 to useful chemicals [114]. They used a basic IL of [BMIm]OH for the reactions of CO_2 and aminobenzonitriles to produce rather complex compounds of quinazoline-2,4(1H,3H)-diones (Scheme 37), which are important because of their biological activity and widely used as key structures in medical drugs. [BMIm]OH gave good to excellent product yields from a broad range of aminobenzonitriles at 120°C and 3 MPa CO_2 for 18 h. This protocol was also applicable for five/six membered *N*-heterocyclic carbonitriles. [BMIm]OH was much more effective for the reaction than organic and inorganic bases of Et_3N and Cs_2CO_3. Other ILs of [BMIm]BF_4 and [BMIm]HSO_4 showed no activity. After the reaction, the product was precipitated by pouring the reaction mixture to water and filtered out. The catalyst in the filtrate was recovered by removing water under vacuum. The catalyst was recycled four times. Unfortunately, gradual decreases in the yield on the recycling was observed; however, no structural change of [BMIm]OH was observed. Loss of the catalyst during the recycling might be the reason for the decrease in the yield. Scheme 38 represents the proposed reaction mechanism in which the deprotonation of the amino group by the OH anion of the IL, the formation of [BMIm] carbamate ester, intermolecular cyclization, and the rearrangement of the resulting cyclic intermediate are involved.

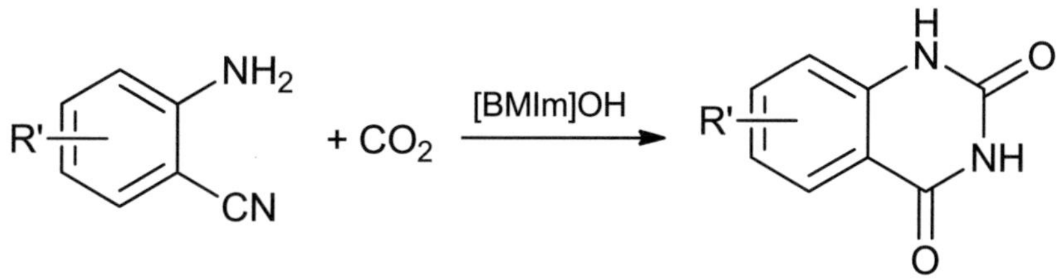

Scheme 37. Synthesis of quinazoline-2,4(1H,3H)-diones from CO_2 and aminobenzonitriles.

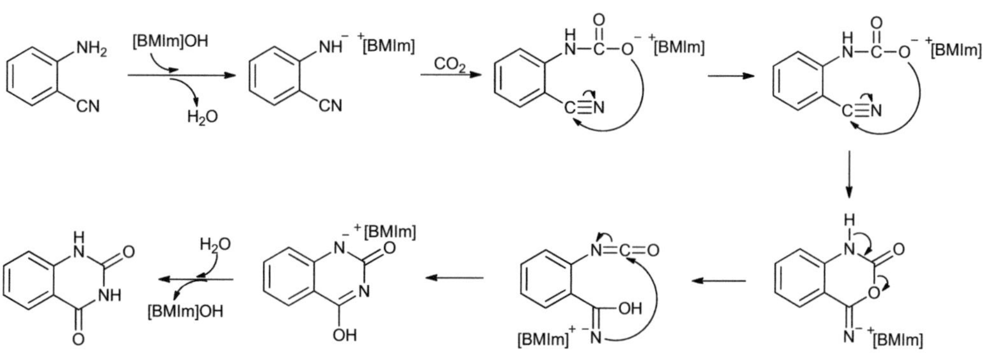

Scheme 38. A proposed mechanism for quinazoline-2,4(1H,3H)-dione from CO_2 and aminobenzonitrile catalyzed by [BMIm]OH [114].

5. Polycarbonate synthesis in/by IL

Polycarbonates, with such features as high ductility, good transparency, high heat and impact resistance, and high biocompatibility, are considered as specialty polymers and find increasing uses in engineering thermoplastics and many different fields [115]. The copolymerization of CO_2 and epoxides to produce polycarbonates is the most promising area of CO_2 utilization as a direct material, which may lead to a solution for resource, energy, and environmental problems. Inoue et al. first reported the synthesis of high molecular weight polycarbonates by alternating copolymerization of CO_2 and epoxides using a catalyst derived from diethyl zinc and water [116]. Since then, several excellent reviews [117–121] have appeared on this interesting topic. It is demonstrated that coordination compounds based on zinc, aluminum, cadmium, chromium, manganese, cobalt, and rare-earth metals are promising candidates for highly active catalytic systems for the copolymerization of epoxides with CO_2. In the following, we will deal with the polycarbonate synthesis from CO_2 and epoxides with ILs as catalysts or cocatalysts.

Park et al. used several imidazolium-based ILs of different alkyl groups (C_2–C_8) and anions as catalysts in the copolymerization of CO_2 with a substituted epoxide of phenyl glycidyl ether (PGE) in a batch or semi-batch reactor without any solvent (Scheme 39) [122–124]. In the presence of the ILs, CO_2 could be effectively added to the epoxide ring of PGE to produce the polycarbonate whose polydispersity was close to unity, i.e. a narrow molecular weight distribution. The activity of IL containing Cl^- was in the order of $[BMIm]^+ > [EMIm]^+ \approx [HMIm]^+ \approx [OMIm]^+$. This was ascribed to the streric hindrance effect and/or the differences in the hydrophobicity and the viscosity. The order of the activity for the different anions in imidazolium-based ILs decreased as $Cl^- > BF_4^- >> PF_6^-$, which was proposed to result from the difference in the solubility of IL in PGE. $[EMIm]PF_6$ had no activity. Every Cl-containing ILs produced polycarbonates with the carbonate content, $f(CO_2)$, of 100% by the copolymerization at 80°C. For the copolymerization at 120°C, the high $f(CO_2)$ value of the polycarbonate was maintained with [BMIm]Cl, whereas those value were about 70% with the other ILs. Higher CO_2 pressure was favorable for the PGE conversion and $f(CO_2)$ [124]. The reaction mode (batch or semi-batch) also affected $f(CO_2)$ [122,123]. The polycarbonates produced in the batch reactor sometimes had lower $f(CO_2)$ than those produced in the semi-batch reactor (under continuous pressurized CO_2 flow) depending on the polymerization conditions (temperature, CO_2 pressure, reaction time). For comparison, $[Bu_4N]Cl$, and [BPy]Cl were also used for the coplymerization [123,124]. Compared to [BMIm]Cl, $[Bu_4N]Cl$, and [BPy]Cl showed lower activities, average molecular weight (Mn), and $f(CO_2)$.

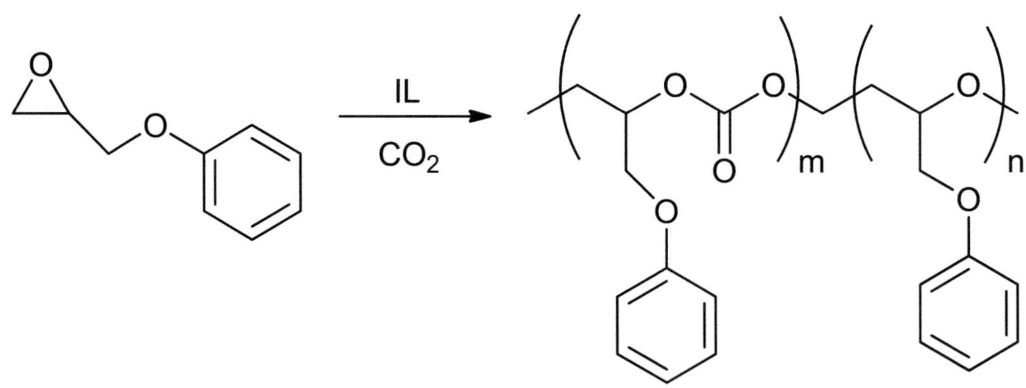

Scheme 39. Copolymerization of PGE and CO_2 in the presence of IL.

As mentioned in section 2.1., there can be some synergistic effect by the combination of IL and zinc halide for the CO_2 cycloaddition to epoxide. Park et al. also investigated the influence of the co-presence of $ZnBr_2$ on the copolymerization of CO_2 and PGE with [BMIm]Cl [124]. $ZnBr_2$ alone showed a low activity and a low Mn, but enhanced the activity of [BMIm]Cl. By the co-presence of $ZnBr_2$, Mn of the copolymer produced was also increased; however, its f(CO_2) was smaller than that of the copolymer produced with the IL alone. For the enhancement of the activity, they proposed the cooperative activation of epoxide by Zn and halide anion of the IL, which was very similar to that proposed for the CO_2 cycloaddition reaction (Scheme 6 in section 2.1.2.).

They also prepared silica-immobilized imidazolium ILs of different alkyl groups and used them as the catalysts for the same copolymerization [125]. Although the immobilized IL catalysts were easily separated by filtration after the polymerization, they produced copolymers whose Mn and f(CO_2) were smaller than those of the copolymers prepared with the corresponding homogeneous IL catalysts. Effects of the alkyl chain length of IL and the reaction conditions on the nature of the copolymer synthesized showed similar trends to those observed with the homogeneous ILs. The immobilized ILs could be reused up to four consecutive runs without considerable loss of activity and f(CO_2).

Several research groups reported that various ILs could be effective cocatalysts in the copolymerization of CO_2 and epoxides with metal salen or metal porphyrin complexes [126–130]. In some cases, it was shown that the activities of theses metal complexes were drastically enhanced by the co-presence of IL, although they had no or a very low activity for the coplymerization in the absence of IL.

Inoue and co-workers reported the successful alternating copolymerization of CO_2 with EO, PO, or CHO (cyclohexene oxide) by the use of Al porphyrin complexes in combination with quaternary phosphonium or ammonium salts [126]. The copolymerization of CO_2 and PO or CHO was conducted at room temperature and 5 MPa CO_2, producing polycarbonates with the perfect alternating structure (f(CO_2) = 100%) and a very narrow molecular weight distribution (Mw/Mn < 1.1). For the polycarbonate produced from EO, f(CO_2) was 70%. When one reaction run with PO was completed, the next run was carried out by just adding CO_2 and PO to the reactor. The molecular weight distribution measurements of the starting and the final copolymers revealed that the average molecular weight was increased, while keeping the narrow distribution. This indicated the "living" nature of the copolymerization with this catalyst system. They also achieved the synthesis of block copolymers.

The same group also used the system of Al salen complex and [Et_4N]OAc for the copolymerization of CO_2 and CHO [127]. This catalyst system required a higher reaction temperature of 80°C to obtain the copolymer with a high yield and high f(CO_2), compared to the above mentioned catalyst system using Al porphyrin complex. In the absence of the IL, the salen Al complex was less active and yielded the polymer mostly containing ether linkage (f(CO_2) = 2%). The lower activity and a very lower f(CO_2) value without IL were also reported by Jung et al. [128] for the coplymerization with Al salen/[Et_4N]Br. Thus, one can say that the cocatalyst of IL enhances not only the polymerization activity of the metal complex catalyst, but also its ability for the CO_2 insertion.

Lu and Wang reported the CO_2 and PO copolymerization over the binary catalysts of (salen)Co^{III}X/[Bu_4N]Y at 25°C and 2 MPa CO_2 for 3 h [129]. All the polycarbonate synthesized with these catalysts had f(CO_2) values more than 99%. However, the selectivity to polycarbonate and cyclic carbonate could be drastically affected by adjusting axial group X of (salen)Co^{III}X and the anion Y of [Bu_4N]Y. It was found that the use of a cobalt complex with an electron-withdrawing axial group X and a quaternary ammonium salt with poor leaving ability anion, increased the selectivity for poly(propylene carbonate). For example, in the presence of the $[Bu_4N]Br$ cocatalyst, a change in the axial group X of chiral $[salenCo^{III}X]$ from acetate to dinitrophenoxy improved the selectivity to polycarbonate from 3 to 78%. On the other hand, with the Co complex of dinitrophenoxy as the catalyst, a change in the anion Y of $[Bu_4N]Y$ from Br^- to Cl^- or CH_3COO^- resulted in an increase in the selectivity for polycarbonate from 78 to 99%. Besides, the catalyst system was found to be applicable to other aliphatic epoxides, providing the corresponding polycarbonates with > 99% carbonate linkages.

Imidazolium salts ILs had also been proved to be effective cocatalysts in the copolymerization of CO_2 with cyclohexene oxide (CHO) catalyzed by $(salen)Cr^{III}Cl$ complex [130]. The loading of ILs had great effect on the copolymerization. Although the Cr complex alone showed no activity, it produced the polycarbonate in the presence of 2.25 equivalent [OMIm]Br with a TOF value of 219 h^{-1} (based on Cr) and $f(CO_2)$ of 98%. The TOF value was slightly increased to 286 h^{-1} by using 13.5 equivalent IL, while keeping the high $f(CO_2)$. However, the co-presence of 20 equivalent IL decreased the TOF to 86 h^{-1} and $f(CO_2)$ to 56%. These results strongly suggest that the activity of [OMIm]Br was low, and there is the synergistic effect for the activity and $f(CO_2)$ in the $(salen)Cr^{III}Cl$/[OMIm]Br system with appropriate ratios, although the authors did not point out these issues. Anions of imidazolium salts with higher nucleophilicity and better leaving ability, such as Cl^- and Br^-, were beneficial for improving the activity, $f(CO_2)$, and Mn than BF_4^- and PF_6^-. In addition, the activity was increased with the alkyl chain length of the IL cation. Dodecylmethylimidazolium bromide ([DDMIm]Br) displayed an enhanced TOF of 242.5 h^{-1}, $f(CO_2)$ > 99%, and, and a lower polydispersity index of 1.1. The enhanced activity might come from the increased nucleophilicity of an anion, the hydrophobicity, and solubility of CO_2 with increasing alkyl chain length.

6. Concluding remarks

This chapter reviews the application of ILs to the chemical and electrochemical synthesis of carbonates and related compounds using CO_2 as a direct or indirect reactant. Several types of ILs are active catalysts and active components in multi-component catalysts for these synthetic reactions, depending on the kind and chemical nature of cations and anions involved. The above-reviewed reactions may proceed under relatively mild conditions, which is one of practical merits of using ILs. In some cases, however, a long reaction time is needed. For one-step reactions such as the synthesis of a cyclic carbonate from an epoxide and CO_2, the selection of active ILs is important and the separation and recycling of ILs are also an issue of practical importance. The ILs may be dispersed and immobilized on the surfaces of suitable solids in order to change the homogeneous catalytic reactions into the heterogeneous ones, in which the post-reaction catalyst separation and recycling would be easier. A few interesting attempts are to conduct multi-step reactions such as the synthesis of a cyclic carbonate from an olefin and CO_2, including the first formation of an epoxide and the second addition of CO_2 to this product, in a single reactor. Multi-component catalysts containing a few different types of ILs should be required, which are respectively and selectively active for certain reactions involved. In addition, if possible, these active components should be effective under the same or very similar reaction conditions, which, however, is not an easy requirement at present. A possible way to cope with these difficulties is the use of different reaction conditions, which are switched at a certain stage of reactions. For this, careful kinetics analysis of the reactions is required in advance.

Note again that ILs are interesting and effective components of catalysts for organic synthetic reactions, as demonstrated with many examples. For the industrial application of ILs in order to realize eco-friendly and effective processes including chemical/catalytic and post-reaction chemical/physical procedures, further work is still of significance to clarify their catalytic, physicochemical, and biological features in more detail.

7. Acknowledgements

One of the authors (JMS) acknowledges the financial support from National Natural Science Foundation of China (21073049), State Key Lab of Urban Water Resource and Environment of Harbin Institute of Technology (HIT2010DX15). This work was supported in part by Japan Society for the Promotion of Science with Grant-in-Aid for Scientific Research (B) 22360327.

8. References

[1] M. Lancaster, Green Chemistry: An Introductory Text, Roy. Soc. Chem., Cambridge, 2002, pp. 154-161.
[2] T. Welton, Chem. Rev. 99 (1999) 2071–2084.
[3] M. A. P. Martins, C. P. Frizzo, D. N. Moreira, N. Zanatta, H. G. Bonacorso, Chem. Rev. 108 (2008) 2015–2050.
[4] M. Haumann, A. Riisager, Chem. Rev. 198 (2008) 1474–1497.
[5] F. Jutz, J.-M. Anderson, A. Baiker, Chem. Rev. 111 (2011) 322-353.
[6] A. G. Shaikh, S. Sivaram, Chem. Rev. 96 (1996) 951–976.
[7] X. Xiaoding, J. A. Moulijn, Energy Fuels 10 (1996) 305-325.
[8] J. Sun, S. Fujita, M. Arai, J. Organomet. Chem. 690 (2005) 3490-3497.
[9] J.J. Peng, Y.Q. Deng, New. J. Chem. 25 (2001) 639–641.
[10] H. Kawanami, A. Sasaki, K. Matsui, Y. Ikushima, Chem. Commun. (2003) 896–897.
[11] T. Seki, J.-D. Grunwaildt, A. Baiker, J. Phys. Chem. B 13 (2009) 114–122.
[12] V. Caló, A. Nacci, A. Monopoli, A. Fanizzi, Org, Lett. 4 (2002) 2561-2563.
[13] Y. Gu, F. Shi, Y. Deng, J. Org. Chem. 69 (2004) 391-394.
[14] H. Kisch, R. Millini, I.-J. Wang, Chem. Ber. 119 (1986) 1090–1094.
[15] H.S. Kim, J.J. Kim, H. Kim, H.G. Jang, J. Catal. 220 (2003) 44–46.
[16] J. Palgunadi, O.S. Kwon, H. Lee, J.Y. Bae, B.S. Ahn, N.-Y. Min, H.S. Kim, Catal. Today 98 (2004) 511-514.
[17] F. Li, L. Xiao, C. Xia, B. Hu, Tetrahedron Lett. 45 (2004) 8307–8310.
[18] J. Sun, S. Fujita, F. Zhao, M. Arai, Green Chem. 6 (2004) 613-616.
[19] J. Sun, S. Fujita, F. Zhao, M. Arai, Appl. Catal. A: Gen. 287 (2005) 221-226.
[20] R. Srivastava, S. Fujita, M. Arai, React. Kinet. Catal. Lett. 96 (2009) 55-64.
[21] S. Fujita, M. Nishiura, M. Arai, Catal. Lett. 135 (2010) 263-268.
[22] J. Sun, L. Wang, S. Zhang, Z. Li, X. Zhang, W. Dai, R. Mori, J. Mol. Catal. A: Chem. 256 (2006) 295–300.
[23] S.-S. Wu, X.-W. Zhang, W.-L. Dai, S.-F. Yin, W,-S, Li, Y,-Q. Ren, C.-T. Au, Appl. Catal. A: Gen. 341 (2008) 106–111.
[24] K. Kossev, N. Koseva, K. Troev, J. Mol. Catal. A: Chem. 194 (2003) 29-37.
[25] Sibaouih, P. Ryan, M. Leskela, B. Rieger, T. Repo, Appl. Catal. A: Gen. 365 (2009) 194–198.
[26] F. Ono, K. Qiao, D. Tomida, C. Yokoyama, J. Mol. Catal. A: Chem. 263 (2007) 223–226.

[27] J. Sun, L. Ren, S. Zhang, W. Cheng, Tetrahedron Lett. 50 (2009) 423–426.
[28] X.-B. Lu, R. He, C.X. Bai, J. Mol. Catal. A: Chem. 186 (2002) 1–11.
[29] X.-B. Lu, X.J. Feng R. He Appl. Catal. A: Gen. 234 (2002) 25–33.
[30] X.-B. Lu, Y.J. Zhang, B. Liang, X. Li, H. Wang, J. Mol. Catal. A: Chem. 210 (2004) 31–34.
[31] M. North, R. Pasquale, Angew. Chem. Int. Ed. 48 (2009) 2946–2948.
[32] J. Sun, S. Zhang, W. Cheng, L. Ren, Tetrahedron Lett. 49 (2008) 3588–3591.
[33] Y. Zhou, S. Hu, X. Ma, S. Liang, T. Jiang, B. Han, J. Mol. Catal. A: Chem. 284 (2008) 52–57.
[34] J.-Q. Wang, D.-L. Kong, J.-Y. Chen, F. Cai, L.-N. He, J. Mol. Catal. A: Chem. 249 (2006) 143–148.
[35] J.-Q. Wang, X.-D. Yue, F. Cai, L.-N. He, Catal. Commun. 8 (2007) 167–172.
[36] Zhu, T. Jiang, B. Han, J. Zhang, Y. Xie, X. Ma, Green Chem. 9 (2007) 169–172.
[37] L.-F. Xiao, F.-W. Li, J,-J, Peng, C.-G. Xia, J. Mol. Catal. 253 (2006) 265–269.
[38] H.-L. Shim, S. Udayakumar, J.-I. Yu, I. Kim, D.-W. Park, Catal. Today, 148 (2009) 350–354.
[39] S. Udayalumar, V. Raman, H.-L. Shim, D.-W. Park, Appl. Catal. A: Gen. 368 (2009) 97–104.
[40] T. Sakai, Y. Tsutsumi, T. Ema, Green. Chem. 10 (2008) 337-341.
[41] X. Zhang, D. Wang, N. Zhao, A.S.N. Al-Arifi, T. Aouak, Z.A. Al-Othman, W. Wei, Y. Sun, Catal. Commun. 11 (2009) 43–46.
[42] T. Takahashi, T. Watahiki, H. Yasuda, T. Sakakura, Chem. Commun. (2006) 1664–1666.
[43] S. Udayalumar, S.-W. Park, D.-W. Park, B.-S. Choi, Catal. Commun. 9 (2008) 1563–1570.
[44] K. Motokura, S. Itagaki, Y. Iwasawa, A. Miyaji, T. Baba, Green Chem. 11 (2009) 1876–1880.
[45] S. Udayalumar, M.-K. Lee, H.-L. Shim, D.-W. Park, Appl. Catal. A: Gen. 365 (2009) 88–95.
[46] W.-L. Dai, L. Chen, S.-F. Yin, S.-L. Luo, C.-T. Au, Catal. Lett. 135 (2009) 295–304.
[47] Y. Xie, Z. Zhang, T. Jiang, J. He, B. Han, T. Wu, K. Ding, Angew. Chem. Int. Ed. 46 (2007) 7225–7258.
[48] K. Qiao, F. Ono, Q. Bao, D. Tomida, C. Yokoyama, J. Mol. Catal. A: Chem. 303 (2009) 30–34.
[49] H.S. Kim, J.J. Kim, H.N. Kwon, M.J. Chung, B.G. Lee, H.G. Jang, J. Catal. 205 (2002) 226–229.
[50] D.W. Park, B.-S. Yu, E.-S. Jeong, I. Kim, M.-I. Kim, K.-J. Oh, S.-W. Park, Catal. Today, 98 (2004) 499–504.
[51] J. Sun, W. Cheng, W. Fei, Y. Wang, Z. Meng, S. Zhang, Catal. Today, 148 (2009) 361–367.
[52] Y. Zhao, J.-S. Tian, H.-X. Qi, Z.-N. Han, Y.-Y. Zhuang, L.-N. He, J. Mol. Catal. A: Chem. 271 (2007) 284–289.
[53] P. Tascedda, E. Duñach, J. Chem. Soc. Chem. Commun. (1995) 43–44.
[54] P. Tascedda, M. Weidmann, E. Dinjus, E. Duñach, Appl. Organometal. Chem. 15 (2001) 141–144.
[55] D.F. Niu, C.T. Xu, L. Zhang, Y.W. Luo, K. Zhang, J.X. Lu, Chin. J. Catal. 28 (2007) 880–884.
[56] K. Otsuka, T. Yagi, I. Yamanaka, Chem. Lett. 3 (1994) 495–498.
[57] M.A. Casadei, S. Cesa, M. Feroci, A. Inesi, L. Rossi, F.M. Moracci, Tetrahedron 53 (1997) 167–176.

[58] M.A. Casadei, A. Inesi, L. Rossi, Tetrahedron Lett. 38 (1997) 3565–3568.
[59] M.A. Casadei, S. Cesa, M. Feroci, A. Inesi, New J. Chem. 23 (1999) 433–436.
[60] M.A. Casadei, S. Cesa, L. Rossi, Eur. J. Org. Chem. 2000 (2000) 2445–2448.
[61] H.Z. Yang, Y.L. Gu, Y.Q. Deng, F. Shi, Chem. Commun. 3 (2002) 274–275.
[62] H.Z. Yang, Y.L. Gu, Y.Q. Deng, J. Org. Chem. 22 (2002) 995–998.
[63] Y. Wang, G.Q.Yuan, Y.C. Zeng, H.F. Jiang, Chin. J. Org. Chem. 27(2007) 1397-1400.
[64] L. Zhang, Y.W. Luo, D.F. Niu, X.D.Yu, J.X. Lu, Chin. J. Catal. 28 (2007) 100–102.
[65] P.A. Wender, F.E. McDonald, Tetrahedron Lett. 31(1990) 3691–3694.
[66] J.A. Verdol (1962) US Pat 3 205 305
[67] J.M. Sun, S. Fujita, B.M. Bhanage,M.Arai, Catal. Commun. 5 (2004) 83–87.
[68] J.M. Sun, S. Fujita, B.M. Bhanage, M. Arai, Catal. Today 93-95 (2004) 383–388.
[69] J.M. Sun, Y.L. Wang, X.J. Qu, D.Z. Jiang, F.-S. Xiao, S. Fujita, M. Arai, Chem. J. Chin. Univ. 27 (2006) 1522–1525.
[70] N. Eghbali, C.J. Li, Green Chem. 9 (2007) 213–215.
[71] M.B. Smith, J. March, March's Advanced Organic Chemistry. Wiley-Interscience, New York, 2001.
[72] J.L. Wang, J.Q. Wang, L.N. He, X.Y. Dou, F. Wu, Green Chem. 10 (2008) 1218-1223.
[73] J.M. Sun, S. Fujita, F.Y. Zhao, M. Arai, J. Catal. 230 (2005) 398–405.
[74] Y.L. Wang, J.H. Sun, D. Xiang, L.Wang, J.M. Sun, F.-S. Xiao, Catal. Lett. 129 (2009) 437–443.
[75] F. Ono, K. Qiao, D. Tomida, C. Yokoyama, Appl. Catal. A Gen. 333 (2007) 107–113.
[76] G.S. Owens, M.M. Abu-Omar, Chem. Commun. (2000) 1165-1166.
[77] M. Aresta, E. Quaranta, J. Mol. Catal. 41 (1987) 355–359.
[78] M. Aresta, A. Dibenedetto, J. Mol. Catal. A: Chem. 182–183 (2002) 399–409.
[79] D.S. Bai, H.W. Jiang, Green Chem. 12 (2010) 39–41.
[80] Y. Ono, Appl. Catal. A: Gen. 155 (1997) 133–166.
[81] T. Sakakura, K. Kohno, Chem. Commun. (2009) 1312-1330.
[82] Shaikh, S. Sivaram, Chem. Rev. 96 (1996) 951–976.
[83] M.A. Pacheco, C.L. Marshall, Energy Fuels 11 (1997) 2–29.
[84] T. Wei, M. H. Wang, W. Wei, Y. H. Sun, B. Zhong. Green Chem. 5 (2003) 343–346.
[85] Behr, Angew. Chem. Int. Ed. Engl. 27 (1988) 661–678.
[86] S.T. King, Catal. Today 33 (1997) 173–182.
[87] S. Fang, K. Fujimoto, Appl. Catal. A: Gen. 142 (19956) L1–L3.
[88] S. Fujita, B. M. Bhanage, Y. Ikushima, M. Arai, Green Chem. 3 (2001) 87–91.
[89] H. Wang, B. Lu, Q. H. Cai, F. Wu, Y. K. Shan, Chin. Chem. Lett., 16 (2005) 1267-1270.
[90] E.-S. Jeong, K.-H. Kim, D.-W. Park, S.-W. Park, J.-W. Lee, React. Kinet. Catal. Lett. 86 (2005) 241–248.
[91] H.-Y. Ju, M.D. Manju, D.-W. Park, Y. Choe, S.-W. Park, React. Kinet. Catal. Lett. 90 (2007) 3–9.
[92] H.-Y. Ju, M.D. Manju, K.-H. Kim, S.-W. Park, D.-W. Park, Korean. J. Chem. Eng., 24 (2007) 917–919.
[93] M.M. Dharman, H.-Y. Ju, H.-L. Shim, M.-K. Lee, K.-H. Kim, D.-W. Park, J. Mol. Catal. A: Chem. 303 (2009) 96–101.
[94] K.-H. Kim, D.-W. Kim, C.-W. Kim, J.-C. Koh, D.-W. Park, Korean J. Chem. Eng., 27 (2010) 1441–1445.
[95] D.-W. Kim, C.-W. Kim, J.-C. Koh, D.-W. Park , J. Ind. Eng. Chem., 16 (2010) 474–478.

[96] D. W. Kim, D.-O. Lim, D.-H. Choc, J.-C. Koh, D.-W. Park, Catal. Today (2010), doi: 10.1016/j.cattod.2010.11.010.
[97] Z. Yang, L.-N. He, X.-Y. Dou, S. Chanfreau, Tetrahedron Lett. 51 (2010) 2931–2934.
[98] J.-S. Tian, J.-Q. Wang, J.-Y. Chen, J.-G. Fan, F. Cai, L.-N. He, Appl. Catal. A: Gen. 301 (2006) 215–221.
[99] B. M. Bhanage, S. Fujita, Y. Ikushima, M. Arai, Appl. Catal. A: Gen. 219 (2001) 259–266.
[100] S. Fujita, B.M. Bhanage, D. Aoki, Y. Ochiai, N. Iwasa, M. Arai, Appl. Catal. A: Gen. 313 (2006) 151–159.
[101] X. Chen, C. Hu, J. Su, T. Yu, Z. Gao, Chin. J. Catal. 27 (2006) 485–488.
[102] L. Zhang, D. Niu, K. Zhang, G. Zhang, Y. Luo, J. Lu, Green Chem. 10 (2008) 202–206.
[103] P. Xiao, G. Zhang, L. Zhang, D. Niu, J. Lu, Chin. J. Catal. 30 (2009) 43–47.
[104] X.Y. Wang, S.Q. Liu , K.L. Huang, Q.J. Feng, D.L. Ye, B. Liu, J.L. Liu, G.H. Jin, Chin. Chem. Lett. 21 (2010) 987–990.
[105] D. Yuan, C. Yan, B. Lu, H. Wang, C. Zhong, Q. Cai, Electrochim. Acta 54 (2009) 2912–2915.
[106] F. Shi, Y. Deng, T. SiMa, J. Peng, Y. Gu, B. Qiao, Angew. Chem. Int. Ed. 42 (2003) 3257–3260.
[107] T. Jiang, X. Ma, Y. Zhou, S. Liang, J. Zhang, B. Han, Green Chem. 10 (2008) 465–469.
[108] R. Nomura, T. Nakao, Y. Nishio, S. Ogawa, A. Ninagawa, H. Matsuda, Chem. Ber. 122 (1989) 2407–2409.
[109] H. Kawanami, H. Matsumoto, Y. Ikushima, Chem. Lett. 34 (2005) 60–61.
[110] S. Fujita, H. Kanamaru, H. Senboku, M. Arai, Int. J. Mol. Sci. 7 (2006) 438–450.
[111] K. Sato, S. Arai, T. Yamagishi, Tetrahedron Lett. 40 (1999) 5219–5222.
[112] J.L. Thomas, J. Howarth, K. Hanlon, D. McGuirk, Tetrahedron Lett. 41 (2000) 413–416.
[113] Y. Gu, Q. Zhang, Z. Duan, J. Zhang, S. Zhang, Y. Deng, J. Org. Chem. 70 (2005) 7376–7380.
[114] Y.P. Patil, P.J. Tambede, K.M. Deshmukh, B.M. Bhanage, Catal. Today 148 (2009) 355–360.
[115] W. Kuran, Polymeric Material Encyclopedia, Vol. 9, CRC Press, 1996.
[116] S. Inoue, T. Tsuruta, H. Koinuma, J. Polym. Sci. Polym. Lett. 7 (1969) 287–292.
[117] D.J. Darensbourg, Chem. Rev. 107 (2007) 2388–2410.
[118] M.S. Super, E.J. Beckman, Trends Polym. Sci. 5(7) (1997) 236–240.
[119] D.J. Darensbourg, R.M. Mackiewiicz, A.L. Phelps, D.R. Billodeaux, Acc. Chem. Res. 37 (2004) 836–844.
[120] G.W. Coates, D.R. Moore, Angew. Chem. Int. Ed. 43 (2004) 6618–6639.
[121] H. Sugimoto, S. Inoue, J. Polym. Sci. Part A: Polym. Chem. 42 (2004) 5561–5573.
[122] N.Y. Mun, K.H. Kim, D.W. Park, Y. Choe, I. Kim, Korean J. Chem. Eng. 22(2005) 556–559.
[123] D.W. Park, N.Y. Mun, E.H. Lee, Y. Choe, S.W. Park, React. Kinet. Catal. Lett. 89 (2006) 149–156.
[124] J.Y. Ahn, H.L. Shim, K.H. Kim, I. Kim, S.W. Park, D.W. Park, Korean. J. Chem. Eng. 25 (2008) 693–696.
[125] H.L. Shim, M.K. Lee, K.H. Kim, D.W. Park, S.W. Park, Polym. Adv. Technol. 19 (2008) 1436–1440.
[126] T. Aida, M. Ishikawa, S. Inoue, Macromolecules 19 (1986) 8–13.
[127] H. Sugimoto, H. Ohtsuka, S. Inoue, J. Polym. Sci. Part A: Polym. Chem. 43 (2005) 4172–4186.
[128] J.H. Jung, M. Ree, T. Chang, J. Polym. Sci. Part A: Polym. Chem. 37 (1999) 3329–3336.
[129] X.B. Lu, Y. Wang, Angew. Chem. Int. Ed. 43 (2004) 3574–3577.
[130] X.Q. Xu, C.M. Wang, H.R. Li, Y. Wang, W.L. Sun, Z.Q. Shen, Polymer 48 (2007) 3921–3924.

13

Ionic Liquids in Polar Diels-Alder Reactions Using Carbocycles and Heterocycles as Dienophiles

Mancini Pedro M.E., Kneeteman María,
Della Rosa Claudia, Bravo Virginia and Adam Claudia
Área de Química Orgánica- Departamento de Química- Facultad de Ingeniería Química
Universidad Nacional del Litoral Santa Fe
Argentina

1. Introduction

The design and discovery of Ionic Liquids (ILs) displaying melting point lower than 100 °C, in particular Room Temperature Ionic Liquids (RTILs) have been the subject of considerable efforts over the past decade. In particular, RTILs attract considerable attention because they are expected ideal solvents that provide novel reactions fields in the green chemistry. (Hitchcock, et al, 1986; Welton, 1999)

The interest of this class of molecules arises in part from their use as liquid media for a variety of chemical transformations as a substituted for molecular solvents. Moreover it is possible think on ILs as reagents in chemical reactions. (Adam, et al, 2009)

ILs have generated a large amount of interest from the scientific and academic communities due to their surprising and unique properties and their potential use in areas as diverse as synthesis, biocatalysis, electrochemistry, etc. Thus, these classes of molecules are increasingly used in organic chemistry, material sciences and physical chemistry. (Wasserscheid & Keim, 2000; Welton, 2004)

ILs are salts -substances composed exclusively of cations and anions-. This fact differentiates them from simple ionic solutions, where ions are dissolved in a molecular medium. They are also distinct from inorganic molten salts because their melting points are lever than 100 °C (most of them exist in the liquid form at or near room temperature).

RTILs exhibits a variety of desirable properties, such as negligible vapour pressure, which make them interesting for various applications. In particular, the option of fine-tuning chemical and physical properties by an appropriate choice of cations and anions has stimulated much of the current excitement with respect to these compounds and has led to the term "*designer solvents*". An alternative, complementary strategy to tailor properties is to mix two (or more) ionic liquids, including binary mixtures of ionic liquids and molecular solvent.

The majority of ILs belongs to the *N,N*-methyl-(alkyl)-imidazolium, alkyl-pyridinium or ammonium cations associated with a variety of anions such as X^-, BF_4^-, BPh_4^-, PF_6^-, NO_3^-, $CF3SO_3^-$ NCS^-, $N(CN)_2^-$, TiO^-, NTf_2^-, $AlCl_4^-$, etc. Recently a few amidinium-based ILs have been also reported. (Dechambenait, et al, 2010)

The many opportunities arising from all possible combination of cations and anions allows to tune with good precision the characteristics of ILs such as melting point, viscosity, hygroscopy, dielectric constant and microscopic molecular properties. (Rogers & Seddon, 2003; Forsyth, et al, 2004; Kulkarni, et al, 2007; Wilkes, 2002; Yamada, et al, 2007) Due to the large variety of ionic liquids, the imidazolium ILs constitute a large family of anion-cation couples considered as model system which have been the most intensively studied. Specifically, the imidazolium ILs formed by 1-alkyl-3-methyl imidazolium cations which different type of anions, exhibit nano scale structural organization in polar and non polar region for which the spatial distribution evolve with the alkyl chain length. This phenomenon results from the structural asymmetry of cation composed of a polar head group and a non polar alkyl side chain leading to a balance between domains constituted by the non polar alkyl chain interacting through van der Waals forces. The existence of the micro-phase segregation in polar/non polar domains provides new perspectives to elaborate high-performance materials or to consider new selective salvation abilities in imidazolium ILs with respect to non polar solutes and/or associating species. (Martins, et al, 2008; Mac Farlane & Seddon, 2007; Plechkova & Seddon, 2008; Weingärtner, 2008; Yamada, et al, 2007)

In base to their ionic nature, the structure of ILs incorporates different level of complexity. First and in order to maintain local electro neutrality, the high-charge density parts of cations and anions must create a three-dimensional network where the nearest neighbors of a given ion are of opposite sing. Second, the low-charge density residues that are often presents in the ions (generally as alkyl side chain) are segregated from the polar network, forming non polar domains. This nano-segregation/structuration between polar and non polar regions, first predicted by molecular dynamics simulation studies and later corroborated by diffraction techniques, implies the existence of differentiated and complex interactions not only in pure ionic liquid but also in their mixtures with molecular solvents or even other ionic liquids.

Various physical and chemical properties useful to understand the solubility of RTIL´s have been studied, among which dielectric properties are crucially important. However, there are, at least, two problems in the study of dielectric properties. One concerns with experimental techniques and the other concerns with scientific aspects. Furthermore, there arises a basic question of how the permittivity derived assuming that the IL´s are homogeneous is related to the interconnection polar to non polar domains predicted by computer simulation and evidenced by experiments. In addition, anomalous phase separation behavior was reported for binary system of RTILs with some organic compounds.

The specific properties of the RTILs mostly result from having both a very low vapour pressure gaseous phase and a liquid phase of significant polarity, high thermal conductivity offering wide electrochemical windows. ILs have high solvation ability for a wide variety of polar, non polar, organic and inorganic molecules as well as organometallic compounds. Moreover, the possibility of changing their properties allowing the selective solvation of solutes and thus control the mutual miscibility of particular organic compounds such as for instance alcohols and water. As a consequence, the characterization of the properties of different classes of ILs used as solvent for specific applications and for chemical reactions and catalysis, have been intensively investigated. (Mancini et al, 2008, 2004; Fortunato, et al, 2010). Nevertheless, many questions are still open and more fundamental studies remain to be done before achieving a better understanding of the physical properties of ILs at the molecular level. Experimental and theoretical research has been devoted to determine the connection existing between the nanostructural organization in ILs or in its mixtures with molecular solvents, and the remarkable solvation and/or chemical properties.

The Diels-Alder (DA) reaction is one of the most useful processes in preparative organic chemistry. Its potential in heterocyclic chemistry and natural products synthesis is very well known. It provides the chemist with one of his best tool for the preparation of cyclic compounds having a six-membered ring. The process is in one step inter or intramolecular from a diene and dienophile bearing an almost unlimited number of variants. It is worth noting that these variants exist not only in the substitution of the reaction component but also in the electronic nature of the dienes and dienophiles.

While a great effort has been dedicated to the development and use of aromatic compound in these reactions as diene, in general their use as dienophiles has been considered less probable, mainly because aromatic and heteroaromatic compounds have proved to be relatively unreactive as dienophiles. However, it exist a number of examples of five-membered aromatic heterocycles acting as dienophiles in DA reactions with normal electron demand. For example aromatic pentaheterocycles holding an electron withdrawing group in their structures at α- or β-position have shown to be good dienophiles in the interaction with different dienes at an elevated temperature. In that direction, in the 80's studies of aromatic heterocycles -furans, *N*-tosylpyrroles, and thiophenes- in cycloaddition reactions demonstrated the viability of these systems as dienophiles. (Wenkert, et al, 1988, Wenkert & Piettre, 1988)

While the use of these substrates as dienes in thermal and high-pressure DA reactions has been widely analyzed, the employment of such compounds as dienophile has received relatively little attention in the literature. Due to our interest in the cycloaddition chemistry of substituted aromatic heterocycles with electron withdrawing groups, we have reported that 2-and-3-nitrofurans, 2- and 3-nitrothiophenes, 2- and 3-nitro-*N*-tosylpirroles and 2- and 3-nitroselenophenes react with strong and poor dienes under different thermal conditions. The participation of nitro-*N*-tosylpyrroles in cycloaddition reactions made possible a one spot simple indole synthesis. Further studies focused on the dienophilic character of indoles derivatives seemed to be appealing for the total synthesis of carbazol and *Aspidosperma* alkaloids. (Biolatto, et al, 2001). Moreover, we explored the synthesis of different dibenzofurans using nitrobenzofurans as dienophiles. (Della Rosa, et al, 2011) In addition, we have been realized studies on the dienophilic character in DA reactions of other aromatic systems such as nitronaphthalenes. (Paredes, et al, 2007)

In general, when these nitrodienophiles were exposed to a different dienes yielded the corresponding initial cycloadduct products, which normally suffer the thermal extrusion of the nitro group. Unexpected results were observed when using nitronaphthalenes with the less reactive dienes. Interestingly, when 1- and 2-nitronaphthalenes reacted with isoprene they produced the corresponding *N*-naphthylpyrrole. From the results obtained in all these processes, we can conclude that there are two competitive reactions that might take place on mononitronaphthalenes and related compound: the addition to the nitro group (hetero DA process) and the normal DA reaction on the C1-C2 bond, depending on the strength of the diene partner. This hetero DA behavior can be extended to the nitrothiophenes probably due to its high aromaticity in relation to the others aromatic heterocyclepentadienes. More recently, we have reported the findings on the dienophilic character of nitroquinolines and nitroisoquinolines, particularly 5- and 8-nitroquinolines. (Cancian et al, 2010). These dienophiles were exposed to different dienes under thermal conditions using molecular solvents as reaction media. In these last reactions, the best results were obtained with chloroform as molecular solvent due to its potential character hydrogen bond donors (HBD) which could be influence the reactivity of the reaction systems.

In general, these cited polar cycloadditions are a domino process that is initialized by a DA reaction to give the formally [4 + 2] cycloadduct followed for the subsequent concert irreversible elimination of nitrous acid, which is the factor responsible for the feasibility of the overall process. (Domingo, et al, 2008)
In particular, for polar DA reactions one of the most interesting aspects is its solvent dependence. Moreover, in recent years, this reaction has been subject of several studies in order to enhance the reactivity. For specific DA reactions were demonstrated that the aqueous solutions have remarkable increase in reactivity and selectivity, and these results were discussed in terms of hydrophobic effects. ILs with similar properties to water, such as being highly ordered media and good HBD ability have also been shown to have potential influence the outcome of polar DA reactions.

2. Main objectives

The aim of the present review is twofold. Considering the results obtained in the cited type of thermal polar DA reactions using molecular solvents as reaction media, the first purpose is to analyze the influence of Protic Ionic Liquids (PILs) in this class of polar cycloaddition reactions in which the dienophiles are relatively poor. With this purpose ammonium- and imidazolium-based PILs have been selected because the differences in their HBD acidity. The second purpose concerned with theoretically studies using the Density Functional Theory (DFT) methods. We try to obtain information about the factors affecting reactivity and selectivity. In general, it would be possible demonstrated that the ILs solvent effect in these reactions is in general determined by the solvent hydrogen bond donation ability.

2.1 An overview

At first and with the purpose of comparison and reference, we show a compilation of the results corresponding to thermal cycloaddition reactions that using molecular solvent as reaction media. In each Figure we show the reaction conditions and the results by both thermal reactions in molecular solvents and the reactions development in PILs (described later). In the case of the reactions in PILs only we expressed the results with EAN because the results using

2.1.1 Nitroaromatic pentaheterocycles as dienophiles in thermal DA reactions using benzene as solvent

The study of dienophilic character of monosubstituted nitro-aromatic pentaheterocycles was carried out employing 2-nitrofuran **1a** 2,5-dimethyl-3-nitrofuran **1b**, 1-tosyl-2-nitropyrrole **2a**, 1-tosyl-3-nitropyrrole **2b**, 2- and 3- nitrothiophene **3a**, **3b**, and 2- and 3-nitroselenophene **4a, 4b**. In addition, isoprene **5**, 1-*N*-acetyl-*N*-propyl-1,3-butadiene **6**, 1-methoxy-3-trimethylsiloxy-1,3-butadiene (Danishefsky´s diene) **7** and 1-diethyl-amino-3-*tert*-butyldimethyl-silyloxy-1,3-butadiene (Rawal´s diene) **8** were selected as dienes, covering an attractive spectre of nucleophilic character (Figure 1).
When **1a** was heated with the less reactive isoprene **5** in a sealed ampoule at 150-200°C for 72 h using benzene as solvent, it showed its dienophilic character taking part in the DA cycloaddition reactions. The reactions proceeded to produce the mixture of isomeric benzofurans **10a** and **10b** as the principal products and dihydrobenzofurans **9a** and **9b** (global yield 62%) (Della Rosa, et al, 2005). (Figure 2) In the same manner, in the case of 1-tosyl-2-nitropyrrole **2a**, it reacted with isoprene furnishing dihydroindole isomers **11a,b** and

indole isomers **12a,b** as the principal products in moderate yield (50%). All addition products showed extrusion of the nitro group as nitrous acid. (Della Rosa, et al, 2007)

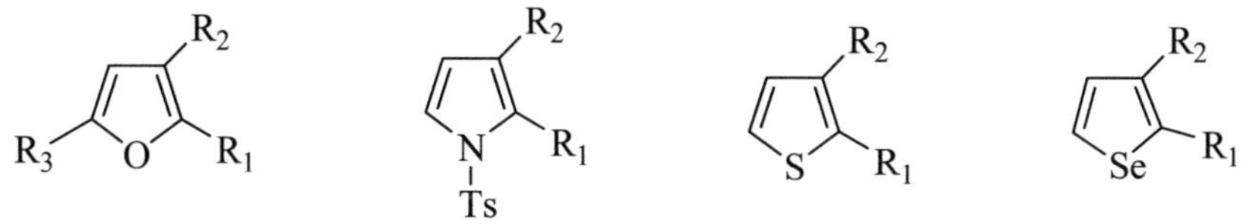

1a R 1= NO_2, R_2=R_3=H **b** R_2= NO_2, R_1=R_3=Me

2a R_1= NO_2 , R_2 = H **b** R_1= H, R_2= NO_2

3a R_1= NO_2, R_2=H **b** R_1 = H, R_2= NO_2

4a R_1= NO_2 , R_2 = H **b** R_1= H, R_2= NO_2

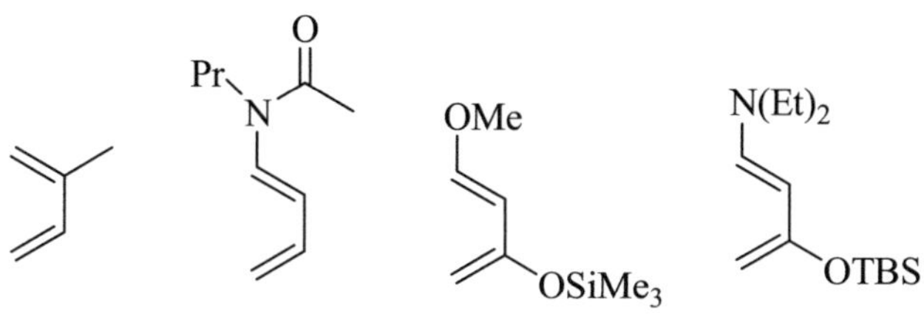

5 **6** **7** **8**

Fig. 1. Dienophiles and dienes used in the different experiences

In contrast with the above mentioned behavior, when 2-nitrothiophene **3a** was tested with **5**, it gave traces of pyrrolyl-thiophene **13** formed by a heterocycloaddition followed by thermal rearrangement. (Della Rosa, et al, 2004). This unexpected behavior was also found with nitronaphthalenes as dienophiles.

Quite unexpectedly, in the reaction of 2-nitroselenophene **4a** with the same diene, we obtained a mixture of aromatic cycloadducts **14a** and **14b** with moderate yields (55%). (Della Rosa, et al, 2007). No pyrrolyl-derivatives from hetero DA were detected.

Then, the dienophilicity of **1a, 2a, 3a** and **4a** was investigated in the reactions with 1-*N*-acetyl-*N*-propyl-1,3-butadiene **6**, Danishefsky´s diene **7** and Rawal´s diene **8**. The use of dienes **7** and **8** allowed analyze the influence of bulky substituents on the yields and regiochemistry of the cycloadditions under thermal conditions.

Exposure of **1a, 2a** and **4a** to dienamide at 120 °C for 72 h led to benzofuran **15**, tosylindole **18** and benzoselenophene **23**, with moderate yield (58%, 53%, and 48%, respectively) (Figure 3). The results of this type of DA reactions indicate a possible sequential pathway where the ease of thermal extrusion of nitrous acid accompanying the DA reaction would lead to the 4-substituted-dihydro adduct, which would undergo thermal aromatization by losing the *N*-acetyl-*N*-alkylamino substituent. On the other hand, when 2-nitrothiophene reacted with **6**, it generated only pyrrolyl-thiophene **21**. To confirm this result the reaction was performed using 2-nitrothiophene and 1-methoxy-1,3-butadiene, thus obtaining pyrrole derivative **21**. With this dienophile, the reaction way was determined by the nucleophilicity of diene as a consequence of the similar energy between the processes, normal DA and hetero DA. (Della Rosa, et al, 2007, 2004, 2007)

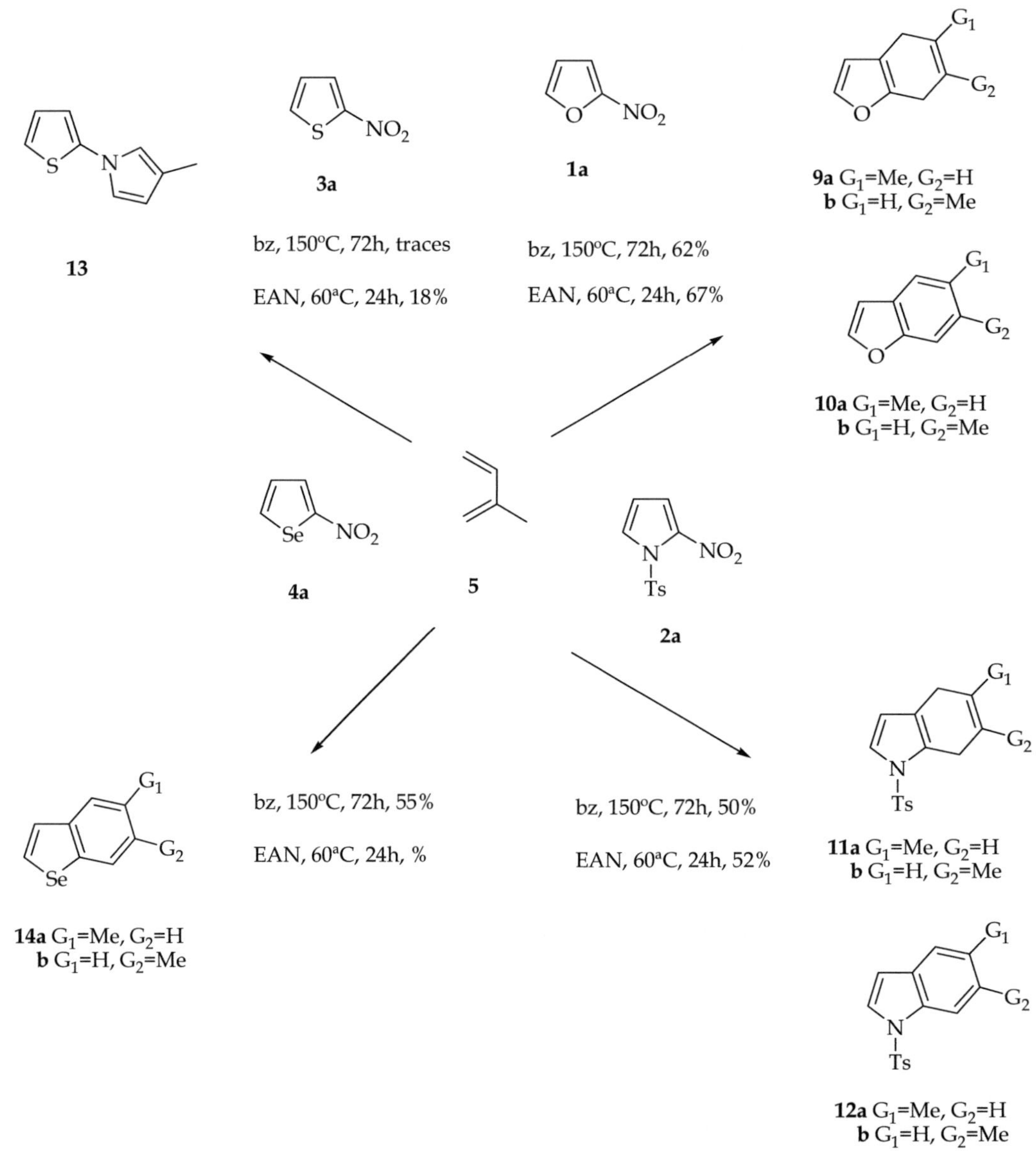

Fig. 2. Diels-Alder reactions of aromatic 2- nitroheterocycles with isoprene

The reaction of Danishefsky´s diene with **1a** yielded benzofuran **16**. Similarly, in the reaction of **2a** and **3a** with diene **7** all they provided normal DA products **19** and **22** (Figure 4), respectively. The products in the reactions with diene **7** resulted from the aromatization of the nitro-adducts promoted by the loss of the nitro and methoxyl groups as nitrous acid and methanol, respectively. The intermediate that suffered nitrous acid extrusion and retained the methoxy group was not detected in any case. (Della Rosa, et al, 2005, 2007)
The reactions of **1a**,. **2a, 4a** with Rawal´s diene afforded benzofuran **17** with very good yield, tosylindole **20,** and benzoselenophene **24** (82%, 50%, and 50%, respectively), (Figure 5). (Della Rosa, et al, 2007, 2005)

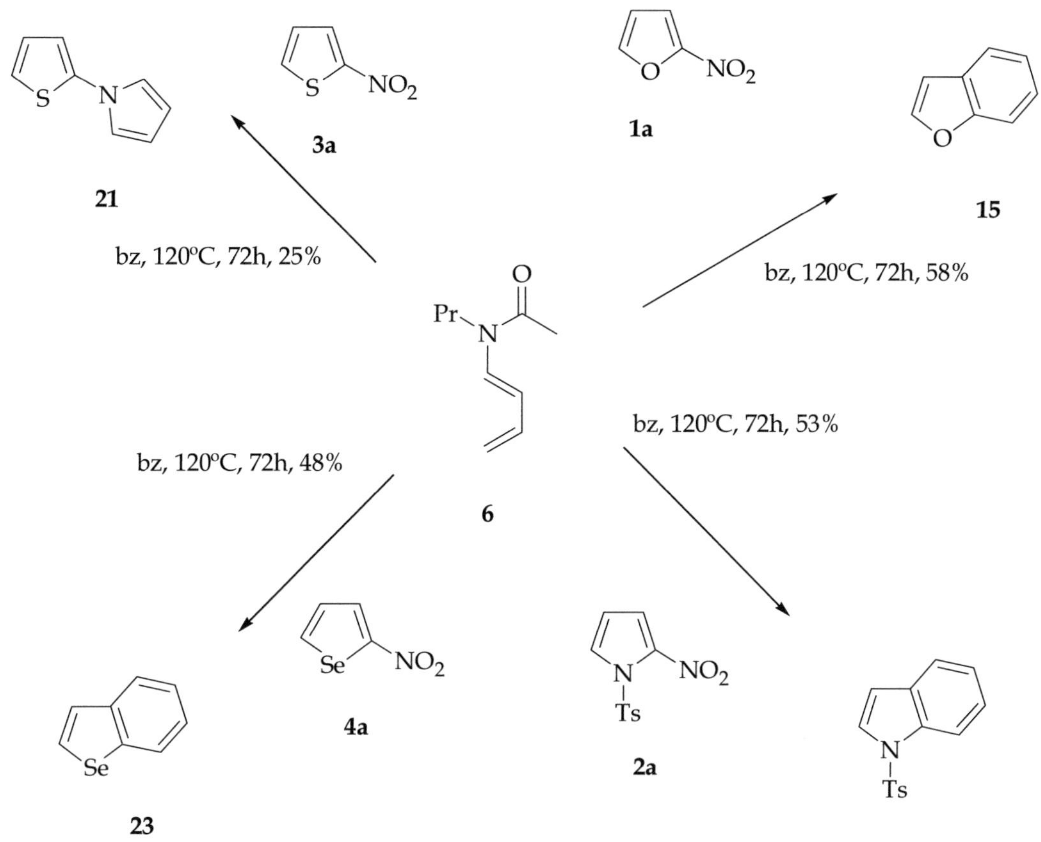

Fig. 3. Diels-Alder reactions of aromatic 2- nitroheterocycles with 1-*N*-acetyl-*N*-propyl-1,3-butadiene.

When 1-tosyl-3-nitropyrrole **2b** and 3-nitroselenophene **4b** were tested with isoprene as diene under thermal conditions, the reactions afforded a mixture of regioisomeric cycloadducts previously informed: **11 a,b; 12 a,b** and **14 a,b,** respectively (Figure 6). In the same direction and due to the impossibility of obtaining 3-nitrofuran, when 2,5-dimethyl-3-nitrofuran **1b** was exposed to **5**, the DA reaction at 150ºC and 200ºC proceeded to furnish the mixture of regioisomers **25a** and **25b** (Figure 6). (Della Rosa, et al, 2007)

Exposure of 1-tosyl-3-nitropyrrole **2b** to dienes **6** and **7** yielded 1-tosyl-indole **18** and 1-tosyl-1-indole-6-ol **26,** respectively with moderate to high yield (Table 1). The selenophene derivative **4b** reacted with diene **8** yielding **28**. At the same time, mononitrated substrate **3b** with diene **6** afforded the pyrrolyl derivative **27´** and **3b** did not undergo cycloaddition with diene **7**. Probably this behavior is a consequence of the special reactivity of this substrate connected with its aromaticity. (See the results with 1-nitronaphthalene)

2.1.2 Disubstituted nitroaromatic pentaheterocycles as dienophiles

Complementary, the behavior of aromatic pentaheterocycles disubstituted with electron withdrawing groups, a nitro group being one of the substituents, was studied. The influence of the disubstitution was first considered for the substrates indicated below.

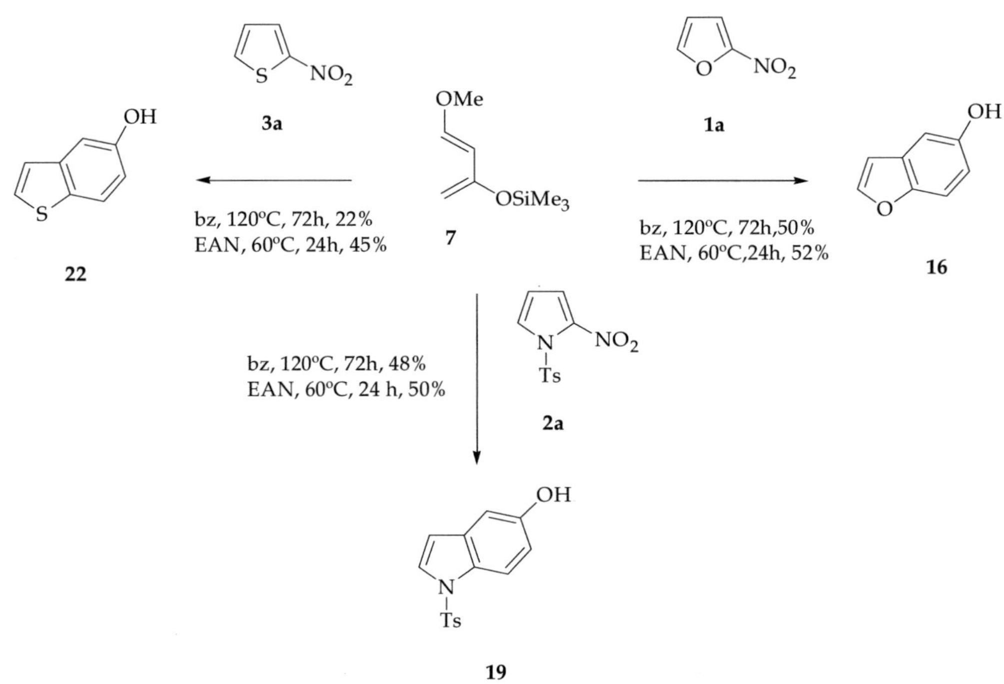

Fig. 4. Diels-Alder reactions of aromatic 2- nitroheterocycles with Danishefsky´s diene.

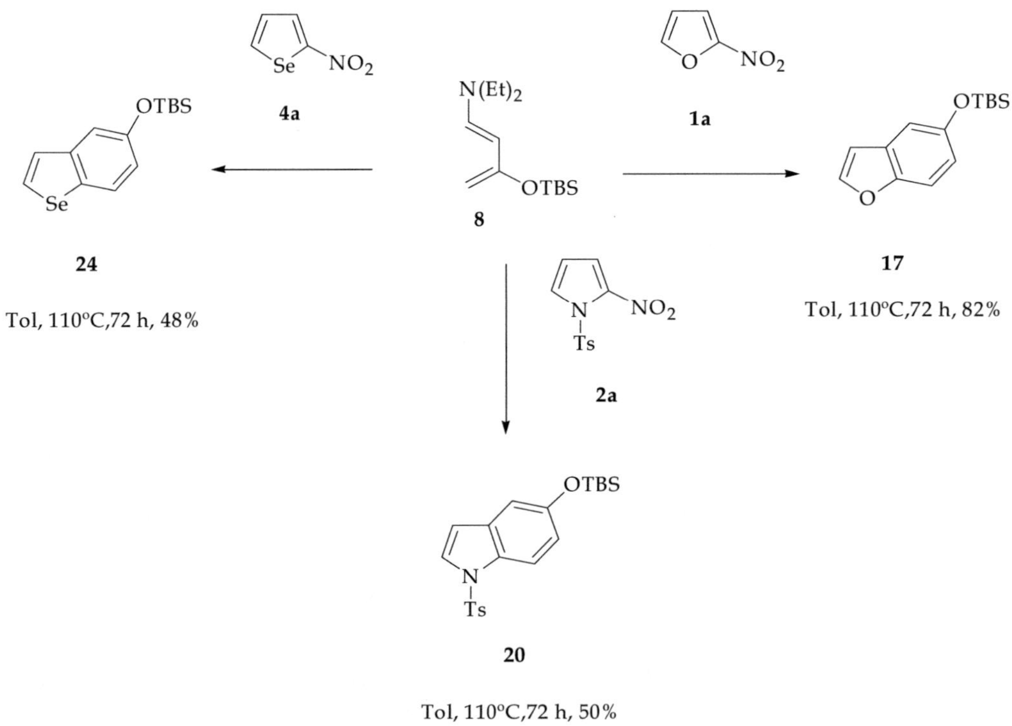

Fig. 5. Diels-Alder reactions of aromatic 2- nitroheterocycles with Rawal´s diene.

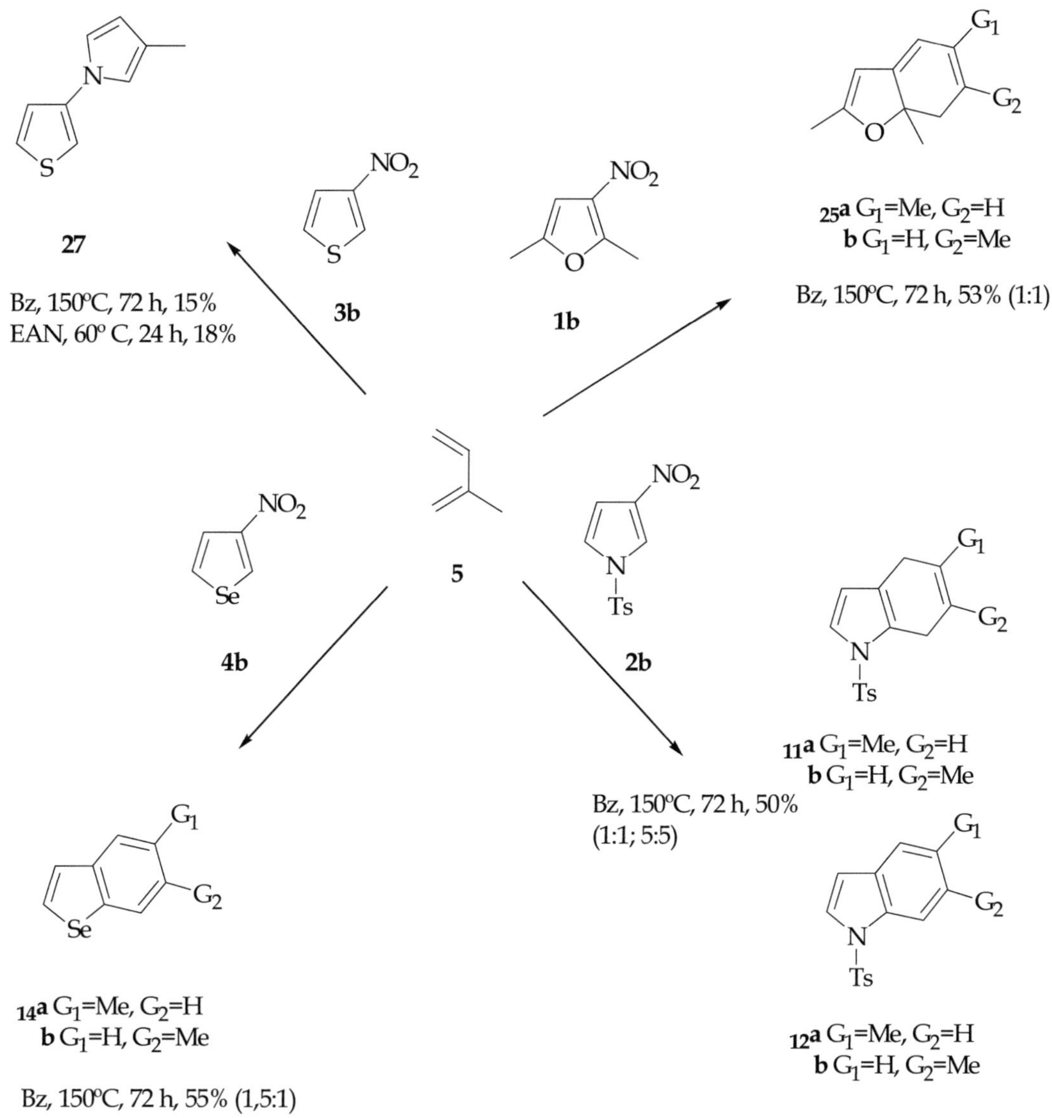

Fig. 6. Diels-Alder reactions of aromatic 3-nitroheterocycles with isoprene

The DA behavior of methyl 5-nitrofuran-3-carboxylate **29a** with isoprene was studied under similar conditions than before (Figure 8).[30] These reactions yielded the mixture of isomeric cycloadducts **30a** and **30b,** as principal products (60%) and a mixture of double addition adducts **31a-d** (10%) and **32a-d** (10%), in both cases as regioisomer mixtures. When 1-*N*-acetyl-*N*-propyl-1,3-butadiene (**6**) reacted with **29a**, it afforded benzofuran **33**. Meanwhile, exposure of **29a** to **7** and **8** yielded benzofurans **34** and **35**, respectively. In these reactions, only 1:1 adducts whose structure revealed site selectivity and regioselectivity were obtained.

Entry	Dienophile	Diene	Solvent, T, Time	Product	Yield
1	2b	6	Bz, 150º C, 72 h	18	55
2	2b	7c	Bz, 150º C, 72 h	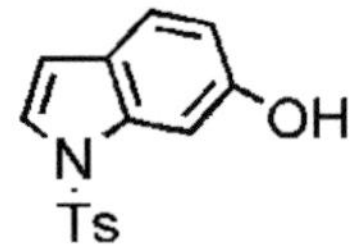 26	48
3	2b	7	EAN, 60º C, 24 h	26	50
4	3b	6	Bz, 150º C, 72 h	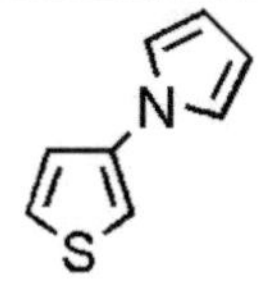 27′	20
5	4b	8	Bz, 150º C, 72 h	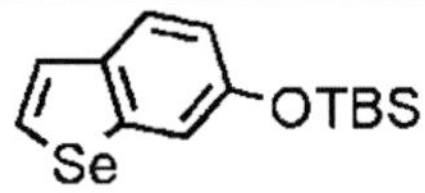 28	54

Table 1. Diels-Alder reactions of aromatic 3-nitroheterocycles with dienes **6-8**

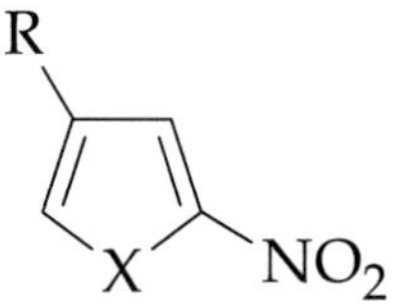

29 a X= O, R= CO_2CH_3
b X= N-Ts, R= CO_2CH_3
c X= S, R= CO_2CH_3

Fig. 7. Dienophiles nitroaromatic pentaheterocycles disubstituted

On the other hand, the treatment of methyl 5-nitro-1-tosyl-pyrrole-3-carboxylate **29b** with **5** afforded a mixture of isomeric indoles **36a** and **36b**, as the principal products (45%) and a mixture of double addition adducts **37a-d** and **38a-d** (global 15%), in both cases as regioisomeric mixtures (Figure 9). (Della Rosa, et al, 2007)

Exposure of 29b to diene 6 produced indole 39, with moderate yield (48%) and with loss of nitro group and N-acetyl-N-alkylamino substituents. The cycloaddition between diene 7 and 29b proceeded to produce indole 40 (52%). (Figure 9) (Della Rosa, et al, 2007)

When methyl 5-nitrothiophene-3-carboxylate 29c was heated with less reactive isoprene, it gave pyrrolyl-thiophene 42 as the principal product formed (32%) by a hetero DA followed by thermal rearrangement (Figure 9). (Della Rosa, et al, 2004). We assumedacycloaddition of the diene to the heterodienophilic fragment of the nitro group forming the oxazine *N*-oxide (Paredes, et al, 2007; Della Rosa, et al, 2004).. According to the normal hetero DA mechanism, a second electron-withdrawing group placed on thiophene nucleus in **29c** caused an increase in pyrrole yields. Therefore, this behavior would confirm the preference of these substrates for the hetero DA process, as a consequence of a minor reaction energy requirement.

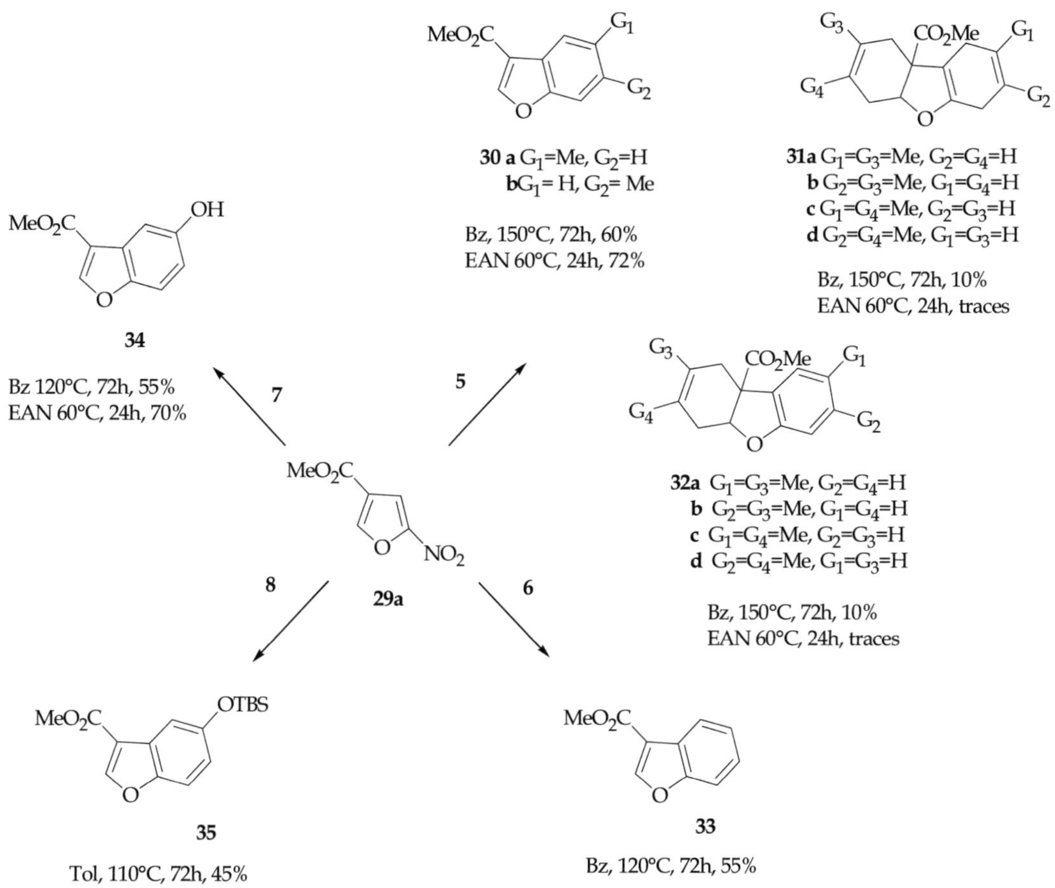

Fig. 8. Diels-Alder reactions of methyl 5-nitrofuran-3-carboxylate with different dienes

The products of normal cycloaddition to the C4-C5 bond of thiophene **41a** and **41b** were isolated as well although in minor proportion. In all cases, the reactive double bond of thiophene nucleus was the nitrated one. In any case the bis-adduct from double cycloaddition of diene was detected. When the diene 1-*N*-acetyl-*N*-propyl-1,3-butadiene was used the loss of the substituents was observed in the resultant pyrrolyl-thiophene **43** (Figure 10). With a highly activated diene such as Danishefsky´s diene react with **29c** (like **29a** and **29b**) afforded **44** (Figure 9). With Danishefsky´s diene, **29a**, **29b** and **29c** gave the nitro-adducts. These cycloadducts suffer aromatization providing the corresponding methyl 5-hydroxybenzofuran-3-carboxylate, methyl 5-hydroxybenzothiophene-3-carboxylate and methyl 5-hydroxy-1-tosylindole-3-carboxylate, respectively. In these reactions, only 1:1 adducts were obtained. The products structure revealed site selectivity and regioselectivity. This behavior could be ascribed to the stronger electron-withdrawing effect of the nitro group.

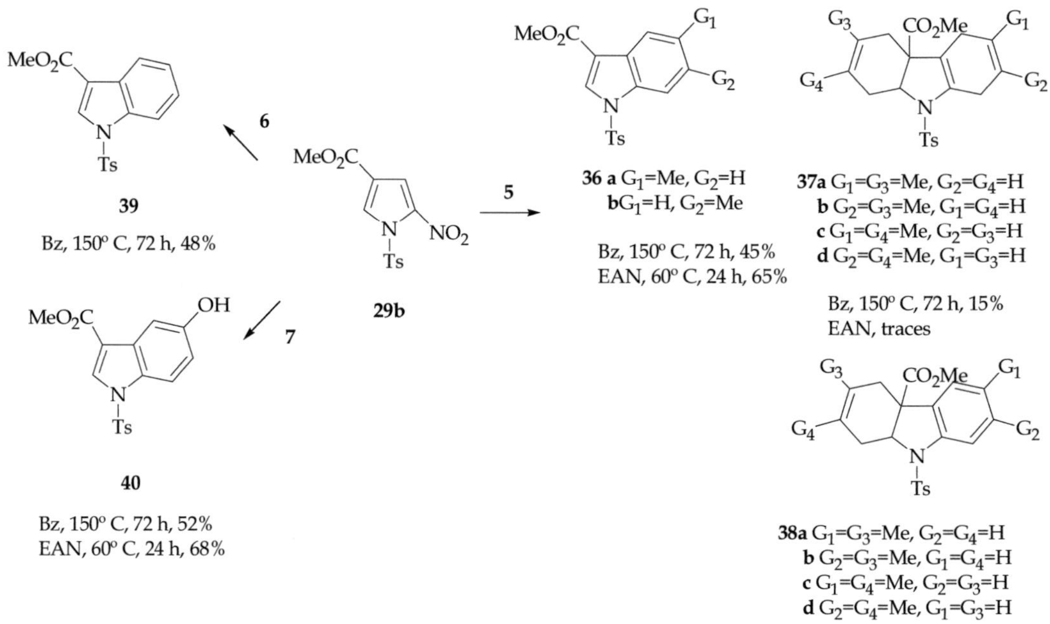

Fig. 9. Diels-Alder reactions of methyl 5-nitro-1-tosyl-pyrrole-3-carboxylate with diverse dienes

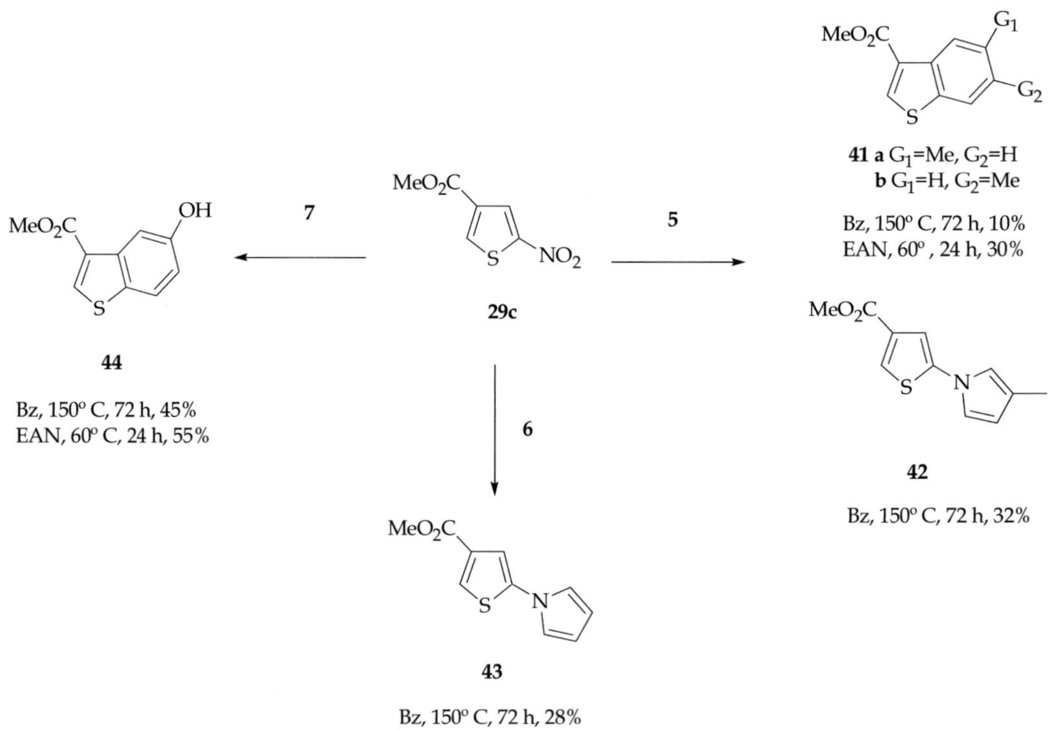

Fig. 10. Diels-Alder reactions of methyl 5-nitrothiophene-3-carboxylate with different dienes.

In the Figure 11 we show others dienophiles pentaheterocicles aromatic disastituted.
In the case of methyl 5-nitrofuran-2-carboxylate (**45a**), the thermal reaction with isoprene gave a mixture of aromatic cycloadducts of simple addition **47a** and **47b** with moderate yields (Figure 12). (Della Rosa, et al, 2005) The reactivity and regioselectivity of methyl 5-nitrofuran-3-carboxylate, which were higher than those methyl 5-nitro-2-carboxylate compounds is in consonance with the different behavior informed for β-acylfurans and their α-acyl isomers. (Wenkert, et al, 1988)

45 a X=O, R=CO_2CH_3
b X=N-Ts, R=CO_2CH_3
c X=S, R=CO_2CH_3
d X=Se, R=$COCH_3$

46 a X= N-Ts, R= CO_2CH_3
b X=S, R= CO_2CH_3
c X= Se, R= $COCH_3$

Fig. 11. Dienophiles nitroaromatic pentaheterocycles disubstituted

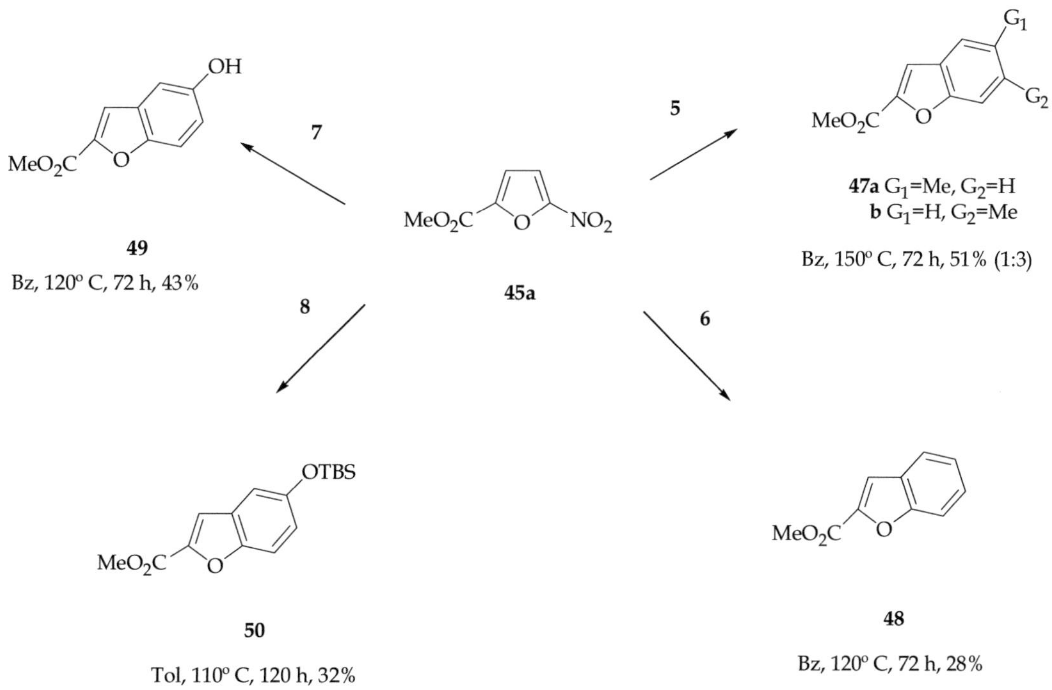

Fig. 12. Diels-Alder reactions of methyl 5-nitrofuran-2-carboxylate with diverse dienes

Treatment of **45a** with dienes **6, 7, and 8** yielded benzofused compounds **48, 49** and **50,** respectively (Figure 12). On the other hands, the reactions of **45b** with dienes **5, 6** and **7** yielded a mixture of dihydrotosylindoles **51a**, **51b**, and the indoles derivatives **53** and **54a**. (Figure 13)

Again, these reactions proceeded by the selective addition of the diene to the nitro-substituted double bond of the dienophile (Della Rosa, et al, 2007, 2007)

The reactive behavior of 4-nitro-2-carboxylate compounds was also studied under similar conditions. Reactions of nitropyrrole **46a** with isoprene proceeded to produce the regioisomer mixtures of the dihydroindoles **51a,b** and the indoles **52a,b**. The reactions of **46a** whit the dienes **6** and **7** afforded with the reasonable yield the tosyl indole derivatives **53** and **54b** (Figure 13).

After 72 h at 150 °C, the reaction of methyl 5-nitrothiophene-2-carboxylate **(45c)** with isoprene afforded pyrrolyl-thiophene **55**. This dienophile yielded the pyrrolyl-thiophene **56** when reacted with diene **6,** and in the reactions with diene **7** afforded the benzothiophene derivative **57,** in both cases with good yield. (Figure 14)

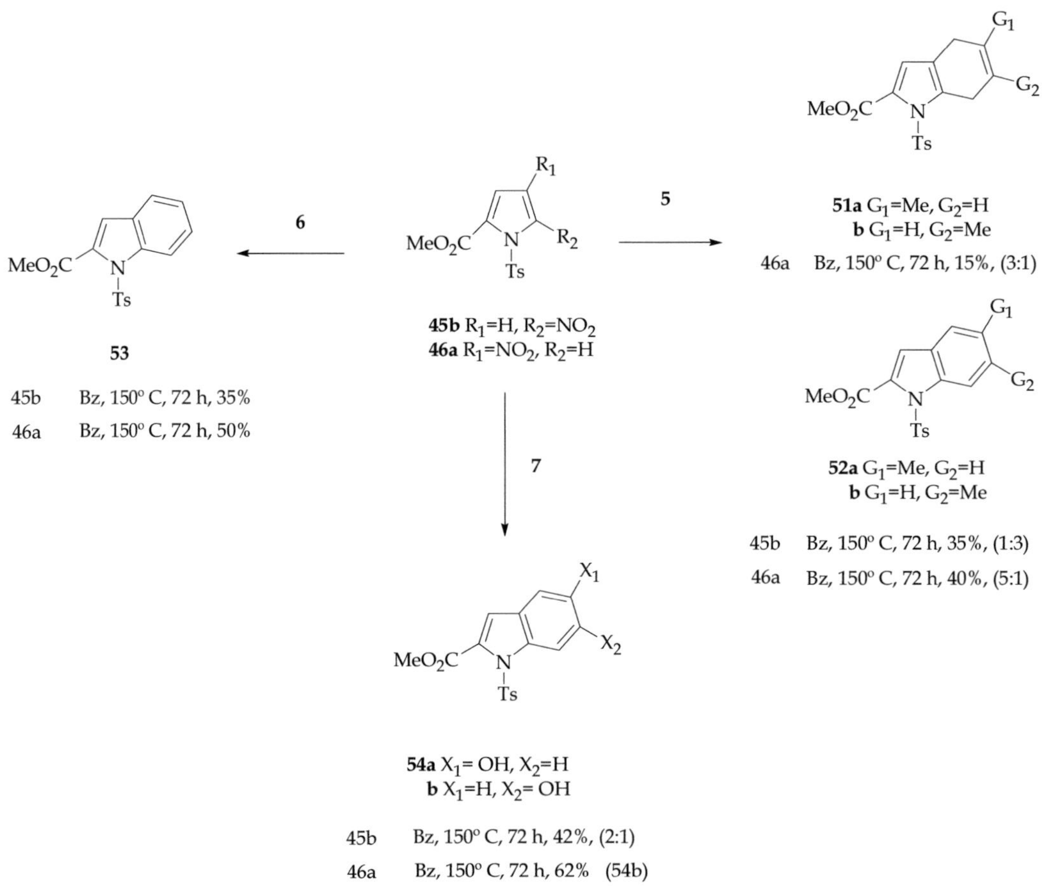

Fig. 13. Thermal reactions of disubstituted pyrroles with dienes **5**, **6** and **7**

2.1.3 Nitroindoles as dienophiles

When 1-tosyl-3-nitroindole **58** was reacted with Danishefsky´s diene under thermal conditions (120 °C, 72 h) the reaction surprinsingly afforded 37% of a mixture of

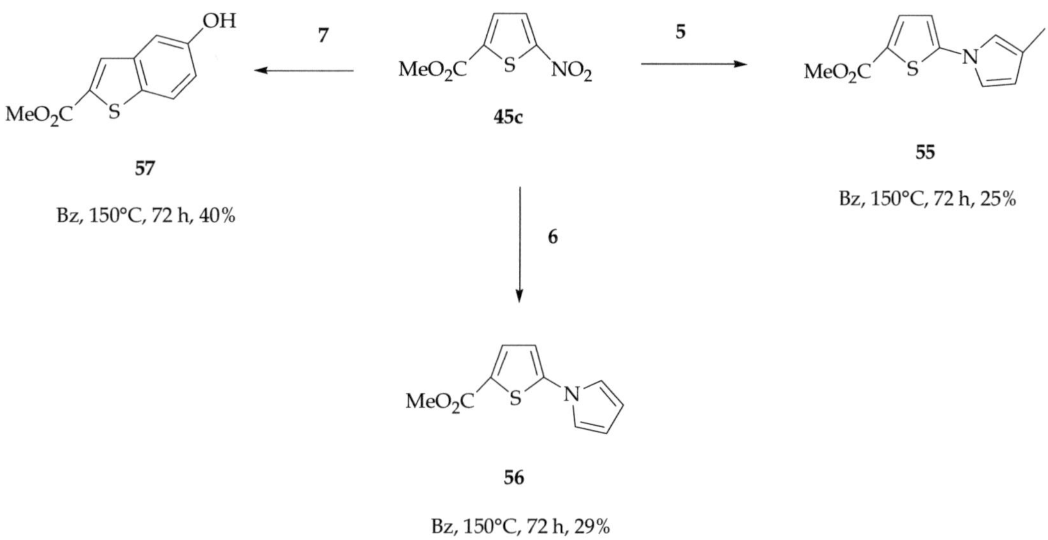

Fig. 14. Thermal reaction of methyl 5-nitrothiophene-2-carboxylate **45c** with dienes **5**, **6** and **7**

diasteroisomeric cycloadducts **60** and **61** in a 30:1 ratio after hydrolysis, with 48% of hydroxycarbazole **59** (Figure 15). When the reaction temperature was lowered to 70°C the isomeric ratio change to ca. 1:1 (64% yield), with only 11% to hydroxycarbazole **59**. This result demonstrated that the isomer **61** is more likely to suffer thermal aromatization because of the *trans* arrangement of its nitro and methoxy substituents. The intermediate that suffered nitrous acid extrusion and retained the methoxy group was not detected in either case. (Biolatto, et al, 2001)

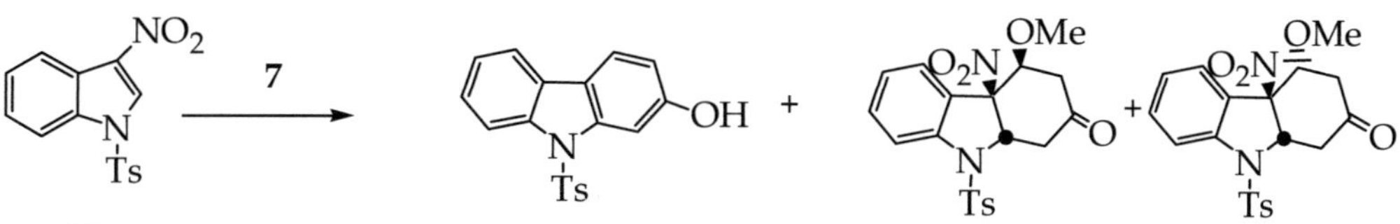

58 59 60 61

Bz, 120°C, 72 h, 48% Bz, 120°C, 72 h, 37%, (30:1)
Bz, 70°C, 48 h, 11% Bz, 70°C, 48 h, 64%, (1:1)

Fig. 15. Thermal reaction of 1-tosyl-3-nitroindol with Danishefsky´s diene

2.1.4 Nitrobenzofurans as dienophiles

To explore the normal electron demand DA dienophilicity of nitrobenzofurans **62a-b** we choose 1-trimethylsilyloxy-1,3-butadiene (**63**), as diene (Figure 16) .

The thermal reactions of **62a** with isoprene (150 °C, 72 h) afforded the mixture of aromatic regioisomeric cycloadducts **64a** as principal products and dihydrodibenzofurans **65a** and **65b** with reasonable yield (Figure 17). On the other hand, reactions of 62**a** with 1-trimethylsilyloxy-1,3-butadiene (120-140°C, 72 h) yielded dibenzofuran **66** with loss of trimethylsilyloxy and nitro groups. The yield is good. In the same way, in the thermal reaction of 62**a** with the Danishefsky`s diene aromatic cycloadduct **67a** was obtained with very good yield and complete regioselectivity (Figure 17). (Della Rosa, et al, 2011)

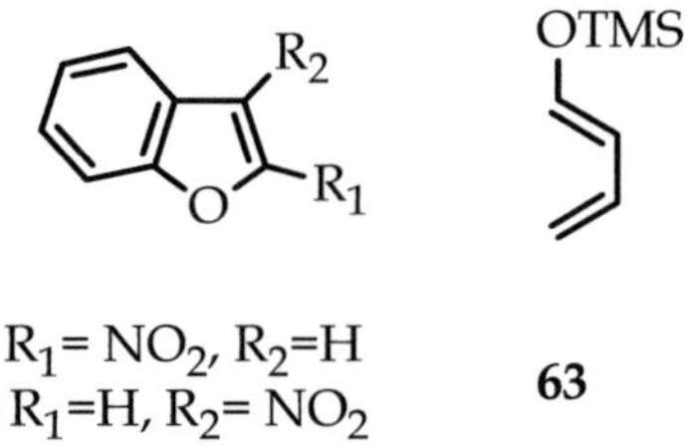

Fig. 16. 2- and 3- nitrobenzofurans, and 1-trimethylsilyloxy-1,3butadiene

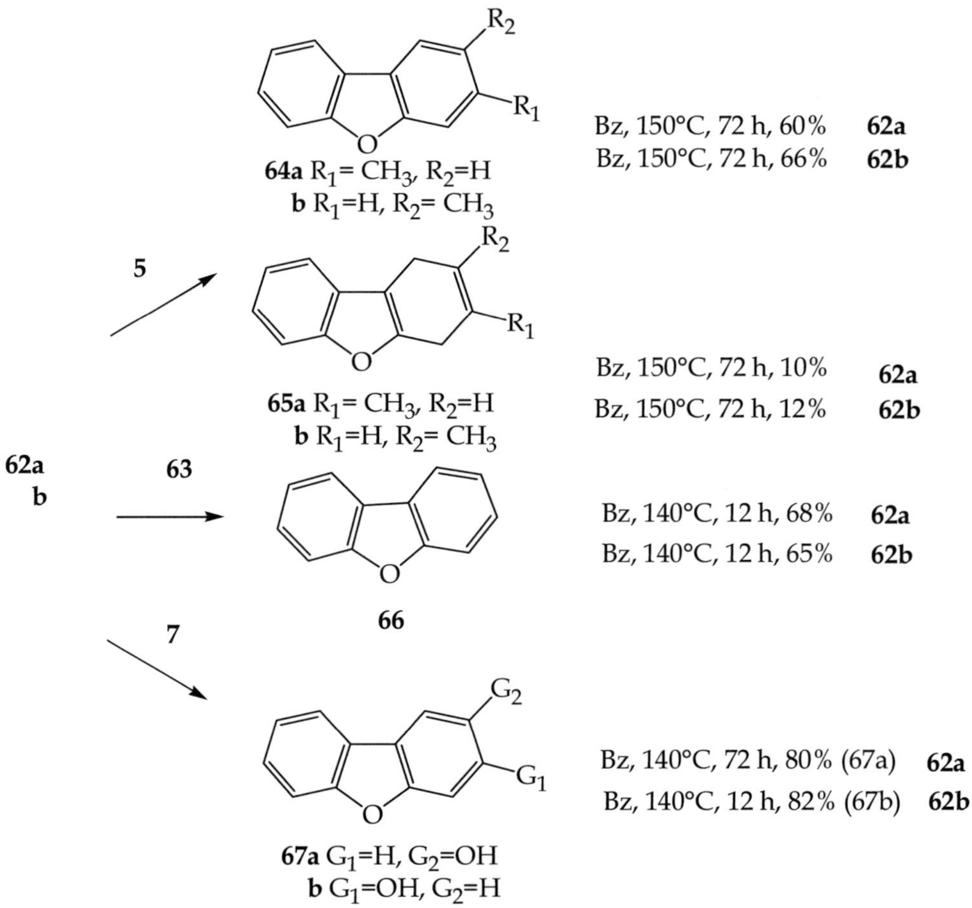

Fig. 17. Thermal reactions of 2- and 3-nitrofurans with different dienes

Then, we analyzed the reactivity of 3-nitrobenzofuran **62b** with the dienes **5**, **63** and **7**. This allowed us to compare not only the relative reactivity of the 2- and 3-substitution of the aromatic ring but also the regioselectivities in the reactions with dienes properly substituted. When **62b** reacted with isoprene (150 °C, 72 h) we observed the same mixture of regioisomeric products **64a, 64b**, **65a** and **65b** obtained in the reaction with the substrate **62a.** The yield of these reactions was reasonable. The reaction of 62b with the diene **5** (120-140 °C, 72 h) offer the dibenzofuran **66** in a similar yield that before. On the other hand, the thermal reaction of **62b** with the Danishefsky´s diene yielded the aromatic cycloadducts **67b** with high yield and complete regioselectivity, which is controlled by the nitro and methoxy groups. (Figure 17)

2.1.5 Nitronaphthalenes as dienophiles

To explore the normal electron-demand DA dienophilicity of 1-nitronaphthalenes (**1-NN**) **68**, 2-nitronaphthalene (**2-NN**) **69**, 1,3-dinitronaphthalene (**1,3-DNN**) **70**, 1,4-dinitronaphthalene (**1,4-DNN**) **71**, 1,5-dinitronahpthalene (**1,5-DNN**) **72** and 1,8-dinitronaphthalene (**1,8-DNN**) **73**, the dienes **5**, **6**, **7 and** 1-methoxi-1,3-butadiene **74** were selected.

We started the study by testing the reactivity of 1-nitronaphthalene with Danishefsky's diene since it is one of the strongest dienes and is appropriate to test these aromatic substrates. Thus, when **68** and **7** were heated in sealed ampoule (150 ºC, 72 h) using benzene as solvent, 62% of 2-hydroxy-phenanthrene **75** was regioselectively produced. (Figure 18)

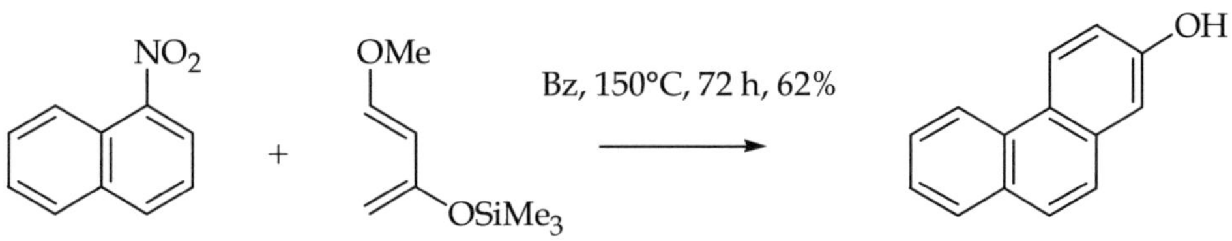

68 **7** **75**

Fig. 18. Diels- Alder reaction of 1-nitronaphthalene with Danishefsky's diene

In an analogous fashion to **68**, 2-nitronaphthalene **69** reacted whit diene **7** to give 45% of hydroxyphenanthrene **76**. The regioselectivity of these reactions was controlled by both the nitro group of the dienophile and the methoxyl group of Danishefsky´s diene.

In the case of 1,4-dinitronaphthalene, due to the enhanced reactivity of this dienophile, the DA reaction with **7** proceeded to furnish 75% of the single regioisomer 2-hydroxy-9-nitrophenanthrene **77**. On the other hand, the reaction of 1,3-dinitronaphthalene with the same diene produced the two isomers 2-hydroxi-10-nitrophenantrene and 3-hydroxi-9-nitrophenantrene resulting from the cycloaddition to C1-C2 and C3-C4 bonds (1:4,6 ratio, respectively), with a total yield of 88%. The reactions of 1,5-dinitronaphthalene and 1,8-dinitronaphthalene yielded in both cases the corresponding phenanthrenes derivative. With **72** the yield is similar to mononitronaphthalenes. In the case of **73**, the low yields obtained could be explained by steric effects derived from the proximity of the nitro groups, which caused considerable overcrowding in the molecule. (Paredes, et al, 2007)

Unexpected results were observed when using nitronaphthalenes with less reactive dienes such as **5** and **6**. Interestingly, when **68** and **69** reacted with these dienes, they produced *N*-(1-naphthyl)-3-methylpyrrole **78** and *N*-(2-naphthyl)-3-methylpyrrole **79** with yields according to the electron-donor properties of the diene. Moreover, when the dinitronaphthalenes reacted with these dienes they generated a mixture of *N*-naphthylpyrroles and phenanthrenes with different yield ratios. In the case of 1,3-dinitronaphthalene, a clear tendency towards the DA cycloaddition to the C3-C4 bond was observed with these dienes. With isoprene, 1,3-dinitronaphthalene gave the corresponding 2-metil-10-nitro-phenanthrene **80**, together with traces of [3-methyl-1-(1-nitronaphthalen-3-

yl)-pyrrole] **81a** and [3-methyl-1-(2-nitronaphthalen-4-yl)-pyrrole] **81b**. The reactions with **6** yielded 3-methyl-9-nitrophenantrene **82** and traces of corresponding *N*-naphthylpyrroles **83a** and **83b**.

2.1.6 Nitroquinolines as dienophiles

Considering that with benzene or toluene -molecular solvent frequently used in this type of process- were observed solubility problem we decided employ chloroform as reaction media. This molecular solvent has potential HBD character which could be influence the reactivity of the reaction systems. The reaction temperature was in the range 120-150 °C, and we used different times of reaction.

Because 2-nitroquinoline and 4-nitroquinoline were no reactive in all the experiences development in this reaction conditions, the study started by testing the reactivity of 5-nitroquinoline **84** and 8-nitroquinoline **85**, with the dienes **5**,**7**, **8** and 1-trimethylsyliloxy-1,3-butadiene **63**. When **84** was heated with the less reactive isoprene **5** it gave as principal product the 5-(3-methyl-1*H*-pyrrolil)-quinoline **86**, through the participation of the nitro group in a hetero DA process followed by a thermal rearrangement -hetero DA pathway-. (Cancian, et al, 2010)

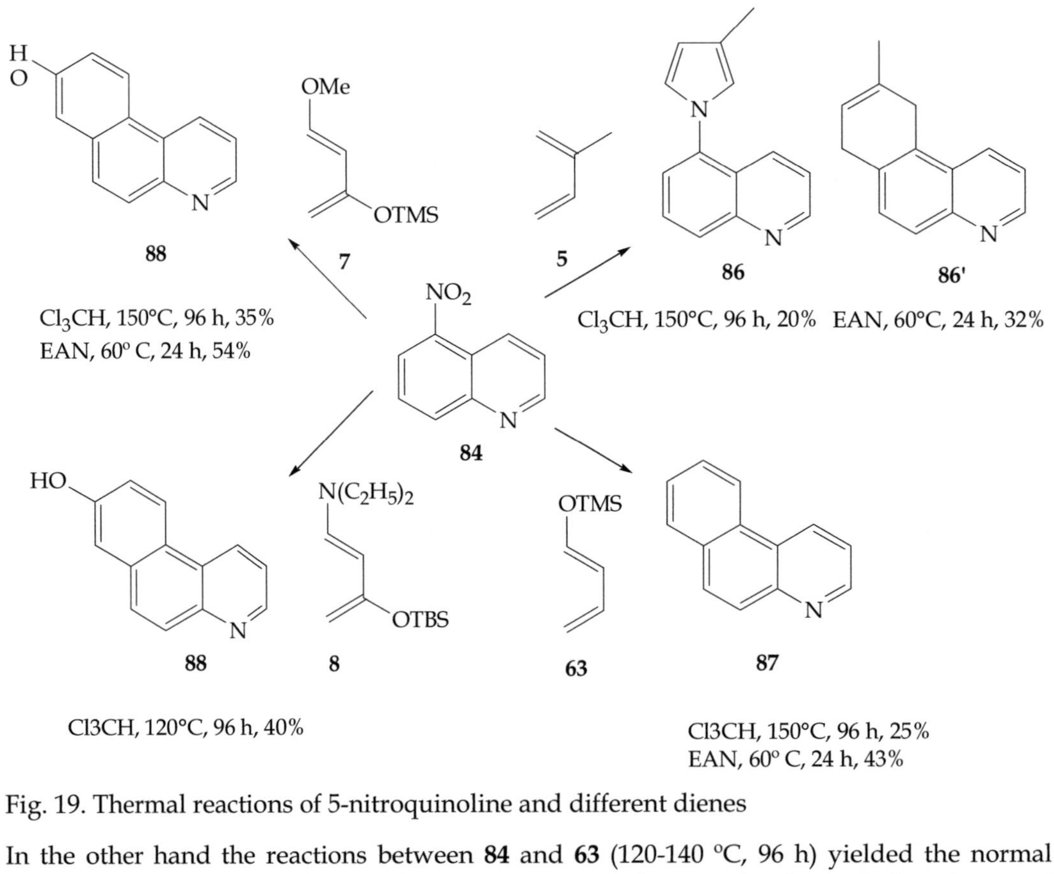

Fig. 19. Thermal reactions of 5-nitroquinoline and different dienes

In the other hand the reactions between **84** and **63** (120-140 °C, 96 h) yielded the normal addition product **87** with complete aromatization due to the loss of the nitro and

trimethylsilykoxy groups. In the same way, in the thermal reactions of **84** with the diene **7** (120-140ºC, 96 h), aromatic cycloadduct **88** was obtained with very good yield and complete regioselectivity. Then was explored the reaction between **84** and **8** observing the same product that in the cycloaddition using the diene **7** although with higher yield. (Figure 19)
In turn the reactions of **85** with the dienes **5**, **7**, and **63**, yielded in all cases the normal addition products (Figure 20).

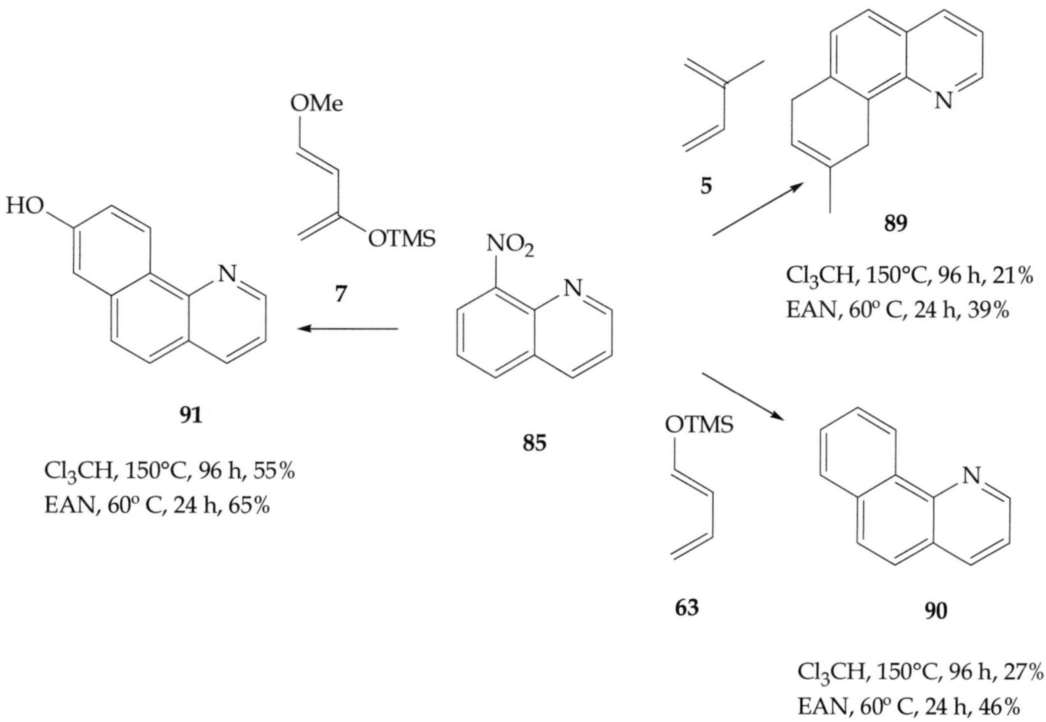

Fig. 20. Thermal reactions of 8-nitroquinoline and different dienes

The reaction between **85** with **5** shown as principal product el dihydro derivative 7,10-dihydro-9-methyl-benzo[f]quinoline **89**, joint to traces of its regioisomer. In this case the normal addition process observed could be attributed to the near position and consequent interaction of the nitro group and the heteroatom. The employ of **63** as diene yielded the aromatic cycloadddition product **90** with a hight yield that in the reaction using **84**. The reaction of **85** and diene **7** yielded the aromatic heterotricyclic product **91** with very high yield and complete regioselectivity. (Figure 20)

2.2 Ionic liquids as solvent in polar DA reactions

Many reports refer to the DA cycloaddition as a typical example of a reaction that is indifferent towards the choice of the solvent. This is only strictly true for the very special case of DA reactions between two purely hydrocarbon reactants, such as cyclopentadiene dimerization. Actually, DA reactions proceed at an appreciable rate only when either the diene or the dienophile are activated by an electron donating or electron withdrawing group, normally characterized by the presence of a heteroatom that can therefore efficiently

interact with the solvent. The influence of the solvent on these latter reactions has been extensively investigated, in particular after Breslow and Rideout in 1980 evidenced the dramatic accelerating effect exerted by water. (Riodeout, & Breslow, 1980)
PILs, with their peculiar properties such as high polarizability/dipolarity, good hydrogen bond donor ability, etc., were straight away considered to have the potential to influence the outcome of DA reactions. The first investigation on the reaction between cyclopentadiene and alkyl acrylatesin an a IL was performed using ethylammonium nitrate which, surprisingly, gave a mixture of *endo* and *exo* products in a ratio of 6,7 : 1 (Jaeger & Tucker, 1989). Subsequently, Welton have investigated the influence of ILs based on the 1-butyl-3-methylimidazolium cation, $[BMIM]^+$ in DA reactions. The *endo* : *exo* ratio and associated acceleration observed in the DA cycloaddition of cyclopentadiene with methyl acrylate was attributed to the ability of the IL to hydrogen bond to the dienophile (methyl acrylate).
In that direction and to complete the reaction approach of aromatic carbocyclic and heterocyclic substituted with electron with-drawing groups as dienophiles in cycloaddition reactions, we have investigated PILs solvent effects in polar DA reactions using ethylammoniun nitrate (EAN), 1-methylimidazolium tetrafluoroborate ($[HMIM][BF_4]$), and 1-butyl-3-methylimidazolium tetrafluoroborate ($[BMIM][BF_4]$). The dienophiles studied were 2-nitrofuran, 1-tosyl-2-nitropyrrole, 1-tosyl-3-nitropyrrole, 2-nitrothiophene, 3-nitrothiophene, methyl-5-nitrofuran-3-carboxylate, methyl 5-nitro-1-tosylindole-3-carboxylate, methyl-5-nitrothiophene-3-carboxylate, 1-nitronaphthalene, 2-nitronaphthalene, 1,3-dinitronaphthalene, 1,4-dinitronaphthalene, and 5- and 8-nitroquinolines. These dienophiles were exposed to different dienes with variable nucleophilicity. This allowed us compared in thermal cycloaddition reaction conditions not only the relative reactivity of these substrates also the regioselectivities when we change molecular solvents for ILs.

2.2.1 2-nitrofuran (1.a) as dienophile

When **1a** was reacted with the less reactive isoprene **5** in a sealed ampoule at 60°C for 24 h using EAN as solvent, the reactions proceeded to produce a mixture of isomeric benzofurans **10a** and **10b** (1:1) as the principal products with reasonable yield (40%) and dihydrobenzofurans **9a** and **9b** (1:1) (10%). (Figure 2). If the time of the reaction increased to 24 h we observe a 1:1 mixture of isomeric benzofurans **10a** and **10b** in 55% yield and trazes of the isomeric dihydrobenzofurans **9a** and **9b**. Similar results were observed when the reaction was development in $[HMIM][BF_4]$ although the yields were slightly lower.
The reactions of **1a** with 1-trimethylsilyloxy-1,3-butadiene using EAN, $[HMIM][BF_4]$, and $[BMIM][BF_4]$, respectively as solvents, in seals ampoule at 60°C (12 and 24 h, respectively) offered in all cases good yield in benzofuran. The best yield (ca. 60%) was obtained with EAN as solvent working 24 h. On the other hand $[HMIM][BF_4]$ is a better solvent than $[BMIM][BF_4]$, probable due to its character HBD.
The reaction of Danishefsky´s diene (60°C, 12 and 24 h, respectively) with **1a** using the three ILs cited in the before paragraph yielded 5-hydroxyibenzofuran **16** in reasonable yield (Figure 4). Similarly to the reactions with isoprene and 1-trimethylsilyloxy-1,3-butadiene, the best yield was observed with EAN (ca. 65%). The product obtained in the reactions with diene **7** resulted from the aromatization of the nitro-adducts promoted by the loss of the nitro and methoxyl groups as nitrous acid and methanol, respectively. The intermediate that suffered nitrous acid extrusion and retained the methoxy group was not detected in any case.

In all cases the presence of PIL´s as reaction media improve the yields respect to use of molecular solvent, while the temperature and reaction time decrease.

2.2.2 2-and 3-nitro-1-tosylpyrrole (2a and 2b)

In the reactions of **2a** and **2b** with diene **7** in EAN and [HMIM][BF_4] (60°C, 24 h), all they provided normal DA products **19** (Figure 4) and its isomer 1-tosyl-6-hydroxyindole **26**, respectively, with moderate yield (ca 50%). The products in the reactions with diene **7** resulted from the aromatization of the nitro-adducts promoted by the loss of the nitro and methoxyl groups as nitrous acid and methanol, respectively. Again, the intermediate that suffered nitrous acid extrusion and retained the methoxy group was not detected in any case. In these reactions the regioselectivity was controlled by the nitro and the methoxy group of the diene.

2.2.3 2-and 3-nitrothiophenes (3a and 3b)) as dienophile

When 2-nitrothiophene **3a** was tested with isoprene using EAN, and [HMIM][BF_4], respectively as solvent (60°C, 12, 24 and 48 h) , afforded pyrrolyl-thiophene **13** formed by an heterocycloaddition followed by thermal rearrangement, in relatively low yield (ca de 13, 18 and 25%, depending on the temperature) (Figure 2). The yields using EAN were lightly better. This behavior was also found with molecular solvents and probably is due to the aromatic character of the thiophene ring. Similar results were obtained when we use the dienophile **3b** (Figure 6). In this case the observed product was **27** and the yields were lighly lower than when the dienophile was **3a**.
On the other hand, the reactions of **3a** with diene **7** using both IL´s, provided the normal DA product **22** (Figure 4) with moderate yield (ca de 45%) (60°C, 24h). This result is compatible with those corresponding to the use of molecular solvent. Only the yields are better in presence of ILs.

2.2.4 Disubstituted nitroaromatic heterocyclepentadienes as dienophiles

In this case the behavior of aromatic heterocyclopentadienes disubstituted with electron withdrawing groups (a nitro group being one of the substituents) as dienophile in the presence of IL´s (EAN and [HMIM][BF_4]) was studied. With this purpose we choose the following dienophiles: methyl 5-nitrofuran-3-carboxylate **29a**, methyl 5-nitro-1-tosylindole-3-carboxylate **29b**, and methyl 5-nitrothiophene-3-carboxylate **29c**.
The reaction of methyl 5-nitrofuran-3-carboxylate **29a** with isoprene was studied in presence of both PILs at 60°C, 24 h (Figure 8). These reactions yielded a mixture of isomeric cycloadducts **30a** and **30b**, as principal products (72%) and traces of double addition adducts **31a-d** and **32a-d** in both cases as regioisomer mixtures. Meanwhile, exposure of **29a** to diene **7** yielded benzofurans **34** (70%). In this last reaction, only 1:1 adducts whose structure revealed site selectivity and regioselectivity were obtained. In all cases the presence of EAN produce lightly better yields than the use of [HMIM][BF_4].
The treatment of methyl 5-nitro-1-tosylpyrrole-3-carboxylate **29b** with **5** in both PILs afforded a mixture of isomeric indoles **36a** and **36b**, as the principal products (65%) and traces of a mixture of double addition adducts **37a-d** and **38a-d**, in both cases as regioisomeric mixtures (Figure 9). The cycloaddition between diene **7** and **29b** proceeded to produce indole **40** (Figure 10) with good yield (68%). Once again EAN is better solvent.
When methyl 5-nitrothiophene-3-carboxylate **29c** was heated with less reactive isoprene in both PIL´s at 60 °C, 24 h, it gave a mixture 1:1 of methyl 5-methylbenzothiphene-3-

carboxylate and methyl 6-methylbenzothiophene-3-carboxylate (30%). In this condition we observe differences respect to the reactions in molecular solvents where the process is hetero DA. When a highly activated diene, such as Danishefsky´s diene, react with **29c** (like **29a** and **29b**) in the presence of PILs yielded **44** (Figure 10) (55%). Once again EAN is a better reaction media.

2.2.5 Nitronaphthalenes as dienophiles

To explore the normal electron-demand DA dienophilicity of nitronaphtahlenes in presence both PILs (EAN and [HMIM][BF_4]) we selected 1-nitronaphthalenes **68**, 2-nitronaphthalene **69**, 1,3-dinitronaphthalene **70**, and 1,4-dinitronaphthalene **71.**

When **68** and **7** were heated in a sealed ampoule (60ºC, 24 h) using EAN or [HMIM][BF_4] as solvent, 65% of 2-hydroxy-phenanthrene **75** was regioselectively produced. In a similar way 2-nitronaphthalene **69** reacted in the same conditions to give 43% of 3-hydroxyphenanthrene. The regioselectivity of both reactions was controlled by both the nitro group of the dienophile and the methoxyl group of Danishefsky´s diene.

On the other hand, when 1-nitronaphthalenes reacted with isoprene in the presence of the cited PILs (60ºC, 24 h), it produced a mixture of *N*-naphtyl-2-methylpyrrole (30%) and 2-methylphenantrene (12%). Similar result was obtained using 2-nitronaphthalene as dienophile although the yields were lightly lower. It is interesting cite that the presence of PILs modify the products obtained in these reactions. With molecular solvents we observed only the hetero DA product, meanwhile using IL´s appear in the mixture the normal DA product.

Moreover, when the dinitronaphthalenes **70** and **71** reacted with isoprene (60ºC, 24 h) in the presence of EAN or [HMIM][BF_4], they generated a mixture of phenanthrenes and *N*-naphthylpyrroles with different yield ratios and a moderate predominance of the normal DA product. In the case of 1,3-dinitronaphthalene **70**, newly a clear tendency towards the DA cycloaddition to the C3-C4 bond was observed. In the reactions with isoprene, 1,3-dinitronaphthalene gave the corresponding 2-metil-10-nitro-phenanthrene (32%) , together with [3-methyl-1-(1-nitronaphthalen-3-yl)-pyrrole] and [3-methyl-1-(2-nitronaphthalen-4-yl)-pyrrole]. On the other hand, the reactions of 1,4-dinitronaphthalene with isoprene afforded 2-methyl-9-nitrophenantrene (27%) and 3-methyl-1-(4´-nitronaphtalene-1´-iy)-pyrrole.

EAN was a better reaction media in all cases discussed. The preference for the normal DA products in the presence of PILs respect to the use of molecular solvent, probably is due to the increase of the electrophilicity of the dinenophile as a consequence of HBD interactions

2.2.6 5-and 8-nitroquinolines as dienophiles

It was explored the cycloaddition reactions between 5-nitro and 8-nitroquinolines with the dienes 1-trimethylsiyloxy-1,3-butadiene and Danishefsky´s diene, respectively, in presence of EAN and [HMIM][BF_4]. With both PIL´s the reaction systems improved the yields (approximately 20%) and the reaction time and temperature were considerably smaller (60ºC, 24 h) than similar reactions development in molecular solvents. In all reaction systems, EAN make possible major yields than [HMIM][BF_4]. The cycloaddition products were the same than those obtained using chloroform as reaction media, except for the reaction of **84** whit **5** in which the observed product was consequence of normal DA process (8-Methyl-7,10-dihydro-benzo[h]quinoline) **86**. (Cancian, et al, 2010)

2.3 Theoretical studies

In the last years the DFT has been successful in explaining the reactivity and regioselectivity of cycloaddition reactions. In this direction there are several parameters which can be used as global or local reactivity descriptors. For instance, the chemical hardness (η) describes the resistance of the chemical potential to a change in the number of electrons. The electronic chemical potential (μ) it is usually associated with the charge-transfer ability of the system in its ground state geometry. Both quantities can be approximated in terms of the energies of the HOMO and LUMO frontier molecular orbitals (Eqs. 1 and 2) (Domingo, et al, 2002; Domingo & Aurell, 2002)

$$\eta = \left(\varepsilon_{LUMO} - \varepsilon_{HOMO}\right) \tag{1}$$

$$\mu = \frac{\left(\varepsilon_{LUMO} + \varepsilon_{HOMO}\right)}{2} \tag{2}$$

The global electrophilicity index (ω), introduced by Parr (Parr, et al, 1999) is a useful descriptor of the reactivity that allows a quantitative classification of the global electrophilicity character of a molecule within a unique scale. Current studies based on the DFT and applied to DA reactions, have shown this classification of the diene/dienophile pair is a powerful tool to predict the feasibility of the process and the type of mechanism involved. This index defined as Eq. 3

$$\omega = \frac{\mu^2}{2\eta} \tag{3}$$

Useful information about polarity of de DA processes may be obtained from the difference in the global electrophilicity power of the reactants. This difference has been proposed as a measure of the polar character of the reaction. On the other hand, local reactivity indexes are associated with site selectivity in a chemical reaction. These descriptors should reflect the sites in a molecule where the reactivity pattern stated by the global quantities should take place. For instance, an important local reactivity parameter was introduced by Parr *et al.*, and it was defined as the Fukui function (Domingo, et al, 2002)

Eq. (4) provides a simple and direct formalism to obtain the Fukui function from an approach based on a relationship with the FMO´s. The condensed Fukui function for electrophilic (nucleophilic) attack involves the HOMO (LUMO) FMO coefficients (c) and the atomic overlap matrix elements (S).

$$f_k^a = \sum_{\mu \in k} | c_{\mu a} |^2 + \sum_{\upsilon \neq \mu} c_{\mu a} c_{\upsilon a} S_{\mu \upsilon} \tag{4}$$

This scheme has been corroborated for several reactions that are well documented.

Eq. (5) has been introduced to analyze at which atomic site of a molecule the maximum electrophilicity power will be developed.

$$\omega_k = \omega f_k^+ \tag{5}$$

Furthermore, the first approaches toward a quantitative description of nucleophilicity, in the form of a regional reactivity index, have also been reported. Eq (6) has been developed by

Domingo (Domingo, et al, 2008) with the purpose of identifying the most nucleophilic site of a molecule and assessing the activation/deactivation caused by different substituents on the electrophilic aromatic substitution reactions of aromatic compounds.

$$N_k = N f_k^- \quad (6)$$

$$N = \left(\varepsilon_{HOMO,Nu} - \varepsilon_{HOMO,TCE}\right) \quad (7)$$

Where $e_{HOMO,TCE}$ is the HOMO energy of tetracyanoethylene (TCE) taken as a reference molecule because it exhibits the lowest HOMO energy in a large series of molecules previously considered in the framework of polar DA cycloadditions). N is the global nucleophilicity index and N_k is its local counterpart. This nucleophilicity index has been useful to explain the nucleophilic reactivity of some molecules towards electrophiles in cycloaddition as well as substitution reactions. (Domingo, et al, 2008)

Normally the polarity of the normal electron demand DA process has been studied by means of global electrophilicity index difference between reactants and the regioselectivity of the normal electron demand DA reaction using the local electrophilicity index for dienophiles (electrophiles in the reaction) and the local nucleophilicity index for dienes (nucleophiles in the reaction).

In the direction cited before we show different theoretical studies related to the polar DA reactions described experimentally, in which the dienophiles are aromatic heterocycles or carbocycles. In same cases, the mechanism of these reactions, specially respect to regio-, site- and stereochemistry were analyzed in detail.

2.3.1 Five-membered heterocycles: General

In Table 2 we present a classification of the dienophiles and dienes in decreasing order of the electrophilicity power (ω). We can assume that high nucleophilicity and high electrophilicity corresponds to opposite extremes of this scale. (Della Rosa, et al, 2011; Brasca, et al, 2009). In the table we also included some global properties such as the chemical potential and the chemical hardness. A good electrophile is characterized by a high value of μ and a low value of η.

In order to verify the validity of this scale we carried out the corresponding thermal DA reactions of some dienophiles with dienes of different nucleophilicity such as isoprene and Danishefsky's diene.

- The substitution of one hydrogen atom in all the dienophiles by one of the most powerful electron-withdrawing groups (nitro group) produces an increment in the electrophilicity character and therefore an increase in the reaction rate is expected. The 2-nitro-substituted heterocycles show high electrophilicity power respect to the 3-nitro-substituted ones. Experimentally we obtained higher yields when the nitro group is place in the 2-position of the thiophene's ring than when it is in the 3-position. So these last results support the tendency observed in the tables.
- The dienophiles substituted by two different electron-withdrawing groups (methyl carboxylate and nitro groups) show the highest values in electrophilicity power, indicating that these disubstitutions are suitable in order to increment the reactivity of the dienophiles. Moreover, the yields corresponding to the DA reactions of the disusbstituted heterecycles with the dienes showed an increase respect of that corresponding to the monosusbstituted heterecycles.

I. Dienes

Molecule	Global properties		
	μ (a.u.)	η (a.u.)	ω (eV)
Isoprene	-0.1209	0.1962	1.01
Danishefsky's diene	-0.0945	0.1851	0.66

II. Dienophiles

Molecule	Global properties		
	μ (a.u.)	η (a.u.)	ω (eV)
Methyl 5-nitrofuran-3-carboxylate	-0.1897	0.1814	2.70
2-Nitrofuran	-0.1810	0.1775	2.51
3-Nitrofuran	-0.1767	0.1808	2.35
Furan	-0.1024	0.2441	0.58

a. Furan and derivatives.

Molecule	Global properties		
	μ (a.u.)	η (a.u.)	ω (eV)
Methyl 5-nitrothiophene-3-carboxylate	-0.1922	0.1767	2.84
2-Nitrothiophene	-0.1845	0.1738	2.66
3-nitrothiophene	-0.1794	0.1821	2.40
Thiophene	-0.1545	0.1566	2.07

b. Thiophene and derivatives.

Molecule	Global properties		
	μ (a.u.)	η (a.u.)	ω (eV)
1-Tosyl-methyl 5-nitropyrrole-3-carboxylate	-0.1734	0.1776	2.30
1-Tosyl-2-nitropyrrole	-0.1655	0.1739	2.14
1-Tosyl-3-nitropyrrole	-0.1668	0.1765	2.14
1-Tosylpyrrole	-0.1348	0.1752	1.41

c. 1-Tosyl-pyrrole and derivatives.

Molecule	Global properties		
	μ (a.u.)	η (a.u.)	ω (eV)
Methyl 5-nitroselenophene-3-carboxylate	-0.1899	0.1717	2.85
2-Nitroselenophene	-0.1829	0.1695	2.68
3-Nitroselenophene	-0.1776	0.1803	2.38
Selenophene	-0.1220	0.2195	0.92

d. Selenophene and derivatives.

Table 2. Global properties for some common reagents participating in Diels-Alder reactions

On the other hand, the differences in the global electrophilicity power between the dienophile/diene pair (Δω) are higher for the Danishefsky's diene than for isoprene. Therefore, we can expect a high reactivity and a high regioselectivity for the pair dienophile/Danishefsky's diene. This fact is also consistent with the experimental researches in both molecular solvent and IL´s.

2.3.2 2-and 3-nitrobenzofurans as dienophiles

When 2-nitrobenzofuran (**2-NBF**) and 3-nitrobenzofuran (**3-NBF**) were reacted with isoprene, 1-trimethylsilyloxy-1,3-butadiene and de Danishefsky´s diene, under different reaction conditions they showed their dienophilic character taking part in a normal demand polar DA cycloaddition reactions. These reactions could be considered a domino process that is initialized by a polar DA reaction, and the latter concerted elimination of nitrous acid from the [4+2] cycloadduct yields the corresponding dibenzofurans. (Della Rosa, et al, 2011)
The electrophilicity of isoprene falls in the range of moderate electrophiles within of the electrophilicity scale proposed by Domingo *et al.* When electron-donating substituents, -OCH_3 and -$OSi(CH_3)_3$, are incorporated into the structure of butadiene, a decrease in the electrophilicity power is observed. Therefore, the electrophilicity of Danishefsky´s diene falls in the range of marginal electrophiles, good nucleophiles, within of the electrophilicity scale. This behavior indicates that the nucleophilic activation in Danishefsky´s diene is greater than in isoprene, in clear agreement with the high nucleophilicity index of the diene. 1-trimethylsilyloxy-1,3-butadiene have a intermediate behavior.
As consequence of the high electrophilic character of these substituted dienophiles and the high nucleophilic character of the dienes, it is expected that these DA reactions proceed with polar character. The polarity of the process is assessed comparing the electrophilicity index of the diene/dienophile interacting pairs, that is $\Delta\omega$. Evidently, the differences in the global electrophilicity power ($\Delta\omega$ are higher for the Danishefsky´s diene than 1-trimethylsilyloxy-1,3-butadiene and isoprene. Therefore, a high reactivity and high regioselectivity are expected for the pair dienophile/Danishefsky´s diene.
Finally, the flux of the electron-density in these polar cycloaddition reactions is also supported by means of a DFT analysis based on the electronic chemical potentials of the reagents. The electronic chemical potentials of the substituted heterocyclic dienophiles (nearly -5 eV), are higher than those of the dienes (nearly -3 eV), thereby suggesting that the net charge transfer will take place from these electron-rich dienes towards the aromatic dienophiles. (Table 3 and 4)

Compound	ω(eV)
2-NBF	3.3275
3-NBF	3.0002

Table 3. Global electrophilicity indexes for dienophiles

Dienophile	Site	ω_K(eV)
2-NBF	2	0.1571
	3	0.5359
3-NBF	2	0.7263
	3	0.0482

Table 4. Local electrophilicity indexes for dienophiles

According to the global electrophilicity index ω shown, the dienes will act as nucleophiles and the dienophiles as electrophiles. To study the regioselectivity we used the local

electrophilicity and nucleophilicity indexes for dienophiles and dienes respectively. The sites of study were C2 and C3 in **2-NBF** and **3-NBF** and C1 and C4 in the dienes. The more favourable adducts are the ones where the most electrophilic and nucleophilic sites interact first. In the reactions in which it is possible discussed the regioselectivity the experimental data agree with the computational results. (Della Rosa, et al, 2011)
The 2-nitrosubstituted benzofuran show higher electrophilicity power than the 3-nitrosubstituted benzofuran probably due to the proximity of the nitro group with the heteroatom

2.3.3 Nitrofurans as dienophiles: Theoretical mechanistic approach

Specifically the reactions of 2-nitrofuran, methyl 5-nitro-3-furancarboxylate, and methyl 5-nitrofuran-2-carboxylate, respectively, and Danishefsky´s diene were studied using the hardness, the polarizability and the electrophilicity of the corresponding DA primary adducts as global reactivity indexes. The experimentally observed products for these DA reactions using different conditions were indicated in the Figure 21, and related experiments. It has been demonstrated that both the hardness as well as the electrophilicity power of adducts are appropriate descriptors for predicting the major product of the reactions at least in the cases study here. (Brasca, et al, 2011)

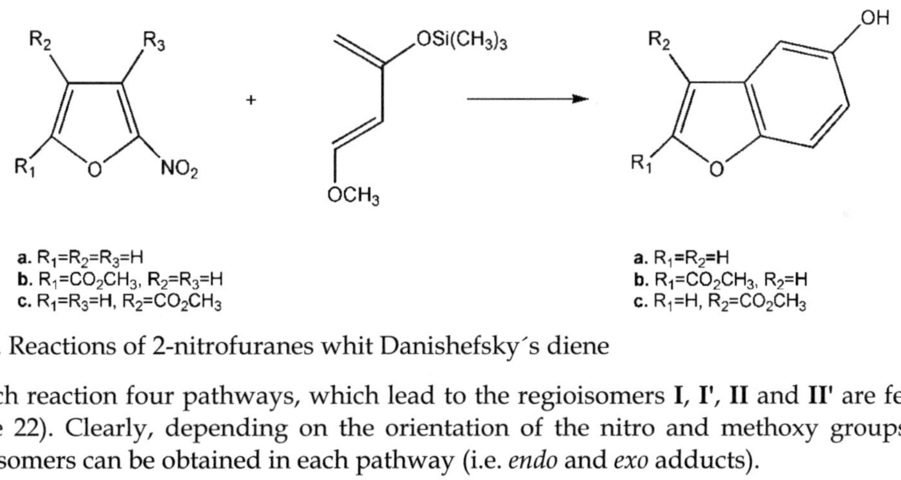

a. R_1=R_2=R_3=H
b. R_1=CO_2CH_3, R_2=R_3=H
c. R_1=R_3=H, R_2=CO_2CH_3

a. R_1=R_2=H
b. R_1=CO_2CH_3, R_2=H
c. R_1=H, R_2=CO_2CH_3

Fig. 21. Reactions of 2-nitrofuranes whit Danishefsky´s diene

For each reaction four pathways, which lead to the regioisomers **I**, **I'**, **II** and **II'** are feasible (Figure 22). Clearly, depending on the orientation of the nitro and methoxy groups, two stereoisomers can be obtained in each pathway (i.e. *endo* and *exo* adducts).
The regioisomer that have the higher value of η and the lower values of ZPE, α and ω, should correspond to the major product.[i] The calculated hardness and electrophilicity power correctly predict the regioisomers **I.a-I.c** as the main adducts of the DA reactions. However, the calculated total energies and polarizabilities show a random behaviour. Therefore, neither of these two parameters can predict the predominant regioisomer of the DA reactions in a correct way.
The results obtained in gas phase revealed the same tendency as in molecular solvent and in ILs. The chemical hardness and the electrophilicity power are useful parameters to predict which regioisomers will lead to the main product of the DA reactions.
We can conclude that the predominant regioisomeric adduct of the reactions between furan derivatives and Danishefsky's diene has always the less electrophilicity and high hardness values. Therefore, the regioselectivity experimentally observed can be confirmed by this approach.

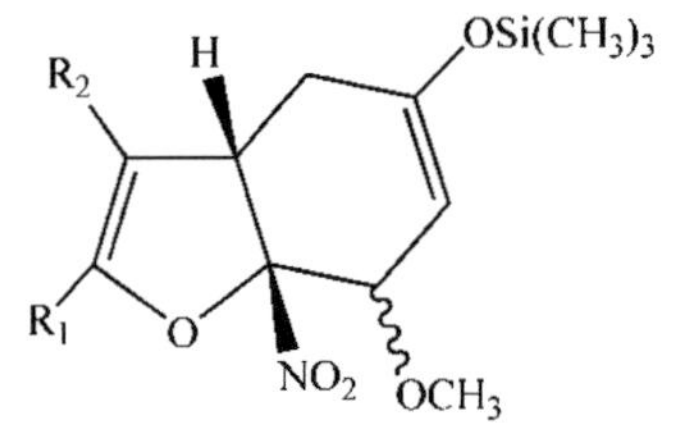

I.a. $R_1=R_2=H$
b. $R_1=CO_2CH_3$, $R_2=H$
c. $R_1=H$, $R_2=CO_2CH_3$

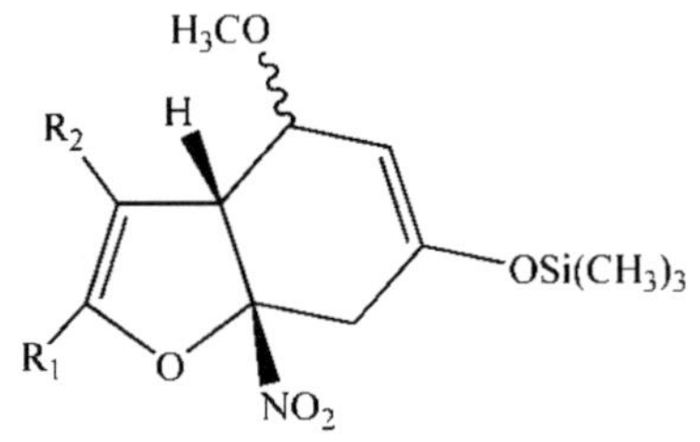

I'.a. $R_1=R_2=H$
b. $R_1=CO_2CH_3$, $R_2=H$
c. $R_1=H$, $R_2=CO_2CH_3$

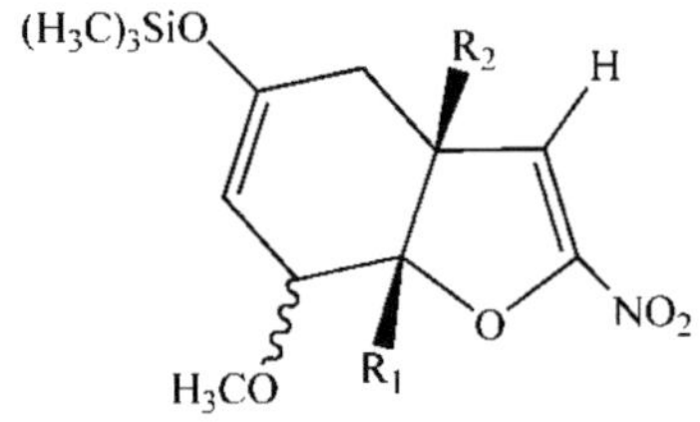

II.a. $R_1=R_2=H$
b. $R_1=CO_2CH_3$, $R_2=H$
c. $R_1=H$, $R_2=CO_2CH_3$

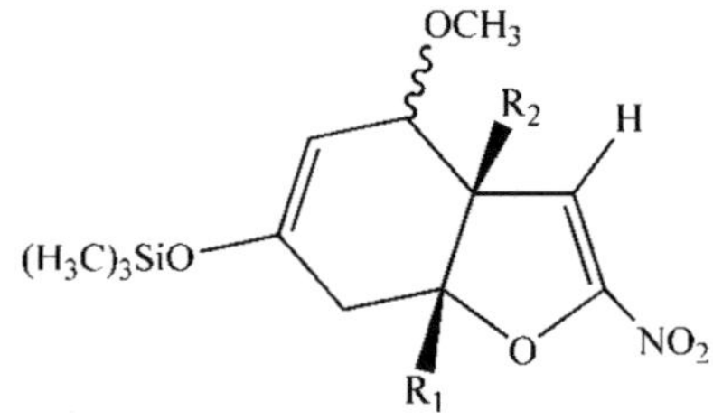

II'.a. $R_1=R_2=H$
b. $R_1=CO_2CH_3$, $R_2=H$
c. $R_1=H$, $R_2=CO_2CH_3$

Fig. 22. Possible regioisomeric Diels-Alder adducts

The obtained energies show that the I.*endo* isomer is more stable than the I.*exo* one. Moreover, the more stable I.*endo* isomer has lower electrophilicity value than the I.*exo* isomer in all cases. Regarding the hardness, we found that only in the presence of solvent the maximum hardness principle is obeyed.

The investigation by DFT theory, including solvent effects (molecular solvents and ILs), shows that these cycloadditions proceed by a concerted but asynchronus reaction mechanism. The lowest activation energies for concerted reactions are obtained. However, the stepwise additions have significantly lower activation energy lead to substantially less stable products. Moreover, the primary cycloadducts could never be isolated but were converted into 5-hydroxybenzofurans by subsequent extrusion of nitrous acid, hydrolysis of the silyl enolether, and elimination of methanol. Elimination of nitrous acid is calculated to have lower overall barriers than cycloaddition reactions and is strongly exothermic, thus explaining the preferred reaction pathway.

2.3.4 Nitronaphthalenes as dienophiles

The reactions of naphthalenes properly mono- and disubstituted with an electron-withdrawing and a series of dienes in thermal conditions, were evaluated with the frontier molecular orbitals (FMO) theory which provide qualitative information about the feasibility of these DA reactions. Besides, the global electrophilicity index (ω) is employed to estimate the electrophilic character of the dienophiles used in the cycloaddition reactions.

Taking into account that the reactivity of a DA reaction depends on the HOMO-LUMO energy separation of the reactants, and that in a normal electron demand DA reaction the strongest interaction takes place between the HOMO of the diene and the LUMO of the dienophile, we compared the corresponding energies of the reacting partners in order to explain the experimental tendency observed. (Table 5).

Molecule	FMO Energy
1,4-DNN	-3.3223
1,3-DNN	-3.1313
1,5-DNN	-3.1022
1,8-DNN	-2.7392
Danishefsky's diene	-5.0883

Table 5. Calculated LUMO energies for the dinitronaphthalenes and calculated HOMO energy for Danishefsky´s diene

When the FMO of the reacting pairs are closer in energies, the interaction is higher. Therefore, we expected higher reactivities for **1,4-DNN** and **1,3-DNN**, which is consistent with the experimental results

Since all the reactions involving **1,3-DNN** produce only one product in appreciable amounts, we selected this dienophile to explain the difference in reactivity towards the different dienes. Thus, the FMO energies of the reactants were evaluated and calculations of the electrophilicity index were performed. Table 6 shows how Danishefsky's diene leads to the minimum HOMO-LUMO energy difference compared to the other dienes, so that an increase in the reactivity of the system is to be expected. This effect is revealed by an increase in the yield of the DA reaction. (Paredes, et al, 2007; Domingo, et al, 2008)

Molecule	Energy	μ	η	ω
1,3-DNN	-3.1313	-5.11	3.97	3.30
Isoprene	-5.9563	-3.29	5.34	1.01
1-(*N*-acetyl-*N*-propylamine)-1,3-Butadiene	-5.3250	-2.99	4.67	0.96
1-Methoxy-1,3-butadiene	-5.3032	-2.83	4.95	0.81
Danishefsky's diene	-5.0883	-2.57	5.03	0.66

Table 6. Global properties values and global electrophilicity scale for reagents of DA reaction. The calculated LUMO energy for **1,3-DNN** and the calculated HOMO energies for the dienes are also shown.

It has been demonstrated a good correlation between the difference in the global electrophilicity of the diene/dienophile interacting pair ($\Delta\omega$) and the feasibility of the cycloaddition.

2.3.5 Study of the domino reaction of 1-nitronaphthalene with the Danishefsky´s diene: Theoretical mechanistic approach

The reaction of 1-nitronaphthalene with the Danishefsky´s diene to give 3-hydroxyphenantrene has been theoretically studied using DFT methods. This reaction is a domino process that is initialized by a polar DA reaction between the par dienophile/diene to give the formally [2 + 4] cycloadduct . The subsequent concerted elimination of nitrous acid from the primary adduct yields the precursor of the phenatrene derivative. Analysis of the global reactivity indixes as well as the thermodynamic data for this domino process indicates that while the large electrophilic character of 1-nitronaphthalene together with the large nucleophilic character of Danishefsky´s diene are responsible for the participation of these reagents in a polar DA reaction. The DA reaction has a two-step non-intermediate mechanism characterized by the nucleophilic attack of the non-substituted methylene of the diene to the electrophilically activated C2 position of 1-nitronaphtalene. The subsequent ring closure affords the primary cycloadduct. The latter concerted elimination of nitrous acid yielded the precursor of the tricyclic aromatic final product. Spite of the large activation free energy associated with the DA reaction and the endergonic in the primary adduct, the irreversible extrusion of nitrous acid makes feasible thermodynamically the domino reaction. (Domingo, et al, 2008)

2.3.6 Nitroquinolines as dienophiles

Although the global electrophilicity for the 5-nitroquinoline indicated a lightly major reactivity than the 8-nitro isomer (Figure 23). This result does not agree with the experimental data. In this respect it is possible think that the attack of the dienophile to the "*para*" position would be a reversible process meanwhile the attack to the "*orto*" position to the nitro group evolves in form irreversible to the cycloaddition product. The major reactivity of the 8-nitro derivative could be occur due to the presence of electroelectronic factors more favourables, which are produced during the nucleophilic attack of the diene, for instance a better stabilization of the negative charge in the nitro group. These effects were does not considered in the reagents. (Cancian, et al, 2010)

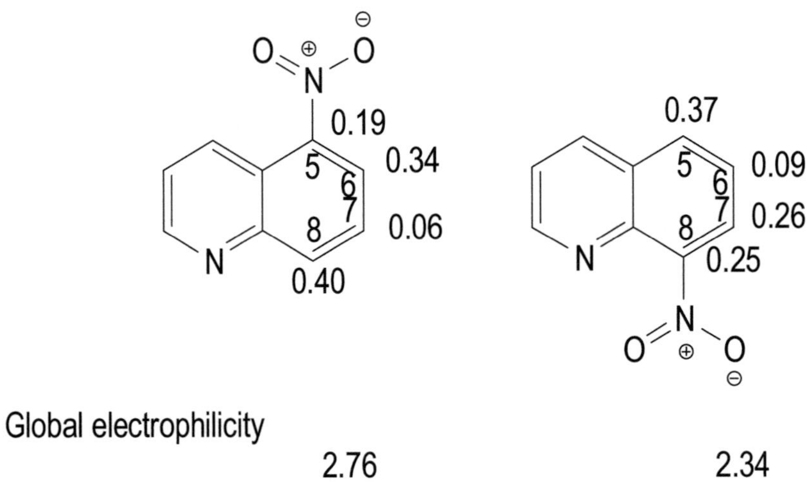

Fig. 23. Global and local electrophilic indexes for 5- and 8-nitroquinolines

Using the 8-nitroquinoline we showed specifically the interactions models with the lowest energy between the dienophile and the solvents (chloroform, EAN and HMIM) (Figure 24). For these structures were calculated the global electrophilicity and nucleophilicity values (Table 7).

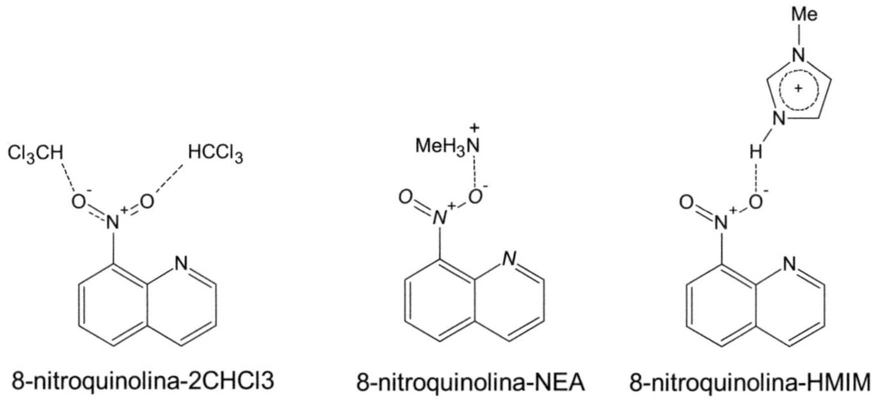

Fig. 24. Theoretical models for the interaction between 8-nitroquinoline and solvent molecules

	HOMO	LUMO	μ	η	ω	N
8-nitroquinolina	-0,2497	-0,0945	-0,1721	0,1553	2,59	2,32
8-nitroquinolina-2CHCl$_3$	-0,2677	-0,1060	-0,1868	0,1617	2,94	1,84
8-nitroquinolina-HMIM	-0,3615	-0,2242	-0,2929	0,1373	8,50	-0,72
8-nitroquinolina-NEA	-0,3482	-0,2044	-0,2763	0,1438	7,22	-0,35
Dieno Danishefsky	-0,1705	0,0125	-0,0790	0,1830	0,46	4,48

Table 7. Global electrophilicity and nucleophilicity indexes

3. Conclusions

It was possible demonstrate the influence of the solvent in these particular type of DA reactions. A series of aromatic carbocyclic and hetrocyclic substituted by electron withdrawing groups can act as dienophiles in polar cycloaddition reactions joint to a different dienes in the presence of PILs. However, DA reactions proceed at an appreciable rate only when either the diene or the dienophile are activated by an electron donating or electron withdrawing group, normally characterized by the presence of a heteroatom that can therefore efficiently interact with the solvent. ILs, with their peculiar properties such as high polarizability/dipolarity, good hydrogen bond donor ability, were straight away considered to have the potential to influence the outcome of these DA reactions, accelerating them.

In general, the products of the reactions development in PILs are similar to those in molecular solvents. However, the presence of PILs improved the reaction rate probably due to the hydrogen bonding interactions between the neoteric solvent and the dienophile. Only in a few cases we can note differences in the product distribution. For the reactions in which are possible to observe a competition between normal and hetero DA process, the PILs favor the normal pathway because they improve the electrophilicity of the dienophiles.

The DFT analysis of the global properties of the interacting pair diene/dienophile illustrates the normal electron demand character of these DA reactions. It is possible to show that the local indexes provide useful clues about the regiodirector effects, particularly of the nitro group. The presence of a solvent (molecular or neoteric) as the reaction media does not impart a prominent influence on the relative reactive sites. In few cases among those studied we can note that the relative reactive sites are affected by the solvent and the basis set (e.g. methyl 5-nitrofuran-2-carboxylate).

The site-, regio-, and stereochemistry of some of these DA reactions has been investigated by the density functional theory, including solvent effects. Generally, these cycloadditions proceed by a concerted but asynchronous reaction mechanism. The *endo* stereochemistry is in most case preferred.

In general, the normal DA reaction mechanism is a domino process that is initialized by the polar reaction between the diene and the dienophile to give the primary cicloadduct. These DA reactions have a two-step non-intermediate mechanism characterized by the nucleophilic attack on the non-substituted methylene of the diene to the electrophilically activated position of the dienophile. The subsequent ring-closure affords the primary cicloadduct. This behavior makes the reaction to be regioselective. The latter concerted elimination of the nitrous acid from the primary cicloadduct yields the precursor of the final aromatic product. Spite of the large activation free energy associated with the DA reaction and the endergonic character of formation of the primary cicloadduct, the irreversible extrusion of the nitrous acid make feasible thermodynamically the domino reaction.

DFT calculations of the electrophilicity and nucleophilicity indexes in general agree with the experimental results and they are a good reactivity and regioselectivity predictors in this type of polar cycloaddition reactions

The presence of a PILs in the reaction media improves significantly the electrophilic character of the dienophile. However, the differences between the experimental results using PILs or molecular solvents are not so bigger how the calculated electrophilic values indicated.

With the present results in mint, in a second step would be desirable development others DA reactions using new dienes with a strong projection in organic synthesis.

4. Acknowledgement

This research was supported by PICT 2008, N° 1214 at Agencia Nacional de Promoción Científico Tecnológica de la República Argentina (ANCyT). We thanks so much to ANCyT

5. References

Adam, C., Fortunato, G., Mancini, P.M.E. (2009) Nucleophilicity and acid catalyst behavior of a protic ionic liquis in a molecular reaction media. Part 1. *Journal of Physical Organic Chemistry*, 22, 460-465, ISSN 1099-1395

Biolatto, B., Kneeteman, M., Paredes,E., Mancini, P.M.E., (2001). Reactions of 1-tosyl-3-substituted indoles with conjugated dienes under thermal and/or hyperbaric conditions. *Journal of Organic Chemistry*, 66, 3906-3912, ISSN 0022-3263

Brasca, R., Della Rosa, C., Kneeteman, M., Mancini, P.M.E. (2011). Five-membered aromatic heterocycles in Diels-Alder cycloaddition reactions: Theoretical studies as a complement ofthe experimental research. *Letters in Organic Chemistry*. 8, 2, 82-87, 2011, ISSN 1687-6865

Brasca, R., Kneeteman, M., Mancini, P.M.E., Fabian, W.M.F. (2011). Comprehensive DFT Study on site-, regio-, and stereoselectivity of Diels-Alder Reactions Leading to 5-Hydroxybenzofurans. European Journal of Organic Chemistry, 721-729, 2011, ISSN 1434-193x

Brasca, R., Kneeteman, M., Mancini, P.M.E., Fabian, W.M.F. (2009) Theoretical explanation of the regioselectivity of polar cycloadition reactions between furan derivatives and

Danishefsky diene. *Journal of Molecular Structure,THEOCHEM*, 911, 134-131, ISSN 0166-1280

Cancian, S., Kneeteman, M., Mancini, P.M.E. (2010). Nitroquinolines as dienophiles in Polar Diels-Alder reactions. Influence of molecular solvent and ionic liquids. 14th International Electronic Conference on Synthetic Organic Chemistry (ECSOC-14), ISSN 0095-2338

Dechambenait, P., Ferlay, S., Kyrtsakas, N., Hosseini, M.W. (2010). Amidinium based ionic liquids. *New Journal of Chemistry*, 34, 1184-1194. ISSN 1144-0546

Della Rosa, C., Kneeteman, M., Mancini, P.M.E. (2005). 2-Nitrofurans as dienophiles in Diels-Alder reactions. *Tetrahedron Letters*, 46, 8711-8714, ISSN 0040-4039

Della Rosa, C., Kneeteman, M., Mancini, P.M.E. (2007). Behavior of selenophenes substituted with the electron withdrawing groups in polar Diels-Alder reactions. *Tetrahedron Letters*, 48, 7075-7078, ISSN 0040-4039

Della Rosa, C., Kneeteman, M., Mancini, P.M.E. (2007). Comparison of the reactivity between 2- and 3-nitropyrroles in cycloaddition reactions. A simple indole synthesis. *Tetrahedron Letters*, 48, 1435-1438, ISSN 0040-4039

Della Rosa, C., Paredes, E., Kneeteman, M., Mancini, P.M.E. (2004). Behavior of thiophenes substituted with electron-withdrawing groups in cycloaddition reactions. *Letters in Organic Chemistry*, 1, 148-150, ISSN 1687-6865

Della Rosa, C., Sanchez, J.P., Kneeteman, M., Mancini, P.M.E. (2011). Diels-Alder reactions of nitrobenzofurans: A simple dibenzofurans synthesis. Theoretical studies using DFT methods.*Tetrahedron Letters*, 52, 2316-2319, ISSN 0040-4039

Domingo L. R., Aurell M. J. (2002). Density Functional Theory Study of the Cycloaddition Reaction of Furan Derivatives wth masked o-Benzoquinones. Does the Furan Act as a Dienophile in the Cycloaddition Reaction?.*Journal of Organic Chemistry*, 67, 959-965, ISSN 0022-3263

Domingo L. R., Aurell M. J., Pérez P., Contreras R. (2002). Quantitative characterization of the local electrophilicity of organic molecules. Understanding the regioselectivity on Diels-Alder reactions. *Journal Physical Chemistry A*, 106, 6871-6875, ISSN 1089-5647

Domingo L.R. (2002). A density functional theory study for the Diels-Alder reaction between N-acyl-1-aza-1,3-butadienes and vinylamines. *Tetrahedron*, 58, 3765-3774, ISSN 0040-4020

Domingo, L.R., Aurell, M.J., Kneeteman, M., Mancini, P.M.E. (2008). Toward an understanding of the participation of nitronaphtalenes in polar Diels-Alder reactions *Journal Molecular Structure: THEOCHEM*, 853, 68-76, ISSN 0166-1280

Domingo, L.R., Chamorro, E., Pérez, P. (2008). Understanding the reactivity of captodative ethylenes in polar Cycloaddition reactions. A theoretical study. *Journal of Organic Chemistry*, 73, 4615-4624. ISSN 0022-3263

Forsyth, S.A., Pringle, J.M., Mac Farlane, D.R. (2004). Ionic liquids - An overview. *Australian Journal Chemistry*, 57, 113-119, ISSN 0004-9425

Fortunato, G., Mancini, P.M.E., Bravo, V., Adam, C. (2010). New solvent designed on the basis of molecular microscopic properties of binary mixtures of the type: protic molecular solvent + 1-butyl-3-methylimidazolium-based ionic liquids. *The Journal of Physical Chemistry B*, 114, 11804-11817, ISSN 1520-5707

Hitchcock, P.B., Mohammed, T.J., Seddon K.R., Zora, J.A., Hussey, C.L., Ward, E.H. (1986). 1-methyl-3-ethylimidazolium hexachlorouranate(IV) and 1-methyl-3-ethylimidazolium tetrachlorodioxo-uranate(VI): Synthesis, structure, and electrochemistry in a room temperature ionic liquid. I*norganic Chimica Acta*, 113, L25-L26, 19-191, ISSN 0020-1693

Jaeger, D.A., Tucker, C.E., (1989). Diels-Alder reactions in ethylammonium nitrate, a low-melting salt. *Tetrahedron Letters*, 30, 1785-1788, ISSN 0040-4039

Kulkarni, P.S., Branco,L.C., Crespo, J.G., Nunes,M.C., Raymundo, A., Afonso, C.A.M. (2007). Comparison of physicochemical properties of new ionic liquids based on imidazolium, quaternary ammonium, and guanidinium cations. *Chemistry - A Euopean Journal*, 13, 8478-8486, ISSN 1521-3765

Mac Farlane, D.R., Seddon, K.R. (2007). Ionic liquids-progress on the fundamental issues. *Australian Journal Chemistry*, 60, 3-5, ISSN 0004-9425

Mancini, P.M.E., Fortunato, G., Adam, C., Vottero, L.R. (2008) Solvent effects on chemical process. New solvent designed on the basis of molecular microscopic properties of binary mixtures: molecular solvents + 1,3-dialkylimidazolium ionic liquids. *Journal of Physical Organic Chemistry*, 21, 87-95, ISSN 1099-1395

Mancini, PM.E., Fortunato, G., Vottero, L.R. (2004). Molecular solvents/ionic liquids binary mixtures: designed solvents basis on the determination of their miscroscopic properties. *Physical of Chemistry Liquids*, 4, 6, 625-632, ISNN 1029-0451

Martins, M.A.P, Frizzo, C.P., Moreira, D.N., Zanatta, N., Bonacorso, H.G.(2008). Ionic liquids in heterocyclic synthesis. *Chemical Review*, 108, 2015-2050, ISNN 0010-8545

Paredes, E., Brasca, R., Kneeteman, M., Mancini, P.M.E. (2007). A novel application of the Diels-Alder reaction: nitronaphthalene as normal electron demand dienophiles. *Tetrahedron*, 63, 3790-3799, ISSN 0040-4020

Plechkova,N.V., Seddon, K.R. (2008).*Chemical Society Review*, 37, 1, 123-150

Parr, R.G., Von Szentpaly L., Liu, S. (1999). Electrophilicity Index. *Journal of the American Chemical Society*, 121, 1922-1924, ISSN 0002-7863

Rideout, C.D., Breslow, R., (1980). Hydrophobic acceleration of Diels-Alder reactions. *Journal of the American Chemical Society*, 102, 7816-7820, ISSN 0002-7863

Rogers, R.D., Seddon, K.R. (2003). Ionic liquids. Solvents of the future?. *Science*, 302, 792-793, ISSN 0036-8075

Wasserscheid, P., Keim, W, (2000) Ionic liquids - New 'solutions' for transition metal catalysis *Angewandte Chemie, International Eddition*, 39, 3772, ISSN 1433-7851

Weingärtner, H (2008). Understanding ionic liquids at the molecular level: Facts, problems, and controversies., *Angewandte. Chemie, International Eddition*. 47, 654-670, ISSN 1433-7851

Welton, T. (1999). Room-Temperature Ionic Liquids. Solvents for Synthesis and Catalysis. *Chemical Review*, 99, 2071, ISSN 0010-8545

Welton, T., (2004). Ionic liquids in catalysis. *Coordination Chemistry.Review*, 248, 2459-2477, ISSN 0010-8545

Wenkert, E., Moeller, P.D., Piettre, S.R. (1988). Five-membered aromatic heterocycles as dienophiles in Diels-Alder reactions. Furan, pyrrole and indole *Journal of the American Chemical Society*, 110, 21, 7188-7194, ISSN 0002-7863

Wenkert, E., Piettre, S.R. (1988). Reaction of α–and β–acylated furans with conjugated dienes. *Journal of Organic Chemistry*, 53, 5850-5853, ISSN 0022-3263

Wilkes, J.S. (2002). A short history of ionic liquids - From molten salts to neoteric solvents. *Green Chem*istry, 4, 73-80, ISSN 1463-9270

Yamada, T., Lukac, P.J., M. George, M., Weiss, R.G. (2007). Reversible, room-temperature ionic liquids. Amidinium carbamates derived from amidines and aliphatic primary amines with carbon dioxide. *Chemistry of Material*, 19, 967-969, ISSN 0897-4756

Yamada, T., Lukac, P.J., Yu, T., Weiss, R.G. (2007). Reversible, room-temperature, chiral ionic liquids. Amidinium carbamates derived from amidines and amino-acid esters with carbon dioxide. *Chemistry of Material*, 19, 4761-4768, ISSN 0897-4756